D0990347

Survey Nonresponse

WILEY SERIES IN SURVEY METHODOLOGY
Established in Part by WALTER A. SHEWHART AND SAMUEL S. WILKS

A complete list of the titles in this series appears at the end of this volume.

Survey Nonresponse

Edited by

ROBERT M. GROVES
University of Michigan and Joint Program in Survey Methodology, USA

DON A. DILLMAN
Washington State University, USA

JOHN L. ELTINGE
U.S. Bureau of Labor Statistics

RODERICK J. A. LITTLE
University of Michigan, USA

A Wiley-Interscience Publication
JOHN WILEY & SONS, INC.

This text is printed on acid-free paper. ♾

Published simultaneously in Canada.

For ordering and customer service, call 1-800-CALL-WILEY

Library of Congress Cataloging-in-Publication Data

Survey nonresponse / editors, Robert M. Groves . . . [et al.].
p. cm. — (Wiley series in probability and statistics)
"A Wiley-Interscience publication."
Includes bibliographical references and index.
ISBN 0-471-39627-3 (cloth: alk. paper)
1. Social surveys—Response rate. I. Groves, Robert M. II. Series.
HN29 .S727 2001
300'.723—dc21 2001045522

Printed in the United States of America

10 9 8 7 6 5 4 3 2 1

Contents

II. IMPACTS OF SURVEY DESIGN ON NONRESPONSE

III. NONRESPONSE IN DIVERSE TYPES OF SURVEYS

IV. STATISTICAL INFERENCE ACCOUNTING FOR NONRESPONSE

Contributors

Zenel Batagelj, University of Ljubljana, Ljubljana, Slovenia

Paul Beatty, National Center for Health Statistics, Hyattsville, Maryland, USA

Jelke G. Bethlehem, Statistics Netherlands, Voorburg, The Netherlands

David Brownstone, University of California, Irvine, California, USA

Jane Burris, University of Illinois, Chicago, Illinois, USA

John B. Carlin, Royal Children's Hospital and University of Melbourne, Melbourne, Australia

Paul Clarke, University College, London, England

Mick P. Couper, University of Michigan, Ann Arbor, Michigan and Joint Program in Survey Methodology, College Park, Maryland, USA

Wim de Heer, Statistics Netherlands, Heerlen, The Netherlands

Edith de Leeuw, MethodikA, Amsterdam, The Netherlands

Wil Dijkstra, Free University, Amsterdam, The Netherlands

Don A. Dillman, Washington State University, Pullman, Washington, USA

Murray Edelman, Voters New Service, New York, New York, USA

John L. Eltinge, U.S. Bureau of Labor Statistics, Washington, D.C and Texas A&M University, College Station, Texas, USA

Andrew Gelman, Columbia University, New York, New York, USA

Thomas F. Golob, University of California, Irvine, California, USA

Robert M. Groves, University of Michigan, Ann Arbor, Michigan and the Joint Program in Survey Methodology, College Park, Maryland, USA

Steven G. Heeringa, University of Michigan, Ann Arbor, Michigan, USA

Douglas Herrmann, Indiana State University, Terre Haute, Indiana, USA

Joop Hox, Ultrecht University, Amsterdam, The Netherlands

Timothy P. Johnson, University of Illinois, Chicago, Illinois, USA

David R. Judkins, Westat, Rockville, Maryland, USA

Camilla Kazimi, San Diego State University, San Diego, California, USA

Jon A. Krosnick, Ohio State University, Columbus, Ohio, USA

Hyunshik Lee, Westat, Rockville, Maryland, USA

James M. Lepkowski, Survey Research Center, University of Michigan, Ann Arbor, Michigan

Virginia Lesser, Oregon State University, Corvallis, Oregon, USA

Roderick J. A. Little, University of Michigan, Ann Arbor, Michigan, USA

Katja Lozar Manfreda, University of Ljubljana, Ljubljana, Slovenia

Peter Lynn, University of Essex, Colchester, England

David A. Marker, Westat, Rockville, Maryland, USA

Jean Martin, Office for National Statistics, London, England

Robert Mason, Oregon State University, Corvallis, Oregon, USA

Xiao-Li Meng, Harvard University, Cambridge, Massachusetts, USA

Daniel M. Merkle, ABC News, New York, New York, USA

Danna L. Moore, Washington State University, Pullman, Washington, USA

Gad Nathan, Hebrew University, Jerusalem, Israel

Elizabeth Nichols, U.S. Bureau of the Census, Washington, D.C., USA

Diane O'Rourke, University of Illinois, Chicago, Illinois, USA

Linda Owens, University of Illinois, Chicago, Illinois, USA

Danny Pfeffermann, Hebrew University, Jerusalem, Israel

Trivellore E. Raghunathan, University of Michigan, Ann Arbor, Michigan, USA

Eric Rancourt, Statistics Canada, Ottawa, Canada

Cleo D. Redline, U.S. Bureau of the Census, Washington, D.C., USA

Donald B. Rubin, Harvard University, Cambridge, Massachusetts, USA

Carl E. Särndal, Statistics Canada, Ottawa, Canada

Jun Shao, University of Wisconsin, Madison, Wisconsin, USA

Eleanor Singer, University of Michigan, Ann Arbor, Michigan, USA

Johannes H. Smit, Free University, Amsterdam, The Netherlands

Tom W. Smith, National Opinion Research Center, Chicago, Illinois, USA

Patrick Sturgis, Department of Statistics, University of Surrey, Surrey, England

Seymour Sudman, (deceased), University of Illinois, Urbana, Illinois, USA

John Tarnai, Washington State University, Pullman, Washington, USA

Michael W. Traugott, University of Michigan, Ann Arbor, Michigan, USA

Vasja Vehovar, University of Ljubljana, Ljubljana, Slovenia

Diane K. Willimack, U.S. Bureau of the Census, Washington, D.C., USA

Marianne Winglee, Westat, Rockville, Maryland, USA

Metka Zaletel, University of Ljubljana, Ljubljana, Slovenia

Elaine Zanutto, Wharton School of Business, University of Pennsylvania, Philadelphia, Pennsylvania, USA

Alan Zaslavsky, Harvard Medical School, Cambridge, Massachusetts, USA

Preface

As statistical surveys enter their second century of existence, they simultaneously exhibit evidence of success and face threats of failure. The evidence of success is that surveys are ubiquitous. Almost all countries in the world use them to measure the socioeconomic status, health, and well-being of their populations. Leaders in government use them to guide important policy, and industry leaders use them to make capital investment decisions. Scientists test theories about human behavior using their observations. The complex set of survey design steps that require both statistical and social science expertise are practiced thousands of times daily by government, academic scientists, and commercial researchers. However, all is not well. In most countries of the world, surveys depend on the voluntary participation of sample households and businesses. In a fundamental sense, surveys work because the samples drawn into them want them to work. Without the active participation of sample persons, few of the valuable statistical properties of sample surveys survive. Over the past two decades, in the developed countries of the world, cooperation rates in sample surveys, especially household surveys, appear to have declined.

Declining cooperation rates increase the cost of conducting surveys, as repeated attempts to seek information are made to reluctant sample members. Declining cooperation rates can also damage the ability of the survey statistics to reflect the corresponding characteristics of the target population. Lower rates *can* do this, but they do not necessarily do so. One of the important scientific challenges facing survey methodology at the beginning of this century is determining the circumstances under which nonresponse damages inference to the target population. A second challenge is the identification of methods to alter the estimation process in the face of nonresponse to improve the quality of the sample statistics.

This book was written to provide a review of the current state of the field in survey nonresponse. It was stimulated by the International Conference in Survey Nonresponse, held in Portland, Oregon, USA, October 28–31, 1999. The conference was sponsored by a consortium of professional organizations interested in statistical surveys—the American Statistical Association (Survey Research Methods Section), the American Association for Public Opinion Research, the Council of American Survey Research Organizations, the Council of Marketing and Opinion Research, and the International Association of Survey Statisticians. These organizations offered seed money for the organizing committee and advertised the conference among their memberships. More than 500 researchers from throughout the world attended the conference.

Several organizations offered financial support for the conference and monograph activities. There were three levels of financial contributions. The highest-level contributors included the U.S. Bureau of the Census, U.S. Bureau of Justice Statistics, U.S. Bureau of Labor Statistics, U.S. National Agricultural Statistics Service, and Nielsen Media Research. The next-highest level of contributors were the Australian Bureau of Statistics, U.S. Bureau of Transportation Statistics, The Gallup Organization, Mathematica Policy Research, and the U.S. Substance Abuse and Mental Health Services Administration. The third group of contributors were the U.S. Energy Information Administration and Westat. We are deeply grateful for the altruism and support for scientific activities implied by these organizations' financial support.

An editorial committee guided the work that led to this book. It consisted of Don A. Dillman, John L. Eltinge, Robert M. Groves (chair), and Roderick J. A. Little. These four divided up the editorial oversight.

The conference had an organizing committee—Donald Bay (National Agricultural Statistics Service), Diane Bowers (Council of American Survey Research Organizations), Donald Clifton (The Gallup Organization), Cynthia Clark (U.S. Bureau of the Census), Lee Giesbrecht (U.S. Bureau of Transportation Statistics), Joseph Gfroerer (Substance Abuse and Mental Health Services Administration), Bruce Hoynoski (Nielsen Media Research), Michael Rand (U.S. Bureau of Justice Statistics), Siu Ming Tam (Australian Bureau of Statistics, and Clyde Tucker (U.S. Bureau of Labor Statistics). In addition, there was a very active committee that organized the contributed paper sessions at the conference, chaired by Lilli Japec (Statistics Sweden), who devoted hundreds of hours to the task, and Kari Djerf (Statistics Finland), Brian Harris-Kojetin (Arbitron, Inc.), Abby Isräels (Statistics Netherlands), Sylvie Michaud (Statistics Canada), Jane Sheppard (Council for Marketing and Opinion Research), and Clyde Tucker (U.S. Bureau of Labor Statistics). This committee handled the hundreds of contributed paper sessions of the conference and oversaw each of the sessions in the conference. Further, the Joint Program in Survey Methodology contributed support through a conference Web page, and arranged for the services of Pamela Ainsworth and Karen Kane, who worked with Lee Decker (American Statistical Association) to handle the many conference logistical issues. Finally, Pamela Fennell provided invaluable assistance in designing and updating the conference Web page.

This book was written for survey professionals. It deliberately combines literatures in social science behavioral approaches to survey nonresponse and literatures in statistical approaches to design and estimation in the presence of missing data. We hope that it may be used as a supplementary text in graduate seminars in survey methodology. The book is divided into four parts:

I. Perspectives on Nonresponse, Chapters 1–6, providing the conceptual frameworks useful to understanding how survey researchers approach nonresponse

II. Impacts of Survey Design on Nonresponse, Chapters 7–12, providing current thinking and practice on features of survey protocols and target populations that are related to response rates

III. Nonresponse in Diverse Types of Surveys, Chapters 13–17, describing how nonresponse characteristics differ by mode of data collection, household or business target populations, and longitudinality of the survey

IV. Statistical Inference Accounting for Nonresponse, Chapters 18–29, describing weighting, imputation, substitutions, and estimation procedures that attempt to reduce the effects of nonresponse on inference to population attributes.

The inclusion of both social science approaches and statistical approaches to nonresponse was deliberate. The editors and authors believe that only by a blending of these two perspectives will issues of survey nonresponse be understood and controlled.

The book's preparation was greatly aided by Karen Kane, Amy Chuang Luo, Duncan Wurm, and Jui Zheng. We thank them for their attention to detail.

Finally, we, the editors, thank the authors of this monograph for their diligence and support of the goal of providing an overview of a dynamic research field. We are confident that their continued diligence will lead to discoveries of new insights into the dynamics of survey participation, new methods to increase participation in surveys, and innovative estimation tools to improve survey statistics from data subject to nonresponse.

ROBERT M. GROVES
DON A. DILLMAN
JOHN L. ELTINGE
RODERICK J. A. LITTLE

August 2001

PART I

Perspectives on Nonresponse

CHAPTER 1

Survey Nonresponse in Design, Data Collection, and Analysis

Don A. Dillman, *Washington State University*
John L. Eltinge, *U.S. Bureau of Labor Statistics*
Robert M. Groves, *University of Michigan and Joint Program in Survey Methodology*
Roderick J. A. Little, *University of Michigan*

Nonresponse occurs when a sampled unit does not respond to the request to be surveyed or to particular survey questions. Error caused by nonresponse is only one of several sources of potential error in surveys—others include coverage, measurement, and sampling error (Groves, 1989)—but it is one that has attracted much interest in recent years, as response rates to certain surveys appear to have been declining, and this is of much concern to social scientists and statisticians throughout the world.

In this volume, we consider the causes and consequences of two types of nonresponse behavior. One type is unit response, which occurs when the person or organization that constitutes the sample unit, fails to respond to a survey. We also consider item nonresponse, the situation in which a unit response is obtained, but the respondent does not answer all of the questions.

In this chapter, we provide an overview of relevant concepts and nonresponse issues, the aim of which is to serve as an introduction to the chapters that follow.

1.1 DESIGN STRUCTURES AND NONRESPONSE

The design of the sample selection process, data collection instruments, field rules, and editing methods can lead to several related forms of incomplete data. First, considerations of cost or burden may lead to deliberate "designed missingness" in the final collected data. Examples include rotation sample designs in panel studies (e.g., Lent et al., 1999); two-phase sample designs (Särndal and Swensson, 1987); and split questionnaire designs (Raghunathan and Grizzle, 1995; Renssen et al., 1997). Second, for sensitive or potentially burdensome items, questionnaires sometimes include a series of related nested questions that attempt to capture an item of interest in progressively more refined forms. As one moves to higher levels of refinement, more units will drop out of the question series due to lack of information or perceived sensitivity. See, for example, the "bracketed income questions" study in Chapter 24 of this volume and the general discussion of relationships between coarsened data and missing data in Heitjan (1989). Third, responses to preliminary items may render some subsequent items irrelevant. The resulting "skip patterns" in questionnaires produce nonrectangular data patterns that can require specialized analyses. Fourth, data collection field rules frequently affect both unit and item nonresponse rates. For example, one frequently seeks to increase response rates by requiring multiple callbacks for noncontacted sample units and extensive conversion efforts for reluctant sample subjects. Chapters 8 through 11 of this book consider callback conversion and incentive work in further detail. In addition, field rules that allow proxy reporting may lead to increased nominal unit response rates, but may have less of an impact on response rates for some items, and nonrespondent conversion and proxy reporting may lead to increases in reporting error, thus possibly inflating the overall mean squared error of the resulting survey estimator. Fifth, data editing rules sometimes require the effective deletion of an item or an entire unit record if the collected items do not satisfy specified internal or external consistency rules. Thus, the editing process itself can reduce response rates.

1.2 EFFECTS OF NONRESPONSE ON SURVEY ESTIMATION

In work with population means, totals, and related parameters, survey nonresponse is of practical concern for several reasons, including (1) biases in point estimators, (2) inflation of the variances of point estimators, and (3) biases in customary estimators of precision.

The point-estimation bias issue (1) principally receives attention in the survey literature and is the major reason that survey organizations devote extensive efforts to reduction of, and adjustment for, nonresponse. However, the efficiency and inference issues arising in (2) and (3) are also of importance in sample surveys with nonresponse. Methods that fail to account for the loss of precision due to missing data, such as naive single imputation methods, lead to underestimation of variance associated with point estimates, confidence intervals that have lower than nominal coverage, and tests that have greater than nominal size.

To illustrate the bias issue (1), consider a finite population U containing N units with items Y_i, $i = 1, \ldots, N$, and define the associated population mean $\mu = N^{-1}\Sigma_{i=1}^{N} Y_i$. In addition, let S be the set of indices of n sample units selected through a complex design D. This design may involve a combination of stratification, clustering and unequal probabilities of selection. For each unit i in the population, let π_i

equal the probability that unit i is included in the sample s, and define the associated probability weight $w_i = 1/\pi_i$. Then a standard point estimator of the population mean μ is

$$\hat{\mu}_F = \left(\sum_{i \in S} w_i\right)^{-1} \sum_{i \in S} w_i Y_i \tag{1.1}$$

Under moderate regularity conditions, the full-sample point estimator $\hat{\mu}_F$ is approximately unbiased for μ, where the expectation of $\hat{\mu}_F$ is evaluated with respect to the sample design for fixed characteristics Y_i, $i = 1, \ldots, N$.

In the presence of nonresponse by one or more elements $i \in S$, one cannot compute the full-sample estimator (1). A simple alternative is the unadjusted estimator,

$$\hat{\mu}_{UA} = \left(\sum_{i \in R} w_i\right)^{-1} \sum_{i \in R} w_i Y_i \tag{1.2}$$

where now the sample S is partitioned into subsets R and M containing responding and missing units, respectively. We may rewrite expression (1.2) as,

$$\hat{\mu}_{UA} = \left(\sum_{i \in S} w_i r_i\right)^{-1} \sum_{i \in S} w_i r_i Y_i \tag{1.3}$$

where r_i is the response indicator for element i: $r_i = 1$ if element i responds; $r_i = 0$ otherwise.

The operating characteristics of $\hat{\mu}_{UA}$ depend on the nonresponse process, which can be formalized using the following quasirandomization model (Oh and Scheuren, 1983)

(M.1) For $i = 1, \ldots, N$, assume that the r_i are independent Bernoulli (p_i) random variables.

Note that model (M.1) allows the response probabilities p_i to vary across i. Under the quasirandomization model (M.1) and additional regularity conditions (see, e.g., Eltinge, 1992), the unadjusted point estimator $\hat{\mu}_{UA}$ has a bias approximately proportional to the finite population correlation of Y_i with the response probabilities p_i. In this and all subsequent discussions, expectations are evaluated with respect to both the original sample design and the quasirandomization model (M.1).

One approach to reducing the bias of (1.3) is to apply a nonresponse weighting adjustment. Suppose that unbiased estimates $\hat{p}_i$ can be computed using available information from the survey or external sources. Sections 1.2 and 1.3 below will discuss estimation of p_i from auxiliary data. Then one could compute the approximately unbiased estimator

$$\hat{\mu}_p = \left(\sum_{i \in R} w_{pi}\right)^{-1} \sum_{i \in R} w_{pi} Y_i \tag{1.4}$$

where $w_{pi} = w_i/\hat{p}_i$ is a weight that has been adjusted by the inverse of the estimated selection probability for unit i. See, e.g., Oh and Scheuren (1983) and Särndal and Swensson (1987).

The efficiency issue (2) then relates to the increase of the variance of the estimator $\hat{\mu}_p$ relative to the idealized estimator $\hat{\mu}_F$; and the inference issue (3) depends on the bias and stability properties of variance estimators applied to (1.4), and the coverage of confidence intervals and tests constructed using (1.4) and these variance estimators.

1.3 UNIT NONRESPONSE

Unit nonresponse, the failure to obtain any survey measurements on a sample unit, arises in ways that are specific to a survey design. For example, in interviewer-assisted surveys, contacted sample units may refuse the explicit request of the interviewer to participate. In mail self-administered surveys, some envelopes containing the questionnaire may not even be opened, and thus the sample person may not have actually received a request to participate.

From the perspective of the researcher, designs vary in their tendency toward nonresponse of different types, the costs of reducing a particular source of nonresponse, and the potential nonresponse error in key statistics arising from the nonresponse. Figure 1.1 illustrates how noncontacts, refusals, and incapacities vary in their manifestations over different modes of data collection.

Survey Mode	Cause of Nonresponse	Example	Cost of Reducing Nonresponse
Mail	No Request	Nondelivery	Low
	Refusal	Mailing read, but ignored	High
	Incapacity	Illiteracy	High
Web	No Request	E-mail not received	Low
	Refusal	Access the first page, then abandon site	High
	Incapacity	Computer can't download	High
Telephone	No Request	No one answers the call	Low
	Refusal	Sample person hangs up	High
	Incapacity	Deafness; language problems	High
Face to Face	No Request	No one answers	Low
	Refusal	Sample person declines request	High
	Incapacity	Language problems	High

Figure 1.1. Types of nonresponse by mode of data collection.

The first two modes do not involve a human interviewer as part of the request mechanism. This implies that the researcher (or his/her agent) does not observe the unit nonresponse. In modes without interviewers, the causes of unit nonresponse must be inferred from the absence of a completed survey instrument. In addition, various frame problems are confounded with failure to complete an instrument (e.g., selection of an ineligible unit or of a duplicate frame record of an eligible unit). For that reason, classification of unit nonresponse is more difficult without the assistance of an interviewer.

In self-administered modes, the failure to deliver a request for the survey can arise because the sample person did not receive the sent request or because the person fails to attend to the request (e.g., does not read the email requesting completion of a web survey). Interviewers in the other modes make observations that classify the type of nonresponse.

The costs of reducing nonresponse are partially a function of the cause of the nonresponse. In telephone and face-to-face survey modes, interviewers can sometimes observe attributes of the nonresponding sample unit and apply different followup procedures to different units. Reducing noncontact nonresponse generally relies on repeated efforts at different times of day and days of the week. Reducing refusal is generally less successful, and thus relatively expensive. In modes without use of an interviewer, most inducements for participation must be made part of the written or visual material that is sent to the sample person as part of the request.

Because the causes of unit nonresponse vary across modes, it is possible that the association between nonresponse rates and nonresponse errors also varies across modes.

1.3.1 Theoretical perspectives on Influences toward Unit Nonresponse

The challenge of theories of unit nonresponse is both to describe the dynamics of the process that leads to survey nonparticipation and to identify those circumstances that lead to more or less error due to nonresponse. Chapters 3 and 7 demonstrate that the behaviors underlying noncontact are quite different from those underlying refusal.

Influences on Nonresponse From Failure to Delivery the Survey Request. In interviewer-assisted surveys, failed attempts to deliver the survey request arise when the persons in the sample unit are not present or are inaccessible to the interviewer. Impediments to access abound. On the telephone, these include answering machines, caller identification features, and call blocking. In face-to-face surveys, these include locked central entrances to apartment buildings, walled subdivisions, locked gates, intercoms, etc. Groves and Couper (1998) found empirical support that these access impediments slowed down contact (i.e., required more calls to first contact) but did not otherwise greatly affect the overall probability of contact.

The more calls on sample cases, the higher the likelihood of eventual contact (e.g., Purdon et al., 1999). When the calling times are varied, there is a higher chance of eventual contact; evening and weekend calls are more productive than weekday daytime calls (e.g., Weeks et al., 1987). Wealthier, younger, and single person households invest in answering machines (Tuckel and O'Neil, 1995). Further, there is consistent evidence that at-home patterns vary by household composition—single person households with employed members tend to be home less frequently; elderly single person households and households with children (especially younger children) tend to be at home more frequently (Lievesley, 1988; Groves and Couper, 1998).

In self-administered surveys, the failure to expose the sample person to the survey request can arise because of (a) erroneous contact information (i.e., mailing address, email identification), (b) delivery errors (e.g., postal errors), (c) interception by others followed by failure to forward to the sample person (e.g., the mail handling within the unit does not forward the survey request), and/or (d) the sample person's failure to read or understand the survey request. These factors differ from those in interviewer-assisted surveys in that once delivered, the survey request remains active until the sample person attends to it or disposes of it.

Influences on Survey Participation. Groves and Couper (1998) organized the influences on cooperation into four blocks. Two of them are exogenous to the survey design: (a) social environmental influences (e.g., survey-taking climate, urbanicity effects), and (b) knowledge and social psychological attributes of the sample persons (e.g., civic duty, interest in politics). The other two are features that are chosen by the survey designer: (a) survey protocols (e.g., mode of data collection, incentives, burden of the interview), and (b) the selection and training of interviewers (this factor being absent in self-administered surveys). Together, these four factors influence the sample person's behavior during the few moments of interaction with a survey interviewer, determining what features of the request become salient and what personal knowledge is brought to bear in the judgment about participation. Linked to these are various concepts, proffered as influences on cooperation.

Opportunity Costs. "Opportunity costs" hypotheses posit that a prospective survey respondent weighs all the costs of participation against the benefits of participation. All other things being equal, the burden of providing the interview is larger for those who have little discretionary time. Although there is some empirical support that those who live alone (and thus cannot share household duties with others) have lower cooperation rates, there is not much empirical research on this hypothesis.

Social Exchange. "Social exchange" reflects the ubiquitous norm that favors provided by one actor to another lead to reciprocation. Dillman (1978) focused on social exchange within a relatively closed system (survey organization and householder), with relatively small gestures on the part of the survey organization (personalized letters, token incentives, reminder letters). These efforts, which are accompanied by efforts to reduce costs of responding (e.g., questionnaires that are difficult to understand and appeals that subordinate the respondent) and promote trust, are hypothesized to evoke a reciprocating response from the householder. In order to explain the propensity to respond, Goyder (1987) has expanded the notion of exchange to evoke a wide array of obligations and expectations over an extended period of time between an individual and various institutions of society.

Theories of social connectedness, isolation, or disengagement are related to those of social exchange. Social isolates are out of touch with the mainstream culture of a society and will not be guided by norms of the dominant culture. Survey researchers sometimes have noted that feelings of "civic duty" prompt survey participation. As reviewed in Chapter 4, those who are alienated or isolated from the broader society/polity would be less likely to cooperate with survey requests that represent such interests.

Topic Saliency. There is much speculation that when the sampled persons share the purposes of the conversation, they would tend to cooperate. This speculation is based on hypotheses that surveys on salient topics may offer some chance of personal gain to the respondents because their group might be advantaged by the survey information, and also that the chance to exhibit one's knowledge on the topic would be gratifying. When the topic of the interview is used as an important attribute by the interviewer in persuading the householder, then prior knowledge about the topic and personal relevance to the householder can affect response propensity. This issue is particularly evident in self-administered surveys where the complete content of the questionnaire can be examined before the respondent decides whether or not to respond. This hypothesis offers a direct link between the behavioral and statistical concerns—topic saliency effects are likely to be associated with statistical biases due to nonresponse, reviewed in Section 1.4 (Groves et al., 2000).

Interviewer Effects on Cooperation Propensity. Interviewers vary in their cooperation rates. Drawing on focus groups of interviewers and statistical findings about correlates of cooperation, Groves and Couper (1998) note that interviewers can act to make different features of the survey request more or less salient to the respondent's decision to cooperate (see also Morton-Williams, 1993). This behavior, labeled "tailoring," can effectively address concerns of the sample person, leading to cooperation. Inexperienced interviewers often encounter low cooperation because they make salient survey features less attractive or irrelevant to respondent concerns.

Influences toward Cooperation in Other Designs. Two other survey designs are noteworthy: longitudinal surveys and establishment surveys. In longitudinal surveys (as is described and elaborated on in Chapter 17), the secondary waves of interviewing are affected by the respondent experiences of the prior waves. If the initial waves produce memories of a pleasant event, cooperation propensities rise; if not, they fall. In establishment surveys, two additional influences pertain. First, the cooperation process sometimes involves multiple members of the organization. Some members may have the authority to permit or deny providing survey information; others have access to the desired information and can independently determine the nature of compliance with the request. Second, in contrast to most households, many organizations have made formal policies concerning the release of information to external entities. Corporation attorneys may enforce these policies, with little or no grant of appeal.

1.3.2 Design Impacts on Response Propensity

Design Features Affecting Contact with Sample Units. As implied above, with self-administered surveys, the paucity of information available on the causes of the nonresponse associated with cases that are not returned may result in the designer using followup procedures of a rather generic sort (e.g., mailing reminders, sending a second questionnaire in paper surveys). However, it is also possible for the designer to use knowledge about differences in populations, survey objectives, and delivery situations to tailor procedures in ways that take such differences into account. For example, the knowledge that the respondent to a web survey must look up information prior to responding suggests that sending a printable attachment to help in the collection of such data can greatly reduce the frustration of proceeding several pages into the electronic questionnaire, only to discover that the respondent is not prepared to answer it. Similarly, six to seven contacts within two weeks to respond to a diary that must be filled out for certain dates is not likely to seem unreasonable to respondents, but in other survey situations surely would. Many examples of such tailoring of data collection procedures are provided by Dillman (2000). In interviewer-assisted surveys, a variety of tools have been used to reduce noncontact nonresponse.

Call Scheduling. Some survey designs have specific rules for the calling patterns on households. Some telephone surveys use software in computer-assisted interviewing systems to enforce such rules. For example, surveys may specify that the interviewer make four call attempts over a number of days in order to make first contact with a household, and then obtain a final disposition within four additional calls.

Length of Data Collection Period. It is clear that the length of the data collection period has an effect on contact rates, other things being equal. If a survey were limited to a few hours in a single evening (as in some political attitude polls), all those households who were away from home on that evening remain uncontacted, regardless of efforts made by the interviewers. On the other hand, all households are likely to have someone at home at some point during a survey data collection period lasting several months. There are surveys attaining near 100% contact rates with about a one-week period of data collection (Groves and Couper, 1998).

Interviewer Workload. Another limiting factor on the efforts of interviewers to reach cases is the size of their workload—on average, how many cases they have been assigned to be completed during the data collection period. The size of workload that is manageable is probably a complex function of the mode of data collection, the length of the interview, the compactness of the sample segments (in face-to-face surveys), the nature of the study population, and the period of the data collection.

Design Features Affecting Cooperation
Agency of Data Collection. Sample persons' knowledge and attitudes concerning the sponsor can affect whether they grant an interview or complete a self-administered questionnaire. If the organization acts indirectly to benefit the target persons or their group, then they might be more positively disposed to the request. If the householders know nothing about the sponsor, they may focus more on the interviewer's characteristics or nature of the written request (stationery, personalized features) to aid in their judgments.

Advance Warning of the Survey Request. It has become commonplace in face-to-face surveys to mail a letter to the sample household that alerts the unit to the upcoming call. From a theoretical perspective, this design feature can be used to manipulate a variety of influences known to affect survey participation: (a) by using official stationery of a legitimate sponsoring agency, it communicates that the survey is so authorized; (b) by reviewing the social benefits of the survey, it can communicate, "independently" of the interviewer, the survey's contribution to the common good; and (c) by reviewing the confidentiality provisions of the survey, it can allay fears of undesired distribution of private information. Perhaps for all of these reasons, advance letters increase the interviewers' confidence while seeking cooperation.

Respondent Incentives. Another common design tool to improve participation is to provide an incentive—giving the sample unit something judged by the sponsor of the survey to be valued by the unit. Most incentives take the form of money, either cash, checks, or money orders. These are found to be most effective when sent in advance. Some incentives are objects that have cash value—gift certificates, coffee mugs, books, calculators, jewelry, tool kits, kitchen magnets, American flags, pens, medallions, and a host of other things. Chapter 11 provides a summary of the effectiveness of survey incentives across survey modes.

Follow-up Procedures. If the initial contact reveals that the householder is reluctant to cooperate with the survey protocol, there are a variety of design options. These include switching to another interviewer (matched to the attributes of the respondent), sending persuasion letters or information customized to the respondent concerns, and switching to another mode of contact (usually to one offering more interaction).

Increasing Cooperation in Self-Administered Surveys. Practical tools for increasing response rates in self-administered surveys have been influenced by attempts to apply concepts of social exchange to survey design (Dillman, 2000). These include the use of "tailored design" features that attempt to create respondent trust, reduce costs of responding, and increase benefits of cooperation. The rewards of responding are viewed to be more credible when some degree of trust is generated among sample persons. This trust may be generated by sending some token of appreciation with the request, through sponsorship by a legitimate agent, and by the apparent importance of the request for the group to which the sample person belongs. The rewards for responding may include features as simple as showing a positive regard for the sample persons, thanking them for their cooperation, seeking their advice, supporting their group values, increasing the interest in the questionnaire, giving testimony to others in the group supporting the survey, and communicating the scarcity of opportunities to participate in such research. The costs of cooperating may be reduced by limiting the cognitive burden or threat of embarrassment. For self-administered questionnaires, much attention is currently being given to the use of visual design principles (see Chapter 12) to reduce such burdens.

1.3.3 Other Design Features Chosen to Reduce Unit Nonresponse

Other design features have been used to decrease unit nonresponse. These include sample design that assigns lower probabilities of selection to subpopulations known to produce low response rates. The reduced sample size releases some resources, which are then focused on the achieved smaller sample in those subpopulations. The higher per-unit resources devoted to these sample cases is devoted to achieve higher-than-normal response rates in that group. The response rate computation is then weighted by selection probabilities. A similar scheme, based on two-phase sampling, occurs *after* data collection is finished. Such double sampling schemes draw a probability sample of nonrespondents and then use more expensive measurement techniques (Elliott et al., 2000).

1.4 ITEM NONRESPONSE

1.4.1 Contrasts with Unit Nonresponse

Item nonresponse is the failure to obtain substantive answers to individual survey questions when a unit response is obtained. However, the exact line of demarcation between unit and item nonresponse remains somewhat fuzzy. Survey procedures often require that key questions and/or a certain proportion of questions must be answered before classifying a survey unit as having responded to the survey.

Defining item nonresponse is further complicated by the nature of the response options that are considered nonresponse. Refusing to answer an interviewer's question or leaving a question on a self-administered questionnaire blank both constitute item nonresponse. However, questionnaire designers often build "don't know" or "no opinion" categories into attitude questions, which, if selected, may indicate an unwillingness to state an opinion. Choosing such an option may or may not be interpreted by the surveyor as item nonresponse. For example, in response to the question, "how satisfied are you with your work?" a "no opinion" may be treated as item nonresponse because all respondents are viewed as having the necessary information for choosing a level of satisfaction. However, in response to the question of "to what extent do you feel that management efficiency has been improved by your new e-mail system?" a "no opinion" may be analyzed as a legitimate and useful response because of this phenomenon is not visible to many employees.

In addition, item nonresponses may occur for all items located toward the end of a questionnaire as a result of a sudden termination of an interview, or the failure to provide substantive answers may be sprinkled throughout, and may only occur for certain kinds of questions (income) and not others. Thus, considerable variation exists with regard to how the failure to provide a substantive answer while providing a unit response becomes interpreted as item nonresponse.

1.4.2 The Causes of Item Nonresponse

It would be a mistake to think of the causes of item nonresponse as being the same as those that lead to unit nonresponse. Somewhat different factors play the predominant roles in determining unit nonresponse, though it is important to realize that for some respondents a decision not to respond to a particular question may be the precursor of a unit nonresponse.

Survey Mode. In general, item nonresponse is more likely to occur in self-administered questionnaires than in those conducted by interviewers. (Tourangeau et al., 2000). In self-administered formats, particularly paper questionnaires, it is up to the respondent whether each question gets read, the order in which the questionnaire is read, and whether an answer is recorded. Poor visual layout and design of such questionnaires is a major cause of item nonresponse (Jenkins and Dillman, 1997). In contrast, interviewers are normally instructed to read every question that applies to the respondent. Further, the interviewer can respond to a respondent's inclination not to answer a question by encouraging reconsideration of a previously offered answer, e.g., "Let me read the question to you again. . . ."

Interviewers and Interviewer Training. Interviewers are trained by different organizations to respond in various ways to unanswered questions. Sometimes, interviewers are encouraged to use multiple probes to discourage the offering of no opinions or refusals. In other instances, they are encouraged to accept whatever answer the respondent offers. Also, some interviewers are quite skillful in developing a high level of rapport with respondents, so that answers that might not be given to one interviewer are willingly offered to another.

Question Topics. Item nonresponse is more likely to occur for questions involving some psychological threat. Questions about personal or household income are notorious for eliciting refusals. Also, questions about sexual behavior, drinking of alcoholic beverages, and violations of laws are more likely to elicit item nonresponses in most surveys (Tourangeau et al., 2000).

Question Structures. Certain question structures are more likely to produce a high item nonresponse than are others. In self-administered surveys, open-ended questions are more likely to be left unanswered than are closed-ended questions for which a respondent has only to choose from among the presented alternatives. Multiple-part questions in which the total is expected to sum to 100%, for example, providing the percent of income spent on a variety of household expenses, may yield higher missing data. Also, questions in self-administered questionnaires that appear not to apply to individuals but include as a final category "does not apply to me" are likely to induce more item nonresponse. In addition, self-administered questionnaires that include branching instructions that are needed for directing respondents who choose different response categories in one question to answer different future questions, produce higher item nonresponse rates (Featherston and Moy, 1990; Turner et al., 1992).

Question Difficulty. Sometimes, questions are too difficult for certain respondents to answer. Asking someone how many times they have visited a doctor's office during the last three years may prove impossible for some respondents to answer, whereas a question that asks only about the last three months or most recent visit could be easily answered. In surveys of businesses, questions are sometimes asked that require information that is not collected by the responding organization, or collected in a different way, so that providing a meaningful response is not possible, and high item nonresponse results (Tomaskovic-Devey et al., 1995).

Institutional Requirements and Policies. Institutional review boards for the protection of human subjects often require that respondents be informed that providing answers to each question is voluntary. This may discourage interviewer efforts to convince reluctant respondents to answer certain questions, particularly those deemed most essential to the survey. In self-administered surveys, it may lead to the practice of offering a "prefer not to answer" or "no opinion" category for all questions. A frequently touted advantage of Web or Internet surveys (Vehovar, et al., Chapter 15, this volume) is that respondents can be required to answer every question before moving on to the next question, but that potential advantage may be mitigated by institutional requirements.

Some organizations, particularly businesses, treat certain information available to them as proprietary, and as a matter of policy will not provide it to others. Thus, item nonresponse may reflect an unwillingness to provide the requested information to the surveyor (Dillman, 2000)

Respondent Attributes. Substantial variation exists among respondents with regard to their propensity to offer no opinions, refusals, or no answers. Although the attributes associated with item nonresponse vary by populations and survey topics, older people and those with less education are less likely to provide answers in many surveys. In addition, reluctant respondents or converted refusals, as reported in Chapter 10 by Mason, Lesser, and Traugott, are more likely to refuse to respond to individual items.

1.4.3 Procedures for Reducing Item Nonresponse

Efforts to reduce item nonresponse are contingent on both the nature of the item nonresponse (e.g., no answer versus no opinion) and which of the above factors are the primary causes. Opinion researchers have frequently preceded individual items with a screening question that asks whether or not the respondent has an opinion, in effect encouraging those who do not have one to opt out of answering. Some effects of this procedure have been discussed in detail by Schuman and Presser (1981).

To overcome the high item nonresponse associated with sensitive questions, researchers have also proposed changing the structures of questions. For example, instead of asking the respondent to state or write down household income, the research may ask the respondent to choose from among broad income categories (Dillman, 2000). Converting questions from open-ended to close-ended has been used to reduce item nonresponse to a wide variety of behavioral questions. Bradburn and Sudman (1979) have discussed the use of embedding and numerous other techniques to de-emphasize threatening questions about respondent behavior.

Several chapters in this book discuss various aspects of efforts to reduce item nonresponse. In Chapter 5, Beatty and Herrman raise the fundamental question of whether efforts should be made to reduce item nonresponse for certain questions. They provide a theory of the decision-making process by which respondents decide whether or not to answer individual questions. They also propose and evaluate a methodology that might be used by others to identify questions for which respondents are likely to provide answers even though it is unlikely that a meaningful answer can be provided. In Chapter 6, Krosnick provides a contrasting perspective, though only for opinion questions. He argues that no opinion filters, proposed by Converse (1964) as a way of separating meaningful from unmeaningful opinions, do not work. The reasons are that no opinion responses often result from ambivalence, question ambiguity, satisficing, intimidation, and self-protection. He concludes that there is something meaningful to be learned from pressing respondents to report their opinions.

In Chapter 12, Redline and Dillman focus on a questionnaire structure issue that has perplexed designers of self-administered questionnaires for decades. It is the problem of item nonresponse stemming not from intention, but from unintentional mistakes made in following the branching instructions in questionnaires where the respondent is expected to answer some questions but not others. They propose two quite different designs for branching instructions: a prevention technique and a detection technique. Each depends upon different manipulations of the graphical, symbolic, and verbal languages used to guide respondents to the next appropriate question, and shows promise for reducing inappropriate item nonresponse.

The problem of item nonresponse is linked back to the issue of unit nonresponse by Mason, Lesser, and Traugott in Chapter 10. They examine the empirical consequences of achieving higher response rates to a telephone survey through refusal conversion, a process that produces an item nonresponse rate of 25% compared to 11% for previous respondents. Based on their analysis of these data, the question is raised as to whether refusal conversion is worth the effort. They also discuss alternatives, from imputation for the missing data to collecting data from a larger sample without refusal conversion.

1.5 ESTIMATION AND INFERENCE FOR SURVEY DATA SUBJECT TO UNIT AND ITEM NONRESPONSE

Despite attempts to reduce unit and item nonresponse in surveys, missing values inevitably occur. Analysis methods for data subject to unit and item nonresponse seek to create valid point estimates of population quantities from the incomplete data. Since good statistical analyses usually also require an assessment of statistical uncertainty of estimates through confidence intervals or tests of hypotheses, it is also important to reflect the loss of information arising from missing data, so that, for example, nominal 95% confidence intervals cover the true population quantity at least approximately 95% of the time in repeated sampling.

Approaches to the analysis of incomplete data (e.g., Little, 1997) include:

1. Complete-case (CC) analysis, where complete-data methods are applied to the rectangular data set obtained by deleting the incomplete cases. An extension is weighted CC analysis, where the respondents are weighted to compensate for bias.
2. Imputation, where missing values are replaced by estimates and the filled-in data are analyzed by complete-data methods. Often a single value is imputed (single imputation). Multiple imputation imputes more than one set of imputations of the missing values, allowing the assessment of imputation uncertainty.
3. Analysis of the incomplete data by a method such as maximum likelihood (ML) that does not require a rectangular data set.

The standard approach to unit nonresponse in surveys is weighted CC analysis, as discussed in the next subsection. Item nonresponse is generally handled by either discarding the incomplete cases or imputation, although ML approaches are becoming more common in specific problems, such as mixed-model analysis of repeated-measures data arising from panel surveys (Harville, 1977; Laird and Ware, 1982; SAS Institute, 1992; Hedeker, 1993; Pfeffermann and Nathan, Chapter 28, this volume).

Survey statisticians seek to avoid modeling assumptions, relying on the randomness of the sampling distribution as the basis for the inference. However, assumptions are inevitable when dealing with survey nonresponse; the only way to avoid them is to have no missing data! In particular, imputation methods make assumptions about the predictive distribution of the missing values, whether explicit in a formal statistical model or implicit in the nature of the imputation algorithm. CC analysis does not create a predictive distribution for the missing values but it makes (potentially stronger) assumptions about the representativeness of the complete cases of the underlying target population.

Two types of assumptions are typically evoked in developing survey adjustments for nonresponse, reflecting design and model-based approaches to survey inference. The first is to assume that within subgroups of the population based on available information internal or external to the survey, the probabilities of response are independent of the survey outcomes subject to nonresponse, so that nonresponse can be effectively considered a form of random subsampling of the sampled cases. Oh and Scheuren (1983) used the term "quasirandomization" to describe this assumption. The second is to assume some form of model for the predictive distribution of the missing values given the observed data (Little, 1982; Rubin, 1987). Design-based survey statisticians favor quasirandomization because it avoids models for the population values; however, the predictive modeling approach is more flexible, in that it allows inference for problems involving complex nonmonotone missing data patterns and fuller conditioning on observed data (e.g., Heeringa et al., Chapter 24; Gelman and Carlin, Chapter 19, this volume).

Let $Y = (y_{ij})$ denote an $(n \times p)$ rectangular matrix of the survey data in the absence of missing values, with ith row $y_i = (y_{i1}, \ldots, y_{ip})$, where y_{ij} is the value of variable Y_j for subject i. With missing data, define the *missing-data indicator matrix* $M = (m_{ij})$, such that $m_{ij} = 1$ if y_{ij} is missing and $m_{ij} = 0$ if y_{ij} is present. The matrix M then defines the pattern of missing data. Some methods for handling missing data apply to any pattern of missing data, whereas other methods assume a special pattern. An important example of a special pattern is *univariate* nonresponse, where missingness is confined to a single variable. Another is *monotone* missing data, where the variables can be arranged so that $Y_{j+1}, \ldots, Y_p$ is missing for all cases where Y_j is missing, for all $j = 1, \ldots, p-1$. This is often the predominant pattern in longitudinal survey data subject to attrition.

The missing-data mechanism concerns the reasons why values are missing, and in particular whether these reasons relate to values in the data set. For example, a subject in a longitudinal survey may be more likely to drop out if she or he moved to a different location; drop-out might be related to the variables in an employment survey if relocation is related to a change of job. Any analysis of data involving unit or item nonresponse requires some assumption about the missing-data mechanism, and a variety of such assumptions are made in this book. The mechanism can be formalized in terms of the conditional distribution of M given the survey outcomes Y and the survey design variables Z, which we assume are fully observed (Rubin, 1976, 1987; Little, 1982; Little and Rubin, 1987). Let $f(M|Y, \theta)$ denote this distribution, where θ are unknown parameters. If missingness does not depend on the values of the data Y or Z,

$$f(M|Y, Z, \theta) = f(M|\theta) \qquad \text{for all } Y, Z, \theta \tag{1.5}$$

and the data are called *missing completely at random* (MCAR). Note that this assumption does not mean that the pattern itself is random, but rather that missingness does not depend on the data values. An MCAR mechanism is plausible in planned missing-data designs, for example, a design that follows up a random subset of nonrespondents, but it is usually an unrealistically strong assumption when missing data do not occur by design, because missingness usually does depend on recorded variables. Let Y_{obs} denote the observed values of Y, and Y_{mis} the missing values. A less restrictive assumption than Eq. (1.5) is that missingness depends only on values of the design variables Z, and not on the survey outcomes:

$$f(M|Y, Z, \theta) = f(M|Z, \theta) \qquad \text{for all } Y, \theta \tag{1.6}$$

Rubin (1987) uses the term *unconfounded* for this mechanism, and it is considered in Chapter 21. A less restrictive assumption than Eq. (1.6) partitions the survey variables as $Y = (Y_1, Y_2)$, where Y_1 is a fully observed adjustment cell variable used to classify respondents and nonrespondents for nonresponse weighting or imputation (see Chapter 18). The latter methods assume that

$$f(M|Y, Z, \theta) = f(M|Y_1, Z, \theta) \qquad \text{for all } Y_2, \theta \tag{1.7}$$

where missingness is allowed to depend on Y_1 or Z but not on Y_2. This assumption is closely related to the assumption of quasirandomization (Oh and Scheuren, 1983). A still weaker assumption is

$$f(M|Y, Z, \theta) = f(M|Y_{obs}, Z, \theta) \qquad \text{for all } Y_{mis}, \theta \tag{1.8}$$

where Y_{obs} denotes all the observed survey data, including the fully observed survey variables and the observed values of the incomplete variables. If (1.8) holds, then the missing data mechanism is called missing at random (MAR) (Rubin, 1976). Note that (1.5) to (1.7) are all stronger forms of the MAR assumption (1.8) that differ on the degree of conditioning on observed data. Simple missing data methods, such as discarding incomplete cases or weighting within adjustment cells, often require stronger assumptions about the mechanism than fully model-based methods that condition on all the available data (Y_{obs}, Z), such as the Bayesian methods in Chapter 24, which assume (1.8). Thus there is a trade-off between model complexity and the strength of assumptions about the mechanism.

Most current survey nonrespondent adjustments assume that the missing-data mechanism is MAR. If the distribution of the mechanism depends on missing values after conditioning on the observed data, that is, (1.8) does not hold, then the missing-data mechanism is not missing at random (NMAR). Chapter 25 discusses an NMAR model in a specific application.

1.5.1 Methods for Unit Nonresponse

The standard approach to unit nonresponse in surveys is to discard the unit nonrespondents and assign a nonresponse weight to the respondents to reduce nonresponse bias. In probability sampling, a sampling weight inversely proportional to the probability of selection is used to adjust for differential rates of selection. If nonresponse is viewed as another stage of probabilistic selection of units, then the product of the probability of selection by design and the probability of response given selection is the probability of being observed, and its inverse can be used as a weight in the analysis.

Sampling probabilities are generally known, but nonresponse probabilities are unknown and need to be estimated from the data. A standard approach is to form adjustment cells (or subclasses) based on background variables that have a distribution over respondents and nonrespondents that is known or can be estimated from the sample data. All nonrespondents are given zero weight and the nonresponse weight for all respondents in an adjustment cell is then the inverse of the (estimated) response rate in that cell. If more than one background variable is measured, adjustment cells can be based on a joint classification, collapsing small cells as necessary.

Two approaches to the creation of adjustment cells can be distinguished. First, they can be based on variables measured for respondents and nonrespondents in the survey. For example, the location of nonrespondents and respondents in the sample is generally known, so adjustment cells might be based on geography, for example primary sampling units. Alternatively, adjustment cells can be based on external information from a census or larger survey. For example, if age is known only for respondents but the age distribution of the population is available from census data, then the respondent distribution of age can be weighted to correspond to the population age distribution using a method such as poststratification. Both these approaches yield weighted estimates, but the associated standard errors are different, since the proportions of the population in each adjustment cell are estimated in the first case and known in the second case. Both approaches remove the component of nonresponse bias attributable to differential nonresponse rates across the adjustment cells, and eliminate bias if within each adjustment cell respondents can be regarded as a random subsample of the original sample within that cell (i.e., the data are MAR given indicators for the adjustment cells).

A useful extension of this approach with more extensive background information is *response propensity* stratification, where (a) the indicator for unit nonresponse is regressed on the background variables, using the combined data for respondents and nonrespondents and a method such as logistic regression appropriate for a binary outcome; (b) a predicted response probability is computed for each respondent based on the regression in (a); and (c) adjustment cells are formed based on a categorized version of the predicted response probability. Theory (Rosenbaum and Rubin, 1983; Little, 1986) suggests that this is an effective method for removing nonresponse bias attributable to the background variables.

Inverse probability weighting methods can be useful for removing or reducing nonresponse bias, but they do have serious limitations. First, the method is inefficient, particularly if information ignored in the incomplete cases is extensive (e.g., prior wave information for nonrespondents to one wave of a longitudinal survey). Inverse probability weighting can have unacceptably high variance, as when outlying values of a variable are given large weights. One approach to this problem is to develop weights that correspond to linear or loglinear prediction models based on observed covariates (Deville and Särndal, 1992; Bethlehem, Chapter 18, this volume). This form of weighting does not necessarily yield design-consistent estimates in missing-data settings, but estimates can be adjusted to achieve this (Kott, 1994). Another approach is to reduce the variability of the weights by random effects models (Elliott and Little, 2000; Gelman and Carlin, Chapter 19, this volume).

A second problem is variance estimation for estimates involving estimated weights. Explicit formulas are available for simple estimators such as means under simple random sampling (Oh and Scheuren, 1983), but methods are not well developed for more complex problems, and often ignore the component of variability from estimating the weight from the data. One way to avoid this problem, at the expense of increased computational effort, is to compute sampling errors using a replication method (e.g., the bootstrap or the jackknife) that recomputes the nonresponse weights on each replicate data set, as in the chapters by Shao (Chapter 20) and Lee, Rancourt, and Sarndal (Chapter 21).

A common practical alternative to weighting for unit nonresponse is substitution, where substitute respondents are found for the unit nonrespondents. Analyses typically ignore any differences between nonrespondents and their substitutes, and as such are vulnerable to bias and overstated precision. In this volume, Rubin and Zanutto (Chapter 26) propose development of substitution methods that attempt to address these problems, and we hope that their contribution motivates additional methodological work in this area.

1.5.2 Methods for Item Nonresponse

A common and simple analytic approach to item nonresponse simply discards any cases that involve missing values, a form of CC analysis. This method is simple but entails a loss of information in the discarded cases and the potential for bias if the complete cases are a biased sample, that is, if the missing data are not missing completely at random (MCAR; see Rubin, 1976). The size of the resulting bias depends on the degree of deviation from MCAR, the amount of missing data, and the specifics of the analysis. In particular, the bias in estimating the mean of a variable is easily shown to be the difference in the means for complete and incomplete cases multiplied by the fraction of incomplete cases.

Methods that fill in or impute the missing values have the advantage that observed values in the incomplete cases are retained. Imputations are often developed using information from respondents, but can also be based on external data such as administrative records, as discussed by Zanutto and Zaslavsky (Chapter 27). A common naive approach imputes missing values by their simple unconditional sample means (i.e., marginal means). This can yield satisfactory point estimates of unconditional means and totals, but it yields inconsistent estimates of other parameters, for example, variances or regression coefficients (Kalton and Kasprzyk, 1982; Little and Rubin, 1987). Inferences (tests and confidence intervals) based on the filled-in data are seriously distorted by bias and overstated precision. Thus, the method cannot be generally recommended.

An improvement over unconditional mean imputation is *conditional mean* imputation, in which each missing value is replaced by an estimate of its conditional mean, given the values of observed values. For example, in the case of univariate nonresponse with $Y_1, \ldots, Y_{p-1}$ fully observed and Y_p sometimes missing, one approach is to classify cases into cells based on similar values of observed variables, and then to impute missing values of Y_p by the within-cell mean from the complete cases in that cell. A more general approach is regression imputation, in which the regression of Y_p on $Y_1, \ldots, Y_{p-1}$ is estimated from the complete cases, including interactions as needed, and the resulting prediction equation is used to impute the estimated conditional mean for each missing value of Y_p. Iterative versions of this method lead (with some important adjustments) to ML estimates under multivariate normality (Little and Rubin, 1987, Section 8.2).

Although conditional mean imputation yields best predictions of the missing values in the sense of mean squared error, it leads to distorted estimates of quantities that are not linear in the data, such as percentiles, correlations and other measures of association, and variances and other measures of variability. A solution to this problem is to use random draws rather than best predictions to preserve the distribution of variables in the filled-in data set. An example is *stochastic regression* imputation, in which each missing value is replaced by its regression prediction plus a random error with variance equal to the estimated residual variance.

Other imputation methods impute values observed in the dataset. One such method is the *hot-deck,* as used by the Census Bureau for imputing income in the Current Population Survey (CPS) (Hanson, 1978; David et al., 1986). Each nonrespondent is matched to a respondent based on variables that are observed for both; the missing items for the nonrespondent are then replaced by the respondent's values. A more general approach to hot-deck imputation is to define a distance function based on the variables that are observed for both nonrespondents and respondents. The missing values for each nonrespondent are then imputed from a respondent that is close to the nonrespondent in terms of the distance function (Little, 1988a; Lazzeroni et al., 1990). Hot-deck methods are discussed in this volume by Marker et al. (Chapter 22).

A serious defect of imputation is that it invents data. More specifically, a single imputed value cannot represent all of the uncertainty about which value to impute, so analyses that treat imputed values just like observed values generally underestimate uncertainty, even if nonresponse is modeled correctly and random imputations are created. Large-sample results (Rubin and Schenker, 1986) show that for simple situations with 30% of the data missing, single imputation under the correct model results in nominal 90% confidence intervals having actual coverage below 80%. The inaccuracy of nominal levels is even more extreme in multiparameter testing problems.

Two approaches to this deficiency are considered in this book. The first approach is to apply a replication method of variance estimation and recompute the imputations on each replicate sample (Fay, 1996; Rao, 1996; Shao, Chapter 20, this volume; Lee, Rancourt, and Särndal, Chapter 21, this volume). For example, if imputations are created using regression imputation, and standard errors are computed on the filled-in data using the bootstrap, then the regression imputations for each bootstrap sample are recomputed with regression coefficients estimated from that bootstrap sample, rather than from the overall sample. In simple settings, analytical formulas can be used to create the necessary adjustments of the imputations from the overall sample.

The second approach is multiple imputation (MI) (Rubin 1987, 1996). Instead of imputing a single set of draws for the missing values, a set of M (say $M = 5$) datasets are created, each containing different sets of draws of the missing values from their predictive distribution. We then apply the analysis to each of the M datasets and combine the results in a simple way. In particular, for scalar estimands, the MI estimate is the average of the estimates from the M datasets, and the variance of the estimate is the average of the variances from the five datasets plus $1+1/M$ times the sample variance of the estimates over the M datasets (the factor $1+1/M$ is a small-M correction). The last quantity here estimates the contribution to the variance from imputation uncertainty, which is missed by single imputation methods. The chapters in this volume by Heeringa et al. (Chapter 24), Brownstone, Golub, and Kazimi (Chapter 25), Rubin and Zanutto (Chapter 26), and Zanutto and Zaslavsky (Chapter 27) provide applications of MI in a variety of settings. The relative merits of these alternative approaches to propagating imputation uncertainty are debated in Rubin (1996), Fay (1996), Rao (1996), and the associated discussion of these articles

Likelihood-based methods, in particular, maximum likelihood (ML) or Bayesian inference, avoid imputation by formulating a statistical model and basing inference on the likelihood function of the incomplete data. Define Y, M, and Z as above, and suppose the data and missing-data mechanism are modeled in terms of a joint distribution for Y and M given Z. *Selection models* specify this distribution as:

$$f(Y, M|Z, \theta, \psi) = f(Y|Z, \theta) f(M|Y, Z, \psi) \tag{1.9}$$

where $f(Y|Z, \theta)$ is the model in the absence of missing values, $f(M|Y, Z, \psi)$ is the model for the missing-data mechanism, and θ and ψ are unknown parameters. The likelihood of θ, ψ given the data Y_{obs}, M, Z is then proportional to the density of Y_{obs}, M given Z (regarded as a function of the parameters θ, ψ), and is obtained by integrating out the missing data Y_{mis} from Eq. (1.9), that is:

$$L(\theta, \psi|Y_{obs}, M, Z) = \text{const} \times \int f(Y, M|Z, \theta, \psi) dY_{mis} \tag{1.10}$$

The likelihood of θ ignoring the missing-data mechanism obtained by integrating the missing data from the marginal distribution of Y given Z, that is:

$$L(\theta|Y_{obs}, Z) = \text{const} \times \int f(Y|Z, \theta) dY_{mis} \tag{1.11}$$

The likelihood (1.11) is easier to handle computationally than (1.10), and more important, it avoids the need to specify a model for the missing-data mechanism, about which little is known in many situations. Rubin (1976) showed that valid inferences about θ are obtained from (1.11) when the data are MAR, as in Eq. (1.8). If in addition θ and ψ are distinct in the sense that they have disjoint sample spaces, then likelihood inferences about ψ based on (1.11) are equivalent to inferences based on (1.10); the missing-data mechanism is then called *ignorable* for likelihood inferences. Large-sample inferences about θ based on an ignorable model are based on ML theory, which states that under regularity conditions

$$\theta - \hat{\theta} \sim N_k(0, C), \tag{1.12}$$

where $\hat{\theta}$ is the value of θ that maximizes (1.11), and $N_k(0, C)$ is the k-variate normal distribution with mean zero and covariance matrix C given by the inverse of the observed or expected information matrix. Thus, if the data are MAR, the likelihood approach reduces to developing a suitable model for the data and computing $\hat{\theta}$ and C. In many problems, maximization of the likelihood to compute $\hat{\theta}$ requires numerical methods. Standard optimization methods such as Newton–Raphson or scoring can be used. Alternatively, the expectation–maximization (EM) algorithm (Dempster et al., 1977) or extensions (McLachlan and Krishnan, 1997) can be applied. Little and Rubin (1987) and Meng and Pedlow (1992) provide many applications of EM to particular models.

Nonignorable, non-MAR models apply when missingness depends on the missing values. A correct likelihood analysis must be based on the full likelihood from a model for the joint distribution of Y and M. The standard likelihood asymptotes apply to nonignorable models providing the parameters are identified, and computational tools such as EM also apply to this more general class of models. However, information to estimate both the parameters of the missing-data mechanism and the parameters of the complete-data model is often very limited, and estimates are very sensitive to misspecification of the model. Often, a sensitivity analysis is needed to see how much the answers change for various assumptions about the missing-data mechanism.

Heitjan and Rubin (1991) and Heitjan (1994) extended the formulation of missing-data problems via the joint distribution of Y and M to more general incomplete data problems involving coarsened data. The idea is to replace the binary missing-data indicators by random coarsening variables that map the y_{ij} values to coarsened versions $z_{ij}(y_{ij})$. Full and ignorable likelihood can be defined for this more general setting. Heeringa et al. (Chapter 24) apply this theory when developing MI methods for multivariate coarsened survey data.

Asymptotic standard errors for ML estimates might be based on an estimate of the information matrix or replication methods (Little, 1988b; Efron, 1994; Shao, Chapter 20, this volume; Lee et al., Chapter 21, this volume). An alternative is to switch to a Bayesian simulation method that simulates the posterior distribution of θ. For Bayesian methods, a prior distribution is assumed for the parameters and inference is based on the posterior distribution of the parameters of interest. For ignorable models this is

$$p(\theta|Y_{\text{obs}}, M, Z) \equiv p(\theta|Y_{\text{obs}}, Z) = \text{const}\, p(\theta|Z) \times f(Y_{\text{obs}}|Z, \theta)$$

where $p(\theta|Z)$ is the prior and $f(Y_{\text{obs}}|Z, \theta)$ is the density of the observed data. Since the posterior distribution rarely has a simple analytic form for incomplete-data problems, simulation methods are often used to generate draws of θ from the posterior distribution $p(\theta|Y_{obs}, M, Z)$. Data augmentation (Tanner and Wong, 1987) and the closely related Gibbs' sampler (e.g., Tanner, 1996) are iterative methods of simulating the posterior distribution of θ and the missing values that combine features of the EM algorithm and multiple imputation. The Gibbs' sampler starts with an initial draw $\theta^{(0)}$ from an approximation to the posterior distribution of θ. Given a value $\theta^{(t)}$ of θ drawn at iteration t:

(A) Draw $Y_{mis}^{(t+1)}$ with density $p(Y_{mis}|Y_{obs}, Z, \theta^{(t)})$;
(B) Draw $\theta^{(t+1)}$ with density $p(\theta|Y_{obs}, Y_{mis}^{(t+1)}, Z)$.

The procedure is motivated by the fact that the distributions in (A) and (B) are often much easier to draw from than the correct posterior distributions $p(Y_{mis}|Y_{obs}, Z)$ and $p(\theta|Y_{obs}, Z)$. The iterative procedure can be shown in the limit to yield a draw from the joint posterior distribution of Y_{mis}, θ given Y_{obs}, Z. This algorithm can be run independently K times to generate K_{iid} draws from the approximate joint posterior distribution of θ and Y_{mis}. Schafer (1996) developed algorithms that use iterative Bayesian simulation to multiply impute rectangular data sets with arbitrary patterns of missing values when the missing-data mechanism is ignorable, and the complete-data matrix can be modeled from the multivariate normal, multinomial loglinear, and general location models. For a real survey application, see Ezzati-Rice et al. (1995). Heeringa et al. (Chapter 24) apply the Gibbs' sampler to create multiple imputations under an extension of the general location model. We expect future research to continue to develop a broader range of models tuned for survey applications.

Model-based inferences about parameters θ can be based directly on their simulated posterior distribution, in which case the imputations play a role of stepping stones for computing these distributions. Alternatively, the imputations of the missing values themselves might be regarded as the main output of the method. In particular, an important use of the methods is to create multiple imputations for a public use file, which may be subject to a variety of analyses by survey users. One advantage of this approach is that the imputation model may differ from the analysis model. Thus, the imputer may include in the imputation model information (such as fine-grained geographic data) that is not available to the user because of confidentiality constraints. Another useful feature is that the imputation model is used only to create imputations, so the impact of misspecification of that model is confined to the imputations it produces. If the amount of missing data is very modest, even a simple and limited imputation model may yield serviceable inferences. This is important since building a complicated imputation model may entail an unnecessary use of resources in that setting. Meng (Chapter 23) discusses theoretical and simulation properties of multiple imputation when the imputation model and the analysis model differ.

Most approaches to unit and item nonresponse in surveys to date have assumed that the missing data are MAR. Nonignorable, non-MAR models are needed when missingness depends on the missing values. For example, suppose a subject in an income survey refused to report an income amount because the amount itself is high (or low). If missingness of the income amount is associated with the amount after controlling for observed covariates (such as age, education, or occupation) then the mechanism is not MAR and methods for imputing income based on MAR models are subject to bias. A correct analysis must be based on the full likelihood from a model for the joint distribution of Y and M. The standard likelihood asymptotics apply to nonignorable models, providing the parameters are identified, and computational tools such as the Gibbs' sampler also apply to this more general class of models. Brownstone, Golub, and Kazimi (Chapter 25) discuss one application of a nonignorable model in a survey setting, and this is an active area of research (see for example Amemiya, 1984; Fay, 1986; Glynn et al., 1993; Little and Rubin, 1987, Chapter 11; Little, 1993, 1995; Wu and Carroll, 1988; Wu and Bailey, 1988; Scharfstein et al., 1999).

1.6 BLENDING BEHAVIORAL AND STATISTICAL VIEWPOINTS ON NONRESPONSE

This volume deliberately combines research that asks why nonresponse occurs, how it might be reduced by survey design, and how to analyze data in the presence of nonresponse. These three questions do not share the same research history, but it is becoming clear that they need to be blended in the future.

Although nonresponse increases the cost of data collection, its effects on the error properties of survey statistics are of greater concern. As Sections 1.1 and 1.5 show, the key to reduced nonresponse error, either through nonresponse rate reduction or analytic techniques, is insight into how the influences on nonresponse behavior relate to the survey variables of interest. Whether nonresponse is actually MAR depends on to what extent missingness is caused by missing values and whether the causes of nonresponse are the survey variables themselves.

Most survey design features aimed at reducing nonresponse are based on the implicit assumption that all sample units react similarly to the feature. Because of that, the literature addresses whether unit response rates increase with advance letters, incentives, and persuasion attempts. For item nonresponse, it addresses whether training interviewers to probe inadequate responses or particular visual layouts for questionnaires reduces item missing data. Even when such investigations succeed in determining what produces lower missing data, they may not necessarily address to what extent nonresponse errors of various statistics are affected by the lower missing data.

At this point in the evolution of the field, we see great need for the development of behavioral theories that reveal the influences on the response decision *and* their relationships to the survey variables. Summarizing the literature reviewed above, two influences on unit nonresponse may be linked with the threat of nonignorable nonresponse—topic salience and the effect of sponsorship. For example, if some survey protocols have the effect of strong appeal only to those very interested in the key survey measures, then MAR assumptions are likely to fail. If behavioral research could demonstrate such links between survey variables and the causes of response, then nonresponse error reduction either through increased response rates or properly specified adjustment models could be achieved.

On the estimation side, more work is needed on the development of good imputation models that reflect the main features of the population and the survey design. Nonignorable modeling is still in its relative infancy in the survey field, and work is needed in building realistic nonignorable models that reflect what is known about the missing-data mechanism (e.g., Little, 1995). The question of the most appropriate way to assess imputation uncertainty, either by multiple imputation or by replication methods, is somewhat unresolved at present, and could benefit from additional research. On the model-checking side, more work on diagnostics for imputation and weighting models is needed, and further theoretical and practical exploration of the properties of imputation when the imputer's and analyst's model differ would be useful, along the lines of Chapter 23 of this book.

CHAPTER 2

Developing Nonresponse Standards

Tom W. Smith, *National Opinion Research Center, University of Chicago*

2.1 INTRODUCTION

For over a half century, scholars have been studying survey nonresponse and survey researchers have been discussing standards for their field. This chapter examines the nexus of these two efforts: the development of nonresponse standards. It considers (1) the formulation of definitions and standards concerning nonresponse among (a) scholarly researchers, (b) professional, trade, and academic organizations, and (c) the federal government; (2) the current nonresponse definitions and standards of professional, trade, and academic organizations; (3) how survey nonresponse is actually reported by newspapers, scholarly researchers, survey organizations, and government agencies; and (4) how professionalization has influenced the development of nonresponse standards.

2.2 THE FORMULATION OF DEFINITIONS AND STANDARDS CONCERNING NONRESPONSE

2.2.1 Scholarly Research

Nonresponse has been a major topic of interest in survey methods. Research extends back to the emergence of polling in the 1930s and has been a regular feature in statistical and social science journals since the 1940s. An analysis of JSTOR statistical journals dates the first nonresponse article from 1945 and the *Public Opinion Quarterly* index's earliest reference is from 1948. The index of *Public Opinion Quarterly* contains 125 articles on this topic; a full-text search of journals covered in JSTOR finds the following number of articles, by subject area, that included the word "nonresponse": political science—62, economics—87, sociology—146, and statistics—431. *Sociological Abstracts* lists 219 articles using the word "nonresponse." The vast majority of these articles focus on the main issues in the field: reducing nonresponse, measuring nonresponse bias, and compensating for nonresponse by imputation and/or weighting.

Some scholarly publications have, however, focused on the matter of developing standards for surveys in general and nonresponse in particular. Some have directly argued for the establishment of standards for survey research. A statisticians' early effort was W. Edward Deming's "Principles of Professional Statistical Practice" (1965, pp. 1892–1993). He proposed that "The statistician's report or testimony will deal with statistical reliability of the results," including "nonresponse and illegible or missing units." Similarly, market research calls for nonresponse standards showed in George S. Day's (1975, p. 466) urging that "top priority should be assigned to documenting the seriousness of the problem in terms of accepted and reasonably standardized industry-wide measures of the components of non-response rates." Another line of research tried to bring order to the babel of terms used in descriptions of survey nonresponse by clearly defining terms (e.g., Kviz, 1975; Groves, 1989; Lessler and Kalsbeek, 1992).

2.2.2 Organized Efforts by Professional, Trade, and Academic Organizations and the Federal Government

Many organizations have struggled to develop standards for survey research in general and for nonresponse in particular. The intertwined, major initiatives of professional, trade, and academic organizations and the federal government from the late 1960s on will be considered.

Professional, Trade, and Academic Organizations. Although the American Association for Public Opinion Research (AAPOR) had struggled with the issue of standards from its inception (Smith, 1999), it was not until the 1960s that AAPOR and other professional associations adopted codes and specifically addressed the matter of nonresponse. AAPOR adopted standards for the disclosure of survey findings in 1967. Likewise, in 1968 the National Committee on Published Polls (renamed the National Council on Public Polls in 1969) was formed. This trade organization also adopted standards of disclosure.

In 1973, the American Statistical Association (ASA) with support from the National Science Foundation (NSF) held two Conferences on Surveys of Human Populations (CSHP) to "discuss the problems of present-day surveys . . . [and] explore whether or not these problems may now have reached a level or are growing at a rate that pose a threat to the continued use of surveys as a basic tool of social science research ("Report," 1974, p. 30)." The group found that "confusion" exists "in regard to what is meant by such measures as completion rates, nonresponse, refusals, etc." Since the participants felt that problems in general and nonresponse in particular were on the rise, they recommended that a well-staffed committee be funded to do "a detailed examination of the characteristics and impact of the trends explored in this report and to develop ways and programs for the survey research community to meet these trends" and "to develop adequately comprehensive guidelines or standards for use throughout the survey research profession (p. 33)."

The CSHP suggestions were carried forward indirectly by the establishment in 1974 of the Survey Research Methods (SRM) section of the ASA. At its first meeting, this new group set up a task force on standards (Bailar and Lanphier, 1978, p. vii). Moreover, in 1975 SRM secured funds from NSF to assess survey practices via a developmental and feasibility study of surveys of the federal government. This led to a survey of surveys. Even though the data collectors and/or federal managers were directly contacted for the desired details on survey design and methods, Bailar and Lanphier (1978, p. 13) found that "Survey Response rates were difficult to collect and compare. Response rates have different names and different definitions in different places and circumstances."

This led to incomparability across surveys. For example, "in one survey, a response of .90 was actually .56 when substitutions were properly considered. They were able to get a reported response rate from 26 of their 36 surveys, were able to check the reported response rate for 23 surveys, and found correct calculations for 18, only half of the targeted surveys.

Building on the ASA's NSF-supported efforts were the NAS Panels on Incomplete Data and Survey Measurement of Subjective Phenomena. The Panel on Incomplete Data was formed in 1977 "to make a comprehensive review of the literature on survey incompleteness in sample surveys and to explore ways of improving the methods of dealing with it." In its three volume report and collection of papers (Madow et al., 1983, Vol. 1, p. 8), it made 35 recommendations, including the following:

#4 Compute nonresponse rates, completion rates, and item coverage rates during as well as after the data collection effort for the samples, for important domains, and for important items.

#5 Prepare one or more accountability tables and define nonresponse and completion rates in terms of entries in the accountability table. Provide one or more tables containing the data used in calculating item coverage rates to avoid any misunderstanding of the definition.

#6 Consider preparation of accountability tables for domains and specified number of callbacks as well as for the total sample.

The Survey Measurement Panel was organized in 1980 and suggested (Turner and Martin, 1984, p. 311) that:

> A knowledgeable working party of six or eight people could produce a document specifically oriented to surveys of subjective phenomena comparable to the booklet, "Standards for Discussion and Presentation of Errors in Survey and Census Data" (Gonzalez, et al., 1975), based on practice of the Bureau of the Census and other government agencies engaged in fact-finding surveys. To be helpful such a document need not have the force of law; the cited document has only the force of example in regard to nonfederal survey organizations.

In the commercial sector, the nonresponse problem was at the same time taken up by the Marketing Sciences Institute (MSI) and the Council of American Survey Research Organizations (CASRO). In the late 1970s, in response to industry concerns about threats to market research in general and nonresponse in particular, MSI, in cooperation with CASRO, launched a series of nonresponse studies under the direction of Frederick Wiseman. One (Wiseman and McDonald, 1978) looked at response rates from 182 telephone surveys. It found that comparisons were difficult "because of the lack of uniform definitions" and concluded that "top priority should be given to the establishment of industry-wide standard definitions and reporting procedures to help monitor industry trends, and to establish benchmarks to help research firms and users evaluate response levels in their own surveys (p. vi)."

This led to the establishment in 1979 of a MSI Nonresponse Steering Committee consisting of representatives from the corporations that supported MSI. One of its main activities was to conduct a study of MSI and CASRO members (Wiseman and McDonald, 1980). Firms were asked how often they calculated response rates and were given a set of final outcome figures for a survey and asked to report what the response and other outcome rates were for the example survey. They found that many firms often did not calculate response rates and that few firms calculated rates in the same manner. Of the 55 responses to their survey, 40 calculated response rates from the supplied figures as requested and they produced 29 different rates with no more than three adopting the same method. For the example data, the calculated response rates ranged from 12% to 90%.

In 1982 a CASRO task force's report "On the Definition of Response Rates" repeated the lament that there was no common understanding of nonresponse terms and proposed a basic definition of response rates and examples of how they should be calculated. The task force also seemed to favor "accounting tables" that report the final status of all sampled cases.

It was not until the AAPOR initiatives in the mid 1990s on Best Practices and Standard Definitions that professional, trade, and academic organizations again made serious forward strides. After a failed attempt to gather data on response rates in the late 1980s (Bradburn, 1992), the next major AAPOR initiative involving standards was taken in 1995–1997 when the Council decided to formulate a statement of "Best Practices for Survey and Public Opinion Research and Survey Practices AAPOR Condemns" (AAPOR, 1997). Expanding on the AAPOR Code's Standards of Minimal Disclosure, "Best Practices" indicated that a series of outcome rates should be reported for surveys.

At about the same time (1996–1998), an AAPOR committee developed standard definitions for the final disposition of case codes and of various outcome rates (e.g., response rates and cooperation rates) based on these codes. At their urging, AAPOR's official journal, *Public Opinion Quarterly,* adopted the new standards (Price, 1999, pp. i–ii).

Government Agencies. The Office of Management and Budget (OMB) has the responsibility to approve all U.S. Federal surveys and began issuing various guidelines and admonitions in the mid-1970s. In 1974, OMB (OMB, 1976, pp. 12, 13) indicated that nonresponse error should be carefully considered "in designing the survey, in establishing controls over survey operations, and for the information of users of the data when they are published." It also stated that "strenuous efforts should be made to collect data from every unit in the sample, using follow-ups where necessary."

Also, perhaps as early as 1976, OMB formulated positions on acceptable levels of nonresponse (Wiseman, 1983). In 1979 OMB (OMB, 1979, p. 6) adopted interim guidelines that stated:

> It is expected that data collection based on statistical methods will have a response rate of at least 75 percent. Proposed data collections having an expected response rate of less than 75 percent require special justification. Data collection activities having a response rate of under 50 percent should be terminated. Proposed data collection activities having an expected response rate of less than 50 percent will be disapproved.

No definition of response rates was offered, however.

Current guidelines (OMB, 1999, p. 73) merely prohibit statistical surveys that "do not produce reliable results" and nonresponse is listed among the common sources of failure. In Appendix C: Frequently Asked Statistical Questions, OMB discusses estimating likely response rates for a survey (FASQ #1—Estimating Response Rates) and data quality problems that may arise from nonresponse (FASQ #2—Consequences of Low Response Rates). As before, no definitions are provided.

In the early 1990s the OMB-led Federal Committee on Statistical Methodology (FCSM) formed a subcommittee on nonresponse. It was to "better understand unit nonresponse in surveys, including levels of nonresponse and measures used to compute nonresponse rates" (Shettle et al., 1994, p. 972). The subcommittee reviewed nonresponse trends in federal demographic and establishment surveys for 1981–1991 (Shettle, et al., 1994; Johnson et al., 1994; Osmint et al., 1994). Although not yet adopted as general FCSM guidelines, the subcommittee recommended that (pp. 975–976):

1. Survey staffs should compute response rates in a uniform fashion over time and document response rate components on each edition of the survey.
2. Survey staff for repeat surveys should monitor response rate components (e.g., refusals, not-at-homes, out-of-scopes, address not locatable, postmaster returns) over time, in conjunction with documentation of cost and design changes.
3. Agencies that sponsor surveys should be empowered to report the response rates of these surveys. The sponsoring agency should explain how response rates are computed for each survey it sponsors. Response rates for any one survey should be reported using the same measure over time, so that users may compare the response rates. Response rate components, including actual counts, should be published in survey reports.

Two other interagency efforts dealing with nonresponse were organized in the 1990s—the Interagency Group on Establishment Nonresponse (IGEN), and the Interagency Household Survey Nonresponse Group (IHSNG) (Bay, 1999; Atrostic and Burt, 1999; Kydoniefs and Stanley, 1999). These groups are studying and monitoring nonresponse and developing general, interagency methods to reduce nonresponse, but do not seem to be aimed at formulating standards on nonresponse per se.

In sum, efforts have been underway since the 1940s to develop nonresponse standards. Since the late 1960s, interconnected efforts have been carried out by professional, trade, and academic organizations and various government agencies to explore the problem of nonresponse and forge standards related to it. Different organizations sometimes worked together (e.g., MSI and CASRO; AAPOR and SRM), government agencies funded academic programs (e.g., NSF's support of the ASA/SRM and NAS), several private undertaking were motivated, at least in part, by governmental actions (e.g., CASRO and MSI by OMB guidelines), and many individuals participated across succeeding efforts. (On overlapping memberships, see Smith, 1999.) Through these and other initiatives a series of professional and industrial nonresponse codes were developed.

2.3 CURRENT PROFESSIONAL, ACADEMIC, AND INDUSTRY CODES AND POSITIONS

To ascertain the current situation regarding nonresponse standards, the codes and official positions of 18 major professional, trade, and academic associations involved in survey and marketing research were examined. Coverage is restricted to international organizations and those in the United States, although some foreign organizations are mentioned. For more on standards in other countries see Kasse, 1999; Hidiroglou et al., 1993; and Allen et al., 1997. Trade or industry associations are those to which organizations rather than individuals belong. Professional and academic associations have individuals as members.

The core professional associations on the survey research side are AAPOR and the World Association for Public Opinion Research (WAPOR). The European Society for Opinion and Marketing Research (ESOMAR) is a bridge between survey research and market research. The main market research associations are the American Marketing Association (AMA), the Marketing Research Association (MRA), and the Association for Consumer Research (ACR). The main statistical groups are ASA, the International Association of Survey Statisticians (IASS), and the International Statistical Institute (ISI). The main trade associations are CASRO, the Council for Marketing and Opinion Research (CMOR), the National Council on Public Polls (NCPP), and, on the market research side, the Advertising Research Foundation (ARF) and Association of European Market Research Institutes (AEMRI). Within the market research arena are two associations dealing with audience and readership studies, the Audit Bureau of Circulations (ABC) and the Media Ratings Council (MRC). Finally, there are two associations of associations—the Research

Industry Coalition (RIC), which includes among its members AAPOR, ACR, AMA, ARF, CASRO, MRA, and NCPP, plus the American Psychology Association, the Newspaper Association of America, the National Association of Broadcasters, the Professional Marketing Research Society of Canada, the Qualitative Research Consultants Association, and the Travel and Tourism Research Association; and the European Federation of Associations of Market Research Organizations (EFAMRO). A detailed account of the positions of these organizations is presented in Smith (1999) and summarized below.

Of the 18 professional, trade, and academic organizations examined, three have no codes or any relevant official statements (CMOR, ACR, and IASS, but IASS is covered by the code of ISI). Another three organizations have only brief general statements about doing good, honest research (AMA, ARF, MRA). Yet another three have general pronouncements about being open about methods and sharing technical information with others, but no details on what should be documented (ASA, ISI, RIC). Then there are nine that have some requirement regarding nonresponse (AAPOR, CASRO, ESOMAR, AEMRI, EFAMRO, NCPP, WAPOR, ABC, and MRC, plus ARF, if its guidelines for newspaper audience surveys were counted as opposed to its less detailed, general pronouncements).

Of the eight organizations with an official journal (AAPOR, WAPOR, ESOMAR, AMA, ARF, ISI, IASS, and ASA), one (AAPOR) has a definite standard about reporting and calculating response rates, two have general pronouncements that mention nonresponse bias or response rates (AMA and ARF), and one has a standard on data sharing (ASA).

The nine referring to nonresponse in their codes and statements all require that response rates (or some related outcome rate) be reported. AAPOR, WAPOR, EFAMRO, ABC, and MRC put nonresponse among those facets of survey methodology that must be reported automatically and routinely, whereas CASRO, ESOMAR, AEMRI, and NCPP have less comprehensive reporting rules. However, the ABC and MRC requirements are only studies of media usage. ARF also has more stringent guidelines just for media usage surveys.

Only a subset of the nine mentioning nonresponse require anything beyond reporting. Five organizations provide at least some definition of response and/or related outcome rates and these appear in nonbinding documents and statements and not as part of their codes (AAPOR, CASRO, NCPP, ABC, and MRC) and four provide no definitions (ESOMAR, EFAMRO, AEMRI, and WAPOR). Only the AAPOR, CASRO, and ABC definitions are detailed.

Two organizations deal with the issues of nonresponse bias in their codes. The WAPOR code, right after requiring the reporting of the nonresponse rate, calls for information on the "comparison of the size and characteristics of the actual and anticipated samples," and the ESOMAR code requires in client reports "a discussion of any possible bias due to non-response." Three organizations mention nonresponse bias in official documents. AAPOR in its "Best Practices," but not its code, urges that nonresponse bias be reported. The ASA addresses the matter in its "What is a Survey?" series. The AMA in its publication, the *Journal of Market Research,* requires authors to "not ignore the nonrespondents. They might have different characteristics than the respondents."

Three organizations deal with technical standards. AAPOR, as part of "Best Practices," but not its code, indicates that survey researchers should try to maximize response rates and discusses means to do. ABC and ARF are more precise in specifying minimum number of calls and details on other efforts that should be employed by media evaluation surveys.

Finally, only ABC specifies a minimally acceptable response rate, although even it provides exceptions to its standard.

In brief, only the professional, trade, and academic organizations at the core of survey research and in the subarea of media ratings research take up nonresponse in their codes, official statements, and organizational journals. General market research and statistical organizations do not explicitly deal with nonresponse issues in their codes and standards and only marginally address these in the guidelines of their official journals. Even among those organizations that consider nonresponse, reporting standards are incomplete, technical standards are lacking and/or regulated to less official status, and performance standards are nonexistent.

2.4 REPORTING NONRESPONSE

To see how the disclosure codes and other statements on documenting methods in general and nonresponse in particular have been carried out, the reporting and documentation practices of survey researchers, archives, newspapers, academic journals, and government agencies were examined.

2.4.1 Survey Reports and Releases

First, the current study descriptions and standard methodology statements of 11 major organizations conducting public opinion research in the public sphere were examined (Associated Press, Gallup, Harris, Los Angeles Times, New York Times/CBS News, Pew Research Center/Princeton Survey Research Associates, Roper Starch Worldwide, Wall Street Journal/NBC News, Washington Post/ABC News, Wirthlin, and Yankelovich). Four only mentioned sampling variation as a source of error in their surveys, four mentioned some additional sources of error, but did not mention nonresponse as one of these sources, and three included some statement about nonresponse as a source of error. For example, Harris states that "Unfortunately, there are several other possible sources of error in all polls or surveys that are probably more serious than theoretical calculations of sampling error. They include refusals to be interviewed (non-response), question wording and question order, interviewer bias, weighting by demographic control data and screening." However, no organization routinely reported response rates as part of their standard documentation.

Second, 14 university-based survey research organizations were examined (Indiana University Center for Survey Research, National Opinion Research Center, University of Chicago, Ohio State Social and Behavioral Sciences Survey Research

Center, Rutgers University Eagelton Institute, University of California, Los Angeles, Institute for Social Science Research, University of Cincinnati Institute for Policy Research, University of Maryland Survey Research Center, University of Michigan Institute for Social Research, University of Oregon—Oregon Survey Research Laboratory, University of Virginia Center for Survey Research, University of Wisconsin, Milwaukee, Institute for Survey and Policy Research, University of Wisconsin Survey Center, and Virginia Commonwealth University Survey and Evaluation Research Laboratory). Of these, five regularly issued state-level, public opinion polls roughly akin to the periodic polls conducted by the public pollsters. In their standard methodology statements three mentioned nonsampling sources of error including nonresponse and two routinely reported response rates. For example, the Buckeye State Poll of May, 1997 reported "A total of 2,881 randomly-generated telephone numbers were used in this survey. . . . Of these numbers, 1,508 were known to reach a household in Ohio with an eligible respondent. From these households, interviews were completed in 55% of the cases" (www.survey.sba.ohio-state.edu/bsp0597d.htm). The other nine did not have a regular poll with a standard methodology statement, but did provide a list of their recent surveys on their Web sites. Two provide no information on response rates, two report response rates for some studies, and five generally report response rates (often as part of full-text reports).

In sum, response rates are not routinely reported when survey results are released. Such reports are exceedingly rare in among public pollsters, but more common within the academic sector.

2.4.2 Survey Archives

Information on nonresponse for most surveys cannot be found by going to the survey archives where the data sets and their documentation are stored. The three major survey archives indicate that they rarely have data on nonresponse. The Roper Center, University of Connecticut, reported "There are virtually no data provided on response rates in the datasets archived. . . . Occasionally, reports will provide this information, but that would be the case in probably fewer than 5% of the studies." The archive at the Institute for Social Science Research, University of North Carolina, indicated that "fewer than 2% of the studies . . . include response rate in the stuff they send us (despite the fact that we ask for it)." The Interuniversity Consortium for Political and Social Research, University of Michigan, said that reporting was usual for federal surveys (especially in the last 10–15 years) and flagship surveys like the General Social Survey and American National Election Study, but that "many of the frequently-conducted telephone surveys right up 'til today . . . do not include response rates in their technical documentation. . . ."

2.4.3 Newspapers

Given the infrequency of the reporting of response rates in survey reports and especially the low level among the top public pollsters, it is unsurprising that newspapers rarely report response or related outcome rates. A study in 1980 by the Panel for Survey Measurement of Subjective Phenomenon found that only 2% of poll reports in American newspapers mentioned the nonresponse rate and less than 1% did so in British newspapers (Turner and Martin, 1984). A search of the Nexis/Lexis data base of major newspapers over a six month period from December, 1998 to May, 1999 found 70 articles that mentioned something about response or other outcome rates (e.g., Daily Yomiuri—Tokyo, "poll of traffic controllers has 83% response rate" and Star Tribune—Minnesota, "The cooperation rate among those contacted for this poll is 66% . . ."). This represented about 0.7–1.4% of all articles citing survey data.

2.4.4 Academic Journals

Following the lead of Presser (1984), the top academic journals in sociology, political science, and survey research were examined. Surveys are directly used in 86% of the survey research articles, 45% of sociology articles, and 31% of political science articles (for details see Smith, 1999). Of articles that directly use surveys, response rates were documented in 34% for survey research, 29% for sociology, and 20% for political science. In addition, partial information on response rates (e.g., panel attrition, but not initial nonresponse, and nonresponse for screened in cases, but not screening nonresponse) were provided for 6% of the articles in survey research, 15% in sociology, and 2% in political science. Thus, 60% of articles using surveys in survey research made no mention of response rates, as did 56% in sociology, and 77% in political science. A few articles without response rates did refer to other articles or reports in which technical details and sometimes explicitly response rates were available, but the majority of articles in each field had neither direct nor indirect information on response rates. Even when response rates were given, the information provided was typically meager. In survey research, only 20% provided definitions, as did 8% in sociology, and 2% in political science. The definitions were usually minimal and supporting details were incomplete.

In addition, in a number of cases the reported rates were miscalculated or questionable. One study reported a 62% response rate, but the true rate was about 52% when noncontacts were retained in the base. In another case, uncontacted phone numbers were dropped from the base and an unclear description made it impossible to determine whether they were reporting a response rate or a cooperation rate. Finally, another study reported a suspect response rate of 99%. The omission of and misreporting of nonresponse rates in journals largely comes from the absence of editorial policies dealing with such matters (Smith, 1999).

2.4.5 Government Reports

To gauge current nonresponse reporting practices, papers in the Current Population Report series of the Bureau of the Census were examined. Short reports included a general statement about nonsampling error, a person to contact for more information on data sources and accuracy, and no explicit mention of nonresponse. Moderate-length reports had methods appendixes, usually called something like "Source and Accuracy of Estimates." These generally included a standard table giving the average or general number of noninterview cases for the Current Population Survey (CPS). Definitions of nonresponse or of response rates were not given, nor were rates for particular CPS surveys reported. A few longer reports, especially those dealing with the Survey of Income and Program Participation, included detailed discussion of nonresponse and nonresponse bias and references to relevant methodological papers.

Overall, reporting of nonresponse is a rarity in the mass media and in public polls. Nonresponse is more regularly documented in academic and governmental studies, but is still sporadic at best. Moreover, when reported, response rates are seldom defined. Even rarer are explicit cites of AAPOR, CASRO, or NCPP nonresponse standards. A search of statistical and social science journals on JSTOR found one article that cited the nonresponse rules of these organizations.

2.5 PROFESSIONALIZATION AND INDUSTRY STANDARDS

Standards are developed and enforced by various processes. Any entity (a family, business, voluntary association, government agency, etc.) can establish internal rules and regulations (e.g., corporate dress codes, job performance evaluation procedures, machine precision levels, etc.) and enforce them. More general standards tend to develop and exist in three major types. First, standards may evolve from general practice and be maintained and enforced by convention (e.g., the use of the 0.05 level for accepting statistical significance in the social sciences).

Second, standards may be formally adopted by nongovernmental associations or organizations. Most common are standards adopted by professions and trades. These include rules promulgated by voluntary associations of those engaged in the same occupation to regulate their specialty and trade associations that set standards for an industry.

Third, standards may be set by the government through legislation or administrative rule. They may apply to society in general (e.g., criminal statutes) or to only some subgroup (e.g., affirmative action guidelines for government contractors).

During the last century, an emerging source of standards have been professional associations. More and more occupations have followed the lead of such groups as doctors and lawyers and organized associations for the self-regulation of their professions (Abbott, 1988; Freidson, 1984, 1994; Wilensky, 1964). One of the "necessary elements" of professionalization is the adoption of "formal codes of ethics . . . rules to eliminate the unqualified and unscrupulous, rules to reduce internal competition, and rules to protect clients, and emphasize the service ideal . . . (Wilensky, 1964, p. 145)" and "codes of ethics may be created both to display concern for the issue [good character] and to provide members with guides to proper performance at work (Freidson, 1994, p. 174)."

Survey research has begun to follow the path of professionalization, but has not completed the journey. Wilensky (1964) proposes five sequential steps that occupations go through to professionalization: 1) the emergence of the professional, 2) establishing training schools and ultimately university programs, 3) local and then national associations, 4) state licensing, and 5) formal codes of ethics. Survey research has only partly achieved the second, for although there are some excellent training and university programs, most practitioners are formally trained in other fields (statistics, marketing, sociology, etc.). Survey research has resisted certification and state licensing, although recent support for the proscription of fraudulent practices disguised as surveys (e.g., push polls and sugging—selling under the guise of a survey) have moved the field in that direction. Only in the early formation of professional associations did survey research fully embrace professionalization. Among other things, it has been the failure of survey researchers "to define, maintain, and reenforce standards in their area (Donsbach, 1997, p. 23)" that has deterred full professionalization. As Irving Crespi (1998, p. 77) has noted, " In accordance with precedents set by law and medicine, developing a code of standards has long been central to the professionalization of any occupation (see also Donsbach, 1997 and Smith, 1999)."

2.6 SUMMARY AND CONCLUSION

The impetus for nonresponse standards has come from several sources. First, there is a scientific and professional interest in advancing the field and improving the survey craft. Nonresponse has long been a major topic in the survey methodology literature and widely acknowledged as a major source of error. This has included explicit efforts by individual scholars to establish uniform definitions. Second, there are serious applied concerns about nonresponse. Declining response rates have been a perennial concern of survey researchers since at least the early 1970s ("Report," 1974) and survey practitioners see response-related issues as among the top future challenges (Rudolph and Greenberg, 1994). Third, the adoption of guidelines for surveys done for the federal government have sparked action by professional, trade, and academic associations.

These interests and concerns have generated on-going organized efforts to address the problem and have led to a number of notable standards and official statements on the matter. These are most extensive among organizations at the core of survey research (AAPOR, WAPOR, CASRO, ESOMAR, and NCPP) and in the area of auditing media usage (ABC, ARF, MRC) and less detailed among general market research and statistical organizations less centrally involved. Likewise, survey research journals document nonresponse better than do journals in the other social sciences.

Also, professional, trade, and academic organizations have less directly advanced the cause of standards by their general promotion of research methods. As Hollander (1992) has observed, "the annual AAPOR conference was recognized early on, together with *POQ*, which is older still, as a means of advancing standards. . . ." However, neither the direct nor indirect efforts have in practice led to uniform standards or routine disclosure being adopted.

On the development of nonresponse standards the professional, trade, and academic organizations have so far done the following:

1. Disclosure standards are required by the codes of the major professional, trade, and academic organizations in the field of survey research.
2. Technical standards in terms of the definition of terms and calculation of response rates and other outcome rates are being formulated as part of official statements by the organizations.
3. Performance standards in terms of levels of success that must be achieved (e.g., minimum response rates) or specific procedures that must be followed (e.g., a minimum number of call backs) have not been seriously considered by organizations except for audience validation surveys.

Various agencies of the federal government have in turn carried out the following regarding nonresponse:

1. As part of general disclosure or documentation standards, some parts of the Federal statistical system have adopted guidelines on the reporting of survey results and methodology.
2. Technical standards have generally not been adopted, nor have general definitions, but there are recent calls for consistent reporting of nonresponse within a survey time series.
3. Performance guidelines were adopted by OMB in terms of both minimally acceptable and target response rates, but definitions and technical standards to enable these standards were lacking and these numerical standards were later abandoned.

However, despite the development of these governmental and nongovernmental standards for nonresponse, enforcement has been lacking. Disclosure standards are routinely ignored and technical standards regarding definitions and calculations have not been widely adopted in practice. First, response rates and other outcome rates are usually not reported. Second, when reported, they typically are not defined. Third, when their basis is documented, various meanings and definitions are employed. As Richard Lau (1995, p. 5) has noted, "unfortunately, survey organizations do not typically publish response rates; indeed, there is not even a standard way of determining them" (see also Frey, 1989 and Kasprzyk and Kalton, 1998).

Moreover, it appears that no survey researcher or survey organization has been investigated for failing to disclose required information on nonresponse by any of the professional or trade associations with codes and enforcement mechanisms. If nonenforcement was because the code provisions were being widely adhered to, the lack of reenforcing sanctions would be understandable. But the low level of compliance along with the nonenforcement means that neither the organizations nor their members are taking the nonresponse provisions seriously. This in part is because of incomplete professionalization. Without a professional interest in standards above personal and organizational self-interest, many have opposed standards because they do not want to have to live up to them. Many adopt methods that allow them to claim a much higher response rate than permitted under the stricter professional conventions. For example, Wiseman and McDonald (1978 and 1980) and Bailar and Lanphier (1978) found that when surveys reported response rates that differed from the standard methods, their reported rates were uniformly higher than the stricter, conventional rates.

The development of nonresponse standards has followed a long, difficult, and complex path. Nonresponse has been recognized as a source of survey error for over 50 years and for about as long associations and scholars have been debating professional standards. Slowly, out of this deliberation a set of professional standards on disclosure, definitions, and the calculation of nonresponse have emerged. These standards grew from the efforts of many individual scholars and practitioners, from the organized efforts of several different professional, trade, and academic organizations, and from the federal government's initiatives to support conferences and research projects, establish documentation protocols for reporting their own statistical and survey work, and develop guidelines for survey conducted for the government. Participants in these efforts have fully spanned the various segments of the research community: basic and applied, commercial and noncommercial, and private and governmental. Although the progress towards developing nonresponse standards has been notable, the task remains unfinished. Detailed technical standards are just being established, the disclosure standards have yet to be generally adopted among practitioners and no meaningful enforcement of standards has so far occurred.

ACKNOWLEDGMENTS

I would like to thank the following for their assistance and/or comments on earlier drafts of this chapter: Leo Bogart, Robert Groves, Sidney Hollander, Daniel Kaspryzk, Clyde Tucker, and Barbara Bailar.

CHAPTER 3

Trends in Household Survey Nonresponse: A Longitudinal and International Comparison

Edith de Leeuw, *MethodikA, The Netherlands*
Wim de Heer, *Statistics Netherlands*

3.1 INTRODUCTION

Survey nonresponse is as old as survey research itself. However, trend studies suggest that participation in surveys is declining over time (e.g., Steeh, 1981; Goyder, 1987; Hox and de Leeuw, 1994; Sugiyama, 1992; Kojetin and Tucker, 1999). Brehm (1994) states that all sectors of the survey industry—academic, government, business, and media—are suffering from falling response rates. Groves and Couper (1998) recently devoted an entire book to nonresponse in household surveys. The maintenance of response rates appears to be an increasing problem in the developed world. In the words of Norman Bradburn (1992): "We all believe strongly that response rates are declining and have been declining for some time."

However, some specific studies do not show such clear trends (Smith, 1995; Schnell, 1997). For instance, Smith pointed out that the General Social Survey (GSS) in the United States shows no trend in refusal rate. He also states that they have a commitment to achieve high response rates and show persistence in reaching respondents and gaining cooperation. Schnell concluded that the nonresponse in Germany increased only a few points, but that there were large differences between organizations. This strongly suggests that there may be differences in nonresponse trends (cf. de Heer, 1999).

There is a strong need for clear and comparable data on nonresponse to replace our beliefs and common wisdom by facts. To fulfil this need, the International Workshop on Household Survey Nonresponse was founded in 1990 with main goal the pooling and dissemination of knowledge (de Leeuw, 1999). One of the first initiatives was an international survey on nonresponse in official statistics to establish a dataset with long-term, cross-national data (de Heer, 1999). The accumulated data over time are at the core of our study. We will use these data to analyze what is going on with respect to survey nonresponse trends and to clarify which factors pay a role in the nonresponse process.

We first present an overview of international figures on nonresponse and their trend over time, distinguishing between noncontacts and refusals. In the second part we model differences between countries, using background variables on survey design and fieldwork strategy. We focus on factors that can be influenced by the researcher or agency, only incorporating cultural and sociological differences between countries when they directly reflect on survey design (see Johnson et al., Chapter 4, this volume).

3.2 THEORETICAL BACKGROUND

Over the last two decades, several theoretical models for nonresponse have been proposed (e.g., Dillman, 1978; Groves, Cialdini, and Couper, 1992; Groves and Couper, 1998; Hox et al., 1996). The two main components of nonresponse are noncontact and refusal. Most models focus on factors influencing refusal; Campanelli et al. (1997) incorporate the concept of noncontacts more explicitly.

Most models for nonresponse represents a "multilevel" conceptual framework, including influences from the levels of society and social and cultural context, the sample unit (e.g., the household, potential respondent), the survey organization, the survey design, and the interviewer (cf. Groves and Couper, 1998). From a theoretical point of view, influences on all levels are important. To understand why nonresponse occurs, one should study variables on both "macro level" (e.g., society, culture, economic situation), "mezzo level" (e.g., survey design), and "micro level" (e.g., respondent, interviewer). This knowledge will enable us to handle nonresponse adequately, both by reducing nonresponse and by postsurvey adjustment.

3.3 METHOD

3.3.1 Data Collection Procedure

The need for comparable, international data on household survey nonresponse led to the initiative for an international survey on response trends at the First International Workshop on Household Survey Nonresponse in Stockholm in 1990. Besides questions on response and nonresponse (including refusals, noncontacts and others), questions were asked on survey design, fieldwork strategies, and fieldwork organization. The goal was to establish a dataset with long-term, cross-national, and comparable data on nonresponse and nonresponse variables. Special attention was paid to the definition of response and nonresponse. The number of sampled cases that were actually used in the field, corrected for overcoverage, was the basis for the response and nonresponse figures. The questionnaire was sent to key informants at governmental survey agencies with the request to fill in a questionnaire for each survey that was regularly conducted (e.g., Labour Force Survey, Family Expenditure Survey). To build a database, this was repeated over time and the results were fed back and discussed at the yearly meetings of the International Workshop on Household Survey Nonresponse. The last data collection took place in 1997 asking for the results of 1996. For more details on the development of the questionnaire, the measurement of the nonresponse components, and some preliminary results, see de Heer and Israels (1992); Luiten and de Heer (1994), Maas and de Heer (1995), and de Heer (1999).

The data collected in the international survey on response trends are the core of this study. For each country, the collected response data (i.e., final response rates, noncontact rates, refusal rates) were listed and sent to the representatives in early spring 1999. In the accompanying letter we asked representatives to: (1) carefully check and if necessary to correct the available data, (2) add as much new data as possible, and (3) add any comments thought necessary (e.g., on changes in fieldwork).

3.3.2 Available Data

Data were collected from official statistical offices of 16 countries: Australia, Belgium, Canada, Germany, Denmark, Finland, France, Hungary, Italy, The Netherlands, Poland, Sweden, Slovenia, Spain, United Kingdom, and United States. For Germany, data were available for both the former Eastern Germany and the former Western Germany. To detect potential cultural differences between the former Eastern and Western Germany, these data are analyzed separately.

Ideally, to compare nonresponse trends internationally, a data set should contain very long and detailed time series, it should cover a wide rage of survey types and organizations, and all countries should provide data for all surveys. However, this is the stuff that dreams are made of, and countries do differ in the official surveys they conduct and in the interval between two subsequent surveys. Despite the many practical problems encountered collecting the data, we are amazed by the richness of the data provided by representatives at official statistical agencies around the world.

Available for analysis were time series for 16 countries and 10 different surveys, but not all countries provided data for each survey. For the Labour Force Survey (LFS), data were available for all 16 countries. Fourteen countries (excluding Germany and The Netherlands) provided data for the Family Expenditure Survey (FES). Four countries provided data for surveys on Health, three countries for surveys on National Travel and on Income. Two countries provided data for surveys on Living Conditions, on Consumer Sentiments, and on Victimization. One country had data for a Survey on Housing, and one for a General Household Survey. Although the time periods for which data are available do overlap, they do differ in starting point. Some countries even could provide time series going back as far as the end of the 1970s and the early 1980s. For an overview, see the Appendix.

The international questionnaire also provided us with some data on sampling design, survey design, fieldwork strategy, interviewer corps, and survey climate (cf. de Heer, 1999). Besides the question on survey climate, the questionnaire did not contain questions on the social context in which official surveys are conducted (e.g., economical and political situation of the country). From a separate source (Houtepen, 2000) with additional data for Slovenia added, figures on family size, number of young children, unemployment, and inflation were made available for each country.

3.3.3 Statistical Model

The two most important components of nonresponse are "noncontacts" and "refusals." Countries may show different trends, resulting in an overall stable response; for example, a decreasing noncontact rate and an increasing refusal rate (cf. de Heer, 1999). To investigate the nonresponse process in more detail, three dependent variables are used: proportion response, proportion refusals, and proportion noncontacts.

The data cause problems for standard statistical analysis because the surveys are nested in countries and the time series are of different length (see also the Appendix). To solve these problems, a multilevel logistic regression model will be used with year of study as the lowest level. The multilevel model incorporates the nesting structure, and also accommodates the uneven length of the time series (cf. Goldstein, 1995; Hox, 2000). As Little (1995) shows, multilevel modeling of unequal time series data assumes a missing-at-random mechanism, which is a reasonable mechanism for our data.

We decided to keep as many surveys as possible in the analysis to ensure generalizability over surveys, but we had to eliminate the data for the "general household survey" and the "survey on housing"; for these two surveys, data from only one country were available, making survey and country totally confounded (analyses on the complete data set produce virtually identical results, but the confounding of surveys and countries creates convergence problems with the more complicated models).The final analyses were done on the resulting eight surveys: Labour Force Survey, Family Expenditure Survey, Health Survey, National Travel Survey, Income Survey, Survey on Living Conditions, Consumer Sentiments Survey, and Survey on Victimization. To accommodate for the fact that not all countries had data on all surveys, we used a multilevel model for cross-classified (country∗survey) data. For details on cross-classified models see Goldstein (1995) and Rasbash and Goldstein (1994).

We started the analysis with a multilevel logistic model for cross-classified data, including "year" as explanatory variable. This enabled us to investigate whether in general, over all surveys and countries, response rates are declining and have been declining for some time (cf. Bradburn, 1992). For an easy interpretation of the parameter estimates, year was recoded as follows: 1998 = 0, 1997 = –1, 1996 = –2, etc. In the next step, we focussed on differences between countries, and investigated whether nonresponse trends differ between countries. In the last step, we concentrated on the Labour Force Survey only. This is the only survey for which all countries provided longitudinal data for the overall response, and with the exception of Germany, also provided longitudinal data for noncontacts and refusals. In this step, we addressed the question of whether we can explain differences in nonresponse trends from differences in fieldwork strategies. We started with fitting a multilevel logistic model, using "year" as explanatory variable. Again, we had three dependent variables: proportion response, proportion noncontacts, and proportion refusals. Under the model for each country, residuals were calculated for the intercept and then standardized. The intercept indicates the smoothed (non)response in the last year. The standardized residuals indicate how much a specific country scores above or below the overall mean (z-score). Data on fieldwork procedures in each country were then used to model these residuals.

3.4 RESULTS

3.4.1 Trends over Time

When studying nonresponse over time and over countries, there are three research questions that should be addressed: (1) does nonresponse differ between countries?, (2) does nonresponse increase over time?, and (3) is the increase different between countries? In the analysis, we distinguish between two sources of nonresponse, that is, noncontacts and refusals. Both contribute to the overall nonresponse, but different factors influence each source. To inspect differences between noncontacts and refusals, we present the results for overall response, noncontacts, and refusals, side-by-side in one table.

We start with the basis or "null" model. The null model enables us to answer the first question: does nonresponse differ between countries. The results are summarized in Table 3.1.

When we inspect the column "Response," we note that both the country and the survey variances are significant. This means that the response does differ by country and by survey. The country variance is 0.33, which corresponds to an intraclass correlation or country effect of 0.09. The intraclass correlation in the null model is the proportion of total variance at that level. Thus, the countries differ significantly

Table 3.1. Multilevel logistic analysis for cross-classified data: null model

	Response	Noncontact	Refusal
Fixed effect:			
Intercept	1.48 (.20)	–2.72 (.14)	–2.31 (.31)
Random effect:			
Variance country	0.33 (.12)	0.18 (.07)	0.64 (.24)
Variance survey	0.17 (.09)	.005 (.006)^{ns}	0.44 (.23)

Note: Country by survey: response trends over years. Dependent variables are response rate, noncontact rate, and refusal rate. Standard errors in parentheses. All estimates are significant at $p = 0.05$, unless indicated otherwise with "ns." The parameter estimates are on a logit scale and are not proportions.

in response rates and 9% of the variance is at the country level. These country differences reflect a composite of real country differences like differences in culture and differences between the countries in fieldwork procedures, interviewer corps, etc. (see Hox, Chapter 7, this volume). The survey variance for response is 0.17, which corresponds to an intraclass correlation of 0.04. Some surveys have a lower response than others and 4% of the variance is at the survey level. The estimate for the intercept is 1.48 on a logit scale, which corresponds to an overall response rate over countries, surveys, and years of 0.82.

When we look at the column "Noncontact," the most striking finding is that the survey estimate is not significant; type of survey does not influence the noncontact rate. This fits the theories about nonresponse; noncontact should be attributed to survey calling rules and fieldwork procedures of survey organizations, not to the topic of a survey! The country variance is significant and corresponds to an intraclass correlation or country effect of 0.05. The estimate for the intercept is –2.72 on a logit scale, which corresponds to an overall noncontact rate of 0.06.

Both type of survey and country do influence the refusal rate, as can be seen in the last column. The country variance is 0.64, which corresponds to an intraclass correlation or country effect of 0.15. The survey variance for response is 0.44, or an intraclass correlation of 0.10. The refusal rate for some surveys is higher than for other surveys. The estimate for the intercept is –2.31 on a logit-scale, which corresponds to an overall refusal rate of 0.09.

In sum: (1) countries differ significantly in response rate, noncontact rate and refusal rate; (2) type of survey does influence response rate, and refusal rate; but (3) type of survey does not influence the noncontact rate. This indicates that within a country, there are not many differences in noncontacts between different surveys, probably because fieldwork procedures influencing noncontacts do not differ much between surveys in one country. But fieldwork procedures influencing noncontacts do differ between countries, as a global examination of data on fieldwork confirms (see also Luiten and de Heer, 1994). However, people do react to the topic of a survey, hence the influence of survey on refusal rate.

Survey response differs between countries and between surveys, but is there a downward trend over time? To test this second research question we added "year" as explanatory variable to our model, and in the next step investigated whether "year" had a random slope. The final models for overall response, noncontacts, and refusals, are summarized in Table 3.2.

When we first look at the column "Response," we see that the parameter estimate for year is significant (–0.02). There is a negative trend in response from year to year! If we want to see more clearly what happens, we have to look at the noncontact and refusal rates.

For "Noncontact," "year" is also significant (0.03), corresponding to an increase of about 0.2% per year. Noncontacts are increasing from year to year. When we look into the random part of the model, we see something extremely interesting: the variance for "year" is not significant, nor is the variance for survey; only the variance for country is significant. The noncontact rates differ from country to country, and the noncontacts increase over the years with an average of

Table 3.2. Multilevel logistic analysis for cross-classified data: final model with explanatory variable "year" as random component

	Response	Noncontact	Refusal
Fixed effect:			
Intercept	1.35 (0.21)	–2.56 (0.14)	–2.14 (0.31)
Year	–0.02 (0.01)	0.03 (0.01)	0.03 (0.01)
Random effect:			
Variance country	0.39 (0.14)	0.24 (0.10)	0.57 (0.22)
Variance survey	0.16 (0.09)	0.003 (0.005)ns	0.46 (0.24)
Variance year	0.001 (0.0005)	0.001 (0.001)ns	.002 (0.001)

Note: Country by survey: response trends over years. Dependent variables are Response Rate, Noncontact rate, and Refusal Rate. Standard errors in parentheses. All estimates are significant at $p = 0.05$, unless indicated otherwise with "ns." The parameter estimates are on a logit-scale which is a nonlinear transformation of the proportions.

0.2% per year, but the increase in noncontacts is not different from country to country.

There is one country that has a much higher noncontact rate than the other countries. When we perform a sensitivity analysis and remove this extreme from the data, we nevertheless find the same value for the regression coefficient over the years (0.03). The conclusions remain unaltered: noncontacts are increasing over the years and the trend is the same for each country.

When we look at the refusal rates, a different picture emerges regarding country differences. When we inspect the last column of Table 3.2, we first of all see that the effect of year is significant (0.03). The increase in refusals is on average 0.03 per year on the logit scale, which corresponds to an increase of 0.3% each year or an increase of 1% every three years in ongoing surveys, other things being equal. When we look at the random effects, we note that not only the country and survey variances are significant, but also that the variance of year is statistically significant. Year has random slopes, meaning that not only do the refusals differ from country to country and from survey to survey, but also the increase in refusals over the years differs from country to country and from survey to survey.

In other words, the random effects for country and survey indicate that for some countries and for some surveys the refusal rate is higher than for other countries and surveys. But, most important of all, the variance for year is significant too. Thus, the increase in refusal rate over the years differs from country to country. The fitted regression lines for the countries have different slopes (see Figure 3.1). Meaning that in some countries, the increase in refusals is much steeper over the years than in other countries. Figure 3.1 clearly illustrates that one country has an extreme high overall refusal rate, although it appears to go slightly down. Again, we performed sensitivity analyses by removing extreme cases from the data set. Again, this did not alter the conclusions at all. The estimate for year remains significant with value 0.03; also all the variance components in the random part of the model remain significant. Refusal rates are increasing over time, refusal rates differ from country to

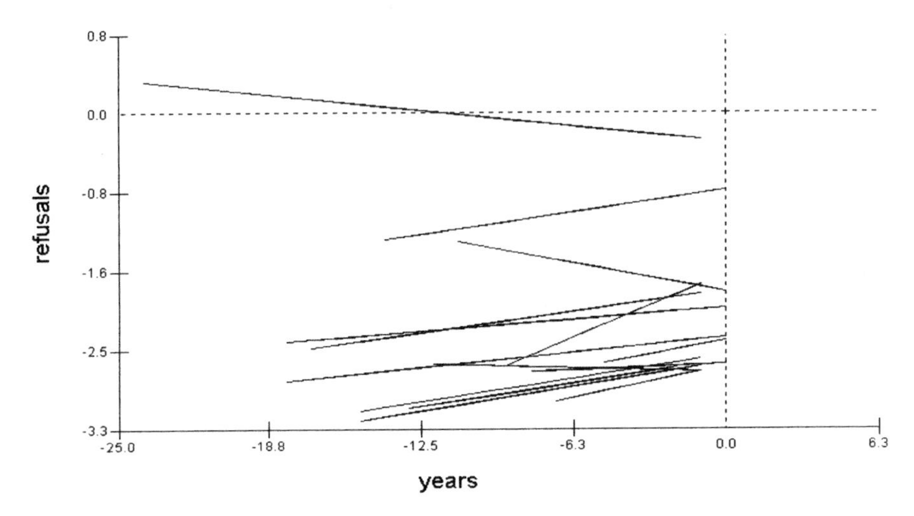

Figure 3.1. Regression lines for refusals across years for different countries (estimates based on the model in Table 3.2; 1998 = 0; refusals on logit scale).

country, and in some countries the increase in refusals is larger than in other countries.

In sum: (1) countries differ in response rate; (2) the response rates have been declining over the years; (3) the trends differ by country; (4) there are no differences between countries in the rate in which the noncontacts are increasing; and (5) the difference in response trends is caused by differences between countries in the rate at which the refusals are increasing.

3.4.2 Differences between Countries

Focus on Labour Force Survey. Countries do differ in response rates. The countries studied here also differ on a number of important points that very well could influence the response (e.g., response climate, fieldwork strategy). In this section, we relate the nonresponse figures to differences between the countries. We concentrate on factors that can be influenced by the researcher; that is, on factors related to survey design and fieldwork strategies. Now we focus on the data for the Labour Force Survey solely. The Labour Force Survey is the only survey for which all countries provided fieldwork data and response figures and, with the exception of Germany, also data on noncontacts and refusals.

For overall response, noncontact, and refusal, the standardized residuals of the intercept were calculated under the multilevel logistic model. The residuals are expressed as standard normal or z-scores, with a mean of zero and a standard deviation of 1. A positive value indicates that a specific country scored above the overall mean in 1998; a negative value indicates that a country scored below average.

For example, for Denmark the standardized residual for noncontact was 2.49, indicating that in Denmark the noncontact rate for the Labour Force Survey is more than two standard deviations above the overall mean of all the countries. Belgium and The Netherlands also have a relative high noncontact rate; for these two countries, the noncontact rate is one standard deviation above the mean. Countries that have a relatively low noncontact rate for the Labour Force Survey are Australia, Slovenia, and the United States; all are one standard deviation below the overall mean. When we look at refusals, we see that for The Netherlands the situation is quite extreme. The standardized residual for refusal in the Labour Force Survey is 3.22. The United Kingdom has a standardized residual of 1.05. The refusal rates of both Denmark and Slovenia are slightly above average. All other countries have lower standardized residuals for the refusal rate in Labour Force Survey.

Compared internationally, The Netherlands is an extreme case, as the z-scores indicate: the Netherlands has both a high noncontact rate and a high refusal rate. For a detailed discussion of international response trends in the Labour Force Survey, see de Heer (1999).

Modeling Differences in Noncontact Rates between Countries. We first focus on factors that can be influenced by the researcher or the agency. Two such factors that may influence the amount of noncontact are the contact strategies used and the respondent selection. Available international data related to contact strategies are mode of data collection, extension of fieldwork period, and interviewer workload. In telephone surveys, more contact attempts can be made than in face-to-face surveys; in addition, in telephone surveys, more often contact attempts are made in the evening. In general, telephone surveys should lead to a lower noncontact rate than face-to-face surveys. A large interviewer workload limits the interviewer in the number of contact attempts that are feasible and, other things being equal, will increase the number of noncontacts. When special attempts are made to contact respondents who could not be contacted during the regular field work period, this should influence the noncontact rate.

Rules for respondent selection also may influence the noncontact rate. Whether or not the household should be interviewed or specific persons, whether or not proxy reporting is allowed, and whether or not substitution of sampling unit is allowed for noncontacts all influence the probability of making a contact.

We correlated the available information on contact strategies with the standardized residuals for the noncontact rate for each country using Spearman's rho, which is a nonparametric measure. The parametric (Pearson) product–moment correlation is rather sensitive for extreme values, and would be misleadingly pulled upwards by the extreme Dutch values. The Spearman correlations were based on 15 countries, as Germany did not provide separate data for noncontacts or refusals.

Mode of data collection does have some effect on the number of noncontacts: when a panel design is used, the number of noncontacts is lower (rho = –0.31), as can be expected. However, telephone interviews do not produce a lower number of noncontacts in the Labour Force Survey. Also, interviewer workload and employment conditions (e.g., free lance, civil servant) are not clearly related to the noncontact rate. What did affect the noncontact rate was whether or not the response rates of the interviewers were monitored: countries that did monitor the interviewers had a lower noncontact rate (rho = –0.36). Finally, rules for respondent selection do have some influence on noncontact rate. Countries that specified that all persons above 16 should be interviewed (individual sample) for the Labour Force Survey have a slightly higher noncontact rate than countries in which the household core should be interviewed (rho = 0.27). It should be noted that all countries allowed some limited form of proxy reporting. Allowing for substitution for noncontacts did not have a strong effect on the noncontact rate.

In sum: survey design (panel), type of sample (household core), and interviewer monitoring had a clear effect in lowering the noncontact rate.

Countries do not only differ in survey design and fieldwork procedures, but also in demographics, economic conditions, and culture. These factors, which are not under the control of the researcher, may influence response (cf. Couper and Groves, 1998; Harris-Kojetin and Tucker, 1999). Differences between societies, such as in average household size and percentage of young children, may influence the at-home patterns. The empirical data support the hypothesis that differences in at-home patterns between the countries may influence the contact rate. Countries with a larger average household size had a slightly lower noncontact rate (rho = –0.37). Countries with a higher percentage of young children also had a lower noncontact rate (rho = –0.60). Of course, survey researchers cannot change the demographics in their country but they can optimize the contact strategy and adapt it to changes in demography. For instance, more contact attempts at different times of the day will also reach those living in small households and working during the day. Strict monitoring of the interviewers has to ensure that the interviewers really attempt to contact a respondent on the prescribed times.

Modeling Differences in Refusal Rates between Countries. Whether or not participation is mandatory is one of the strongest potential determinants of refusal. In addition, survey burden and survey climate may also influence the willingness to participate. For each country, it is known whether or not the participation in the Labour Force Survey is mandatory. As indicator of the burden we used whether or not a panel or a cross-sectional design is used. The informants at each national statistical office also gave an estimate of "survey climate." In addition, we collected data on unemployment and inflation for each country. According to Harris-Kojetin and Tucker (1999) better economic times are associated with lower cooperation with government surveys. Using time series regression, they found that both unemployment rate and inflation rate were associated with refusals to the U.S. Labour Force Survey.

Fieldwork strategies to reduce refusals are under the potential control of the researcher. The use of advance letters and incentives may influence the willingness to participate. In addition, in a face-to-face interview, interviewers have more opportunities to interact with the respondents and convince them to participate than in telephone introductions. Also, the monitoring of interviewers may stimulate the interviewers to put in an extra effort. Finally, whether or not refusal conversion is used and whether or not substitution after a refusal is allowed will influence the final number of refusals.

We correlated the available information on design and fieldwork strategies with the standardized residuals for the refusal rate for each country, again using Spearman's rho. It is not surprising that the mandatory nature of a survey has a strong effect on the refusal rate (rho = –0.72). It did not have any effect on the noncontacts as expected. No clear effects were found of type of survey (panel versus cross-sectional) and the survey climate as seen by the representative at the statistical agency. Since all but one agency for official statistics used advance letters and none used incentives, we could not investigate the effects of these factors on refusals in the Labour Force Survey. Substitution did have the effect of reducing the refusals (rho = –0.35). The refusal rate is the net result of the fieldwork in terms of refusals after substitution. In those countries that did use substitution, apparently not all refusers could be substituted successfully.

Just as with the noncontact rate, the mode of data collection did not have a clear effect but monitoring interviewers did have a slight effect (rho = –0.23). Finally, sending a special letter to the refusers did have some effect in lowering the refusal rate (rho = –0.27)

We found mixed evidence for the hypothesis that better economic times are associated with lower cooperation to official surveys. The percentage unemployed did correlate negatively with the refusal rate (–0.61) but this could be partly attributed to the potential saliency of the topic (Labour Force Survey). Our data also showed a positive correlation between the inflation rate and the refusal rate over countries (rho = 0.44). This indicates that countries with higher inflation also have more refusals on the Labour Force Survey.

Differences between Countries: Main Conclusions. There are relatively large differences between countries in noncontact rate and in refusal rate on the Labour Force Survey as the standardized residuals showed. Denmark has the highest noncontact rate (around 15%) and could gain most by reducing noncontacts. Also, Belgium and The Netherlands (around 11%) can gain in response by improving the contact rate. Slovenia, the United States, and Australia all have an extremely low noncontact rate of around 2%. Other countries are in between.

Regarding refusal rates, The Netherlands is an extreme case (27%) but the United Kingdom also has a relatively high refusal rate (12–13%). In comparison, Denmark, Slovenia, and Hungary have a more moderate refusal rate of 8%, 7%, and 6%, respectively. All other countries have much lower refusal rates.

The strongest predictor of response was the mandatory nature of a survey. All countries in which the Labour Force Survey is voluntary had a much higher refusal rate than countries in which the survey was mandatory. Although very important, this finding cannot be used to reduce the nonresponse actively. Factors that influence nonresponse and can be implemented in fieldwork procedures are monitoring of interviewers and special efforts for refusers. Monitoring the response rate of interviewers has a positive effect in reducing both the noncontact rate and the refusal rate. Sending a special letter to refusers has some effect in reducing the refusals. The last finding can be partially explained by the finding that a special letter is often used in countries in which the Labour Force Survey is mandatory. Finally, countries that use substitution also have a lower refusal rate. We will come back to this in the discussion.

3.5 MAIN CONCLUSIONS AND DISCUSSION

We found that (1) countries differ in response rate, (2) the response rates have been declining over the years, and (3) the nonresponse trends differ by country. It should be noted that these conclusions hold over different types of surveys and are based on time series for 16 countries and 10 different surveys.

Nonresponse has two important components: noncontact and refusal. Further analyses showed that there are differences between countries in noncontact rate and that the noncontact rates are increasing over time, but that there are no differences between the countries in the rate in which the noncontacts are increasing. The difference in the nonresponse trends over the countries is caused by differences between countries in the rate at which the refusals are increasing. For some countries, the increase in refusal rates is much steeper than for other countries. Again these conclusions hold over different types of surveys.

In conclusion: there is ample empirical evidence that response rates are declining internationally. Nonresponse is indeed an increasing problem in the developed world. There are differences between countries in how grave the situation is. Not only do countries differ in their response rate, they also differ in the rate in which the response rates are declining.

To investigate the differences between countries, we analysed differences in survey design and demographic and economic situation. For the Labour Force Survey, auxiliary data were available for all 16 countries.

We found that differences in noncontact rates were associated with differences between countries in average household size and percentage of young children, which indicates potential differences in at-home patterns. Also, differences in survey design were associated with differences in noncontact rates. Panel surveys have, as can be expected, a lower noncontact rate than cross-sectional surveys. Countries that use more lenient rules for sampling and respondent selection (e.g., any member of a household) also have lower noncontact rates than countries that use more strict rules (e.g., the new Eurostat rule: each individual 16 years and older). Finally, it was found that stricter supervision and monitoring of interviewers is also associated with a lower noncontact rate.

Differences in refusal rate were associated with economic indicators of the countries. Countries with a higher percentage unemployed had a lower refusal rate for the Labour Force Survey. Since work-related questions may carry a special significance in countries with a high unemployment rate, this effect could be related to the saliency of the topic. Also, countries with a higher inflation rate had a higher refusal rate. A moderate inflation rate is often seen as a positive sign for the economy. If we adopt this view, the results support the thesis that better economic times are associated with lower cooperation with government surveys (Harris-Kojetin and Tucker, 1999). However, it would be interesting to collect more direct estimators of the economy of different countries and relate those to refusal rates. The strongest correlate of refusal rate was whether or not the survey was mandatory; in countries were the Labour Force Survey was mandatory, the refusal rate was extremely low. However, though extremely interesting, these factors cannot be changed by survey researchers and will be of limited use in the fight against nonresponse.

Several design and fieldwork strategies that are under the control of the researcher were associated with refusal rate. The most important one was the supervision and monitoring of interviewers. Not only is stricter supervision of interviewers associated with a lower noncontact rate, it is also associated with a lower refusal rate! Also, special attempts aimed at refusers, such as a special letter, reduced the refusal rate to some degree. It should be noted that the influence of several potentially effective fieldwork strategies could not be investigated in this study because of lack of variance between the countries. For instance, none of the countries used incentives (for the influence of incentives, see Singer, Chapter 11, this volume).

In conclusion, for the Labour Force Survey, differences in response between countries can be partly explained by differences in survey design and fieldwork strategies between countries.

We showed that response rates have been declining internationally for a large variety of surveys. However, our data are restricted to official government surveys. No comparable data sets are available for commercial or university-based surveys. Although we have a strong suspicion that nonresponse is declining in those fields too, this is only based on limited data from The Netherlands. An international and longitudinal nonresponse study in academics and marketing research would be interesting indeed.

Very high on the research agenda for the coming years should be further investigations into survey design differences and their impact on nonresponse, rather than investigations in country-level cultural differences.

ACKNOWLEDGMENTS

The views expressed in this chapter are those of the authors and do not necessarily reflect the policies of Statistics Netherlands. We sincerely thank the members of the international workshop on household survey nonresponse, who as fairy godmothers advised and commented on earlier versions through the years. We thank Bob Groves, Geert Loosveldt, and Edward Sondik for their helpful comments. We also thank Joop Hox for his help and advice in running complex multilevel models. Above all, we thank the representatives at the statistical agencies, who carefully filled in the questionnaires and provided the data for this study.

APPENDIX—AVAILABLE DATA FOR EACH COUNTRY

Country/Survey	Time period	Country/Survey	Time period
Australia		Netherlands	
Labour Force	1991–1997	Labour Force	1988–1997
Family Expenditure	1991, 1994	Health	1982–1997[2]
Housing	1994	National Travel	1978–1997
Health	1995	Living Condition	1974–1997
Belgium		Consumer Sentiment	1972–1997
Labour Force	1983–1997	Victimization	1992–1997
Family Expenditure	1980–1997	Poland	
Canada		Labour Force	1992–1997
Labour Force	1991–1998	Family Expenditure	1982–1997
Family Expenditure	1990, 1992, 1997	Sweden	
Health	1994, 1996	Labour Force	1980–1998
Germany (West)		Family Expenditure	1985–1996
Labour Force	1990–1997[1]	Income	1980–1998
Germany (East)		Living Conditions	1980–1998
Labour Force	1991–1997	Slovenia	
Denmark		Labour Force	1989–1998
Labour Force	1984–1998	Family Expenditure	1993–1998
Family Expenditure	1981–1998	Consumer Sentiment	1996–1998
National Travel	1991–1998	Spain	
Finland		Labour Force	1980–1998
Labour Force	1983–1997	Family Expenditure	1980, 1990
Family Expenditure	1981–1996	United Kingdom	
Income	1984–1997	Labour Force	1984–1998
France		Family Expenditure	1984–1998
Labour Force	1988–1996	General Household	1984–1998
Family Expenditure	1985, 1989, 1995	National Travel	1990–1998
Hungary		United States	
Labour Force	1992–1997	Labour Force Survey	1983–1997
Family Expenditure	1981–1997	Family Expenditure	1985–1997
Italy		Income	1984–1996
Labour Force	1988–1998	Health	1983–1997
Family Expenditure	1996	Victimization	1983–1997

1. Data are available for both the former Western Germany and the former Eastern Germany.

2. In 1997, a complete redesign took place for the Living Conditions, Health, and Victimization surveys. They were combined into one large survey, POLS (Continuous Survey on Living Condition). In POLS, a general basic question module is used first; after that random, subsamples are asked to answer specific modules (e.g., on Health or on Victimization). See also Akkerboom and deHue (1997).

CHAPTER 4

Culture and Survey Nonresponse

Timothy P. Johnson, Diane O'Rourke, Jane Burris, and Linda Owens, *Survey Research Laboratory, University of Illinois at Chicago*

4.1 INTRODUCTION

Surveys are an inherently social activity. Survey respondents are not merely autonomous information processors, rather, they all exist within complex social matrices that influence their thoughts, feelings, and behaviors. There is now a developing body of evidence regarding avenues by which cultural conditioning may influence the cognitions of survey respondents. Johnson et al. (1997), for example, have reported evidence suggesting cultural variability in survey question comprehension, memory retrieval, judgment formation, and response editing processes in a sample of African American, Mexican American, Puerto Rican, and non-Hispanic White respondents in the United States. Less is known, however, regarding how respondent culture influences survey nonresponse. In this chapter, we review the available empirical literature and discuss theoretical approaches that may be useful in understanding patterns of survey nonresponse across cultures.

In conducting this review, we acknowledge that there are hundreds of definitions of "culture." Hofstede (1980a, p. 21) has defined culture as "the collective programming of the mind which distinguishes the members of one group from another." Triandis (1996, p. 408) defines culture as consisting of "shared elements that provide the standards for perceiving, believing, evaluating, communicating, and acting among those who share a language, a historic period, and a geographic location." Smith and Bond (1998, p. 69) define it more simply as "systems of shared meanings." In our previous work, we have interpreted culture as representing a social group with "a shared language and set of norms, values, beliefs, expectations and life experiences" (Johnson et al., 1997, p. 87). We retain that definition here in modified form that acknowledges Triandis' concern with the temporal boundaries of culture.

4.2 EMPIRICAL EVIDENCE

While there may be hundreds of approaches to conceptualizing culture, measuring it is more problematic. In our review of the available literature, we have identified two forms of empirical measurement that permit some cross-cultural comparisons. Neither of these approaches is completely satisfactory. Each, nonetheless, provides an opportunity to explore the potential effects of culture on nonresponse.

The first approach comes from studies conducted in the United States, where considerable research is available that examines nonresponse behavior across self-identified racial and/or ethnic subgroups within that country. Our review identified 26 comparisons of race and/or ethnic differences in survey nonresponse indicators in U.S. studies conducted since 1975. These studies examined differences in nonresponse to cross-sectional surveys, differences in attrition from panel surveys, and differences in nonresponse to the U.S. Census. Tables 4.1–4.2 summarize this research. In reviewing this information, it is important to note that the outcomes being assessed are not consistent across studies. Some examine respondent refusals only, others assess inability to track or locate respondents, and some examine the resistance of respondents to survey participation. Also, examining nonresponse within a single nation, even one as pluralistic as the United States, will almost certainly find only attenuated cultural effects. The fact that much of the U.S. research is limited to White–Black differences, and that many of that country's ethnic groups are not represented at all, only adds to this problem. We nonetheless treat these empirical studies as an imperfect, but available, data source that is useful in evaluating potential cultural influences on survey nonresponse.

A second, smaller body of research is also available that compares elements of survey nonresponse across nations, permitting cultural comparisons at this broader level of analysis. Unlike the U.S. studies, this approach offers the promise of exploring a wider range of cultural orientations than may be possible within a single country. In doing so, though, it risks masking over subcultural variability, particularly within heterogeneous societies. Variations in the specific survey methodologies used in each country also pose a significant challenge to cross-national nonresponse comparisons.

4.2.1 Studies of Survey Nonresponse in the United States

In Table 4.1, six examinations of nonresponse from cross-sectional U.S surveys are summarized. No differences in nonresponse were found by race in four of them. However, in their assessment of six government-sponsored surveys, Groves and Couper (1998) found that Hispanics were more likely to cooperate than other racial/ethnic groups, even after adjusting for differences in the age and socioeconomic compositions of the groups. Another study that focused on Hispanic groups only found lower response rates among Cubans, relative to Mexican American and Puerto Rican respondents (Rowland and Forthofer, 1993). With the exception of Groves and Couper (1998), these studies in general make few distinctions between refusal and noncontact rates.

Table 4.1. Findings from studies of cultural differences in nonresponse to cross-sectional surveys

Study	Population (study year)	Method	Analysis type	Findings
Casper (1992)	Washington, DC adults (1990)	Face-to-face	Bivariate	No race differences in composition of initial and nonresponse surveys.
DeMaio (1980)	Nationwide adults, CPS (1977)	Face-to-face	Bivariate	No race differences in refusals.
Groves and Couper (1998)	Nationwide samples from six government-sponsored surveys (1990)	Face-to-face	Multivariate	Whites/other races and Blacks less likely to cooperate compared to Hispanics.
Jackson et al. (1997)	Parents of elementary school children in North Carolina	Self-administered	Bivariate	No race differences in nonresponse.
Rowland and Forthofer (1993)	Hispanic adults in Dade County, FL, New York City area, and Southwestern U.S., HHANES (1982–1984)	Face-to-face	Bivariate	Cubans had lower response rates than Mexican-Americans and Puerto Ricans.
Smith (1983)	Nationwide adults, GSS (1980)	Face-to-face	Bivariate	No race differences in cooperation rates.

Table 4.2. Findings from studies of cultural differences in follow-up nonresponse to panel surveys

Study	Population (follow-up year)	Follow-up method	Analysis type	Findings
Aneshensel et al. (1987)	Los Angeles adults (1983)	4 year telephone	Bivariate	Minorities higher attrition
Aneshensel et al. (1989)	Los Angeles adolescent females	2 year face-to-face	Multivariate	Mexican-born higher attrition
Broman et al. (1994)	Michigan autoworkers (1989)	~2 year telephone	Bivariate	Blacks less likely to be reinterviewed
Chen and Kandel (1995)	NY state high school students (1990)	19 year reinterview	Bivariate	Minorities less likely to be reinterviewed
Davidson et al. (1997)	Women from family planning clinics in NY, Dallas, and Pittsburgh (1994–1995)	One year telephone reinterview	Bivariate	No race differences in attrition.
Eaton et al. (1992)	Adults 18–64 in Baltimore, L.A., St. Louis, and Durham, NC, ECA (1979–1983)	1 year face-to-face reinterview	Multivariate	Hispanics less likely to be located for a reinterview
Finkelhor et al. (1995)	Nationwide sample of young people and their caretakers (1993)	~15 month telephone follow-up	Bivariate	Black and Hispanic households less likely to be reinterviewed
Flay et al. (1993)	Adult smokers in Chicago (1989)	Four month telephone	Bivariate	"Negligible" effects of race on attrition
Harris-Kojetin and Tucker (1998)	Nationwide adults, CPS (1995)	8 interviews over a 16 month period	Multivariate	Nonwhites and Hispanics less likely to have completed all eight interviews
Howell and Frese (1983)	Southern U.S. high school students (1979)	13 year fact-to-face reinterview	Bivariate	Blacks "somewhat" less likely to be reinterviewed
Johnson (1988)	Nationwide adults, NSPHPC (1980)	One year telephone	Multivariate	Nonwhites more attrition

Lavrakas et al. (1991)	Two samples of Chicago adults (1984 and 1985)	#1: 12 month phone reinterview #2: 16 month phone reinterview	Bivariate	#1: Blacks less likely to be reinterviewed #2: Asians and Blacks less likely to be reinterviewed
Lepkowski and Couper (this volume)	1. Nationwide adults, ACL (1986) 2. Nationwide adults, NES (1990)	#1: 2.5 & 7.5 years/ face-to-face reinterviews #2: 2 year telephone reinterview	Multivariate	#1: Blacks more difficult to recontact. Race not associated with cooperation #2: Race not associated with recontact. Blacks and Hispanics higher refusal rates
Madans et al. (1986)	Nationwide adults, NHANES (1982–1984)	Face-to-face/phone	Bivariate	Blacks higher attrition
Marcus and Telesky (1983)	Los Angeles adults (1976)	8 telephone	Multivariate	No race differences
		interviews in 1 year		
Menaghan and Merves (1984)	Chicago adults ages 22–69 (1976)	4 year face-to-face	Bivariate	Nonwhites had greater attrition
Patterson et al. (1996)	Washington state adults (1992–1993)	3 year telephone	Bivariate	Nonwhites more attrition
Shettle and Mooney (1999)	College graduates, retired CPS Sample (1992)	3 year mail/ telephone reinterview	Multivariate	Asian, Black, and Hispanic Whites less likely to respond to follow-up survey
Singer et al. (1999)	Detroit adults, DAS (1996)	1 year mail	Multivariate	Blacks higher nonresponse
Vernon et al. (1984)	Alameda County, CA adults (1978)	3-4 year face-to-face reinterview	Bivariate	Mexican Americans and Blacks more likely to refuse; Mexican Americans more likely to be unavailable

In contrast to the cross-sectional research, examination of the effects of respondent culture on panel survey retention suggests greater attrition among minority groups in the United States. Of the 20 panel studies included in Table 4.2, 17 (85%) reported greater minority group attrition. Of the eight studies that reported multivariate findings useful in assessing the effects of culture net of other respondent qualities, seven found differences by race/ethnicity after controlling for other relevant variables. Two of these studies reported findings separately for noncontacts and refusals. Eaton et al. (1992) reported no race or ethnic differences in refusal rates. However, Hispanics, respondents from racial groups other than Whites and Blacks, and persons not initially interviewed in English were less locatable for a one-year follow-up face-to-face interview. Vernon et al.'s (1984) analysis of panel attrition in a face-to-face reinterview found group differences in both refusal and noncontact rates: White respondents were more likely to be contacted and to cooperate.

It is widely acknowledged that the U.S. Census systematically undercounts several minority groups during its decennial Census. It can be documented that for at least 100 years, Blacks have been disproportionately underrepresented in official Census counts (Coale and Rives, 1973). This Census undercount is the result of several factors beyond respondent contact and cooperation, including issues of coverage, sample frame inadequacies, locating problems, and failure to report household members. Some research studies on undercounts have been able to measure these factors and others have not.

What are some of the explanations that might be useful in accounting for the observed undercounts? Choldin (1994, p. 230) suggests that "people in certain situations have good reasons to avoid contact with the government" and that "no amount of advertising and community relations will convince . . . them that it is safe and in their own interest to disclose themselves to the Census." Fear of loss of governmental assistance and/or housing support, and fear of deportation (among illegal U.S. residents) are only a few of the very rational motivations for providing inaccurate household enumeration, or for avoiding the Census altogether, that disproportionately affect minorities in this country.

Numerous ethnographic studies sponsored by the Census Bureau have also sought to identify behavioral causes that influence the undercount of minorities (Bell, 1992; Darden et al., 1992; de la Puente, 1993; Hamid, 1991; and Velasco, 1992). Some of the enumeration barriers suggested by these studies include high mobility patterns, fluid household membership, language barriers, cultural understanding of questionnaire terms such as "household" and "family," distrust of enumerators, and a fear of a loss of resources (Ammar, 1992; Aschenbrenner, 1990; Bell, 1992; Darden et al., 1990; Hamid, 1991; Hudgins et al., 1991; and Velasco, 1992).

4.2.2 Cross-National Studies of Survey Nonresponse

Assessing survey nonresponse across nations provides an alternative approach to evaluating cultural effects. A small number of studies are now available that address this issue and afford us the opportunity to assess nonresponse variability outside of the United States. Goyder (1985b) examined response rates to comparable U.S. and Canadian surveys conducted during the 1960s and 1970s. After controlling for methodological variability across surveys, he reported survey response to face-to-face interviews to be on average 6–7 percentage points lower in Canada. Differences for self-administered questionnaires were only marginally significant, but also favored the United States. The author concluded that socioeconomic differences accounted for about 4 percentage points, with the remaining 3 percentage points attributable to a "residual" cultural difference that reflected greater resistance to survey requests among Canadians.

de Heer (1999) compared the recent survey nonresponse experiences of 16 European and North American countries. Considerable variation in noncontact and refusal rates to labor force surveys were found across these nations. Refusal rates were the highest in The Netherlands and the United Kingdom and lowest in the Australia, Belgium, and Canada. Noncontact rates were highest in Belgium, Denmark, and The Netherlands and lowest in Australia, Slovenia, and the United States. Using the same database, de Leeuw and de Heer (see Chapter 3, this volume) examine cross-national nonresponse differences using multivariate modeling. They report that, net of variations in survey design features and study year, variation remains between these countries in noncontact, refusal, and overall response rates. Research by Hox and de Leeuw (see Chapter 7, this volume) also found national differences in nonresponse rates after controlling for variations in interviewer characteristics, attitudes and behaviors across nine nations.

Both this cross-national literature and the U.S. studies of subcultural differences suggest that some unspecified qualities of respondent culture are probably influencing survey nonresponse. The available evidence suggests that these differences in nonresponse remain even after taking into account cross-cultural variability in socioeconomic status and cross-national differences in survey methods. We now turn to an exploration of cultural elements that may reveal some of the potential mechanisms through which culture exerts its influence.

4.3 THEORETICAL PERSPECTIVES

There are several theoretical traditions that may provide a framework for understanding how culture is linked with survey nonresponse. In this section, we explore a set of psychological, communications, and sociological perspectives that may prove useful. Available empirical assessments within each that may be relevant to the survey nonresponse problem will also be discussed.

4.3.1 Unpackaging Culture

As alluded to earlier, we consider the race/ethnicity indicators employed as part of our review of U.S. findings to be only rough approximations of respondent culture. Cross-national comparisons are also less than adequate measures, as they may be insensitive to subcultural differences. To understand how and why culture influences social behavior, it is necessary to "unpackage" this concept by identifying the specific values and orientations through which it operates. Several dimensions, commonly referred to as cultural "layers" or "orientations," have in fact been identified that have proven useful in developing theories regarding social behavior across broad cultural groups. Researchers have identified a variety of cultural orientations, including uncertainty avoidance, masculine versus feminine orientation, and long- versus short-term time orientations (Hofstede, 1980a; 1991). In this section, we consider three orientations that may be useful in understanding nonresponse differences. Of primary interest will be (1) individualist versus collectivist orientations, (2) power distance, and (3) emphasis on vertical versus horizontal relationships.

Individualist Versus Collectivist Orientation. Cultures vary in the amount of emphasis they place on personal versus group interests or individualism versus collectivism. Self-identity and personal goals are inseparable from, and independent of, the larger social group in collectivist and individualist cultures, respectively. In collectivist cultures, norms, obligations, and duties tend to guide social behavior, whereas personal needs, rights, and contracts guide behavior in individualist cultures. Another important distinction is that collectivist cultures place great emphasis on ingroups versus outgroups. Ingroups are those social groups "about whose welfare one is concerned, and with whom one is willing to cooperate without demanding equitable returns, and separation from whom leads to discomfort or even pain" (Triandis et al., 1984, p. 75).

In general, homogeneous and less complex cultures are more likely to be collectivist, and cultures that are more complex and heterogeneous tend to be individualist (Triandis, 1994). The cultures of Western European nations, and those with strong ties to this region, are generally classified as having strong individualistic orientations. Third World nations and cultures, in contrast, tend to be more collectivistic (Gudykunst and Kim, 1997; Hofstede, 1991). Within the individualistically oriented United States, several minority groups, including those with roots in Latino and Asian cultures, nonetheless tend to be more collectivistic in their outlook (Triandis et al., 1984).

What implications might these cultural qualities hold for survey nonresponse? Gudykunst (1997) argues that collectivists are more likely to view outgroups with suspicion. We believe that survey organizations are more likely to be viewed as outgroups within minority communities. Recognition of this fact has prompted many researchers to employ "local talent" when conducting surveys in minority communities (see Gwiasda et al., 1997; Weinberg, 1971). Persons with individualist orientations, in contrast, would be less likely to make an ingroup–outgroup distinction when evaluating a survey request. Instead, they might be expected to view the request using a simple cost–benefit evaluation, social exchange, or calculus (Groves and Couper, 1998).

Power Distance. Power distance is concerned with social inequality and the degree to which individuals hold power over one another. Hofstede and Bond (1984,

p. 419) define this as "the extent to which the less powerful members of institutions and organizations accept that power is distributed unequally." Its relevance to survey nonresponse is its potential usefulness in interpreting the behavior of strangers during encounters with one another, especially those that may involve varying degrees of power or authority. Persons in low power distance cultures do not accept authority uncritically; they expect it to be legitimated. In cultures with high degrees of power distance, individuals are less likely to question authority, and as Gudykunst (1997, p. 333) points out, "they expect to be told what to do." At first glance, one might expect that such forms of cultural coercion would make it easier to collect survey data from respondents within cultures high in power distance. However, power distance is also strongly correlated with collectivism: that is, cultures high in collectivism also tend to have a great deal of power distance (Smith and Bond, 1998). Conformity and obedience to authority are also strongly associated with power distance and collectivism (Bond and Smith, 1996; Mann, 1980).

Collectivistic cultures treat ingroups and outgroups very differently, and this distinction likely applies to respect for power as well. Consequently, power distance in practice might be expected to improve compliance only to ingroup survey requests in collectivist cultures. Requests from outgroups would likely be met with greater opposition within collectivistic cultures, regardless of the power distance involved. One might also expect less pressure to comply with survey requests in low power distance cultures and this is probably correct. Because low power distance cultures, also tend to be individualistic in orientation, however, outgroup survey requests would also be less likely to be treated differently from ingroup requests. We might thus hypothesize that survey requests would be greeted with: (a) the most compliance within low power distance/high collectivism cultures when the request is made on behalf of an ingroup; (b) the least compliance within low power distance/high collectivism cultures when the request is made on behalf of an outgroup; and (c) intermediate levels of compliance within cultures low on power distance and high on individualism.

Vertical Versus Horizontal Relationships. A third cultural orientation that we briefly examine is concerned with societal emphasis on vertical versus horizontal social relationships. As Triandis (1996) has observed, hierarchy is very important to some cultural groups that place a high emphasis on authority and are willing to accept social inequality. Other societies place less emphasis on hierarchical systems of authority and may be considered more egalitarian in nature. Examples of vertical societies include India and the United States. Horizontal societies include Sweden and Australia. It seems likely that survey nonresponse would be less of a problem among cultural groups with hierarchical orientations, so long as those requesting cooperation are perceived as having legitimate authority to do so. This relationship might be further conditioned by ingroup versus outgroup interaction, however, in that perceived outgroup requests for survey participation may be less respected in vertical societies. Horizontal cultures, in contrast, may make fewer distinctions between ingroup and outgroup requests for survey participation.

4.3.2 Culture and Communication Styles

Communication is perhaps the most fundamental element of the survey process. It has been described as a self-corrective mechanism whereby participants continually check their environment to assure that others have sufficient prior knowledge to reach a mutual understanding (Much, 1991). Effective communication between survey respondents and researchers requires the establishment of a level of intersubjectivity that reflects mutual understanding. Intersubjectivity may be defined as implicit agreement to share the same understanding. The establishment of intersubjectivity allows the formation of a common conceptual framework that permits the respondent to interact within the survey environment without feeling threatened or exposed. This communicative interaction, if established, represents a "partially shared social world" in which each participant implicitly understands the role of the other and cultural differences may be circumvented by the willingness of each to see the other's point of view (Rommetveit, 1974). The extent to which intersubjectivity can be established between the respondent and researcher may influence the respondent's willingness to participate in research and the information they disclose.

Cultures are known to vary along several dimensions of communicative style and several of these have implications for survey nonresponse. Among the communicative norms we explore in this regard are (1) contextual requirements, (2) nonverbal behaviors, and (3) self-disclosure patterns.

Cultural Differences in Context Requirements. Cultures are believed to vary in their reliance on high- versus low-context communication processes. In cultures with low-context requirements, a message's meaning is inferred directly from its content with little regard for context. According to Hecht et al. (1989, p. 176), low context messages require "clear description, unambiguous communication, and a high degree of specificity." Messages of this form are commonly used in situations where precision and clarity are essential, and where double meanings cannot be tolerated, such as in legal communications and computer programming. High-context cultures, by way of comparison, place greater emphasis on the message's context. Thus, messages are interpreted not only from their explicit content but also as a function of nonverbal environmental cues and inferred meanings. Gudykunst (1998) suggests that individualistic cultures rely primarily on low-context communication styles, while collectivistic cultures have higher context requirements. Perhaps not surprising, then, are the findings that North American and North European cultures are classified as low-context whereas East Asian and Latin American cultures are considered to be high-context societies (Hecht et al., 1989).

How might contextual needs influence survey nonresponse? Persons enmeshed in high-context environments may be more cautious in their first interactions with new acquaintances and develop more assumptions about them based upon contextual information (Gudykunst, 1983). Consequently, we speculate that environmental cues may be more likely to influence decisions regarding survey participation among potential respondents with high-context communication styles. In such encounters, interviewers may also be less likely to be taken at "face value" and their "real" motives perhaps more likely to be questioned. Because persons in high-context cultures are also more likely to form assumptions about persons based on their racial or ethnic background (Gudykunst and Kim, 1997), one might also expect to find greater effects of respondent–interviewer social distance on survey nonresponse when surveying persons possessing high-context communication styles. This topic will be explored further in the next section.

Nonverbal Behavior. Nonverbal communication has been referred to as the "hidden dimension" of culture (Hall, 1966). Although more so in high-context cultures, all social groups employ some forms of nonverbal communication. Cross-cultural variability in expectations regarding the appropriateness of eye contact, expressions of emotion, various degrees of interpersonal distance, and sensory involvement during initial encounters may be expected to lead to higher rates of misunderstanding and survey nonresponse.

Self-Disclosure. Revelation of personal information, or self-disclosure, is critical to communication and obviously necessary for survey participation. Research conducted in the United States suggests that Whites disclose information at higher rates than do Blacks (Gudykunst and Kim, 1997). Kochman (1981) has presented considerable information on differences in the communication styles of Blacks and Whites in the United States. These include the acceptability of direct questions, acceptable forms of disagreement, rules for taking turns in conversations, and the importance of various forms of nonverbal communication. Each of these variations in communicative styles may be expected to influence patterns of self-disclosure. Blacks have also been shown to self-disclose more information than do Mexican Americans (Littlefield, 1974). Self-disclosure is most frequently associated with the types of direct, low-context communication patterns observed in individualistic social systems, as opposed to the less direct styles seen in collectivist cultures (Gudykunst, 1998). In our earlier discussion, it was observed that ingroup versus outgroup interactions are qualitatively different in collectivist, but not necessarily within individualistic, cultures. Available evidence suggests that variations in self-disclosure follow a similar pattern. Gudykunst et al. (1992) have observed greater self-disclosure among ingroup members in samples of presumably collectivist Chinese respondents. In contrast, they found few differences in ingroup versus outgroup self-disclosure among individualistic U.S. and Australian samples. Another comparative analysis found similar differences in patterns between respondents in the United States and Poland, a more collectivist country, where a greater distinction was made between willingness to disclose to friends and neighbors, compared to the United States (Derlegh and Stelien, 1977). We suspect that resistance to self-disclosure may be correlated with nonresponse. Although it has not been demonstrated, it seems intuitive that less intention to self-disclose when invited to interact with a stranger should also be predictive of less willingness to engage in any interaction with that individual. It also seems likely that perceptions of the group or institution represented by an interviewer may moderate respondent willingness to participate (Groves and Couper, 1998).

4.3.3 Culture and Social Participation

Patterns of social participation may also be influenced via culture and cultural experiences. In this section, we review three culturally mediated social processes that may influence survey nonresponse: (1) minority group oppression and opposition, (2) social distance perceptions, and (3) helping behavior.

Minority Oppression and Opposition. Interethnic misunderstanding and animosity are unfortunate realities of which we are reminded on an all-too-regular basis. Ethnic minorities in many nations face discrimination, religious and cultural repression, and barriers to social and economic advancement. Alienation from and opposition to the dominant culture are perhaps inevitable consequences of these experiences. A contemporary example is the generalized suspicion and mistrust of Whites by many Black Americans after centuries of race-related mistreatment (Massey and Denton, 1993; Ogbu, 1990; Terrell and Terrell, 1981). There are also multiple historical examples of medical research exploiting Black Americans in the name of scientific advancement dating back centuries (Fry, 1984; Gray, 1998; Humphrey, 1973).

Some survey evidence is available documenting greater levels of mistrust in general among Blacks, compared to other survey respondents. Singer et al. (1993) found that Blacks, but not Hispanics, had higher scores on measures of concern with the privacy and confidentiality of the 1990 U.S. Census when contrasted with non-Hispanic Whites. Aquilino (1994) reported similar findings suggesting that Blacks were less likely than Whites to believe that most people can be trusted. In addition, an experimental assessment of three alternative household rostering methods discovered that an anonymous version added about 30% more Black male residents, on average, per enumerated dwelling (Tourangeau et al., 1997). This finding suggests that deliberate concealment may be a significant reason for the Census undercounts of Blacks described earlier.

Finally, many Black communities resent the intrusion of outside researchers who they believe often benefit professionally by revealing neighborhood problems (Brazziel, 1973; Josephson, 1970; Myers, 1979). Representatives of other minority groups in the United States (Weiss, 1977; Yu, 1982) have expressed similar feelings. These circumstances suggest that many of the same factors that serve as barriers to the socioeconomic advancement of minority populations, such as poor education, poverty, and discrimination, may also serve as barriers to survey participation. We believe the U.S. Black–White experience serves as a useful example of how unique, historical patterns of intercultural relations may influence survey nonresponse patterns worldwide.

Social Distance. Bogardus (1925, p. 299) defined "social distance" as "the degrees and grades of understanding and feeling that persons experience regarding each other." Perceptions of social distance are thought to be influenced by personal similarities in demographics, attitudes and abilities, power differences, and predictability (Triandis, 1994). Interviewers are often demographically "matched" with the race or ethnic composition of neighborhoods where fieldwork is to be done in hopes of minimizing social distance and increasing response rates (Nandi, 1982; Hurh and Kim, 1982). Concern has also been expressed that, because of cultural differences in the social status of women and normative expectations regarding appropriate forms of cross-gender communication, male respondents may be less inclined to comply with an interview request in some cultures when the interviewer is female (Johnson et al., 1997). In other cultures, male interviewers, particularly of different cultural backgrounds, might not be considered appropriate for interviewing female respondents. So confident are researchers of the wisdom of this social distance strategy that virtually no empirical studies have been conducted to confirm its efficacy.

Helping Behavior. Potential cultural differences in willingness to assist a stranger, often referred to as "helping behavior," may also be relevant to survey nonresponse. Variations in helping behavior have been more commonly associated with urbanicity and used to understand nonresponse in that context (Groves and Couper, 1998). A review of the helping literature in the United States by Crosby et al. (1980) found, perhaps not surprisingly, that intraracial helping was more common than interracial helping. Of 13 experiments examined, Whites were significantly more likely to help Whites in 6 (46.1%), Blacks were more likely to be helped by Whites in 2 (15.4%), and 5 differences were not significant (38.5%). Similarly, Blacks were somewhat more likely to help Blacks under identical circumstances (in 7 of 13 experiments; 53.8%). Whites were more likely to be helped by Blacks in 2 experiments (15.4%), and differences were not significant in 4 others (30.8%).

These findings are consistent with Triandis' (1994, p. 221) observation that "in all cultures people are more likely to help an ingroup member than an outgroup member" (although the probabilities may vary between individualist and collectivist cultures). No differences in helping behavior were found by Hedge and Yousif (1992), who reported highly similar rates between England and the Sudan. Miller et al. (1990), in contrast, found considerably greater levels of self-reported helping behavior among respondents in India compared to the United States. A major difficulty in evaluating this body of literature is the wide variety of methods employed to measure helping behavior. Nonetheless, general trends such as the consistent relationship between varied measures of helping behavior and urbanicity indicate that this general construct is robust and may be applicable to the survey nonresponse process, particularly considering the degree to which altruistic motives are commonly cited as a reason for survey participation.

4.4 SUMMARY

Our goal in conducting this review has been to investigate the seldom-studied but very important role that culture may play in the survey response process. In doing so, we have examined the available empirical research literature relevant to the topic. This research suggests cross-cultural variability at both the national and subnational levels. Interpretations of this literature, of course, are complicated by large variations in research context and design, the forms of nonresponse examined, and populations included. It also remains unclear whether the cross-group differences identified might be more appropriately interpreted as being the consequence of racial, ethnic, and national differences in the various socioeconomic status indicators that have been repeatedly linked with survey nonresponse (Groves and Couper, 1998). Of the subset of studies reviewed that introduced multivariate controls, however, most found independent effects of the cultural indicators after holding education and/or income measures constant. The findings support the notion that there are qualitative differences across cultures and subcultures that are influencing nonresponse patterns.

A number of conceptual schemes were presented that may provide a useful framework for interpreting cultural variability in survey nonresponse. In Figure 4.1, we have integrated these into a conceptual model that shows the means by which a respondent's cultural and social characteristics may influence and interact with survey process to influence nonresponse. Recognizing the complexity and pervasiveness of culture and the processes through which it influences human cognition and behavior, we make no claims regarding the adequacy of this model. We nonetheless believe that the mechanisms identified represent a useful first step toward understanding these processes.

As can be seen in Figure 4.1, we believe that cultural values such as those reviewed in Section 4.3.1 influence both respondent accessibility and cooperation indirectly through their effects on patterns of social participation, preferred styles of communication, and socioeconomic opportunities (note that each of these processes may also influence cultural values). Social participation patterns, communication styles, and socioeconomic status, in turn, likely influence and reinforce one another and also directly influence survey nonresponse. Currently, we believe there is very little evidence available with which to directly test this conceptual model.

Perhaps the greatest weakness of the available empirical literature is its reliance on measures of country of residence and racial or ethnic status that at best can only

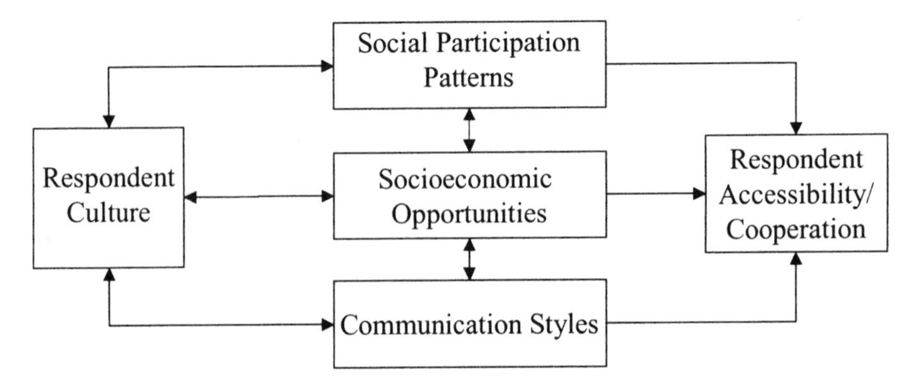

Figure 4.1. Proposed conceptual model of cultural influences on survey nonresponse.

roughly approximate culture. We must investigate the limits of these measures and develop a more sophisticated understanding of those dimensions of culture that are most critical to the conduct of survey research. Determining how these various dimensions map onto the commonly used racial, ethnic, and national indicators will also provide valuable information with which to more precisely interpret the empirical data reviewed in this chapter. Identifying cultural orientations, such as the degree to which a respondent considers him/herself an individualist or part of a collective, have the potential of helping us better understand survey nonresponse.

CHAPTER 5

To Answer or Not to Answer: Decision Processes Related to Survey Item Nonresponse

Paul Beatty, *National Center for Health Statistics*
Douglas Herrmann, *Indiana State University*

5.1 INTRODUCTION

Survey questionnaires are used to ask respondents about behaviors, attitudes, and beliefs covering virtually every facet of their lives. For the most part, we accept respondents' answers as accurate, i.e., reflecting something they did, an event they are aware of, or an attitude that existed in some form prior to our asking the question. If respondents do not have the information we are looking for, we assume that they report that as well.

Yet we now understand a great deal about the cognitive complexities of survey response (Sudman et al., 1996; Schwarz and Sudman, 1996; Tourangeau, 1984) and can see that answers may be based on very different degrees of knowledge. Consider, for example, a survey question asking whether respondents had a routine medical examination within the past 12 months. Statistically, all positive responses appear to be equivalent. However, some respondents might answer based on memories of specific times and places of medical exams, while others more or less guess, based on vague or incomplete memories. Alternately, consider two respondents who have similar but limited memories about their recent medical exams. One may decide that her best guess is sufficient to justify answering the question, while the other does not. Or, one respondent may decide to answer carefully with as much accuracy as possible, while another, lacking such motivation, may answer with little concern for accuracy. Thus, the assumption that substantive answers are of equal quality and reflect some "real" level of knowledge is overly simplistic. Rather, when respondents are faced with a survey question, they have two decisions to make: whether they can respond, and whether they will respond. *Item nonresponse* results when respondents decide negatively in either case.

From a statistical point of view, item nonresponse creates analytic difficulties. The effective sample size is reduced, and item nonresponse should therefore be avoided. From a psychological point of view, things are less clear. We want respondents to provide information that meets a certain level of cognitive precision, while keeping in mind that relatively imprecise answers may be sufficient for purposes of the survey. But wherever we set the threshold for minimal acceptable level of quality, we may be reluctant to accept responses beneath that level. This requires a conscious decision on the part of researchers regarding how much cognitive clarity is enough. Making this decision wisely requires a better understanding of how respondents decide whether or not to actually answer survey questions.

The purpose of this chapter is twofold. The first is to provide a theoretical framework explaining the cognitive processes that lead to either substantive answers or item nonresponse. The second is to illustrate use of a possible evaluation methodology for assessing the likely nature of item nonresponse for particular survey questions, i.e., whether responses are likely to be sufficient for particular research needs. Our primary focus in this chapter is questions that draw on autobiographical memory, as opposed to attitudes. However, our theoretical framework is sufficiently general to be adaptable to attitudinal questions, and some of our empirical data touch on problems of attitude measurement as well.

5.2 BROAD INFLUENCES ON THE DECISION TO RESPOND

Item nonresponse results from two fundamental decisions made by respondents: whether they can answer, and whether they will answer. The first decision is one of *cognition,* involving the ability to remember and apply relevant information to the response task. The second decision is one of *motivation,* which can be low due to sensitivity (if they feel information presents them in an unflattering light), burden (if the information is difficult to provide), or conflict of interest (if they perceive that the information could be used against them, or is requested by an untrustworthy sponsor).

Features of the questions themselves can influence whether respondents choose to answer. The *clarity* of the question and *complexity* of the response task can determine the degree of cognitive challenge posed by the question and affect motivation to respond. Questionnaires also include cues about the acceptability of item nonresponse, such as the explicit availability of "don't know" (DK) categories. A framework that adequately explains response decisions must take into account three factors—what respondents actually know, their interpretations of what the questions are asking them to provide, and their motivations for responding.

More succinctly, we propose that the decision to respond or not respond to a question is essentially driven by the following factors (initially proposed in Beatty and Herrmann, 1995): *cognitive state* (the availability of the information requested); *adequacy judgments* (the respondent's perception of the level of accuracy required by the questioner); and *communicative intent* (the respondent's motivation to provide the information requested). Each of these is considered in turn.

5.2.1 The Role of Cognitive State

Psychological research has consistently demonstrated that respondent knowledge is a matter of degree rather than a dichotomy of knowing and not knowing (Hasher and Griffin, 1978; Reder, 1988). Subsequent research has suggested that respondent knowledge can be classified in terms of four "cognitive states" (Herrmann, 1995; Beatty, Herrmann et al., 1998):

1. Available: The requested information can be retrieved with minimal effort
2. Accessible: The requested information can be retrieved with effort or prompts
3. Generatable: The requested information is not exactly known, but may be estimated using other information in memory
4. Inestimable: The requested information is not known and there is virtually no basis for estimation

Cognitive state is the most obvious determinant of whether or not respondents provide answers. This is especially clear in the "extreme" cognitive states. When information is *available,* the respondent knows the answer to the question by definition; when information is *inestimable,* the respondent does not know the answer. In either case, it is presumably clear to respondents what they "should" do—either answer or admit to ignorance. Still, there is not necessarily a simple connection between what respondents know and what they actually report (Eisenhower et al., 1991; Bradburn et al., 1987).

As Figure 5.1. shows, *any* cognitive state can lead to either a substantive response or item nonresponse. In the extreme cognitive states, such as available, respondents may choose to report or not report what they know. If a respondent knows the answer to a question but chooses not to report it, that constitutes an error of omission (labeled "O"). Alternately, in the inestimable state, a respondent may choose to either admit ignorance or to fabricate information. The latter choice would constitute an error of commission (labeled "C"). Obviously, both errors of omission and commission are undesirable in survey research, reflecting blatant misrepresentations of respondent knowledge.

Generally, however, respondent knowledge is not such an all or nothing proposition. For example, frequency of everyday behaviors may not be known exactly, but respondents may have some relevant information that could be used to estimate them. Thus, a substantive response in the intermediate cognitive states may be seen as correct or incorrect, depending upon the precision expected by the researcher and the precision that the respondent believes is expected.

5.2.2 The Role of Adequacy Judgment

An adequacy judgment is necessary when a potential response contains uncertainty, estimation, or guessing (Jobe and Herrmann, 1996; Schlenker, 1986). In the intermediate cognitive states, a respondent can provide some information, but it may not

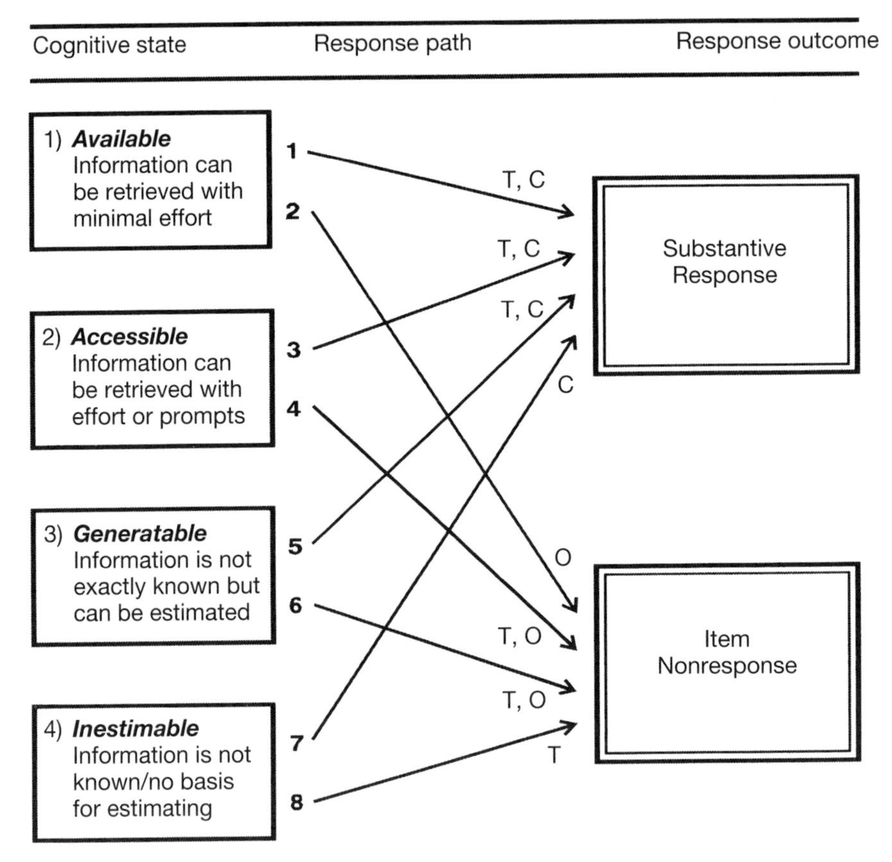

Figure 5.1. Four-state mapping of cognitive states to response outcomes. T: Potentially truthful response. O: Potential error of omission. C: Potential error of commission.

be clear whether this information meets the requirements of the question. Consider, for example, a question that asks "In the last seven days, how many miles did you drive your car?" Most respondents could not answer this precisely (even if we ignore the obvious definitional problems within this question). A respondent may know, however, that her daily commute is approximately 20 miles round trip, 5 days a week, but this varies somewhat. She may also know that local errands are performed on evenings and weekends. Using this information, she figures 110 miles per week, realizing that this may be off by 10 or 20 miles. She must now decide whether this is close enough to report, or whether it would be better to admit that she does not know.

It is likely that this decision will be based in part on the respondent's judgment of the level of precision that is called for by the survey. The difficulty of this judgment will also depend on the degree of uncertainty involved—for example, mild uncertainty would probably not create much of a dilemma for respondents. It may also depend on the cognitive complexity of the question. Judgments may be more complicated when questions provide insufficient guidance regarding the level of certainty that qualifies as a legitimate answer. Surveyors sometimes suggest how much certainty is called for through the use of "filters" that give respondents an explicit opportunity to admit to a lack of knowledge (Schuman and Presser, 1981).

Respondents with equal levels of certainty about their answer could make different decisions about whether an answered should be provided. Some may feel that nonresponse is more appropriate, while others may feel that any substantive answer is better than none, even if it contains imprecision. If a respondent believes that only precise information should be reported, then item nonresponse may be viewed as the most correct action that can be taken. Another respondent might think that a "best guess" is acceptable.

5.2.3 The Role of Communicative Intent

Regardless of how a respondent feels about her or his ability to answer, a decision must be made on what to report (DeMaio, 1984; Hippler et al., 1987). We refer to the decision of what to report as communicative intent.

Respondents may elect not to provide substantive responses for a number of reasons. When they initially receive the survey question, they may choose not to spend the effort necessary to remember and report information even though they technically could do so (Krosnick, 1991). Willingness to spend effort is likely to depend upon their interest in the subject matter and the complexity of the task (Cannell et al., 1981; Eisenhower et al., 1991). Also, since they may feel more of a social obligation to a live interviewer on their doorstep than to one over the telephone (or to a mailed questionnaire), mode could play a factor (Dillman, 2000).

In other cases, respondents may figure a reasonable response to the question but choose not to provide it. One of the most common reasons for such a decision is the belief that their behaviors or attitudes are socially undesirable (DeMaio, 1984). For example, they may be reluctant to admit that they engage in smoking and other behaviors associated with health risks. In other cases, respondents may believe that reporting could put them at risk, particularly in the case of illegal activities (Turner, 1982; Gfroerer et al., 1992).

Respondents with negative communicative intent have several choices. They may elect to plead ignorance (provide a "DK" response) or refuse to answer. In other cases, respondents may provide a substantive answer that they know to be incorrect. The first two avoidance strategies constitute errors of omission, while the latter is an error of commission. In either situation, the respondent's choice of avoidance strategy is largely determined by communicative intent. If a respondent wishes to hide the use of illegal drugs, he would probably deny using them rather than refusing to answer, because the latter might be interpreted as a tacit admission of the behavior. On the other hand, a respondent who wishes to keep his personal finances private may simply refuse to provide such information.

Thus, communicative intent can be an important determinant of both errors of commission and omission. In fact, communicative intent is the *only* reasonable explanation for errors of omission and commission in the extreme cognitive states of available and inestimable. A variety of factors could influence communicative intent, including interest in survey content, complexity of the question, length of the instrument, and perceived risk in answering or not answering.

5.3 RESPONSE DECISION MODEL

The concepts of cognitive states, adequacy judgments, and communicative intent have been proposed before (sometimes with different names, e.g., Cannell et al., 1981; Jobe and Herrmann, 1996). However, they have not been placed explicitly into a model that reveals how all these concepts relate to item nonresponse. We summarize this process in the response decision model shown in Figure 5.2.

The process begins when a respondent first receives the survey question. At this point, she or he must come up with a viable interpretation of what is being asked. The first opportunity for item nonresponse occurs here, if the respondent does not understand the response task (although the respondent may also query further about the intent of the question). If the respondent believes that she or he has a viable interpretation, either initially or following additional guidance, then the response process can continue.

The respondent must then decide whether to make the effort required to respond (Cannell and Henson, 1974). This is the respondent's first decision regarding communicative intent: if his motivation is low, he or she may opt out of continuing the response process. This decision may be influenced by length of the survey and complexity of the question.

The respondent who continues must next retrieve whatever information is relevant to the question and available. If this retrieval reflects information from extreme cognitive states (available or inestimable) communicative intent is the only remaining determinant of item nonresponse. In the available cognitive state, the respondent knows the information that is requested; thus the only explanation for item nonresponse is a communicative intent of nondisclosure. In contrast, a respondent in cognitive state 4 does *not* know the information. In that case, item nonresponse is the most accurate possible outcome.

More commonly, respondent information will fall into the intermediate states (accessible or generatable); the respondent must first evaluate the quality of the potential response. It is not always obvious whether accessible or generatable information is appropriate as a response. Respondents with the same degree of knowledge could choose to respond or not to respond, and both may be correct based on their understanding of the question's intent. Following this judgment, respondents continue with communicative intent decisions, ultimately leading to either substantive responses or item nonresponse.

The model shows four general explanations for item nonresponse: (a) inadequate understanding of the question; (b) low motivation; (c) a decision to withhold information even though it would meet the assumed objectives of the question; or (d) a belief that item nonresponse is more accurate given the precision called for by the question (a judgment which may or may not match the intentions of the researcher). An under-

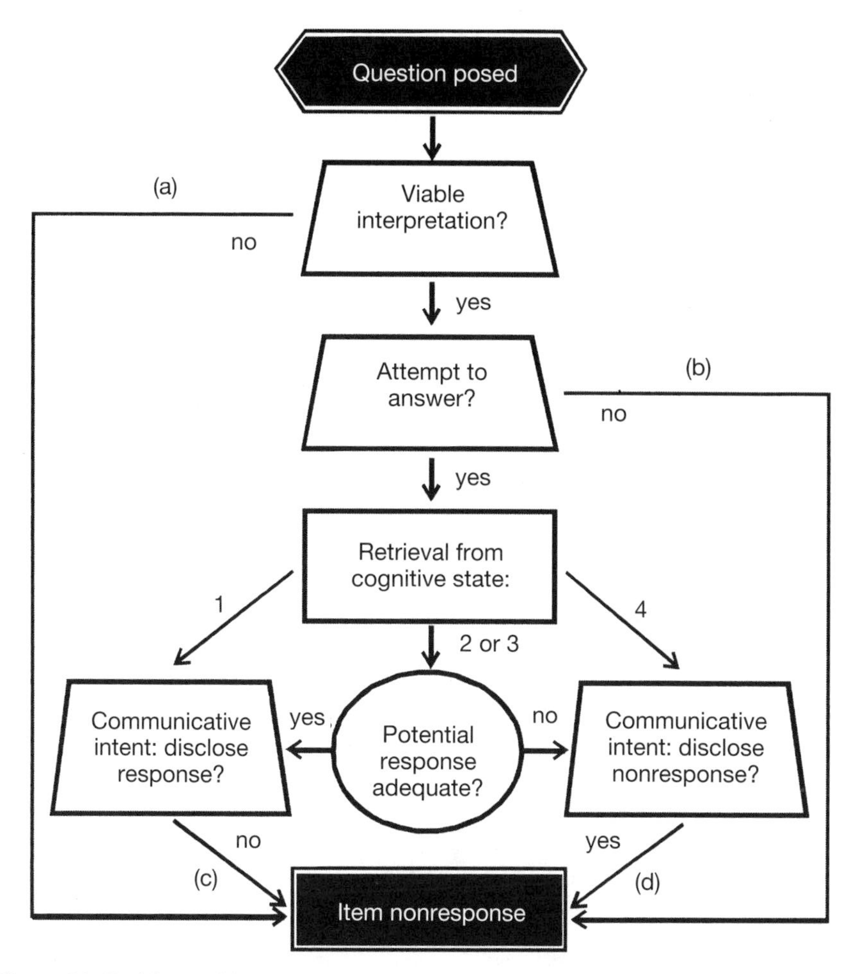

Figure 5.2. Decision model for item nonresponse. (a) Inadequate understanding/comprehension. (b) Low motivation/effort. (c) Withholding of available information. (d) Belief in the inadequacy of response (may or may not agree with researcher objectives).

standing of these explanations for item nonresponse can help us decide on design strategies to reduce item nonresponse (or perhaps, in some cases, to encourage it).

5.4 ERROR CONTROL STRATEGIES—ENCOURAGING OR DISCOURAGING ITEM NONRESPONSE

Researchers can make a number of design decisions that will affect rates of item nonresponse. Consider the following strategies:

1. Do not explicitly offer options for item nonresponse and resist such responses with probes
2. Do not explicitly offer DK's, but accept them when provided
3. Offer DK's as an explicit choice
4. Encourage respondents not to answer questions unless they are confident in the accuracy of their answers

Surveyors may attempt to minimize errors of omission, minimize errors of commission, or do something in between. We refer to these choices as "error control strategies."

Researchers have traditionally been more concerned about errors of omission, since these errors reduce effective sample sizes and create analytic complications. Furthermore, several studies suggest that item nonresponses can often be avoided. Cannell et al. (1979) directed respondents who gave DK answers to try harder, which did increase substantive responses and is used as a rationale for interviewer instructions to always probe a "don't know" response at least once (Guenzel et al., 1983, p. 181). In another study, Poe et al. (1988) argued against providing explicit DK boxes for self-administered factual questionnaires, which reduced effective sample sizes without improving data quality.

Errors of commission have been of greater concern for attitudinal questions, and DK's are more often tolerated for attitudinal than behavioral questions (Groves, 1989, pp. 468–471; Krosnick, Chapter 6, this volume). This mindset partially stems from Converse's (1964) work on "nonattitudes," which suggested that respondents were often unfamiliar with the subject matter of attitudinal questions and provided essentially random responses. Similarly, Schuman and Presser (1981) reported that up to 30% of respondents reported attitudes about highly obscure laws when they were not given an explicit chance to say that they had no opinion. Bishop et al. (1986) interpreted such findings as evidence that people feel "pressured" to provide substantive responses even if they have no basis for doing so. They suggested that it is valuable to provide explicit "outs" for respondents to express lack of an opinion.

Yet the majority of recent studies advise against explicitly accepting DK's for attitudinal questions as well. Smith (1984) argued that item nonresponse reflected not only nonattitudes but also ambivalence. Therefore, probing could encourage respondents to express "leanings" that would still be valuable. Feick (1989) added that item nonresponse could be caused by problems understanding questions and difficulty choosing from available response options. Thus, improvements in questionnaire design would lead to greater gains in data than providing explicit DK options. Also, Gilljam and Granberg (1993) showed that attitudes obtained by probing after initial DK's successfully predicted behaviors. More recently, Visser et al. (2000) confirmed that respondents' "leanings," even when expressed reluctantly, could be used to accurately predict poll results.

Thus, for both factual and attitudinal questions, the preponderance of advice leans toward minimizing item nonresponse (DK's in particular). Most often, this advice is based on the assumption that even weakly held knowledge or attitudes are likely to be of use to researchers. Some researchers have qualified this position—for example, Fowler (1995) points out that the value of what respondents report depends upon how directly questions pertain to their firsthand experiences. However, many social scientists offer general agreement with the dominant statistical position that item nonresponse is rarely, if ever, preferable to responses indicating specific attitudes, beliefs, or behaviors.

Yet, blanket advice such as this is increasingly difficult to defend, for several reasons. Recent years have seen a rapid increase in our knowledge about the inaccuracy, distortions, and biases that may occur in autobiographical recall (Schwarz and Sudman, 1994; Bradburn et al., 1987; Herrmann 1986, 1994, 1995). Similarly, attitudes may be constructed in response to particular words and contexts of survey questions, especially if they did not exist prior to measurement attempts (Feldman and Lynch, 1988; Tourangeau, 1992). It is very clear that, from a psychological perspective, data quality varies along a continuum. Some researchers may be willing to accept data on a relatively low point of this continuum as adequate for their purposes. Others consider responses to be of minimal use if respondents are not confident in their accuracy. Researchers must decide on an individual basis what is their minimum level of *acceptable certainty.*

Furthermore, broad recommendations for accepting or not accepting item nonresponse do not take into account the reasons that respondents fail to provide answers. If item nonresponse stems from poor understanding of questions and low motivation, then the most useful strategies to improve data quality would be to write better questions, offer response incentives, or reduce the burden of questionnaires. Simply accepting item nonresponse, or insisting that respondents give any answer that appear to be substantive, does not address the real problems.

Design decisions can be assisted by knowledge of what cognitive states respondents tend to fall into regarding specific questions. Knowledge of respondent cognitive states could suggest the strategies based on data needs. For example, if respondents tend to fall into the accessible state, specially designed probes may generate acceptable responses.

For other cognitive states, it may be necessary to define the acceptable level of precision for answers. If the researcher does not define the "acceptable" level, respondents will have to do so themselves. Consequently, respondents with the same level of knowledge may make different judgments about the acceptability of their responses. Yet, both could believe that they answered correctly. Instructions for error control can be given according to desired precision levels and to clarify question requirements.

Survey designers can create surveys that maximize truthful communicative intent by developing realistic assessments of sensitivity and burden. Researchers may need to justify the need for sensitive information, offer rewards or other incentives for participation, or provide proof of confidentiality (Willis et al., 1994). And although there is little empirical evidence thus far that survey length decreases data quality, there are ample theoretical reasons to believe that respondent motivation can be taxed to the point that they cooperate only nominally with requests for additional information (Bradburn and Sudman, 1991).

A few additional points bear consideration. First, it is important to consider that error control strategies involve trade-offs. For example, strong attempts to discourage DK's might inadvertently restrict truthful admissions of ignorance (i.e., respondents may provide fabricated responses). Alternately, strong attempts to discourage reporting of imprecise estimates might inadvertently discourage reporting of legitimate information as well, as DK's may appear to be an "easy out." Attempts to eliminate errors of omission could create errors of commission; the reverse is also true. Perhaps more importantly, such strategies may overlook the real problems leading to item nonresponse (such as poorly worded questions, unreasonable response burden, or lack of clarity about what qualifies as a legitimate response).

5.5 USE OF THE THEORETICAL FRAMEWORK TO EVALUATE POTENTIAL SURVEY QUESTIONS

The usefulness of the preceding theoretical framework depends partially upon our ability to collect meaningful data about cognitive states, adequacy judgments, and communicative intent. We present here a test evaluation instrument designed to collect such information for a set of survey questions, and show how this information can be used to assist in questionnaire design. Some research has cast doubt on the feasibility of measuring introspective variables such as cognitive states, since such reports may require access to "metacognitions" that respondents are unlikely to have. For example, Kahneman and Tversky (1973) noted that metacognitive judgments are made with the use of heuristics that can lead to considerable reporting of errors. Nisbett and Wilson (1977) argued that people have very little introspective access to their higher-order cognitive processes. Also, Feldman and Lynch (1988) point out that when beliefs do not exist in long-term memory, or are inaccessible via introspection, the act of measurement itself can create them. For example, respondents' reports about their cognitive states may be based on the *responses themselves* rather than actual awareness of their cognitive states (e.g., "I answered the question, therefore I must actually have information about it.").

In spite of their potential limitations, we believe that respondents are able to provide useful information that pertains to our theoretical model. Most of the thought processes about which Nisbett and Wilson raise concerns (such as respondent explanations of why they behaved in a certain manner) are considerably more complex than the cognitive variables of interest to us. We accept the likelihood that reports about cognitive states contain some error, as many survey responses do. Yet, findings from an earlier study (Beatty et al., 1998) suggest that such reports can still be quite useful. In that study, respondents reported lower cognitive states for questions that were judged to be relatively difficult, and lower cognitive states for proxy responses than for self-reports. Results also suggested that the four cognitive states could be viewed as a scalar of *knowableness,* i.e., likelihood that a respondent knows, or is capable of knowing, the information requested by a survey question.

Our current study is an expanded effort to collect and apply cognitive self-report data and to address the following questions:

1. Can respondents tell us about what cognitive states they fall into on questions that invoke various types of autobiographical memory?

2. Can respondents also provide information about their adequacy judgments and communicative intent?
3. Can this information increase our understanding of decision processes involved in choosing whether or not to provide substantive responses?
4. Can we use these insights to develop instruments for item evaluations?

To answer the questions raised above, we gave two experimental self-administered questionnaires to 218 undergraduate students at Indiana State University in two classrooms. The first questionnaire began with nine survey questions, which varied in terms of type of information solicited, answer format, and complexity. The second questionnaire contained seven of these questions and added four new questions. Ten of the survey questions involved autobiographical memory, ranging from the absurdly simple (date of birth), to somewhat more complicated (number of visits to a doctor in the last year), to the impossible (number of apples eaten in a lifetime). Five of the autobiographical questions dealt with dating of events (four of them asked for an open-ended date, and one used closed answer categories to indicate when an event had taken place). Three questions asked for frequencies (two of objective behavior, one of subjective feelings of poor health). The remaining two questions were dichotomous—one about behavior (whether the respondent had smoked 100 cigarettes), and another about having been talked to by a doctor about high cholesterol. Three questions were nonautobiographical: one asked for beliefs of others' behavior (doctor visits per year for the "average" person); another asked about beliefs of future events (the winner of the 2000 U.S. Presidential election). There was also a factual question (asking for the date of the explosion aboard the space shuttle Challenger).

The questionnaires were administered in three forms, distributed randomly, with different instructions regarding the handling of uncertainty. The first form encouraged respondents to answer every question, even if it was difficult to do so; the second gave minimal guidance regarding item nonresponse, leaving adequacy judgments to respondents' discretion; the third discouraged respondents from providing responses by providing a DK category with each item and encouraging respondents to use it if they thought it was the best answer. Immediately after answering all questions, respondents were asked to assess their cognitive state for each item on a four-point scale: 1 = I knew the answer immediately; 2 = It took me a little while to remember the answer; 3 = I didn't know the exact answer, but knew I could make a guess; and 4 = I had no idea. Then, respondents answered a short follow-up questionnaire asking 1) which questions, if any, they considered not answering but answered anyway (and why); 2) which questions, if any, they did not answer but considered doing so (and why); and 3) which questions, if any, were overly personal or embarrassing.

5.6 RESULTS

Table 5.1 lists characteristics, item nonresponse, and other information for all 13 questions, including cognitive states, based on the 1–4 scale. As shown, the self-

Table 5.1. Item nonresponse under varied conditions for questions with listed characteristics

Item No.	N	Topic[a]	Type of question[b]	Cognitive state[c]	Total item nonresponse (%)	Answered: considered not (%)	Did not answer: considered	Item Nonresponse: Discourage/neutral forms	Item Nonresponse: Encourage form
4	105	Birthday	AED	1.0	0.0	0.9	0.0	0.0	0.0
11	111	Cholesterol	AEDi	1.1	0.0	0.0	0.0	0.0	0.0
8	217	Cigarettes	AEDi	1.2	0.5	0.5	0.5	0.0	1.3
6	106	Music	ABD	1.6	3.7	6.6	0.9	0.0	9.8
12	111	Eye exam	ABD	1.8	1.8	3.6	0.0	0.0	5.7
1	210	Doctor Visit	ABD	1.8	4.8	1.9	1.4	0.7	12.2
10	110	Health	ASpF	1.8	0.0	0.9	0.0	0.0	0.0
2	217	X/year, doctor	ABF	1.9	3.7	3.7	1.4	0.0	10.4
3	217	X/year average	SBF	2.9	5.1	8.7	1.4	0.7	13.1
13	111	Election	SEC	2.9	12.6	9.9	7.2	1.3	37.1
5	216	Bicycle	ABD	3.0	25.0	19.0	6.9	5.8	59.7
9	218	Apples	ABF	3.3	12.8	24.3	3.2	1.4	33.8
7	218	Challenger	FED	3.3	35.3	24.3	12.4	24.8	54.5

[a]Detail on topic: 4, month/year of birthday; 11, been warned of high cholesterol; 8, had smoked ≥ 100 cigarettes in life; 6, month/year last tape/CD purchased; 12, category when last eye exam; 1, month/year most recent visit to doctor; 10, days in last 30 health not good; 2, times/year to doctor; 3, times/year average person to doctor; 13, who will win election; 5, month/year first bicycle ride; 9, number of apples (100's) eaten in life; 7, month/year Challenger exploded.

[b]Type of question coded as follows: *Cognition:* A = autiobiographical; S = speculative; F = factual. *Classification:* E = event; B = behavior; Sp = speculative. *Answer format:* D = date; Di = dichotomous choice; F = frequency; C = categorical.

[c]Cognitive coded on 1–4 scale, where 1 = available; 2 = accessible; 3 = generatable; 4 = inestimatable.

assessed cognitive states covered the full range from available to inestimable, with question means ranging from 1.01 to 3.31 (column 5). Thus, we could conclude that on average, respondent knowledge about whether they have smoked 100 cigarettes in their lives (Q8) is close to *available,* but the number of personal doctor visits made in the last year (Q2) is merely *accessible.* The date of a relatively obscure autobiographical event, the first bicycle ride (Q5), was mostly *generatable,* but the number of apples eaten in a lifetime (Q7) was closer to *inestimable.* Judging by the range of responses, and the correspondence between mean cognitive states and apparent question difficulty, self-assessed cognitive state appears to be a reasonable indicator of respondent knowledge for particular questions. We could conclude with reasonable assurance that respondent knowledge is sufficient for answering the first eight questions listed in Table 5.1. We might be more concerned about responses to the final five questions, and the final three in particular, which have an average cognitive state at less than the "generatable" level. We are also interested in how cognitive states relate to item nonresponse rates. In this study, we considered item nonresponse to have occurred if the respondent checked a "DK" box (when offered), wrote "?" or some other expression of uncertainty, or left an item blank. We also computed the percentage of respondents who reported that they had *considered* item-nonresponse even though they actually chose to answer (column 7, Table 5.1.). These individuals might be referred to as "tentative responders." For the first eight questions, we can see that both item nonresponse rates and the percent who considered not responding are quite low. But for the remaining five questions, the item nonresponse rates are quite high. For example, 25% of respondents did not answer the question about the date of their first bicycle ride (Q5), reflecting the relatively low cognitive states reported for this question; and furthermore, an additional 19% of respondents *considered* reporting that they did not know the answer, for a total of 44% potential nonresponders.

Column 8 of Table 5.1. lists the percent of respondents who did not answer particular questions but considered doing so. Relatively few respondents fall into this category; apparently, this phenomenon is not as common as giving an answer while considering not doing so. Thus, we can evaluate potential item nonresponse problems using several statistics. Self-rated cognitive states provide researchers with a sense of the amount of knowledge that respondents can actually bring to the response process. Rates for ambiguous responses (columns 7 and 8) reveal the degree of effort involved for respondents' adequacy judgments about the quality of their responses. Rates of actual and potential item nonresponse (columns 6 and 7) are alternative ways to evaluate the quality of responses that each survey question might generate. We are not proposing that a particular "cut-off point" separates adequate from inadequate questions for all purposes. However, researchers can use such statistics to select their own comfort level with respondent knowledge levels, and use this information to rewrite questions or devise error-control strategies as appropriate. Furthermore, all of these statistics are easily collectable through pretest questionnaires.

Other items from the evaluation instrument provide additional insights into the decision processes involved in item nonresponse. Of the tentative responders (column 7), 91.7% indicated that they decided to provide an answer (as opposed to a DK) based on an adequacy judgment that their response was "close enough" based on their understanding of question requirements. Fewer of these responders (67.9%) indicated that they answered based on perceived expectations from researchers that they provide a response. Still fewer tentative respondents (43.0%) gave answers because it might appear "silly" to do otherwise. Thus, response decisions seem to be largely based on positive adequacy judgments, suggesting that researchers may want to be explicit about what qualifies as an "adequate" response for questions with highest rates of tentative responding. Other questions were administered to those who considered giving answers but did not. Most of these respondents also reacted on the basis of adequacy judgment, with 80% agreeing that DK was a better choice unless they "knew the answer for sure." Only 31% cited low motivation as having importance in their decision not to respond. However, this was a relatively short questionnaire; we suspect that motivation becomes a *much* more serious issue in longer surveys.

Finally, we considered the effect of respondent instructions on rates of item nonresponse. Item nonresponse was virtually nonexistent on the questionnaire form that discouraged it, and was also very infrequent on the neutral questionnaire form, rising above 10% for only two questions (date of first bicycle ride, and date of the Challenger explosion). Yet, consistent with previous research, item nonresponse rates rose considerably when questionnaire forms included explicit opportunities to provide DK's. It is clear that item-nonresponse rates are related to cognitive states, although not completely determined by them. For example, mean cognitive states are higher for Q5 than for Q9, although Q5 has a much higher rate of item nonresponse. This can probably be explained by question type. Q5 asks for the date of an actual autobiographical event that could be known, while few could answer Q9 about apple consumption literally—a best guess is all that researchers could hope to obtain, whereas the same degree of imprecision for the date of a bicycle ride (Q5) might appear to be inadequate. Also interesting are the nonresponse rates for Q7, concerning the Challenger explosion, which are high when DK's were encouraged but also high even when DK's were discouraged. Of course, this was the only question asked with an externally verifiable, factual answer. It seems reasonable that respondents are generally willing to estimate information that is not literally known, but the likelihood of doing so is reduced if such estimates are clearly a wild guess.

5.7 IMPLICATIONS AND CONCLUSIONS

Traditional perspectives on item nonresponse have been that *any* response is preferable to nonresponse, and that nonresponse should be actively blocked. Alternately, some have argued that such responses should be readily accepted at face value. Our perspective is that these decisions must be based on specific research goals, and that we should try to evaluate the extent and likely causes of potential item nonresponse, and to attack such problems at the source.

With this perspective in mind, the research reported here had two objectives:

first, to develop a theoretical framework explaining the cognitive processes that lead to either substantive answers or item nonresponse, and second, to develop a methodology that might be used to assess the likely nature of item nonresponse for particular questions.

Our theoretical model sought to identify when respondents were likely to give an answer even though they lacked the knowledge to do so, an error of commission, and when they were likely to do the opposite, i.e., not give an answer when they should, an error of omission. Our evaluation methodology sought to obtain information about survey questions by asking potential respondents to report on their own mental processes while answering them.

The proposed evaluation methodology sought to understand item nonresponse by examining the item nonresponse rate (under conditions of encouraging or not encouraging the use of don't know answers), and reports about whether people considered responding or not responding while actually doing the opposite. Based on results from the evaluation questionnaire, we are optimistic that respondents can provide useful information along these lines. Furthermore, this sort of data can be collected relatively easily. An analysis of such data may suggest several courses of action. Some situations may call for simplification of key concepts in the questions; others may call for simplification of the response task (e.g., using response categories rather than asking for specific dates or frequencies). Further instructions about what qualifies as an adequate response may also be helpful. Regardless of the specific solution, it is our hope that this theoretical framework helps researchers approach problems of item nonresponse in a new light.

CHAPTER 6

The Causes of No-Opinion Responses to Attitude Measures in Surveys: They Are Rarely What They Appear to Be

Jon A. Krosnick, *The Ohio State University*

6.1 INTRODUCTION

When survey researchers ask respondents about their attitudes, we usually presume that their answers reflect information or opinions that they previously had stored in memory, and if a person does not have a preexisting opinion about the object of interest, the question itself presumably prompts him or her to draw on relevant beliefs or attitudes in order to concoct a reasonable, albeit new, belief or evaluation (see, e.g., Zaller and Feldman, 1992). Consequently, whether based upon a preexisting judgment or a newly formulated one, responses presumably reflect the individual's beliefs about or orientation toward the object.

What happens when people are asked about an object regarding which they have no knowledge and no opinion? We hope that in such cases, respondents would say that they have no opinion or aren't familiar with the object or don't know how they feel about it (in this chapter, we refer to all such responses as *no-opinion* or *NO* responses). But when respondents are asked a question in such a way as to suggest that they ought to have opinions on the matter, they may wish not to appear foolishly uninformed and may therefore give arbitrary answers. In order to reduce the likelihood of such behavior, some survey experts have recommended that no-opinion options routinely be included in questions (e.g., Bogart, 1972; Converse and Presser, 1986; Payne, 1950; Vaillancourt, 1973). In essence, this tells respondents that it is acceptable to say they have no belief or attitude on a matter.

Do no-opinion filters work? Do they successfully encourage people without meaningful opinions to admit that? Might they go too far and discourage people who have meaningful opinions from expressing them? These are the focal questions considered in this chapter.

6.2 THE NONATTITUDE HYPOTHESIS

Although earlier work on the stability of opinions had been published in various social science disciplines, Converse's (1964) was to become the most frequently cited and widely influential. He happened upon data patterns that raised grave concerns: many of the opinions people expressed on well-known public issues shifted apparently haphazardly from one interview to the next. Converse dubbed these *nonattitudes* and suggested that they were answers masquerading as real opinions, generated by mentally flipping coins and selecting among the offered response alternatives purely randomly. Converse asserted that respondents feel pressure to appear opinionated in surveys even when they are not, and they respond to this pressure by fabricating.

How are we to minimize reporting of nonattitudes? Remarkably, only one method has thus far been proposed: no-opinion filtering. The notion here is that respondents may report nonattitudes partly because survey question wording encourages them to do so. As Schuman and Presser (1981) pointed out, respondents generally "play by the rules of the game" (p. 299), meaning that they choose among the response alternatives offered by a closed question rather than offering reasonable answers outside the offered set. If a question does not explicitly include a NO option, that might imply to respondents that they are expected to have opinions and therefore encourage them to report nonattitudes. Thus, when such a response option is explicitly legitimated via a filter, significantly larger proportions of respondents might admit having no opinion.

The logic underlying this perspective is that when people have a real opinion toward an object, they know it and can readily report it, and whenever people say they do not have an opinion, they truly do not. In situations where people feel pressed to offer an opinion and discouraged from saying they have none, they will make up an answer. But simply legitimating a no-opinion response is enough to eliminate most or all nonattitude reporting, and all or most people who say they have no opinion indeed have none. The division between having an opinion and not having one is presumed to be clear to people, and people are presumed to use knowledge of it to decide when to report opinions and when to say they have none. Therefore, offering a no-opinion response option should increase the number of respondents who say they have no opinion, and saying so should be largely accurate.

Many studies have reported evidence consistent with the first of these two expectations (e.g., Schuman and Presser, 1981). For example, in a relatively early study, Ehrlich (1964) had undergraduates complete 29-item self-administered questionnaires measuring stereotypes of various nationalities and ethnic groups. Respondents were randomly assigned to one of two forms of the questionnaire, one of which explicitly offered "no opinion" and "can't decide" response options, and the other of which did not. The proportion of NO's increased from 0 for the first form to a mean of 22% (range 5–57%) for the filtered items. For all 29 of the items, the proportion of respondents who offered substantive opinions dropped significantly when the filters were included.

6.3 THE VALIDITY OF NO RESPONSES

Are the NO answers that respondents provide to survey interviewers valid? That is, do people who say NO in fact have no opinions on the issues in question? One useful way to address this matter is to examine the correlates of NO responding. If this behavior is more common among people who, a priori, seem less likely to have opinions, that would attest to the validity of these reports, and indeed, there is a good deal of such evidence. This evidence comes from two sorts of studies: nonexperimental (correlating the frequency with which people said NO in answering a single question) and experimental (manipulating the presence or absence of a NO filter in a question and assessing the predictors that could identify individuals most susceptible to the filter's effect).

Among the factors that can identify individuals least likely to say NO are education, knowledge about a question's topic, interest in the topic, exposure to information on the topic, affective involvement in the topic, confidence in one's ability to form an opinion on the topic, and perceived utility of forming an opinion on the topic. Specifically, NO responses are least commonly offered by people who have more formal education (Bishop, Oldendick, and Tuchfarber, 1980; Schuman and Presser, 1981), who are higher in cognitive skills (Colsher and Wallace, 1989; Sigelman et al., 1982), who know more about the topic in question (e.g., Converse, 1976; Faulkenberry and Mason, 1978; Rapoport, 1981; 1982), who are more interested in the topic (Krosnick and Milburn, 1990; Rapoport, 1982; Wright and Niemi, 1983), who are more exposed to information on the topic (Krosnick and Milburn, 1990; Wright and Niemi, 1983), among people with more behavioral experience relevant to the topic (Durand and Lambert, 1988; Krosnick and Milburn, 1990), who feel they have a greater ability to understand the topic (Krosnick and Milburn, 1990), and who feel others are interested in knowing their opinions on the topic (Francis and Busch, 1975; Krosnick and Milburn, 1990).

In experimental studies manipulating the presence or absence of NO options, attraction to such options is greatest among respondents with the lowest levels of education and cognitive skills (Bishop, Oldendick, and Tuchfarber, 1980; Bishop, Oldendick, Tuchfarber, and Bennett, 1980; Narayan and Krosnick, 1996). And people who consider a particular issue to be less of personal interest or importance are more attracted to NO filters (Bishop, Oldendick, and Tuchfarber, 1980; Schuman and Presser, 1981, pp. 142–143). Similarly, NO filters are most likely to attract respondents who otherwise would express moderate attitudes (Ehrlich, 1964).

One final bit of evidence attesting to the validity of NO rates was reported by Converse and Schuman (1984), who compared the percentages of NO responses to various questions asked at approximately the same time by two different survey firms: the Gallup Organization and the National Opinion Research Center (NORC).

Although Gallup generally found higher NO rates than NORC, the correlation between the rates obtained by the two houses across items was a remarkable 0.67. Thus, these rates do appear to reveal something stable and meaningful instead of being wholly arbitrary. Another set of evidence attesting to the validity of NO responses treats them as indicators of the strength of public opinion on issues. Dodd and Svalastoga (1952) proposed that issues that have high rates of NO responses are likely to be ones on which the substantive opinions that are offered are particularly weak. This could occur when, for example, the public has little information about an issue, so many people are reluctant to express opinions at all, and the opinions that are expressed are not especially grounded in confidence, personal importance, or any other aspect of strength. In support of this notion, Dodd and Svalastoga (1952) reported that the percentage of respondents saying NO to each item in a set was strongly and negatively correlated with the consistency over time of the substantive opinions offered by other respondents: $r = -0.91$. Eisenberg and Wesman (1941) also found that for items with higher NO rates, substantive opinions offered were less consistent over time. And Bishop, Oldendick, and Tuchfarber (1980) found that the larger the proportion of people who volunteered a NO response on an unfiltered opinion question, the greater the proportion of respondents attracted by a NO filter when it was included in the question. Also, Page and Shapiro (1983, p. 181) found greater correspondence between public opinion and public policy when smaller proportions of the public declined to report preferences on an issue, although Brooks (1990) found no such relation.

In sum, these studies generally support the notion that NO responses are most likely to be reported when individuals are indeed least likely to have meaningful attitudes toward an object.

6.4 EFFECTS OF NO FILTERS ON DATA QUALITY

Given all the above evidence, it seems likely that offering NO options would increase the quality of data obtained by a questionnaire. That is, respondents who might otherwise offer meaningless opinions would be discouraged from doing so by a filter. But do NO filters work effectively in this sense? That is, is the overall quality of data obtained by a filtered question better than the overall quality of data obtained by an unfiltered question? A variety of evidence addresses this issue. I begin by considering a series of nonexperimental studies and then turn to experimental ones that systematically varied the presence of NO options. The criteria used include the impact of filtering on reliability, correlational validity, and susceptibility to response effects.

In one nonexperimental study, Gilljam and Granberg (1993) asked respondents three questions tapping attitudes toward building nuclear power plants. The first of these questions offered a NO option, and 15% of respondents selected it. The other two questions, asked later in the interview, did not offer NO options, and only 3% and 4% of respondents, respectively, failed to offer substantive responses to them. Thus, the majority of respondents who initially said NO offered opinions in answering the later two questions. At issue, then, is whether these later responses reflected meaningful opinions or were nonattitudes.

To address this question, Gilljam and Granberg (1993) examined two indicators: the strength of the correlation between the two latter attitude reports, and their ability to predict people's votes on an actual nuclear power referendum in a subsequent election. The correlation between answers to the latter two items was 0.41 ($p < 0.001$) among individuals who said NO to the first item, as compared to a correlation of 0.82 ($p < 0.001$) among individuals who answered the first item substantively. Similarly, answers to the second two items correctly predicted an average of 76% of subsequent votes by people who initially said NO, as compared to a 94% accuracy rate among individuals who answered the first item substantively. Thus, the filter apparently separated out people whose expressed opinions were, on average, less predictive than others' opinions. However, the filter also separated out people whose opinions were meaningful to some degree as well. Three other nonexperimental studies taking a different investigative approach produced similarly mixed evidence. Andrews (1984) and Alwin and Krosnick (1991) metaanalyzed the correlates of the amount of random measurement error in numerous survey items, some of which offered NO options and others that did not. In a similar study, Bishop et al. (1979) used existing surveys to assess associations between items asked in either filtered or unfiltered forms and other criterion items. Andrews (1984) found less random error when NO options were offered than when it was not, Bishop et al. (1979) found slightly stronger associations between variables when NO options were offered, but Alwin and Krosnick (1991) found more random error in items that offered NO options.

Experimental studies of NO filters have examined five criteria with which data quality can be inferred: reliability, correlational validity, susceptibility to response effects, measurements of knowledge levels, predictive accuracy, and preventing reporting of opinions toward obscure or fictitious objects.

Reliability. Three studies have explored the impact of experimental variations in the presence or absence of NO filters on reliability. McClendon and Alwin (1993) had respondents answer sets of questions measuring an attitude (e.g., toward lawyers) in either a filtered or an unfiltered form. These investigators then estimated the reliability of the items via structural equation modeling and found no greater reliability when NO filters were included in questions than when they were not.

Krosnick and Berent (1990) reported similar results involving longitudinal data. For their study, respondents were asked about various attitudes on two occasions separated by 2 months, using questions either including or omitting NO response options. No significant change in the over-time consistency of attitude reports appeared depending upon whether NO filters were present or absent. Poe et al. (1988) found this same result in a panel survey of factual matters: longitudinal reliability of responses was equivalent regardless of whether NO opinion options were offered or not.

Correlational Validity. If NO filers improve data quality, they should strengthen associations between variables. Yet in more than 20 experiments, Schuman and

Presser (1981) found that varying the presence of a NO option altered associations between attitudes significantly in only three cases. In two cases, offering the NO option strengthened an observed relation between attitudes, but in the other case, offering the NO option weakened the observed relation. Furthermore, Schuman and Presser (1981; see also, Presser, 1977) found no cases in which filtering altered relations between attitudes and respondent education, interest in politics, age, or gender. Likewise, Krosnick et al. (1999) found no weakening of relationships between attitudes and various attitudinal and demographic predictors of them when the attitudes were measured with questions not including a NO option as compared to questions including that option. Similar results were reported by Sanchez and Morchio (1992), who examined questions tapping beliefs about factual matters (e.g., "which political party had the most members in the U.S. House of Representatives?"). These investigators compared two sets of interviews that differed in NO rates due to differential interviewer probing: in one set, interviewers had probed NO responses more often, thus yielding a lower final rate of NO's. The probing increased the number of correct and incorrect answers given by respondents about equally (which would be expected if respondents were guessing), but it did not significantly alter the relation between measured knowledge and various predictors (e.g., frequency of exposure to political news). Thus, data quality, as indexed in this fashion, was again not compromised by a technique that decreased NO rates (i.e., interviewer probing).

Susceptibility to Response Effects and Manipulations. If items including NO options yield higher quality data, then responses to them should presumably be less susceptible to response effects caused by nonsubstantive changes in question design. McClendon (1991) investigated this possibility by assessing the magnitude of response order effects and acquiescence when NO options were offered and omitted. Although the NO options did reduce acquiescence for one set of items examined, they did not do so for other items, and they had no impact on the magnitude of response order effects.

Krosnick et al. (1999) examined the impact of NO options on data quality by assessing whether responses were equally responsive to manipulations that should have affected them. Specifically, respondents in their study were told about a program that would prevent future oil spills and were asked whether they would be willing to pay a specified amount for it in additional taxes. Different respondents were told different prices, and one would expect fewer people to be willing to pay for the program as the price escalated. In fact, this is what happened. If pressing NO responses into substantive ones creates nonattitudes, then one might imagine that sensitivity to the price of the program would be less among people pressed to offer substantive opinions than among people offered a NO option. But in fact, sensitivity to price was the same in both groups.

Knowledge Accuracy. If a person is accurate when responding "don't know" to a multiple choice or true/false question assessing accuracy of factual knowledge, then he or she should do no better than chance at answering the question if pressed to do so, but Dunlap et al. (1929) did not find support for this assertion. In their study, respondents answered a set of true/false knowledge questions twice, once when instructed to leave blank any questions to which they did not know the answer, and once when they were instructed to guess when answering such questions. Using the grading method of (number correct – number incorrect), guessing should not improve people's scores if they did not in fact know the correct answer to a question. But in fact, guessing did improve scores, which meant that people had more knowledge than their instincts detected when given the opportunity to leave questions blank.

Yet another context for assessing the accuracy of NO responses is preelection polls designed to forecast election outcomes. In these surveys, many respondents say they don't know which candidate they will vote for. But interviewers can press these individuals to indicate which candidate they lean toward in a race. Visser et al. (2000) compared the accuracy of poll data in predicting the actual outcomes of various Ohio elections when treating the data two ways: (1) treating "don't know" responses as valid, and therefore treating these respondents as having no candidate preferences; or (2) treating respondents' indications of which candidates they lean toward as valid measures of their preferences. Collapsing across a series of races, they found the polls were more accurate when using the latter method than when using the former, suggesting that there was validity to the "leaning" responses provided by people who initially said "don't know."

Taken together, the literature on how filters affect data quality suggests that NO filters do not remove all people without meaningful opinions and only people without such opinions. Thus, we see here reason to hesitate about using such filters.

6.5 REASONS FOR NO RESPONSES

In order to make sense of this surprising evidence, it is useful to turn to studies by cognitive psychologists of the process by which people decide that they do not know something. Specifically, Norman (1973) proposed a two-step model that seems to account for observed data quite well. If asked a question such as "Do you favor or opposed U.S. government aid to Nicaragua?" a respondent's first step would be to search for any information in memory relevant to the objects mentioned: U.S. foreign aid and Nicaragua. If no information about either is recalled, the individual can quickly respond by saying he or she has no opinion. But if some information is located about either object, the person must then retrieve that information and decide whether it can be used to formulate a reasonable opinion. If not, he or she presumably replies "don't know," but the required search time makes this a relative slow response. Glucksberg and McCloskey (1981) reported a series of studies demonstrating that "don't know" responses can indeed occur either quickly or slowly, the difference resulting from whether or not any relevant information can be retrieved in memory.

This distinction between first-stage and second-stage NO responses suggests different reasons for them. According to the proponents of NO filters, the reason presumed to be most common is that the respondent lacks the necessary information and/or experience with which to form an attitude. Such circumstances would presumably yield quick first-stage NO responses. In contrast, second-stage NO responses could occur because of ambivalence. That is, some respondents may know a great deal about an object and/or have strong feelings toward it, but their thoughts and/or feelings may be highly contradictory, making it difficult to select a single response.

It also seems possible that NO responses can result at what might be considered a third stage, the point at which respondents attempt to translate their retrieved judgments onto the response choices offered by a question. For example, a respondent may know approximately where he or she falls on an attitude scale (e.g., around 6 or 7 on a 1–7 scale), but because of ambiguity in the meaning of the scale points or of his or her internal attitudinal cues, he or she may be unsure of exactly which point to choose, yielding a NO response. Or a respondent who has some information about an object, has a neutral overall orientation toward it, and is asked a question without a neutral response option might say NO because the answer he or she would like to give has not been conferred legitimacy. A respondent might also realize that the answer implied by the retrieved information will portray an undesirable image of himself or herself, so he or she will choose to say NO instead. Or a respondent may be concerned that he or she does not know enough about the object to defend an opinion toward it, so that opinion may be withheld rather than reported.

Finally, it seems possible that some NO responses occur at a prefirst stage, before respondents have even begun to attempt to retrieve relevant information. For example, if a respondent does not understand the question being asked and is unwilling to answer until its meaning is clarified, he or she might respond "I don't know" (see, e.g., Fonda, 1951). Or if a person is unwilling to expend the cognitive effort required by a memory search, he or she may choose to satisfice by selecting a NO response option (Krosnick, 1991). There is in fact evidence that some NO responses occur for all of these reasons. But as we shall see, NO responses are apparently only very rarely due to complete lack of information and indeed are rarely due to lacking an opinion. So reviewing this literature will show why legitimating NO responses is not a desirable way to improve data quality.

Ambivalence and Question Ambiguity. A number of studies have attempted to identify the reasons for NO responses and have found that genuine lack of opinion apparently predominates. For example, Smith (1984) examined two surveys in which respondents were offered opportunities to say that they had no opinion on a political issue, that they had an opinion but that they were "not sure/it depends," or that they had an opinion but "didn't know" how to express it. Of the responses to 15 items in these categories in one survey, 61% were "no opinion," 34% were "not sure/it depends," and 5% were "don't know." In another survey, the comparable figures were 53%, 41%, and 6%, respectively. Duncan and Stenbeck (1988) found the ratios of "no opinion" to "not sure/it depends" answers in surveys varied significantly across items from 3:1 to 1:1. But "not sure/it depends" never outnumbered "no opinion" responses. Ehrlich (1964) found that 77% of NOs occurred because of lack of interest in or information or thought about the topic, 7% of nonsubstantive responses indicated ambivalence, and 16% indicated that the survey question was too crude to capture the complexity of respondents' views. And Klopfer and Madden (1980) found that lack of engagement was more commonly responsible for NO responses regarding going to church on Sundays than were ambivalence or uncertainty.

However, in other instances, ambivalence or expression problems apparently generated the most NO responses. For example, Klopfer and Madden (1980) found that ambivalence was more often responsible for NOs regarding capital punishment than were lack of engagement or uncertainty, as did Coombs and Coombs (1976).

Faulkenberry and Mason (1978) found that NO responses can sometimes predominantly reflect lack of understanding of the question being asked. These investigators had interviewers note whether nonsubstantive responses occurred when a respondent either understood or did not understand a question about energy generation. Fully 55% of the NO responses were said to have occurred when respondents did not understand the question. Along similar lines, Schaeffer and Bradburn (1989) found that some respondents gave NO responses to a question (about how stressful it would be for them to put another person in a nursing home) because they could not accept the premise of the question (i.e., they would ever do such a thing). Thus, if questions were written so that respondents could better understand them or if they did not require inappropriate presumptions, NO rates might be decreased.

In sum, NOs can sometimes reflect lack of information about or interest in an issue, just as one would hope, but such responses can also occur because respondents have ambivalent feelings on an issue or because the question being asked is not presented sufficiently clearly. Perhaps omitting NO options from questions, thereby compelling people to offer opinions, yields answers that are reasonably stable and meaningful.

Satisficing. Another possible explanation for the fact that NO filters do not consistently improve data quality is satisficing (Krosnick, 1991). According to this perspective, people have many attitudes that are best labeled "latent," meaning that they are not immediately aware of holding those opinions when asked. Instead, the bases of those opinions reside in memory, and people can retrieve those bases and integrate them to yield overall attitude reports (a process called "optimizing"), but doing so requires significant cognitive effort. When people are disposed not to do this work and instead prefer to shortcut the effort they devote in generating answers, they will attempt to satisfice by looking for cues in a question that point to an answer that will appear to be acceptable and sensible but that requires little effort to select. A NO option constitutes just such a cue and may therefore encourage satisficing, whereas omission of the NO option might instead lead respondents to do the cognitive work necessary to retrieve relevant information from memory and report their "latent" opinions.

This perspective suggests that NO options should be especially likely to attract respondents under the conditions thought to foster satisficing: low ability to optimize, low motivation to do so, or high task difficulty. And consistent with this reasoning, as was discussed above, NO filters attract respondents low in educational attainment (an indirect index of cognitive skills) and low on more direct assessments of cognitive skills, as well as respondents with relatively little knowledge and exposure to information on the issue. Thus, evidence that can be viewed as consistent with the notion that NO responses are valid is also consistent with the notion that NO responses reflect satisficing.

Other evidence reviewed above can also be viewed in this light as well. For example, NO responses are especially common among people for whom an issue is low in personal importance, of little interest, and arouses little affective involvement, and this may be because of lowered motivation to optimize under these conditions. Furthermore, people are especially likely to say NO when they feel they lack the ability to formulate informed opinions (i.e., subjective competence), and when they feel there is little value in formulating such opinions (i.e., demand for opinionation). These associations can be conceived of as arising at the time of attitude measurement: low motivation inhibits a person from drawing on knowledge available in memory to formulate and carefully report a substantive opinion on an issue.

All of this evidence is consistent with the satisficing view of NO responses, but it is also consistent with the notion that these responses reflect optimizing. More difficult to interpret in this way, however, is evidence that NO responses are more likely when questions appear later in a questionnaire, at which point motivation is presumably waning. For example, Ferber (1966), Dickinson and Kirzner (1985), Ying (1989), and Culpepper et al. (1992) found that failure to answer an item increased significantly for later questionnaire items (c.f., Craig and McCann, 1978). Also consistent with this perspective are demonstrations that NO responses become increasingly common as questions become more difficult to answer. Although Nuckols (1949) found that questions that were more difficult to understand (because of language complexity) were no more likely to attract NO responses, Klare (1950) and Converse (1976) did find more NO responses for more difficult questions. Furthermore, Converse (1976) found that questions containing long explanations of an issue or requiring respondents to predict the future had higher NO rates than questions with shorter explanations of the issue and ones requiring descriptions of the past or present, especially for respondents with less education.

Converse (1976) also found that NO rates were higher for dichotomous questions than for politimous questions, presumably because the former did not allow respondents an easy opportunity to describe moderate or neutral opinions. In a study that supported this interpretation, Kalton et al. (1980) asked some respondents dichotomous opinion questions (e.g., "Compared with most other people, do you know more or do you know less about how to treat minor ailments?"), where a middle alternative was sensible (e.g., "about the same as most people") but not offered. Not surprisingly, a notable number of respondents declined to provide substantive opinions, and offering the middle alternative explicitly to other respondents signifi-cantly decreased the frequency of NO responses.

Additional evidence consistent with the satisficing perspective comes from a study by Houston and Nevin (1977). These investigators experimentally manipulated the apparent sponsor of a mail questionnaire, either the University of Wisconsin or a small local market research firm. Also, respondents received one of three different appeals for participation, emphasizing enhanced understanding, helping the sponsor, or personal gain for the respondent. NO responses were equivalently frequent across the conditions except when the sponsor was the University and the instructions emphasized understanding, in which case NO responses were notably less frequent. Thus, the match of a prestigious sponsor and a harmonious purpose apparently enhanced respondent motivation and decreased NO responses.

The use of incentives (money or a pen) to enhance response rates for mail questionnaires has also been found to enhance NO rates (Hansen, 1980). Hansen reasoned that when the incentives were not provided, respondents believed that they were intrinsically motivated to complete the questionnaire, whereas people who received the incentives felt they were completing the questionnaire only because they had been given the gifts. A study by McDaniel and Rao (1980) suggests that this effect can be eliminated and in fact reversed by a slight shift in the wording of the explanation for the gift. Instead of simply offering it, these investigators emphasized how minimal the gift was: "I know it's not much, but please accept this new quarter as just a small token of my appreciation for your assistance." This approach presumably minimizes a person's ability to attribute completing the questionnaire to the reward, because it is so small and therefore appears to a genuine expression of gratitude.

A final set of evidence consistent with the satisficing perspective involves mode effects. It seems likely that interviewers can motivate respondents to optimize by creating a sense of accountability and by modeling their professional commitment to the task, and this seems more likely to occur during face-to-face interactions than during telephone interviews (where sense of accountability is likely to be lower, and nonverbal modeling is less likely to occur). Therefore, satisficing may be more likely to occur during telephone interviews. Consistent with this logic, various studies found that respondents said "don't know" significantly more often in telephone interviews than in face-to-face interviews (Aneshensel et al., 1982; Aquilino, 1992; Groves and Kahn, 1979; Herzog et al., 1983; Hochstim, 1962; Jordan et al., 1980; Kormendi, 1988; Locander and Burton, 1976; Schmiedeskamp, 1962; Siemietycki, 1979), though one found no significant mode difference (Rogers, 1976).

Finally, an initially puzzling finding regarding the relation of personality and NO responses is understandable in light of the satisficing perspective. People who are especially trusting of others might seem inclined to conform to the format of a question, offering a NO response only when it is offered. Surprisingly, however, the impact of NO filters is greater for people low in interpersonal trust (Bishop, Oldendick, Tuchfarber, and Bennett, 1980). This may be because the deceit involved in satisficing (i.e., pretending that a response is meaningful when it is not) is most likely to be comfortable to individuals who generally do not trust others to be honest with them, either.

Intimidation. Another reason why NO filters discourage reporting of real attitudes was identified by Hippler and Schwarz (1989). These investigators proposed that strongly worded NO filters might suggest to respondents that a great deal of knowledge is required to answer an attitude question and thereby intimidate people who feel they might not be able to adequately justify their opinions. Consistent with this reasoning, Hippler and Schwarz (1989) found that respondents inferred from the presence and strength of a NO filter that follow-up questioning would be more extensive, would require more knowledge, and would be more difficult. If respondents were motivated to avoid extensive questioning or were concerned that they couldn't defend whatever opinions they might offer, this might bias them toward a NO response.

Further evidence in line with this view was reported by McClendon (1986), who asked respondents how strongly they felt on each of three issues after asking their opinions on these issues using either filtered or unfiltered questions. Offering a NO option lowered reported attitude intensity, suggesting that the filter led people to express more tentative feelings on the issue. This effect carried over to subsequent items in the questionnaire that did not explicitly include a NO option: expressed strength of feeling was weaker for these items as the result of NO options having been offered in previous questions. Furthermore, including NO options in preceding questions increased the number of respondents who selected more tentative substantive opinion options in answering later questions as well (e.g., saying that the state and federal governments should be equally responsible for solving certain problems, instead of choosing one branch of government to have primary responsibility). Thus, NO filters may induce a temporary state of tentativeness in answers that reduces data quality.

A final set of evidence consistent with this perspective was reported by Berger and Sullivan (1970). These investigators gave some respondents special instructions intended to induce careful responding, while other respondents received no such instructions. These instructions stressed that respondents had been carefully selected as a part of a representative sample and that it was "very important that you answer each question." Surprisingly, these instructions were actually associated with increased NO rates, presumably because the stressed importance of the study led to greater reluctance to express opinions that might not be fully informed.

Self-Image Protection. Another reason why people might prefer to select NO options rather than offering meaningful opinions is the desire not to present a socially undesirable or unflattering image of oneself. Fonda (1951) found that people who frequently selected "?" responses in answering questions about their own personalities tended to evidence neurotic tendencies in answering Rorschach (1942) inkblot questions. Similarly, Rosenberg et al. (1955) found that people who selected "?" options more often in personality questionnaires characterized themselves as less agreeable, less cooperative, less self-confident, less free from neurotic tendencies, and possessing other more negative qualities on items they did answer substantively. Forty years later, Johanson et al. (1993) reported a comparable finding. Thus, the self-evaluations respondents declined to make would presumably have been relatively unflattering as well (for similar results, see Chronbach, 1950, p.15; Kahn and Hadley, 1949).

Also consistent with this perspective are studies on mode differences in NO rates. Being interviewed by a person presumably creates a greater sense of social accountability than completing self-administered anonymous questionnaires, and people have been shown in much research to be more willing to disclose embarrassing or undesirable facts about themselves in anonymous self-administered questionnaires. Not surprisingly, then, Newton et al. (1982) found that NO responses were more common when respondents had to answer questions aloud to interviewers than when they could simply write their answers down anonymously. Similarly, Berger and Sullivan (1970) found higher NO rates when respondents were interviewed face-to-face than when they filled out self-administered questionnaires. And in a study by Houston and Jefferson (1975), NOs were more common when self-administered questionnaires identified the name of the respondent than when they did not. These results might have occurred because answering aloud to interviewers or in identified ways on paper would have revealed unflattering or undesirable views.

Taken together, these studies suggest that NO responses often result not from genuine lack of opinion but rather from ambivalence, question ambiguity, satisficing, intimidation, and self-protection. In each of these cases, there is something meaningful to be learned from pressing respondents to report their opinions. NO response options discourage people from doing so under these circumstances. This explains why data quality is not improved when such options are explicitly included in questions.

6.6 CONCLUSION

The essence of Converse's (1964) nonattitudes hypothesis seems unquestionable: Many people who report attitudes in surveys do not have deeply rooted preferences that shape their thinking and behavior. But offering a no-opinion response option does not seem to be an effective way to prevent reporting of weak opinions. In fact, because many real attitudes are apparently missed by offering such options, it seems unwise to use them. This is because the vast majority of NO responses are *not* due to completely lacking an attitude and instead result from a decision not to do the cognitive work necessary to report it, a decision not to reveal a potentially embarrassing attitude, ambivalence, or question ambiguity. This conclusion resonates loosely with a sizable literature in cognitive psychology on the "feeling of knowing" (e.g., Nelson et al., 1984; Schacter, 1983). This phenomenon occurs when a person fails to recall the answer to a question (e.g., "What is the capital of North Dakota?") but claims to be able to recognize the correct answer among an offered set of choices. Indeed, when people have this feeling of knowing, and the related "tip-of-the-tongue" phenomenon (Brown and McNeill, 1977), they indeed often do possess the required information in memory but temporarily cannot gain conscious access to it (see also Koriat and Lieblich, 1974). Thus, failure to give an answer does not mean people do not possess the answer, just as saying NO in answering a questionnaire often does not mean that the person possesses no information with which to make the required judgment. It therefore seems wise to encourage respondents to report whatever opinions they can.

PART II

Impacts of Survey Design on Nonresponse

CHAPTER 7

The Influence of Interviewers' Attitude and Behavior on Household Survey Nonresponse: An International Comparison

Joop Hox, *Utrecht University, The Netherlands*
Edith de Leeuw, *MethodikA, The Netherlands*

7.1 INTRODUCTION

Several studies have addressed the role of the interviewer in nonresponse. There is little evidence that interviewer attributes, such as age and sex, influence response rates. In addition there is hardly any consistent evidence that personality traits or characteristics play any role (cf. Groves and Couper, 1998), but there is some evidence that social skills do play a role (Morton-Williams, 1993). Social skills are associated with knowing and applying rules of accepted behavior and communication, and they can be can be trained (Argyle, 1969; Morton-Williams, 1993). Nonetheless, interviewer training devotes little time to skills or tactics for fighting nonresponse; an exception is Statistics Sweden, which allocates 35% of its training to nonresponse (Luiten and de Heer, 1994). To attain satisfactory response rates, interviewers must learn how to convince reluctant respondents in practice or by informal exchanges with other interviewers. It is therefore not surprising that interviewer experience positively influences response (Durbin and Stuart, 1951; Groves and Fultz, 1985; Couper and Groves, 1992; de Leeuw and Hox, 1996; Singer et al., 1983). What makes these experienced interviewers achieve higher response rates?

Morton-Williams (1993) analyzed tape recordings of survey introductions and identified successful strategies for obtaining respondent cooperation. Important factors were: appear trustworthy (e.g., always identify yourself immediately), appear friendly (e.g., smile, make a compliment), adapt to the situation at the doorstep, and react to the respondent. Successful interviewers use their cultural knowledge and their local knowledge of the sample neighborhood (Groves and Couper, 1998) to optimize their approach and behavior.

Interviewer-respondent interaction is a central concept in the theoretical work of Groves et al. (1992). Using a completely different research method, Snijkers et al. (1999) were able to replicate the main conclusions of Morton-Williams (1993) and of Groves and Couper (1992, 1996): professional competence, social skills, tailoring of the introduction, and maintaining the interaction were all named as good tactics by the more successful interviewers.

A different perspective was introduced by Lehtonen (1996), who concentrated on interviewers' attitude towards persuasion strategies and the role of the interviewer. Those interviewers who have a strong belief in the importance of the voluntary nature of participation and feel negative towards strong persuasion strategies also had a higher probability of nonresponse in face-to-face interviews. The early work of Singer et al. (1983) found that interviewers' stated expectation about the ease of persuading respondents to agree to an interview correlated with interviewers' response rate in a telephone survey. Groves and Couper (1998) find a positive relationship between interviewers' confidence ("can convince almost anyone to respond") and response rate.

Building on the two perspectives—attitude and behavior—de Leeuw et al. (1997) investigated the influence of the interviewer on survey response in a face-to-face interview. They replicated Lehtonen (1996) and showed that interviewer attitude and response rate were correlated in face-to-face interviews. Interviewers with a positive attitude towards persuasion strategies attain a higher response rate. No significant differences between interviewers were found regarding self-reported doorstep behavior.

The questionnaire used by de Leeuw et al. (1997) was partly based on questionnaires used by Campanelli et al. (1997) in the United Kingdom and Couper and Groves (1992) in the United States. To broaden the scope of the study, an international research project was started in 1996 at the international workshop on household survey nonresponse in Mannheim, Germany. Three research questions are central in this international comparison: (1) "Do interviewers in different countries have different attitudes towards the interviewer role?"; (2) "Does interviewer attitude predict interviewer response rate within and across different countries?"; and (3) Does (self-reported) interviewer behavior add to the predictive power for interviewer response rate within and across countries?

7.2 METHOD

7.2.1 Data Collection Procedure

Ideally, a comparative study would use exactly the same survey topics and data collection procedures in all of the collaborating countries. Since this is impossible to achieve, we asked the contributors to provide data from several different surveys, with the intention that the final collection of data sets shows sufficient variation on key characteristics so that wider generalization is warranted. The required data files were on interviewer level. The main dependent variable is the response on the interviewer level for a particular survey; the main independent variables are interviewer attitude and behavior (interviewer questionnaire); background variables are interviewer age, sex, and experience. Most participants provided response data from several surveys. In a few cases, interviewers participated in more than one survey. For interviewer level comparisons, the double interviewer records were removed. When analyzing response rates, they were included in the analysis.

7.2.2 Data Sources

The core of the data set were the data collected by de Leeuw et al. (1997), and Campanelli et al. (1997), who used comparable interviewer questionnaires. Couper and Groves provided data that omitted some attitude questions; Lehtonen provided his original data set (Lehtonen 1996), which did not contain questions on interviewer behavior. New data were collected in the following countries: Belgium, Canada, Finland, Germany, Sweden, Slovenia, and the United Kingdom. In all these cases, the standardized version of the interviewer questionnaire was used. (For a concise description of the available data sets, see the Appendix.) In sum, data was available from nine different countries and 32 surveys. The data came from both official statistics and research institutes, and both face-to-face and telephone surveys were included. Overall, 3064 interviewers approached 32,1947 potential respondents. Country, agency, and interview mode were added as background variables. The response rates were cooperation rates, corrected for noncontacts.

For most studies (except for the United States), we had also data on interviewer background. The majority of the interviewers (87%) were female. There were not many differences between countries regarding interviewer sex, with the exception of the United Kingdom and Germany, which had relatively more male interviewers. The average interviewer age was 47.6 with a standard deviation of 1.8 year. The United Kingdom and Germany had relatively older interviewers, whereas Slovenia had relatively young interviewers. Finland and Holland had the interviewers with most years worked (respectively 12 and 11 years on average at agency), whereas Slovenia had the youngest and least experienced interviewers (on average 3 years at agency).

7.2.3 Instrumentation

We started with the construction of indices for interviewer attitude and for avowed doorstep behavior. To avoid capitalization on chance, we randomly split the total data file into a file for exploration and a file for cross-validation. Step 1 was an exploratory factor analysis using standard techniques and Varimax rotation. The exploratory factor analysis was followed by a confirmatory factor analysis on the same data file. In step 2, the final confirmatory factor analysis model of step 1 was tested on the data of the validation file. In all cases, the final model had a satisfactory fit in the cross-validation. In the third and last step, the complete file was used to calculate factor scores for each interviewer. Some data sets do not contain all the questions, which led to missing data. These missing data were assumed to be missing at random, and they were handled by direct estimation using maximum likelihood (Arbuckle, 1996). The variables were originally scored on a five point scale (1 = always to 5 = never). The final factor scores were calculated in such a way that a high value indicates positive agreement with or frequent behavior on the specific factor.

Exploratory analysis on the ten available attitude questions indicated three distinct factors or groups of related questions. One factor indicates the relevance of "persuasion" (e.g., reluctant respondents can be persuaded), the second factor emphasizes the "voluntary" nature of the interview (e.g., accept refusal of a reluctant respondent), and the last factor stresses the usefulness of sending another interviewer (e.g., when a respondent has refused, it is better to send a different interviewer). One question (more important to gain interest than seek a quick decision), did not have significant loadings on any factor and was dropped from the model. In his original study, Lehtonen (1996) distinguished the same two factors "voluntary" and "persuasion," which was replicated on a much larger sample of interviewers in our study.

The exploratory factor analysis was followed by a confirmatory factor analysis (Bollen, 1989). Since the sample size was large, the assessment of model fit was based on two goodness-of-fit indices that are less sensitive to sample size. The following fit indices for the model were used: the comparative fit index (CFI), the root mean square error of approximation (RMSEA), and a test of close fit (p_{close}), which is a chi-square test of the hypothesis that the RMSEA is not larger than 0.05 (Arbuckle, 1996). For the validation file, the three-factor model also described the data well. Therefore, based on this three-factor model, we calculated factor scores for each interviewer (step 3). A high score on the factor "persuasion" indicates that an interviewer is persuasion-oriented. This factor includes questions that Groves and Couper interpret as interviewer confidence, which we view as a necessary component of persuasion orientation. A high score on "voluntariness" means that the interviewer is more oriented towards the acceptance of refusals and emphasizes the voluntary nature of participation. A high score on "send other" indicates that the interviewer does not prefer to try again, but thinks it best to send another interviewer. This can be seen as a form of "stepping back" (cf. Groves and Couper, 1998). It should be noted that the correlation between the three factors is extremely low; the highest correlation was between "voluntary" and "send other" (–0.14), which is not significant. The attitudes are essentially uncorrelated. The factor matrix for the total sample and the fit indices are depicted in Table 7.1. Factor loadings not present in the table have been fixed to zero.

The same three-step procedure was used for the ten questions on avowed doorstep behavior. Exploratory analysis indicated three distinct factors or groups of related questions. The first factor indicates the avowed use of "social validation" arguments (e.g., mention that most people participate, and emphasize the positive as-

Table 7.1. Standardized factor loadings, attitude questions

	I Persuasion	II Voluntary	III Send other	
C1	—	—	—	Gain interest, not quick decision
C2	—	—	0.52	Send other interviewer if no time
C3	—	—	0.99	Send other interviewer if no cooperation
C4	0.45	—	—	Always persuade reluctant respondents
C5	0.53	—	—	Reluctant respondents can be persuaded
C6	—	0.41	—	Respect privacy of respondent
C7	—	0.64	—	Accept refusals of reluctant respondents
C8	0.22	0.44	—	Emphasize voluntary nature of participation
C9	—	—	0.10	If reluctant, withdraw to return later
C10	0.44	—	—	Caught at right time, most people respond

Exploration	Validation	Total file
CFI = 0.88	CFI = 0.86	CFI = 0.87
RMSEA = 0.04	RMSEA = 0.05	RMSEA = 0.04
(p_{close} = 0.95)	(p_{close} = 0.82)	(p_{close} = 0.95)

pects of participation). The second factor stresses the importance of obtaining a representative sample and uses "scarcity" arguments to persuade (e.g., this is the chance to give your opinion). The last factor concerns a "foot-in-the-door" tactic (e.g., ask to enter the home). The factor matrix for the total sample and the fit indices are depicted in Table 7.2.

The three-factor model gave a satisfactory fit on the validation sample. Again, factor scores were calculated for each interviewer on the total file. A high score on the factor "social validation" indicates that interviewers report that they often use arguments regarding social validation in their persuasion attempts. The factor scores on "scarcity" and "foot-in-the-door" can be interpreted in a similar way. There were substantial positive correlations between the factors. The correlation between social validation and scarcity was 0.41, between social validation and foot-in-door was 0.33, and between scarcity and foot-in-door was 0.54.

7.2.4 Analysis

We started our analyses with a comparison of interviewer attitude and avowed "doorstep" behavior across countries, using analysis of covariance. As there were some differences between countries in interviewers' age, sex, and experience, we used these variables as covariates in the analysis (The data sets from Groves and Couper and Lehtonen lacked some items. If there was at least one question for a factor, the factor score was estimated using direct maximum likelihood estimation. If all questions for a factor were missing, the factor score was set to missing and not used in the comparison. Since the multilevel procedures presented later use the factor scores as predictors, for these analyses the factor scores were also estimated if

Table 7.2. Standardized factor loadings, avowed doorstep behavior

	I Social validation	II Scarcity	III Foot-in-door	
B3A	0.65	—	—	Say topic should interest them
B3B	0.28	—	—	Say you are not salesperson
B3C	0.74	—	—	Say most people enjoy interview
B3D	0.74	—	—	Say most people participate
B3E	0.45	0.23	—	Say this is the chance to give opinion
B3F	—	0.45	0.12	Explain how household was selected
B3G	—	0.98	—	Mention they represent other people
B3H	—	—	0.53	Make a compliment (only face-to-face)
B3I	—	—	0.47	Ask to go into the home (only face-to-face)
B3J	—	—	0.38	Begin asking a question

Exploration	Validation	Total file
CFI = 0.97	CFI = 0.95	CFI = 0.96
RMSEA = 0.04	RMSEA = 0.05	RMSEA = 0.04
(p_{close} = 0.98)	(p_{close} = 0.68)	(p_{close} = 0.96)

all questions for a specific factor were missing. This plug-in estimate is very close to overall mean substitution.)

In the next analysis step, we concentrated on the second and third research questions: which interviewer variables (attitude and behavior) predict interviewer-level response rate. As surveys were nested within countries (see the Appendix), we used a multilevel logistic regression model (cf. Goldstein, 1995). The separate levels in the analysis were interviewers (lowest level), surveys (second level), and countries (third level). In total, we had data from 3064 interviewers, employed in 32 surveys in nine countries. The logistic regression model analyzes response rates as proportions of the interviewer workload; differences among interviewers in workload are automatically included in the model as a weight factor. The estimation method employed was restricted maximum likelihood using second-order Taylor linearization and penalized maximum likelihood estimation (Goldstein, 1995).

7.3 RESULTS

7.3.1 Interviewer Attitudes in Different Countries

Three distinct factors were found that describe interviewer attitudes toward the interviewer role in response or nonresponse. The first is persuasion-oriented, the second reflects the interviewers belief in privacy and voluntariness, and the third reflects the specific feeling that it is better to send a different interviewer than have the same return (see Section 7.2.3).

When we compare interviewer attitudes over the nine countries on which data are available, we see some striking differences. First of all countries do differ on "persuasion" with The Netherlands as the extreme low case ($F = 36.2$, $df = 8$, $p = 0.00$). Ranking countries from high to low on persuasion orientation gives the following order: Germany, Slovenia, United States, United Kingdom, Canada, Finland, Sweden, Belgium, and The Netherlands. Correcting for interviewer characteristics (age, sex, and experience) and for survey organization (government versus nongovernment), the same conclusions hold. The only significant covariate is experience ($p = 0.01$); experienced interviewers tend to be more persuasion-oriented than inexperienced interviewers.

A different picture emerges when we look at "voluntariness." Countries do differ ($F = 73.2$, $df = 7$; $p = 0.00$), but now the most extreme case is Slovenia. Interviewers in Slovenia are strongly persuasion-oriented, and they also value voluntariness. When we rank the countries from high to low on voluntariness, we find the following order: Slovenia, Germany, Sweden, Belgium, The Netherlands, Finland, United Kingdom, and Canada. Correcting for interviewer variables does not change the conclusions. The only significant covariate is again experience ($p = 0.01$). Experienced interviewers tend to be less voluntary-oriented than inexperienced interviewers.

As Table 7.1 shows, the factor "voluntariness" is based on respecting respondents' privacy, accepting a refusal, and emphasizing voluntary nature of participation. Privacy and data protection has been the topic of much discussion in several European countries (e.g., Germany and Sweden); also, refusal conversion is still not generally accepted in Europe. The extreme position of Slovenia favoring privacy and voluntariness can be attributed to the fact that Slovenia is one of the emerging new countries, formerly belonging to Eastern Europe. Under the old communist regime in Eastern Europe, there was not much room for privacy. Therefore, it seems reasonable to assume that at present voluntariness and privacy are highly valued in former Eastern European countries both by interviewers and respondents. It would be interesting to investigate whether other former Eastern European countries show the same trend of valuing privacy and voluntariness.

Finally, we focussed on "send other." Interviewers who score high on send other, think it is a good thing to let another interviewer contact a reluctant respondent (no time, no interest) the second time. Again, there are significant differences between countries ($F = 60.6$, $df = 7$; $p = 0.00$). No significant effects were found for experience. The only significant covariates are age and government. Elder interviewers and interviewers working for official (government) statistical offices are more inclined to favor a renewed try by another interviewer. The rank order of countries is Sweden, Canada, Slovenia, United Kingdom, Belgium, Germany, Finland, and The Netherlands. Interviewers in Sweden are extremely favorable towards a recontact by another interviewer.

Figure 7.1 shows the distribution of interviewer attitudes across countries. Of course, there is also a large variation between interviewers within countries. For, instance an individual Canadian interviewer can score high on voluntariness and an individual Swedish interviewer high on persuasion.

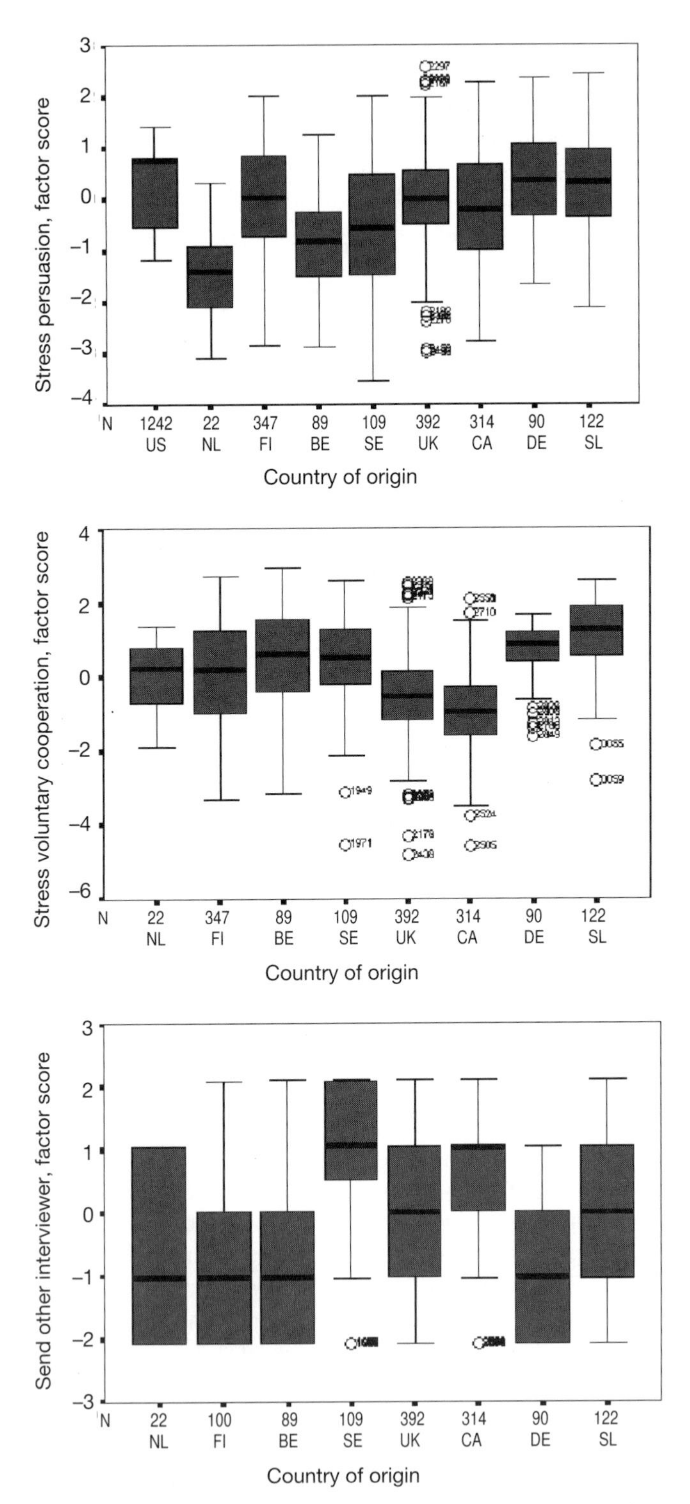

Figure 7.1 Distribution of interviewer attitudes across different countries.

7.3.2 Avowed Interviewer Behavior in Different Countries

What interviewers do or say in their first contact with potential respondents will reflect organizational differences and individual differences. Available are self-reports of avowed behavior on average, not actual observations. But, both methods indicate the same trends, although the intermethod reliability is rather low (Campanelli et al., 1997). We found three distinct factors describing avowed interviewer behavior: social validation, scarcity, and foot-in-door (see Section 7.2.3).

Although there is an overall significant difference between countries on "social validation" ($F = 8.7$, $df = 8$; $p = 0.00$), this is mainly caused by the relatively high value of Canada and the low value of The Netherlands. Interviewers in Canada report that they often use arguments such as, "most people enjoy the interview," "most people participate," "the topic would interest you." When we correct for differences in interviewer characteristics (age, sex, and experience) and organization (government versus nongovernment) the differences between countries become somewhat greater. The only significant covariate is survey organization ($p = 0.00$); interviewers working in official statistics use fewer social validation arguments than interviewers in nongovernmental agencies. Ranking countries from high to low on avowed use of social validation arguments gives the following order: Canada, United Kingdom, Sweden, Finland, United States, Slovenia, Germany, Belgium, The Netherlands.

Looking at the use of "scarcity" arguments (mentioning they represent other people, this is *the* chance to give opinion, explain household selection), we again find a significant difference between countries ($F = 21.5$, $df = 8$, $p = 0.00$). In Canada, these arguments are relatively often used and in The Netherlands, Belgium, and Slovenia, relatively seldom. The only significant covariate is interviewers' sex ($p = 0.00$); female interviewers reported that they use the scarcity arguments more. Ordering the countries from high to low on the use of scarcity arguments gives: Canada, Sweden, Germany, United Kingdom, Finland, United States, Slovenia, Belgium, The Netherlands.

The use of foot-in-the-door or consistency arguments (e.g., begin asking questions) also differs between countries ($F = 24.3$, $df = 8$, $p = 0.00$). There are no significant covariates. The rank order of the countries from high (often use arguments) to low is Canada, Slovenia, Sweden, United States, Germany, Finland, United Kingdom, Belgium, and The Netherlands.

Figure 7.2 shows the distribution of avowed interviewer behavior across the countries. It clearly shows that although countries do differ in avowed interviewer behavior, there is also in some cases a large variation within countries.

7.3.3 Response Rates

The response rates differ considerably across interviewers. The average response rate is 0.82, with a standard deviation of 0.18. The country with the highest average response rate is the United States, at 0.91, and the country with the lowest average

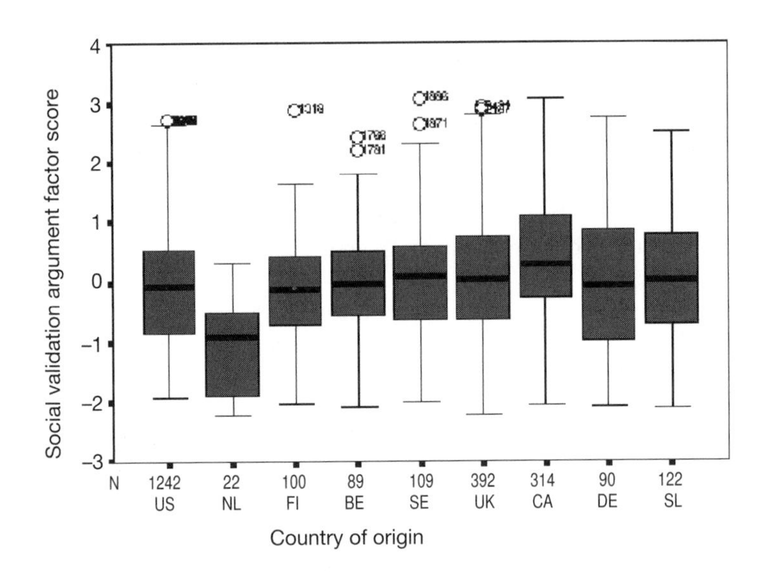

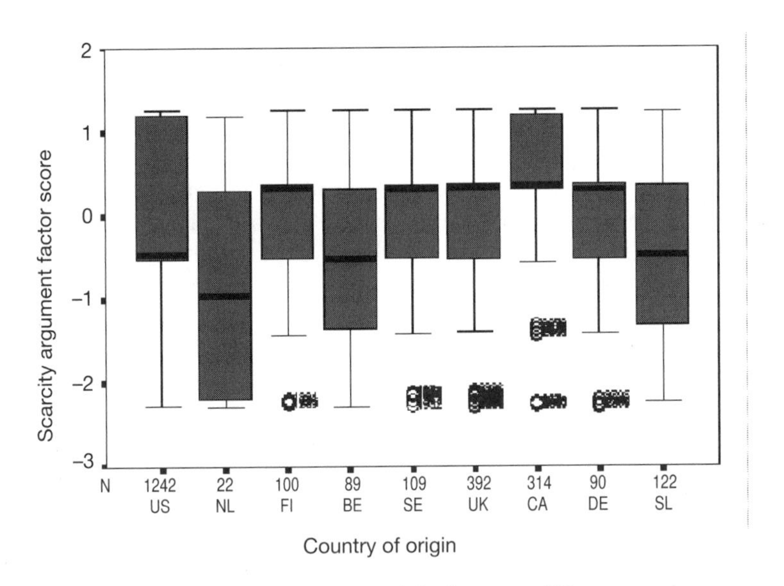

Figure 7.2 Distribution of interviewer behaviors across different countries.

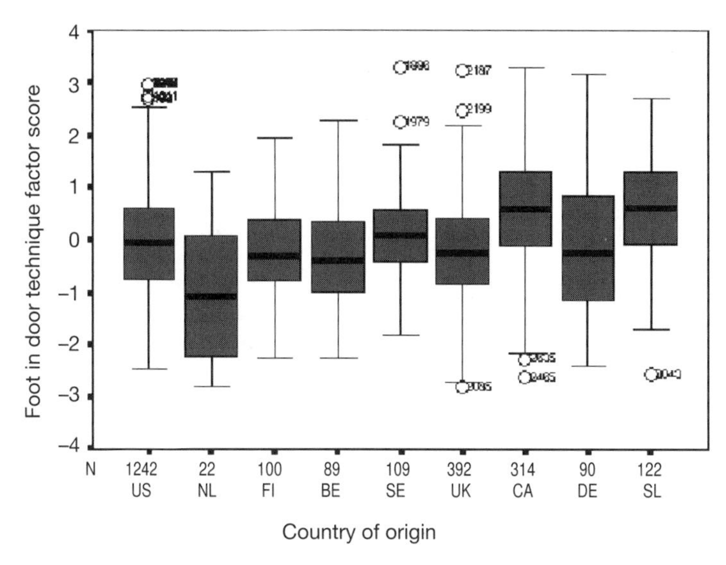

Figure 7.2 Distribution of interviewer behaviors across different countries *(continued)*.

response rate is Germany at 0.52. The highest scoring individual study is the US Census health study, and the lowest is the face-to-face condition in the German social-economic study. However, these figures are not directly comparable because the studies differ in various characteristics. Therefore, we used multilevel logistic regression models to analyze the response rates, including available interviewer characteristics as explanatory variables and study characteristics as covariates.

7.3.4 Predicting Nonresponse Within and Across Countries

For predicting nonresponse, three sets of interviewer-related variables are available. The first set is individual interviewer attributes: age, sex, and amount of interviewing experience. The second set is avowed interviewer behavior: "social validation," "scarcity," and "foot-in-the-door." The third set consists of interviewer attitudes: "persuasion," "voluntariness," and "send other interviewer." Table 7.3 presents the results of four separate models. Each model uses one of the variable sets as predictors of nonresponse. The first model is the null model, which serves as a baseline for model comparison.

In logistic regression modeling, the variance at the lowest level (the interviewer level) is scaled to the variance of the standard logistic distribution, which is 3.29. The null model decomposes the total variance in response into three components. The proportion variance at the country level is 0.14, the proportion variance at the survey level is 0.10, and the proportion variance at the interviewer level is 0.76. Although the country level variance appears to be relatively small, it is not negligible. A country that is one standard deviation below the average has an expected response rate that is 15.9 percentage points lower than average.

Table 7.3. Multilevel logistic regressions on interviewer response rates

Model/ predictor	Null model	Interviewer attributes	Interviewer behavior	Interviewer attitude
Constant	1.25 (0.30)	0.79 (.30)	1.26 (.29)	1.29 (.29)
Age (years)		0.01 (0.001)		
Sex (1 = female)		0.03 (0.015)		
Experience (years)		0.01 (0.001)		
Factor Scores				
Social validation			–0.02 (0.01)	
Scarcity			0.003 (0.01)^{ns}	
Foot-in-door			0.03 (0.01)	
Persuasion				0.10 (0.01)
Voluntariness				–0.02 (0.01)
Send other				–0.02 (0.01)
$\sigma^2_{country}$	0.59 (0.37)	0.62 (0.38)	0.58 (0.37)^{ns}	0.55 (0.35)^{ns}
σ^2_{survey}	0.41 (0.13)	0.40 (0.12)	0.41 (0.12)	0.41 (0.12)
Deviance	557.8	443.0	547.0	225.6

Note: ns = not significant.

If we consider the deviance of each model, which is a measure of misfit, we see that the model with the interviewer attitude variables has the lowest deviance. Thus, interviewer attitudes are the best predictors of response. Interviewer attributes are the second important set of predictors. The effect of avowed interviewer behavior is significant, but small.

Table 7.4 presents a multilevel regression model that contains as predictors all variables that have a significant contribution in Table 7.3. The regression coefficients are defined on the logistic scale. To facilitate interpretation, the last column in Table 7.4 indicates how many percentage points the average response changes if the predictor variable changes one unit.

The deviance of the final model indicates a good fit. All three attitudes are significant, but persuasion makes the largest contribution. Interviewers who are persuasion oriented achieve higher response rates. If interviewers go from one standard deviation below the average to one standard deviation above the average, their response increases on average by $2 \times 1.8 = 3.6$ percentage points.

Interviewer attributes do not appear to be very important, but age and experience are counted in years. Older interviewers have a somewhat higher response rate: a difference of 10 years results in 2 percentage points higher response rates. Experience counts for less: a difference of 10 years results in a predicted increase of 1 percentage point. Age and experience are, of course, correlated, but not perfectly, so the simultaneous effect of age and experience would be slightly higher. Sex does not have a strong influence: women have on average a 0.8 percentage point higher rate.

Table 7.4. Final multilevel model for interviewer response rates

Model/ predictor	Null model	Final model	Effect in percentage points
Constant	1.25 (0.30)	.80 (0.40)	
Age (years)		0.01 (0.001)	0.2
Sex (1 = female)		0.05 (0.02)	0.8
Experience (years)		0.01 (0.001)	0.1
Factor (Z-) Scores			
Social value		–0.02 (0.01)	–0.3
Foot-in-door		0.01 (0.01)[ns]	0.1
Persuasion		0.10 (0.01)	1.8
Voluntariness		–0.02 (0.01)	–0.4
Send other		–0.01 (0.005)	–0.2
$\sigma^2_{country}$	0.59 (0.37)	0.58 (0.36)	
σ^2_{survey}	0.41 (0.13)	0.39 (0.12)	
Deviance	557.8	81.9	

Note: ns = not significant.

Avowed doorstep behavior hardly has any effect. It is interesting to see that the use of social-validation arguments even accounts for a slightly lower response. We will come back to this in the discussion.

The country level variance in the null model is 0.59. Translated back to percentages, this means that a country that is 1 standard deviation below the average in response rate achieves a 15.9 percentage points below average response. Given the large differences between countries, in our data as well as in the data analyzed by de Leeuw and de Heer (Chapter 3, this volume), it is interesting to compare the country variance to the effect of our strongest interviewer variables: interviewer age, experience, and persuasion. For this, the group interviewers is split on the median value for these variables into groups of interviewers that are young versus old, not experienced versus experienced, and low on persuasion versus high on persuasion. On average, the older interviewers have a response rate that is 2.8 percentage points higher than the younger interviewers. The more experienced interviewers have a response rate that is 0.2 percentage point higher than the less experienced interviewers. Finally, the more persuasion-oriented interviewers have a response rate that is 3.6 percentage points higher than the less persuasion-oriented interviewers. Although these differences are not totally negligible, it is clear that they cannot explain the much larger differences between the nine countries.

Finally, there are no significant variances for the regression slopes of the various predictor variables. This implies that there are no large differences in the effectiveness of the predictors across countries.

7.4 SUMMARY AND CONCLUSIONS

Two main questions in this study are whether interviewers in different countries have different attitudes and behaviors, and if such differences can explain differences in response rates within and between countries. Our results show that there are large differences between interviewers in all three attitude dimensions (persuasion orientation, stress voluntariness, and send other interviewer) and in all three doorstep behavior dimensions (social validation argument, scarcity argument, foot-in-door technique). If we inspect the patterns of the differences closely, there are some consistent results. In their attitudes, the United Kingdom and Canada cluster together, including the United States for the one dimension on which the U.S. interviewers are scored. Germany and Slovenia are also close to each other in their attitude scores. Extreme scores were found for The Netherlands, which is very low on persuasion, Belgium, which is simultaneously relatively low on persuasion and relatively high on voluntariness, and Sweden, which is extremely high in sending a replacement interviewer. In the avowed behaviors, most countries are very similar, except for The Netherlands, which scores on average below all other countries for all three dimensions, and Belgium and Slovenia, which score below average in using scarcity arguments. These differences are significant, and remain so if we control for available interviewer characteristics (i.e., age, sex, and experience) and for survey organization (i.e., government versus nongovernment).

We find that the interviewer attributes, attitudes, and avowed behaviors explain only a small part of the variation among countries. Part of the problem may be that there are differences between the studies, which confound the interviewer results. To assess the importance of this problem, we have done follow-up analyses of the response rates. In these follow-up studies, we included the interviewer characteristics reported in Table 7.4 as explanatory variables, with selected study characteristics added to the model as covariates. In general, these follow-up analyses corroborate our results. That is, response differences between countries continue to be important, and the available interviewer characteristics explain only part of them.

We also performed a sensitivity analysis to investigate the importance of the size of the contributed data set. There are large differences between the contributions, with as extremes the United States, with usable data from 1242 interviewers, and The Netherlands with 22 interviewers. We have reanalyzed the model reported in Table 7.4, with weights to compensate for the different sizes of the contributed data sets. The results of the follow-up and sensitivity analyses give us much confidence in the conclusions of our study. We conclude that the generalizability of our results is not impaired by our sample of countries.

The effect sizes we find for the interviewer variables are comparable to the interviewer effects found in other studies. For instance, Hox et al. (1991) report a small interviewer effect (intraclass correlation 0.02) on the response rate in telephone and face-to-face interviews. The small differences between interviewers were not related to interviewer characteristics, which included five personality measures. Groves and Couper (1998, Chapter 7) discuss several studies that relate interviewer characteristics to the interviewer-level response rates. They conclude that there is no strong evidence for a relation between interviewer-level response rates and personality factors, but that interviewer experience and attitudes do have an effect. The effects we find are similar in magnitude, which corroborates their validity. The small but negative effect of using social validity arguments is contrary to expectance (Groves et al., 1992). However, Dijkstra and Smith (Chapter 8, this volume) also find a small negative effect for using social validation arguments. Social validation arguments may remind people too much of a sales pitch and may invoke the wrong ("oh no, they want to sell something") cognitive script. This association will lower the trust in the legitimacy of the interviewer and the survey and may result in the opposite of what is intended: a refusal (cf. van Leeuwen and de Leeuw, 1999).

A limitation of our study is that the studies differ in a number of important characteristics, such as the survey organization, the topic (or mix of topics in an omnibus survey), fieldwork conditions, and so on. This confounding inflates the between-country variance, and at the same time makes it more difficult to find strong interviewer effects. Thus, we suspect that country explains less variance than our analyses suggest, and that the interviewer effects should probably be somewhat larger than we report here. However, weighting the countries differently has hardly an effect on the regression coefficients. Also, none of the regression coefficients shows a significant variation across the nine countries, and at the country level we find no large residuals. The lack of variation in the regression coefficients across countries is reassuring because it indicates that the cultural setting (cf. Johnson et al., Chapter 4, this volume) does not have a strong effect on the efficacy of the interviewer characteristics in our analysis. So, the effect of interviewer experience and persuasion orientation is similar in the different countries.

ACKNOWLEDGMENTS

This chapter was written in collaboration with Mick Couper and Bob Groves (ISR), Wim de Heer (Statistics Netherlands), Vesa Kuusela (Statistics Finland), Risto Lehtonen (Statistics Finland), Geert Loosveldt (KU Leuven), Peter Lundqvist and Lilli Japec (Statistics Sweden), Jean Martin and Roeland Beertens (ONS), Sylvie Michaud and Tamara Knighton (Statistics Canada), Peter Mohler (ZUMA), Patrick Sturgis and Pamela Campanelli (LSE), Vasja Vehovar (University of Ljubljana), and Metka Zaletel and Eva Belak (Statistical Office of the Republic of Slovenia).

This project was possible thanks to the enthusiasm and work of researchers all over the world. The coordinators of the data collection in each country, who contributed and commented on this text, are acknowledged as coauthors. We also sincerely thank all others who in some stage of this project contributed with comments, support, and data. We especially thank Bonnie Brandreth, Pamela

Campanelli, Bob Groves, Elizabeth Martin, and Jean Martin for their helpful suggestions. The usual disclaimers of responsibility apply.

We kindly acknowledge the permission of the health care services in Finland for using their data, and especially the members of the research team: Sirkka-Sisko Arinen (Social Insurance Institute of Finland), Unto Häkkinen (Research and Development Centre of Welfare and Health), Timo Klaukka (Social Insurance Institute of Finland), and Jan Klavus (Research and Development Centre of Welfare and Health).

APPENDIX

Overview of Surveys In Data File

Contact	Country	Study topic	Organization	Main mode	Reference	Number of interviewers	Number of responses	Year
Couper	United States	Expenditure (CE)	Census	Face-to-face	Groves and Couper (1998)	206	10,634	1990
Couper	United States	Health (HIS)	Census	Face-to-face	Groves and Couper (1998)	139	18,322	1990
Couper	United States	Crime (NCS)	Census	Face-to-face	Groves and Couper (1998)	366	47,979	1990
Couper	United States	Census particip.	NORC	Face-to-face	Groves and Couper (1998)	106	2483	1990
Couper	United States	Drug abuse	RTI	Face-to-face	Groves and Couper (1998)	273	11,073	1990
Couper	United States	Comorbidity	Michigan SRC	Face-to-face		152	5642	1990
De Heer	The Netherlands	Living conditions (POLS)	Stat. Netherl	CAPI	de Leeuw et al. (1997)	22	13,066	1996
Kuusela	Finland	Health elderly	Stat. Finland	CAPI		118	1446	1998
Kuusela	Finland	Labor force (LFS)	Stat. Finland	CATI		129	8462	1998
Lehtonen	Finland	Health security	Stat. Finland	CAPI	Lehtonen (1996)	122	1699	1995
Lehtonen	Finland	Health security	Soc. Insurance	Face-to-face	Lehtonen (1996)	95	534	1995
Loosveldt	Belgium	Tourism/recreation	ISPO	Face-to-face		88	2364	1997
Loosveldt	Belgium	Religion/moral	ISPO	Face-to-face		74	1445	1998
Lundqvist	Sweden	Labor force (LFS)	Stat. Sweden	CATI	Japec et al. (1998)	98	24,397	1997
Lundqvist	Sweden	Living cond.(SLC)	Stat. Sweden	CATI	Japec et al. (1998)	92	4667	1997
Martin	United Kingdom	Labour force (LFS)	ONS	CAPI		122	40,483	1998
Martin	United Kingdom	Housing (SEH)	ONS	CAPI		228	17,355	1998

(continued)

Overview of Surveys In Data File *(continued)*

Contact	Country	Study topic	Organization	Main mode	Reference	Number of interviewers	Number of responses	Year
Michaud	Canada	Children (NLSCY)	Stat. Canada	CATI		314	39,427	1999
Mohler	Germany	Social/economic	Infratest	CAPI (panel 1st wave)		87	1475	1998
Mohler	Germany	Social/economic	Infratest	Face-to-face (panel 1st wave)		6	84	1998
Sturgis	United Kingdom	Political tracking	NOP-research	Face-to-face	Campanelli et al. (1997)	16	510	1996
Sturgis	United Kingdom	Family resources	NatCen	CAPI	Campanelli et al. (1997)	16	345	1996
Vehovar	Slovenia	Nat. media (BGP)	Mediana	Face-to-face		22	609	1998
Vehovar	Slovenia	Adult literacy (IALS)	Mediana	Face-to-face		56	4053	1998
Vehovar	Slovenia	TV advertising (ODMEV)	Graliteo	Telephone		27	19,749	1998
Vehovar	Slovenia	Brands (UA)	Graliteo	Telephone		25	36,903	1998
Zaletel	Slovenia	Labor Force	Official Stat.	CAPI		14	1303	1997
Zaletel	Slovenia	Labor Force	Official Stat.	Face-to-face		16	1292	1997
Zaletel	Slovenia	Labor Force	Official Stat.	CATI		8	1453	1997
Zaletel	Slovenia	Household budget	Official Stat.	CAPI		10	605	1997
Zaletel	Slovenia	Household budget	Official Stat.	Face-to-face		8	539	1997
Zaletel	Slovenia	Consumer	Official Stat.	CATI		9	1555	1997

CHAPTER 8

Persuading Reluctant Recipients in Telephone Surveys

Wil Dijkstra and Johannes H. Smit, *Free University, Amsterdam, The Netherlands*

8.1 INTRODUCTION

A major cause of unit nonresponse is the reluctance to cooperate. Especially in telephone surveys, persuasion attempts by the interviewer are the main tools to reduce refusal rates. It is well known that differences exist between interviewers with respect to their effectiveness in persuading recipients to be interviewed. Little is known, however, about the kind of persuasion attempts that are actually employed by interviewers, and about the effectiveness of such persuasive messages. In this chapter, we examine the effects of naturally appearing persuasion attempts by interviewers in telephone surveys, seeking to answer these questions: Which techniques do interviewers use to persuade people to participate in a survey interview? How effective are these techniques? Which interviewer behaviors have negative effects on participation? Can cues from the recipient be helpful in deciding on whether or not to continue the interaction in order to persuade recipients?

8.2 PERSUASION STRATEGIES

We are guided by the distinction made between the "heuristic" and the "systematic" approaches from decision-making theory (Petty and Cacioppo, 1986). In the systematic approach, the person carefully considers all relevant factors. This might lead to questions about the purpose of the survey, anonymity issues, and the difficulty of the questions. In the heuristic approach, the respondent bases his decision to respond or not respond on single, prominent features related to the request. The heuristic approach would lead to straightforward objections, early in the interaction, like "no time," "not interested," or too early compliance with the request to be interviewed. We assume that most objections reflect the heuristic approach to making decisions.

Cialdini (1987, 1988) formulated six compliance principles on which a person may base his decision to yield to a request if following the heuristic approach. They include: reciprocation, social validation, authority, consistency, scarcity, and liking. For a more thorough discussion of the use of these principles in a tailored strategy to encourage survey response see, e.g., Groves and Couper (1998).

8.3 METHOD

8.3.1 The Data

Interviews. The study was a CATI telephone survey about commercials on radio and television and advertisements in magazines and newspapers, conducted by the University of Amsterdam and the Free University of Amsterdam. It was preceded by two pilot studies (Pondman, 1998) to generate tentative hypotheses and refine the coding scheme.

The interview consisted of 49 questions about behavior and attitudes with respect to commercials and advertisements. Interviews took place between six o'clock and ten o'clock in the evening. All interviewers worked three evenings. The interviewer introduced the interview as follows:

> Good evening, this is (name) of the University of Amsterdam. We are conducting a study on how people handle commercials on television and radio and advertisements in newspapers and magazines. The interview will take about 15 minutes. Can I ask you some questions?

Recipients were contacted using random digit dialing. The generated telephone numbers were assigned at random to the interviewers. A total of 2740 contacts were made, some of which were excluded from the analyses. In 71 such cases, the recipient broke off contact during the introductory statements, without any verbal response. Serious language problems occurred in 74 cases. Nonhouseholds (274 cases) were excluded too. In 166 cases, the introduction was not taped or the quality was too bad. A total of 2155 introductions remained for further analysis, leading to 1189 refusals and 966 permissions.

To keep the introductions as similar as possible, no selection procedure for a particular household member was applied, except age: any person of 18 years and over was considered an eligible respondent. The interviews took about 12 minutes on average, the shortest one about 6 minutes, the longest one 22 minutes.

Interviewers. With respect to the interviewers, there were two main concerns. First, for statistical reasons (see Section 8.3.2), the interviewers should be as homogeneous as possible. Second, a situation with either a very high or a very low response rate should be avoided, to ensure enough variance in the dependent variable (refuse versus participate). In view of these two considerations, we decided to use relatively inexperienced interviewers.

All 36 interviewers were recruited from students in the Department of Communication Studies of the University of Amsterdam. All were females, ages 19–22, and none of them had previous interviewer experience. They received one day of training to become acquainted with the CATI procedure and the questionnaire. We found that asking, "Why won't you participate?" in the pilot studies negatively affected the response rate. Hence, interviewers were trained to avoid such behavior. No special persuasion tactics were taught.

The interviewers were fairly atypical with respect to experience. Experienced interviewers (e.g., from a research agency), would have been much more heterogeneous (e.g., with respect to age). Moreover, we also wanted to identify ineffective interviewer behaviors. If we had used very experienced, successful interviewers, such behavior might not have occurred. The use of inexperienced interviewers in this study, however, may limit the generalizability of our results.

Introductions. All verbal utterances of interviewer and recipient during the contact phase of the interview were taped. The introductions were transcribed and divided into meaningful utterances. For example, the introductory statements were split into four different utterances; the interviewer introducing herself, the topic of the interview, the length of the interview, and the request to participate. Each utterance was coded, using a multivariate coding scheme, details of which are provided elsewhere (Dijkstra, 1999a). The eventual data consisted of sequences of codes. Each sequence starts with the code that describes the first introductory statement. A sequence ends with one of three types of codes:

1. A code signifying the interviewer's acceptance of a refusal
2. A code meaning that the recipient grants the interview
3. A code for cases in which the recipient hangs up before the interviewer can express acceptance of a refusal

Our data consist of 2155 coded sequences. The total number of codes (utterances) was 21,790. The mean length of a sequence was 10.1 utterances, with a minimum of 3 utterances. The longest sequence counted 78 utterances and took nearly 3.7 minutes.

In this analysis, the terms "comply" and "object" refer to actions of the recipient during the interaction, expressing willingness to participate or not. An "objection" has to be distinguished from an eventual decision to refuse (a "refusal"). Similarly, we will use the term "grant" for the eventual decision to participate. "Persuasion attempt" refers to deliberately providing arguments or information with the intention to persuade the recipient to participate, e.g., "It is very important for the University that everyone participates" or "Your answers remain anonymous". However, the latter answer given in response to a question like "Are the answers anonymous?" is not conceived as an explicit persuasion attempt. The interviewer provides information because the recipient explicitly asks for that information. Of course such an action can be highly effective, but it is not a deliberately provided argument intended to persuade the recipient.

A "request to participate" refers to an explicit request (e.g., "Would you cooperate?"). An utterance like "I would appreciate it very much if you could cooperate" is a "persuasion attempt" (an argument given in favor of participating), not an explicit request. Many persuasion attempts are, of course, implicit requests.

8.3.2 Analyses

These sequences of codes were analyzed using the Sequence program (Dijkstra, 1999b). This program was used to find particular patterns of interviewer and recipient behavior in a sequence. For example, in which sequences is a particular persuasion attempt by the interviewer followed by compliance, given that the persuasion attempt is preceded by a particular type of objection of the recipient. In this way, frequency tables are produced of, for example, the type of persuasion attempt by its apparent effect (e.g., object or comply). The types of analyses that can be performed by the program are extensively discussed in Dijkstra (1999a).

To determine the significance of such an effect, we have to account for the fact that the observations are not independent, but clustered by interviewers. We used the WesVarPC program) (Brick and Morgenstein, 1996) to calculate a chi-square statistic that takes this design effect into account. This chi-square statistic relies on the modification of the Pearson chi-square by using an estimated design effect based on replication methods (as suggested by Rao and Scott, 1981). A prerequisite for this procedure is that the variance in the dependent variable for each interviewer and each condition (e.g., the type of persuasion attempt) is larger than zero. This condition is usually not met in our data; often there are one or more interviewers with empty cells in the two-dimensional table for that interviewer. To nevertheless obtain an estimate for the design effect, we took two or more interviewers together in order to create nonempty cells. Although this procedure slightly underestimates the design effect, some testing with different numbers of interviewers taken together showed that this hardly affected the significance level (unless one takes all interviewers together, which yields the common Pearson chi-square). However, with two of our tables this procedure did not lead to a solution. In these two cases, we calculated the usual Pearson chi-square, but set the degrees of freedom to the number of interviewers involved. This may overestimate the design effect. In the tables presented, we will give both the common Pearson chi-square, as well as the modified chi-square, indicated as chi-square RS2. RS2 (after Rao–Scott) refers to the way the chi-square is adjusted for the design effect.

8.4 TYPES OF OBJECTIONS

Before an eventual refusal (or grant), recipients have often uttered a number of objections. The more interviewers "maintain interaction," the more objections can be expected from the recipients in a single sequence.

Consideration of the frequency of offering various types of objections by whether they were offered first, second, etc., up to seventh in a specific sequence, revealed that a total of 1296 first objections were offered. The number of objections offered second through seventh were 969, 653, 394, 243, 142, and 81, respectively.

Plain "No's," "Not interested," and "No time" are the most frequent types of objection. The percentage of plain "No's" gradually increases as the interaction continues, from 28% of the first objections to 62% of the seventh objections, whereas "Not interested" and "No time" are most likely to be offered as a first objection, 26% and 30% compared to less than 20% of the objections offered in second through seventh position. It appears that "No time" and "Not interested" are more or less polite reactions intended to break off the interaction as soon as possible, whereas "No" is a more explicit response to persuasion attempts from the interviewer during the course of the interaction.

More substantive objections including "dislike polls," "too old," "too long," "dislike commercials," and "sick" occurred infrequently and with about the same frequency in the first through seventh objection categories. Recipients appear to confine themselves to the same kind of substantive objection during an interaction but alternate this objection with the three predominant types of objections.

8.5 COSTS OF PARTICIPATION

A respondent who takes part in an interview offers much and gains little, except a polite "thank you" from the interviewer after the interview is finished. As far as recipients follow heuristic reasoning, it is not surprising that the cost of participating in an interview is the first thing that comes to their mind. The interviewer might affect this cost in a number of ways.

8.5.1 Length of Interview

Mentioning the Length of the Interview in the Introduction. Most interviewers read the introduction statements as scripted. There are two exceptions, however. Sometimes, interviewers said the interview would take 10 minutes rather than 15 minutes, whereas in other cases the time statement was omitted altogether. Table 8.1 shows whether the recipient immediately complied or objected after the question "Can I ask you some questions?"; sequences wherein the recipient objected or complied before this question and sequences with other types of reactions after this question were excluded.

One interviewer was responsible for 70% of the "ten minutes" statements. Consequently, the other interviewers had a large number of zero cells and chi-square RS2 could not be calculated. However, if we set the degrees of freedom to 36, the effect remains significant.

Apparently, not mentioning the length of the interview is very effective: 66% of the recipients immediately gave permission. Mentioning 15 minutes yielded 36% compliance and mentioning 10 minutes 43% compliance. In more detailed analy-

Table 8.1. Initial reactions of the recipient after the introductory statements

	Object	Comply	Total
15 minutes	847 (64%)	481 (36%)	1328 (100%)
10 minutes	23 (58%)	17 (43%)	40 (100%)
No time statement	87 (34%)	168 (66%)	255 (100%)
Total	957	666	1623

Note: Chi-square = 77.83; $df = 2$; $p < 0.001$. Chi-square RS2 could not be calculated; chi-square = 77.83; $df = 36$; $p < 0.001$.

ses, we also took into account sequences leading to an eventual refusal, notwithstanding an initial comply and sequences leading to interviews that were prematurely broken off. We concluded that not mentioning the length of the interview considerably increased the number of initial permissions, without detrimental effects on the percentage of completed interviews.

Mentioning the Length of the Interview during the Persuasion Process. Quite often, the recipient does not comply immediately with the request to participate. In many cases, his reservation concerned length of the interview. We discerned four different interviewer reactions when the 15 minutes statement was given in the introductory text.

1. Repeat. The interviewer repeated the 15 minutes statement.
2. Repeat/mitigate. The interviewer repeated the 15 minute statement but mitigated its impact, e.g., "Fifteen minutes is much faster than you think" or "Maybe we can do it in less time than 15 minutes."
3. Mitigate. The 15 minute statement was not repeated. Instead a mitigating statement was made; e.g., "It does not take that long" or "Maybe we can do it faster."
4. Decrease. The interviewer explicitly decreased the length by at least three minutes. For example: "I'm sure we can manage it within ten minutes."

Table 8.2 shows the reaction of the respondent immediately following this second time statement of the interviewer. Sequences without such a time statement were excluded.

The more the length of the interview is mitigated, the more recipients comply. The compliance principle of reciprocation may be effective here. The analysis was repeated using the eventual decision of the recipient (refuse or grant) as criterion, which produced nearly the same result.

8.5.2 Psychological Cost

It appeared that a quite common reaction of interviewers to an objection of the recipient is a repetition of the objection, often in a questioning voice: "You won't par-

Table 8.2. Effect of the kind of the second length statement

	Object	Comply	Total
Repeat	49 (77%)	15 (23%)	64 (100%)
Repeat + mitigate	45 (68%)	21 (32%)	66 (100%)
Mitigate	19 (44%)	24 (56%)	43 (100%)
Decrease	6 (32%)	13 (68%)	19 (100%)
Total	119	73	192

Note: Chi-square = 20.08; $df = 3$; $p < 0.001$. Chi-square RS2 = 13.56; $df = 3$; $p < 0.005$.

ticipate?" or "You are not interested?" Such reactions make it very easy for the recipient to persist in his objection and invariably led to another objection, usually a repetition of the first one. We hypothesize that such reactions have a detrimental effect on the chance to eventually persuade the recipient with more convincing arguments. Just consider these two examples:

Example 1
R: "I'm not interested."
I: "You're not interested?"
R: "No."
I: "Are you sure?"
R: "Yes."
I: "What a pity! Your cooperation is very important to us and I would be very grateful if you could participate."

Example 2
R: "I'm not interested."
I: "What a pity! Your cooperation is very important for us and I would be very grateful if you could participate."

We hypothesize that in the first example it is much more difficult for the recipient to comply. After three firm objections the recipient may feel that he will lose face if he complies. Such psychological cost is less in the second example, making it easier for the respondent to change his mind. In Table 8.3, the reactions of the recipient are shown after different numbers of repetitions of objections. Sequences

Table 8.3. Effect of the number of repetitions of objections before a persuasion attempt

	Object	Comply	Total
No repetition	324 (80%)	81 (20%)	405 (100%)
One repetition	135 (92%)	12 (8%)	147 (100%)
More than one repetition	49 (96%)	2 (4%)	51 (100%)
Total	508	95	603

Note: Chi-square = 17.26; $df = 2$; $p < 0.001$. Chi-square RS2 = 9.53; $df = 2$; $p < 0.01$.

wherein the recipient immediately complied and sequences without persuasion attempts were excluded from this analysis.

As can be seen from the table, the probability of a recipient complying decreases from 20% with no repetition of interviewer objections to 4% or less with two or more repetitions of the recipient's objection.

8.6 COMPLIANCE PRINCIPLES

8.6.1 Authority

One of Cialdini's (1987) compliance principles is "authority": people would be more willing to comply to a request if the request came from someone who is perceived as an authority. Interviewers may stress that the survey is conducted by the university and serves scientific purposes. Quite often, the interviewer does not mention the university explicitly but refers to it as "we" or "us"; e.g., "Your help is very important to us." Finally, interviewers sometimes stress that they themselves would appreciate it if the recipient participates; e.g., "I would like it very much if you would participate." The difference between these three kinds of persuasion attempts lies in the use of the words "university" or "scientific research," the more formal "we" or "us," and the more personal "I" and "me." According to the authority principle, we would expect these three kinds of persuasion attempts to yield increasing refusal rates. Table 8.4 shows that this is not quite the case (sequences with two or three different kinds of one of the three types of persuasion attempts and sequences without one of these persuasion attempts are excluded).

It is clear that the "formal" we/us type of persuasion attempt does poorly. In accordance with the authority principle, sequences of the "university" type yield lower refusal rates. Quite unexpectedly, however, the personal appeal seems to be the most effective one, contradicting the authority principle. Because the number of sequences with a personal appeal is fairly small, one might expect that this type of persuasion attempt would be used by only a small number of interviewers. This does not appear to be true. Of the 36 interviewers, 16 interviewers used this technique, usually one or two times. Hence, the positive effect cannot be explained by the actions of two or three otherwise successful interviewers using this technique.

Table 8.4. "Authority" and refusal rate

	Refusal	Grant	Total
University	43 (74%)	15 (26%)	58
Formal	123 (87%)	18 (13%)	141
Personal	11 (46%)	13 (54%)	24
Total	177	46	223

Note: Chi-square = 22.78; $df = 2$; $p < 0.001$. Chi-square RS2 = 15.30; $df = 2$; $p < .001$.

8.6.2 Social Validation

Another compliance principle is "social validation." According to this principle, the recipient would be more willing to comply if he or she believed that other persons would comply. Interviewers often said that they found that most people enjoyed the interview, thus suggesting that people are eager to participate. On the other hand, interviewers also tried to persuade recipients with a statement like "But we need as many people as possible," suggesting that it is not so easy to obtain cooperation from lots of people. According to the social validation principle, the former persuasion attempt should yield lower refusal rates than the latter one. Table 8.5 shows a nonsignificant trend in the reverse direction. Sequences with both types of persuasion attempts and sequences without either one of both types are excluded.

Table 8.5 casts some doubt on the applicability of the social validation principle on survey participation, similar to that reported in Chapter 7. The unexpected results from Tables 8.4 and 8.5 bring into question the usefulness of these two compliance principles for survey participation.

8.7 TAILORING

Tailoring refers to adjustments interviewers make to enhance the probability of obtaining cooperation. In a telephone interview, the interviewer obtains cues from the voice and from information provided by the recipient.

We restrict "tailoring" here to using information provided by the recipient to form a response. For example, after the recipient states: "But, I'm already 87," the interviewer may react with "We are interested in the opinions of older people too." To investigate the effect of tailoring, we looked at the recipient's mention of his or her age, attitudes towards commercials and advertisements, and behavior with respect to watching television, listening to the radio, and reading newspapers and magazines. Sequences wherein the recipient did not provided information about one of these topics where excluded, leaving 186 sequences.

We distinguished between sequences with one or more instances of tailoring; sequences without tailoring, but with other types of persuasion attempts, e.g., "But this research is very important for the university"; and sequences without any persuasion attempt of the interviewer at all. If the interviewer used persuasion attempts, tailoring was a favorite strategy but hardly had the intended effect (Table 8.6).

Table 8.5. "Social validation" and refusal rate

	Refusal	Grant	Total
"Most people enjoy"	105 (87%)	16 (13%)	121
"Need many people"	72 (78%)	20 (22%)	92
Total	177	36	213

Note: Chi-square = 2.70; *df* = 2; *p* = 0.100. Chi-square RS2 = 1.83; *df* = 2; *p* = ns.

Table 8.6. Effect of tailoring on the eventual result

	Refusal	Grant	Total
Tailoring	68 (64%)	38 (36%)	106 (100%)
Other persuasion attempts	25 (69%)	11 (31%)	36 (100%)
No persuasion attempt	39 (89%)	5 (11%)	44 (100%)
Total	132	54	186

Note: Chi-square = 9.10; $df = 2$; $p < 0.05$. Chi-square RS2 = 5.39; $df = 2$; $p < 0.10$.

Tailoring led in 36% of the cases to a grant, whereas persuasion attempts that were not related to the recipient's utterances led to a grant in 31% of the cases. This difference is close to significant.

Our definition of tailoring is rather restrictive. For example, if a recipient states that he or she is too old, the interviewer may react with "But the questions are constructed in such a way that everyone can answer them." We did not consider this persuasion attempt as tailoring because it does not explicitly refer to the recipient's age

We conclude that tailoring may be useful in introductions to telephone interviews. However, the effect is not significantly larger than for other kinds of persuasion attempts, perhaps as a result of our restrictive definition of tailoring

8.8 MAINTAINING INTERACTION

According to Groves and Couper (1992), maintaining interaction has two beneficial effects on the chance that the recipient will participate. First, longer interactions provide the interviewer with more cues that can be used for tailoring. Second, after a long interaction, it becomes more difficult for the recipient to object because of general social norms of social interaction.

8.8.1 Continuers

Maintaining interaction is directed toward keeping the recipient talking, not to obtain quick complies (Groves and Couper, 1992). In conversation analysis, actions that are intended to accomplish this are commonly known as "continuers." Typical examples are repeating part of the other's utterances, preferably in a somewhat questioning voice, and humming and saying "yes" in the sense of "I've heard you, please continue." More explicit continuers are posing questions, requests for repetition, and requests for elucidation. We also view irrelevant utterances (e.g., "Sorry for my voice, I have a cold" or "I went to the dentist this morning, so I am in a somewhat bad temper") as a continuer. Such utterances frequently evoked (irrelevant) reactions from the other person, thus lengthening the interaction. To investigate the impact of such continuers, we took only that part of the interaction into account which starts with the first utterance of the recipient after the introductory statements of the interviewer and ends with the last utterance of the interviewer. We excluded recipients who immediately complied after the introductory statements of the interviewer and granted the interview (666 cases). We also excluded cases in which recipients immediately objected and the interviewer accepted this objection (253 cases) and those in which recipients hung up after their immediate objection (22 cases), thus yielding a refusal. A total of 1214 sequences remained with at least some interaction between interviewer and recipient after the introductory statements.

If interviewers actively try to maintain interaction by using continuers, this should lead to longer sequences, which in turn would lead to fewer refusals. Hence the proportion of continuers (the number of continuers divided by the total number of interviewer utterances in a sequence) should be greater in the granted interviews. This did not appear to be the case, however. The average proportion of interviewer continuers hardly differed for refused (0.118; 915 sequences) and granted (0.126; 299 sequences) interviews ($t = 0.513$; $df = 1212$; $p =$ ns).

Even more surprising was that we found a large difference with respect to the mean proportion of recipient continuers. The mean proportion of recipient continuers was nearly five times as large for the granted interviews (0.332; 915 sequences) than for the refused interviews (0.067; 299 sequences; $t = 22.041$; $df = 1212$; $p < 0.001$). An explanation for this result is that among the recipients who eventually participate, a relatively large number followed the central route (the systematic approach) in the decision making process. Although the most frequent type of continuers consisted of "humming" and "yes" (37%), 28% of the continuers of the recipients were made up of questions (32% for the granted interviews and 22% for the refused interviews). Apparently, these recipients seek information from the interviewer on which to base their decision.

Our results do not clearly support the hypothesis that due to social norms, recipients feel obliged to comply because of the very fact of conversing a while with the interviewer. Rather, prolonged interaction seems to be caused by the recipient himself, seeking information in following the peripheral route of decision making.

8.8.2 When Do Recipients Comply or Object?

Another way to look at the effectiveness of longer interactions is to ask whether effective persuasion takes place relatively early in the interaction, or relatively late. Stated otherwise, it may very well be the case that after a limited length of the interaction the probability of successfully persuading a recipient reduces to near zero. On the other hand, it may also be the case that later in the interaction chances increase that the recipient will comply. A fairly large percentage of the very first reactions of the recipients consists of complies: in 31% of the cases the recipient immediately complies. In the majority of cases however (56%) the recipient immediately reacts with an objection. Thirteen percent of the first reactions concern other kinds of utterances of the recipient, e.g., requests for repetition, asking for information, repeating an utterance of the interviewer, and so on.

The percentages of complies, objections, and other utterances given by recipients from the second through the thirteenth utterances were compared. The percentage of complies at the second utterance of the recipient is quite low (9%), whereas the percentage of objections is quite high (73%). This could be expected: the recipients who comply immediately will be interviewed, thus leaving the more reluctant recipients for further interaction. The percentage of complies remains low until about the tenth utterance, where it begins increasing to 17% for the thirteenth utterance. However, the percentage of objects decreases steadily to 48% for the thirteenth utterance. The difference in percentages is due to the gradual increase in other kinds of utterances, from 18% in the second utterance to 38% for the thirteenth utterance.

Maintaining interaction is a prerequisite for persuading recipients: if there is no interaction, there will be no persuaded recipients. In view of the results of Section 8.8.1, maintaining interaction by using so-called "continuers" is not very effective. Rather, the interviewer should actively try to persuade the recipient or provide relevant information. The probability of persuading a recipient does not decrease with prolonged interaction and may increase a little.

8.8.3 Stop or Continue?

When does it make sense to continue to try to persuade unwilling recipients and when should the interviewer stop? Of the 1296 sequences wherein the recipient objected at least one time, in 761 cases (59%) the interviewer did not make any persuasion attempt, whereas in 232 cases the interviewer undertook only one persuasion attempt. Generally, the success rate increases with the number of persuasion attempts. Explicit persuasion attempts seem to be a necessity to obtain cooperation. In this section, we will discuss behaviors of the recipient that either point to greater or to less susceptibility for persuasion attempts by the interviewer.

"Good Evening." A first cue might be the reaction of the recipient immediately after the first statement of the interviewer: "Good evening, this is . . . of the University of Amsterdam." A number of recipients replied with "Good evening" or a similar statement. We suspected that such an utterance reflected characteristics like sociability, politeness, or adherence to social norms that made them more susceptible to pressures to participate. The refusal rate for these recipients (26% of 159 cases) was significantly lower ($p < 0.001$) than the refusal rate for the recipients who said nothing (62% of 1519). Sequences wherein the recipient gave other utterances after the first statement (e.g., saying "hmm-mm") are excluded.

A similar phenomenon occurs with respect to "humming" and saying "yes" by the recipient after the first and second statement of the introductory text of the interviewer. Although the effect of "humming" is less pronounced than the effect of saying "good evening" (refusal rates are 42% versus 62% after the first statement and 35% versus 61% after the second statement), the number of recipients who show such actions are much larger (429 after the first statement and 460 after the second one).

Premature Objections. A different kind of cue is provided by premature objections—those objections that occur before the interviewer has finished her introductory statements. Premature objections occurred five times after the first statement of the interviewer, 221 times after the second statement, and 51 times after the third statement. We consider only premature objections occurring after the second statement in this analysis. Whereas nearly all (97%) of the recipients giving premature objections refuse, only 36% of those making another utterance and 59% making no utterance at all refuse the interview.

8.9 DISCUSSION

Analysis of the interaction during the introductory phase of the interviews showed the following:

- Telling the recipient the interview takes 10 minutes leads to less refusals than telling the recipient the interview takes 15 minutes.
- Omitting a statement about the length of the interview leads to an even lower refusal rate.
- Mitigating the length of the interview (like "Maybe we can do it faster") leads to less refusals then a plain repetition of the 15 minutes statement.
- Explicitly reducing the length of the interview, e.g., from 15 minutes to 10 minutes, leads to even less refusals.
- Repeating objections of the recipient by the interviewer increases the number of refusals.
- Applying the "authority" compliance principle by explicitly referring to the University is not as effective as expected. Instead, making a personal appeal ("I would like your cooperation very much") is much more effective.
- Applying the "social validation" principle ("Most people enjoy the interview") is not effective either, or may even have a negative effect on the participation rate.
- Tailoring—adjusting the persuasion attempts in view of earlier information or reasons to object by the recipient—may have a positive effect on participation, but the effect is not clear-cut.
- Maintaining interaction by providing information and actively trying to persuade the recipient seems to have more effect than maintaining interaction by using so-called "continuers."
- Maintaining interaction, like saying "good evening" or "hmm-mm" seems to be especially efficient if the recipient shows signs of active listening during the introductory statements of the interviewer.
- Maintaining interaction seems to be less efficient if the recipient reacts with a premature comply.

These findings lend themselves to replication in more controlled experimental studies. For example, it is fairly easy to train interviewers not to repeat the objections of the recipient and then compare their refusal rates with a group of interviewers who are not given that instruction. Similarly, one can train interviewers to use the personal appeal and avoid "we" statements and then compare their results with an appropriate control group. Also, different statements with respect to time and different kinds of mitigation lend themselves to easy experimentation.

Such experimental studies should preferably use experienced interviewers, allowing better generalizability of the findings than our study. In addition, it would be useful to do such research in different countries. Chapter 7 clearly shows cross-cultural differences with respect to behavior and attitudes of interviewers and their effects on refusal rates. In addition, it is important that studies be done on the effect of interaction sequences across multiple contacts with recipients.

Finally, we need theories that apply to persuading unwilling recipients. Our results cast doubt on the applicability of some of the compliance principles. Such compliance principles may work well in other persuasion situations, but for persuading recipients, we probably need some other principles. However, the concept of the heuristic versus the systematic approach seems to have merit when applied to interview introductions.

CHAPTER 9

The Effects of Extended Interviewer Efforts on Nonresponse Bias

Peter Lynn, *University of Essex, UK*
Paul Clarke, *University College, London, UK*
Jean Martin, *Office for National Statistics, UK*
Patrick Sturgis, *University of Surrey, UK*

9.1 INTRODUCTION

Though rarely stated explicitly, an important rationale for wanting to maximize response rate is an assumption that this will bring greater gains in accuracy of estimation than simply increasing the selected sample size. In other words, it is assumed that adding in hard-to-get respondents will not merely improve precision, but will reduce nonresponse bias.

In this chapter, we examine two distinct components of the difficulty of achieving an interview: difficulty of contacting sample members ("ease of contact") and difficulty of obtaining cooperation once contact is made ("reluctance to cooperate") (Groves and Couper, 1998; Smith, Chapter 2, this volume). We assess the separate and combined effects of ease of contact and reluctance on nonresponse bias in order to estimate the impact of extended efforts on survey error and to answer subsidiary questions about the relative effectiveness of different types of extended efforts.

9.2 RESPONSE ELICITATION PROCESSES

Many attempts have been made to measure the difficulty of obtaining response from survey sample members (Cheng, 1998; Drew and Fuller, 1980; Ellis et al., 1970; Filion, 1976a; Fitzgerald and Fuller, 1982; Foster, 1997; Lin and Schaeffer, 1995; Smith, 1984). Such information can be used to aid fieldwork organization and planning, and to estimate the likely extent and direction of survey nonresponse bias (Drew and Fuller, 1980; Ellis et al., 1970; Filion, 1976b; Fitzgerald and Fuller, 1982; King, 1998; Lin and Schaeffer, 1995; Smith, 1984; Stinchcombe et al., 1981). In some cases, this has been extended to the development of nonresponse weighting (Colombo, 1992; Filion, 1976; Fuller, 1974; Lin and Schaeffer, 1995; Scott, 1961). If indicators of ease of contact and/or reluctance could be used to estimate the direction and magnitude of nonresponse bias, this would allow assessment of bias for all survey estimates, rather than just the auxiliary variables available for all selected units.

The first attempts to estimate nonresponse bias using indicators of difficulty of eliciting a response were almost exclusively based on data from postal surveys (Dunkelberg and Day, 1973; Ellis et al., 1970; Ferber, 1948; Filion, 1976b; Hawkins, 1975; Hilgard and Payne, 1944; Larson and Catton, 1959; Pace, 1939; Reid, 1942; Shuttleworth, 1940; Suchman and McCandless, 1940). The indicator was usually the number of mailings prior to response or the number of elapsed days. More recent studies based on interview surveys have typically used the total number of interviewer call attempts as the indicator of difficulty (Cheng, 1998; Drew and Fuller, 1980, 1981; Fitzgerald and Fuller, 1982; Thomsen and Siring, 1983; Traugott, 1987). However, both measures confound the ease of contact and reluctance dimensions. Lin and Schaeffer (1995) recognize this, but still produce a single scale. Other authors have either attempted to isolate ease of contact or reluctance, but not to address the possible interaction between them (O'Neil, 1979; Smith, 1984) or to unite both into a single measure (Stinchcombe et al., 1981).

Thus, a major weakness of much previous research is that it either confounds ease of contact with reluctance or isolates one without considering simultaneously the effect of the other. We therefore develop separate measures of each.

9.3 DATA SOURCES

The analyses of extended interviewer efforts presented in subsequent sections of this chapter utilize data from three large-scale national household surveys in Britain. These surveys were chosen because they have different substantive foci, place different demands on sample members, have different respondent selection criteria, achieve different response rates, and rely on extended efforts to different degrees.

9.3.1 The Family Resources Survey (FRS)

The FRS is commissioned by the (UK) Department of Social Security and carried out jointly by the Office for National Statistics (ONS) and the National Centre for Social Research since 1992. It involves continuous fieldwork with approximately 2000 CAPI interviews per month (Wilmot, 1999). Addresses are sampled with equal probability. Where possible, the interview is carried out with all adult household members present, though some proxy reporting is allowed. The interview mainly concerns income—sources, amounts, regularity—living standards, and characteristics that would determine entitlement to various state benefits. If the household contains more than one "benefit unit" (a single adult or couple living as married and any dependent children), then the income questions module is repeated for each benefit unit. Mean interview length (per household) is about 80 minutes. The data used are from the 1997–1998 FRS (fieldwork April 1997 to March 1998). They relate only to the National Centre half of the sample (PSUs are allocated to the two organizations in a systematic random way each month.) The survey achieved a contact rate of 96.4% and a cooperation rate of 71.7%, resulting in an estimated overall response rate of 69.1%.

9.3.2 The Health Survey for England (HSE)

The HSE is an annual survey commissioned by the Department of Health and carried out by the National Centre for Social Research. It monitors trends in the nation's health and examines the prevalence, nature, and distribution of particular health conditions and associated risk factors. At each sampled household, people aged no less than two are eligible for the survey (Prescott-Clarke and Primatesta, 1998a; 1998b). Those aged 13 and over are interviewed in person, and proxy information is collected from a parent or guardian regarding children aged 2 to 12. After completing the interview(s), the interviewer measures respondents' height and weight and then seeks agreement for a visit by a nurse, to take further measurements. Interviews average around 60 minutes per household. The data analysed here are from 1996 and 1997. The survey achieved contact rates of 97.6% and 97.5%, respectively, in these years, and interview cooperation rates of 73.5% and 69.4%, resulting in estimated overall response rates of 71.7% and 67.7%. (Note: this analysis excludes 1.3% of households where contact was not attempted due to a refusal having been made to the office in response to receiving the advance letter.)

9.3.3 The British Social Attitudes Survey (BSAS)

The BSAS is an annual survey, designed and carried out by the National Centre to investigate the attitudes and views of the British public on a range of social and political topics (Lilley et al., 1997, 1998; Bromley et al., 2000). Funding comes from various sources. The BSAS differs from the FRS and HSE in that, (a) fieldwork is conducted during a limited period in the spring of each year, rather than being continuous throughout the year, and (b) one person aged 18 or over is randomly selected at each address. The data used here are from the 1995, 1996 and 1998 rounds of BSAS. The survey achieved contact rates (with the selected individual) of 97.9%, 97.7%, and 98.2%, respectively, in these three years, and cooperation rates of 70.7%, 69.8%, and 60.0%, giving estimated overall response rates of 69.2%, 68.1%, and 59.0%. For comparison with the other surveys, the proportion of households at which contact was achieved with any adult was 98.9%, 98.4%, and 98.9%, respectively.

9.4 MEASURING THE COMPONENTS OF SURVEY PARTICIPATION

9.4.1 Measures of Ease of Contact

On a face-to-face interview survey, the total number of calls made to the address (TNC) can be used as a proxy measure of ease of contact. This measure was used in some of the previous studies described in section 9.2.2 and is routinely collected for BSAS, HSE, and FRS. However, it is an imperfect measure because (a) the number of calls to make contact depends on interviewer behaviour as well as household characteristics; and (b) TNC also includes calls made subsequent to household contact. For this reason, we undertook a special exercise on BSAS 98 that involved obtaining field documents and coding the number of calls made prior to household contact.

First we look at the distribution of the total number of calls to sample households across the six surveys. Table 9.1 shows that the proportion of addresses requiring more than six calls is lowest on FRS (13.2%), higher on HSE (from 14.8% to 15.4%), and higher still on BSAS (19.7% to 25.9%). The modest difference between FRS and HSE might partly be explained by the admission of proxy responses on the former. The higher number of calls on BSAS is, on the face of it, surprising. The other two surveys require interviews with all adult household members whereas BSAS involves random selection of just one person. Thus, contact should be easier for BSAS. One explanation is that the sample is more highly clustered than for the other two surveys. Another is that the survey may meet with more reluctance than the other two. This seems to be borne out by the lower cooperation rates (see Section 9.3.3). The subject of the interview is much less clear and well defined than for HSE or FRS and this may make it more difficult for interviewers to obtain cooperation. Table 9.1 also demonstrates the limitation of total number of calls as a mea-

Table 9.1. Distribution of number of calls: six surveys

	Total calls at address						Calls until contact
Call number	FRS 1997–1998, %	HSE 1996, %	HSE 1997, %	BSAS 1995, %	BSAS 1996, %	BSAS 1998, %	BSAS 1998, %
1	18.6	11.5	9.9	12.5	12.4	10.5	43.8
2	23.5	25.5	27.8	20.1	22.3	19.3	20.3
3–4	30.1	32.5	33.7	29.1	29.2	27.5	19.7
5–6	14.7	15.2	13.8	15.3	16.3	16.9	7.9
7–9	8.5	9.1	9.3	12.6	9.7	13.6	4.6
10+	4.7	6.3	5.5	10.4	10.0	12.3	3.7
Base[a]	15,450	11,368	9,632	4,969	4,981	4,659	4,659

[a]Eligible sample addresses.

Note: Excluded from the table are sample addresses that were ineligible (not occupied residential dwellings) and those where an interview was not possible for reasons other than refusal or noncontact.

sure of ease of contact. The final column shows the number of calls until household contact for BSAS 98. By comparing this column with the previous one it is obvious that many calls are made beyond those required to make contact. In fact, calls to make household contact account for only 56% of all calls made on BSAS 98.

9.4.2 Measures of Reluctance to Cooperate

We are able to classify sample households into one of three reluctance categories: those who ultimately refused (in this category we have included cases that refused initially, were reissued for a conversion attempt, but no contact was made subsequently), those who refused initially but subsequently agreed to be interviewed ("converted refusals"), and those who agreed to the interview without refusing ("willing respondents"). Table 9.2 shows that at least two-thirds of households do not refuse, even temporarily. This proportion is highest on HSE and lowest on BSAS. The proportion of temporary refusers is highest on BSAS (8.0% in 1995) and lowest on FRS (1.2%). The proportion that ultimately refused to participate is lowest on HSE and highest on BSAS, but it also varies across years of BSAS. BSAS seems to require extended efforts to a greater extent than the other surveys.

9.4.3 Relationship between Ease of Contact and Reluctance

It is of interest to know to what extent households who are difficult to contact are also reluctant to respond. This will aid understanding of the impacts of noncontacts and refusals on nonresponse bias. It will also inform our criticism of many of the earlier studies (see Section 9.2.2 above).

There is no evidence of a relationship between difficulty of contact (as measured by number of calls until household contact was made) and reluctance (as indicated by cooperation rate). The data are from BSAS 98, the only survey for which the number of calls until household contact is known. There is, however, an association between cooperation rate and *total* number of calls for both this and other surveys,

Table 9.2. Distribution of refusals, converted refusals, and willing respondents: six surveys

Calls until contact	FRS 1997–1998, %	HSE 1996, %	HSE 1997, %	BSAS 1995, %	BSAS 1996, %	BSAS 1998, %
Willing respondents	77.1	81.0	82.4	66.1	75.9	63.4
Converted refusals	1.2	2.7	2.1	8.0	2.0	4.1
Refusals	21.6	16.3	15.5	25.9	22.1	32.5
N	14,934	11,177	9,450	4,894	4,693	4,517

Note: Excluded from this table are households for whom we have no direct information about reluctance to cooperate, namely, those where contact was never established, and those who could not be interviewed for reasons of language, temporary absence, or physical or mental health. Collectively, these account for between 1.6% and 6.0% of eligible households.

perhaps reflecting the fact, argued earlier, that total number of calls is imperfect as a measure of ease of contact as it is influenced by reluctance too. In other words, the more reluctant a household, the greater the extent to which total number of calls is likely to exceed number of calls to household contact. So there is no evidence that households who are more difficult to contact are any more or less reluctant than others to cooperate once contacted.

9.5 DO EXTENDED EFFORTS AFFECT SURVEY DATA?

At the start of this chapter, we hypothesised that households interviewed only after extended efforts ("hard-to-get" households) differ systematically from those interviewed without those efforts ("easy-to-get" households). In this section, we test this hypothesis by examining how survey estimates are affected by extended interviewer effort. Furthermore, we compare these effects for the different types of extended interviewer effort, making contact with difficult-to-contact households and converting refusals. Hard-to-get households are defined as those requiring six or more calls or a refusal conversion attempt to achieve an interview. We have compared estimates between these subgroups for a set of demographic variables common to BSAS, HSE, and FRS and also for sets of key variables specific to each survey (Lynn and Clarke, 2001).

9.5.1 Survey Estimates

For the HSE, we examined one general health indicator and four key indicators of health risk factors. These measures are listed in Table 9.3 along with the corresponding estimates from HSE 96 for the total responding sample and each of the four subgroups. It can be seen that respondents for whom extended efforts were necessary were more likely than other respondents to be smokers and drinkers, had lower blood pressure than others, and were less likely to have a longstanding illness. Furthermore, it was the difficult to contact respondents rather than the reluctant respondents who had these characteristics. The reluctant respondents were not significantly different from the "easy" respondents. These patterns were almost identical for HSE 97, except that the reluctant respondents were as likely as the hard to contact to be regular smokers.

A similar analysis was conducted for FRS 97 (data not shown). Hard-to-get households receive a higher proportion of their income from employment (and a lower proportion from state benefits) and have higher housing costs. Employed persons in hard-to-get households work more hours per week than those in "easy" households. Again, these differences are most pronounced for the difficult-to-contact households. Similar analysis was performed for three key attitude scale measures from BSAS. One of the scales exhibited a significant difference in means between the hard-to-get and easy-to-get households for two of the three years, but the other seven comparisons showed no significant differences (data not shown).

Table 9.3. Survey estimates for difficult to contact, reluctant, hard-to-get, and easy-to-get households (HSE 96)

	$\frac{x_{H1}}{n_{H1}}$	$\frac{x_{H2}}{n_{H2}}$	$\frac{x_H}{n_H} = \frac{(x_{H1} + x_{H2})}{(n_{H1} + n_{H2})}$	$\frac{x_E}{n_E}$	$\frac{x_A}{n_A} = \frac{(x_H + x_E)}{(n_H + n_E)}$	
	Difficult to contact (6+ calls)	Reluctant (converted refusal)	Hard-to-get households	Easy-to-get households	All responding households	Nonresponse bias[b]
Regular smokers[a] (%)	28.8	24.5	28.1	23.9	24.6	−1.2
	(0.95)	(1.96)	(0.85)	(0.36)	(0.34)	
Body mass index (mean)	25.3	25.9	25.4	26.0	25.9	0.1
	(0.12)	(0.28)	(0.11)	(0.05)	(0.04)	
Systolic blood pressure (mean)	132.1	138.1	133.0	136.6	136.1	0.6
	(0.40)	(1.19)	(0.39)	(0.19)	(0.17)	
Longstanding illness (%)	35.9	41.9	37.0	43.9	42.8	1.5
	(1.00)	(2.25)	(0.92)	(0.42)	(0.39)	
Heavy drinkers[a] (%)	20.7	18.3	20.2	16.9	17.5	−0.5
	(0.85)	(1.76)	(0.76)	(0.32)	(0.30)	

[a]Regular smokers are respondents who report smoking more than five cigarettes per day on average; heavy drinkers are those who report drinking more than 21 units of alcohol per week on average.

[b]This is an estimate of the (marginal) bias that would have been present in the survey estimate had extended efforts not been made. It is estimated as $\frac{x_E}{n_E} - \frac{x_A}{n_A}$, but only for variables where $\frac{x_E}{n_E} - \frac{x_H}{n_H}$ is significantly different from zero ($p < 0.05$).

9.5.2 Demographic Characteristics

For the same sample subgroups, we produced estimates of five demographic characteristics that are measured consistently across the FRS, HSE, and BSAS. We found systematic differences between the easy-to-get and hard-to-get households. On all six surveys, persons in hard-to-get households were significantly younger and significantly more likely to be employed than those in easy-to-get households. The differences in the proportion employed are particularly striking, ranging from 8.5% to 17.5% across the surveys. Persons in hard-to-get households were also less likely to be white on all six surveys, though this difference only reached statistical significance ($p < 0.05$) for four surveys. In two surveys, persons in hard-to-get households were significantly more likely to be male and less likely to be owner-occupiers. Results from FRS 97 are shown in Table 9.4. The results from the other surveys are broadly similar (see Lynn and Clarke, 2001). Similar patterns have been observed on other studies of the difficulty of response elicitation (Cheng, 1998; Foster, 1997, 1998).

Table 9.4 again reveals that it is the difficult-to-contact households that are most distinct from the easy-to-get households. The proportion in employment, mean age, and proportion white do not differ between reluctant and easy households, only between difficult-to-contact and easy households. For age and employment, this finding holds across all six surveys. Among the four surveys where the proportion white differs between the easy-to-get and hard-to-get households, there are two where only the difficult-to-contact are distinct, but two where the reluctant households are equally distinct.

In addition to these five characteristics, number of persons in the household was examined for BSAS 95 and BSAS 96. The difficult-to-contact households had a significantly smaller mean household size than the easy-to-get households (data not shown).

9.5.3 Discussion

For the variables examined, it is the difficult to contact that are most different from the easy to get. This would appear to suggest that resources for extended efforts might be better concentrated upon making contact with difficult-to-contact households than upon attempting refusal conversions. However, an important distinction must be noted. The final household contact rate on these surveys ranged from 96.3% to 98.2%. On the other hand, the proportion of refusals that are successfully converted is much lower and the final cooperation rates of the surveys range from 60.0% to 73.5%. With around a third or more of the samples remaining as refusals, we cannot be confident that the converted refusals are typical of the remaining refusals. It is possible that although converted refusals are similar to the easy-to-get, the remaining ("harder") refusals are rather different.

There is also some empirical support for the suggestion that extended efforts reduce overall nonresponse bias. United Kingdom population estimates suggest that our survey samples slightly underrepresent small households, persons in employ-

Table 9.4. Demographic characteristics for difficult to contact, reluctant, hard-to-get, and easy-to-get households (FRS 97)

	$\frac{x_{H1}}{n_{H1}}$	$\frac{x_{H2}}{n_{H2}}$	$\frac{x_H}{n_H} = \frac{(x_{H1}+x_{H2})}{(n_{H1}+n_{H2})}$	$\frac{x_E}{n_E}$	$\frac{x_A}{n_A} = \frac{(x_H+x_E)}{(n_H+n_E)}$	
	Difficult to contact (6+ calls)	Reluctant (converted refusal)	Hard-to-get households	Easy-to-get households	All responding households	Nonresponse bias[a]
Male (%)	48.8	47.2	48.7	46.7	47.0	–0.3
	(0.70)	(2.11)	(0.67)	(0.28)	(0.26)	
Age (mean)	43.7	49.1	44.2	48.1	47.5	0.6
	(0.23)	(0.78)	(0.23)	(0.10)	(0.09)	
Owner-occupier (%)	64.5	71.7	65.2	65.7	65.6	—
	(1.17)	(3.33)	(1.11)	(0.48)	(0.44)	
In employment (%)	65.1	57.4	64.3	55.8	57.1	–1.3
	(0.67)	(2.09)	(0.64)	(0.28)	(0.26)	
White (%)	91.3	95.2	91.7	95.0	94.5	0.5
	(0.40)	(0.91)	(0.37)	(0.12)	(0.12)	

[a]This is an estimate of the (marginal) bias that would have been present in the survey estimate had extended efforts not been made. It is estimated as $\frac{x_E}{n_E} - \frac{x_A}{n_A}$, but only for variables where $\frac{x_E}{n_E} - \frac{x_H}{n_H}$ is significantly different from zero ($p < 0.05$).

ment, and nonwhites. Thus, residual nonresponse bias appears to be in the same direction as the bias removed by extended efforts.

9.6 MODELING NONRESPONSE BIAS USING MEASURES OF EASE OF CONTACT AND RELUCTANCE

A set of regression models allows us to assess how survey/demographic variables vary across the two dimensions, as measured by the number of calls prior to contact and respondent reluctance, and to ascertain whether this variation is systematic. Data from BSAS 98 was used in order to take advantage of the superior measures of ease of contact and reluctance that were available. First, we fitted models for respondents. Restricting the analysis to respondents allows us directly to examine variation in the survey variables. Second, we extended the analysis to include nonrespondents. These models allow only an indirect assessment of nonresponse bias using auxiliary variables available for both respondents and nonrespondents.

9.6.1 Models for Respondents

We specified regression models with measures of ease of contact and reluctance as the independent variables, predicting survey and demographic variables. Ease of contact was measured by the number of calls required to make contact with the household. A five-category measure of reluctance was developed, based upon the three-category measure reported in Table 9.2, but with the "willing" category subdivided into three. Thus, for respondents, there are four categories of reluctance: those interviewed at the same call they were contacted on, those for whom one extra call was required, those for whom two or more extra calls were required, and those for whom a refusal conversion was needed. The dependent variables were the same as those reported for BSAS in the previous section. A separate model was fitted to each (a normal linear regression for the continuous measures and a logistic regression for the dichotomous measures).

The results showed no systematic evidence of any interactions between ease of contact and reluctance (data not shown). The main effects were consistent with the observed differences between easy-to-get households and the reluctant and difficult-to-contact households described in Sections 9.5.1 and 9.5.2: the models showed that the three attitude scale scores were not significantly associated with either ease of contact or reluctance. Being in employment was positively associated with both increasing difficulty of contact and increasing reluctance, whereas being an owner-occupier was positively associated with reluctance. The only finding not evident from the analysis in Section 9.5.2 was that age was negatively associated with increasing reluctance as well as increasing difficulty of contact.

9.6.2 Models for Respondents and Nonrespondents

A shortcoming of the models reported in Section 9.6.1 is that they exclude the most reluctant households (refusals) and the most difficult to contact (noncontacts).

Thus, they give only a partial view of the impact of ease of contact and reluctance. To overcome this, we modeled the data for eligible responding and nonresponding households. However, as well as excluding those who do not respond for reasons other than refusal and noncontact, we must also exclude 141 households that were not contacted, even after extended interviewer efforts. This is done essentially for convenience because their position on the reluctance dimension of the scale is unknown. To include them would involve developing a complex model that treats noncontacts as "censored" or "truncated" cases, and is thus beyond the scope of this chapter. It is, however, important to acknowledge that they are potentially different from hard-to-contact households, and that this analysis is incomplete without their inclusion.

It is necessary to use the full five-point reluctance scale to include those cases that eventually refuse. Hence, we are unable to use the survey/demographic variables as dependent variables because they are unavailable for the nonrespondents. Instead, seven auxiliary variables were used (Table 9.5.): six small area estimates from the 1991 population census, and one based upon mid-1997 population estimates. Four of these variables are aggregates of characteristics hypothesized to be correlated with contact and cooperation at the household level. These could therefore be expected to correlate with ease of contact and reluctance at the small area level, though the explanation of the correlation would be via the corresponding household-level characteristic (Groves and Couper, 1998, p. 94). Population density is a characteristic hypothesized to be causal at the small area level (Groves and Couper, 1998, pp. 85, 99, 146, 168, 176). Longstanding illness is an aggregate of a variable found to be affected by extended efforts at the household level (Table 9.3). The proportion of households without a car is correlated (in the United Kingdom) both with economic status/wealth and urbanicity, both factors hypothesized to be causal of nonresponse (Groves and Couper 1998, pp. 127–128, 167–168, 176, 85–87); in addition, it may also be related to social isolation.

As with the models for respondents, no significant interactions were found between the ease of contact and reluctance measures. The model results are summarised in Table 9.5. Population density is found to be positively associated with difficulty of contact, as expected, but the relationship with reluctance is less clear. Being just slightly reluctant (Reluct2) is associated with lower population density, but being a temporary refusal (Reluct4) is associated with higher population density. Complete refusers seem no different from the most willing respondents in terms of population density. At the individual level (in Table 9.3), people with a longstanding illness were found to be less likely than others to be difficult to contact, but no more or less likely to be reluctant. But here we find that people in areas with a high proportion of people with long-term illness are no more or less likely to be difficult to contact, but less likely than others to be reluctant. Presumably, the area-level effect is not just an aggregation of individual-level effects; rather, other factors are at play.

For the other five variables, associations generally correspond with theoretical expectations and/or the individual-level findings reported earlier. There are two exceptions. Slight reluctance (Reluct2) is associated with decreased proportion non-

Table 9.5. Ease of contact and reluctance as predictors of ecological indicators

Dependent variable (y)	β_0	β_1 (# Calls)	β_2 (Reluct2)	β_3 (Reluct3)	β_4 (Reluct4)	β_5 (Reluct5)
Population density	22.91	1.01***	–2.86***	0.49	6.67***	–0.86
	(0.87)	(0.16)	(1.03)	(1.28)	(1.96)	(1.00)
% Owner-occupied	67.71	–0.06	2.45***	0.16	–0.10	1.79***
	(0.60)	(0.11)	(0.71)	(0.86)	(0.94)	(0.69)
% Nonwhite	4.22	0.25***	–0.81*	0.42	1.94***	–0.33
	(0.30)	(0.05)	(0.35)	(0.44)	(0.67)	(0.34)
% Unemployed	5.49	0.06***	–0.47***	0.05	0.09	–0.27*
	(0.10)	(0.02)	(0.11)	(0.14)	(0.22)	(0.11)
% Limiting long-term	16.19	0.01	–0.85***	–0.60***	–0.67	–0.71***
illness	(0.15)	(0.03)	(0.18)	(0.23)	(0.35)	(0.18)
% No car	32.96	0.29***	–3.93***	–1.23	–0.92	–2.40***
	(0.42)	(0.09)	(0.61)	(0.76)	(1.16)	(0.59)
% Socioeconomic	23.31	0.02	1.72***	–2.28	0.84	1.00*
group	(0.38)	(0.07)	(0.45)	(0.56)	(0.86)	(0.44)

Notes: The estimated coefficients shown are from fitting the model $E(y) = \beta_0 + \#\text{ Calls } \beta_1 + \text{Reluct2 } \beta_2 + \text{Reluct3 } \beta_3 + \text{Reluct4 } \beta_4 + \text{Reluct5 } \beta_5$, where # Calls is the number of calls until household contact and Reluct2 to Reluct5 are dummy variables. The reference category is households where an interview was achieved at the same call as contact was first made with the respondent; reluct2 indicates that one extra call was needed to achieve the interview; reluct3 indicates that two or more extra calls were needed; reluct4 indicates that a refusal conversion was needed; reluct5 indicates that the final outcome was a refusal.

$^*p < 0.05$; $^{**}p < 0.025$; $^{***}p < 0.01$.

white and increasing difficulty of contact is associated with an increased proportion unemployed. It is likely that there are uncontrolled factors at play here. It is possible that, for example, areas with high levels of unemployment tend also to have high levels of other barriers to contact, such as entry-phones and security devices. This simply draws our attention to the dangers of assuming that area-level measures will act in the same way as the individual measures of which they are aggregates.

9.7 SUMMARY AND CONCLUSIONS

Our analyses suggest that ease of contact and reluctance to participate are two distinct dimensions on which sample households can be placed. We find evidence, across surveys and across years, that both dimensions are systematically related to substantive survey variables, to demographic survey variables and to auxiliary (small area) variables. We find no evidence of interactions between the effect of ease of contact and that of reluctance.

We find that extended interviewer effort affects both sample composition (distributions of demographic variables) and survey estimates (distributions of substantive variables). These effects are largely due to making contact with difficult-to-contact households, whose characteristics differ from those of easy-to-get households.

Furthermore, the direction of the impact of extended efforts appears to be consistent with nonresponse bias reduction. However, the impact is not of equal magnitude for all variables. It is particularly great for age and employment status. In terms of the survey estimates examined, extended efforts appear to significantly reduce nonresponse bias for certain income and expenditure variables and for measures of health and health risk, but to have no discernable impact upon attitudinal measures. When we fitted models to predict auxiliary (small area) variables, in order to include nonrespondents in the analysis, we found that the direction of the effect of difficulty of contact was sometimes the opposite of that of reluctance. This was not apparent in the analysis of respondents and suggests that converted refusals may not be typical of all refusals

Overall, it seems clear that both noncontacts and refusals contribute to nonresponse bias. Extended interviewer efforts appear to reduce this bias, particularly the component due to noncontact. With considerable efforts, the noncontact rate can be reduced to 3% or less of eligible households. In this situation, there is a limit to the possible extent of residual noncontact bias and it is likely that the "converted" noncontacts are fairly representative of all households that had not been contacted prior to the extended efforts. On the other hand, a relatively small proportion of refusals are converted (typically between 10% and 20% in our experience) and the noncooperation rate remains high (over 25% on all six of our surveys). Thus, converted refusals may not be representative of all refusals. There is a limit to the extent that the refusal rate could be further reduced by greater interviewer efforts and it is likely that survey organisations will have to continue to rely upon judicious weighting to combat refusal bias.

ACKNOWLEDGMENTS

This research was supported by Economic and Social Research Council award R000222938. Most of this research was carried out at the National Centre for Social Research, UK, where the first, second, and fourth authors were employed at the time. We are grateful for the help of many colleagues at the National Centre for Social Research in supplying data, documentation, discussion, and inspiration, and particularly to Tetti Tzamourani for coordinating the coding and analysis of the BSA 98 data.

CHAPTER 10

Effect of Item Nonresponse on Nonresponse Error and Inference

Robert Mason and Virginia Lesser, *Oregon State University*
Michael W. Traugott, *University of Michigan*

10.1 INTRODUCTION

Sampling statisticians and survey methodologists agree that nonresponse in sample surveys is a source of serious error. The error stems from the difference between respondents who participate in a survey and those who either refuse or remain unavailable for interviewing. Assael and Keon (1982) claim that nonsampling error—the error that includes a nonresponse component—is the major contributor to total survey error, whereas sampling error makes a minimal contribution. Data often are not available for nonrespondents, and most studies have focused on ways to increase unit response rates, analyses or comparisons of easy-to-reach versus hard-to-reach respondents, or statistical techniques to estimate differences between respondents and nonrespondents and compensate for them (Groves and Couper, 1998).

Item nonresponse that occurs in practice often results in missing data values and in less efficient estimates because of the reduced size of the data base. In addition, many statistical analysis methods operate on an assumption of complete data sets (Rubin, 1987). Therefore, observations that contain missing values are dropped from the analysis. (See, for example, SAS Version 8, Proc Reg and Proc Logistic procedures.) If the level of item nonresponse increases with efforts to improve unit response rates, researchers may have a false sense that they are increasing precision and power when, in fact, they may not be increasing their sample size at all.

This study has three purposes: The first is to compare the frequency of item missing values for converted and nonrefusal cases. The second is to compare the error associated with a predictive model of voting in which refusal conversion serves as an indicator variable. And the third is to examine the effect of alternative imputations for missing values on survey errors between converted and nonrefusal groups.

The data came from a study of Oregon voters who participated in a statewide mail election on January 30, 1996 to replace Bob Packwood, who had resigned. Ron Wyden, the Democrat, narrowly defeated Gordon Smith, the Republican.

10.2 RESEARCH ON THE EFFECTS OF ITEM NONRESPONSE

10.2.1 The Importance of Nonresponse

Unit nonresponse—the failure to complete an interview with an eligible sampling unit—is not the same thing as item nonresponse—the failure to answer a specific survey item. Examples are refusals to answer or "Don't Know" (DK) responses to a specific question. Rules for the calculation of response rates are equivocal about "partial" or item nonresponse. The AAPOR Standard Definitions (2001) for calculating rates includes item nonresponses as an interview completion; other definitions, however, exclude these cases from the response rate estimate. Item nonresponses are "nonsubstantive" responses (Francis and Busch, 1975) and are among the truly uninformed or unknowledgeable cases that frequently are found in sample surveys (Press and Yang, 1974; Bishop et al., 1980; Hawkins and Coney, 1981). Additional sources include selective refusal to answer a sensitive question as well as other forms of social environmental characteristics that lead to missing responses (Guadagnoli and Cleary, 1992; Groves et al., 1992; Dillman, 2000). Pohl and Bruno (1978) consider item nonresponse to be a serious source of unrecognized nonsampling error. Ferber (1966), one of the first to investigate item nonresponse, reports that only 37% of the returns in a nationwide mail study of consumer buying plans was completely filled out. He concluded that bias from missing observations can be substantial and may foster misleadingly low estimates of standard errors, especially for large samples.

Item nonresponse has produced negative biases in surveys of medical use by the elderly. Grotzinger et al. (1994) report that compared to true population levels the elderly nonrespondents consistently underreported prescription drug use, based on a random sample of 6,500 Pennsylvania Medicare enrollees. Seventy percent were respondents and 30% nonrespondents. True values were obtained from Medicare Part A and Part B use of inpatient hospital, physician, and Medicare-covered nursing home services. Colsher and Wallace (1989) examined item nonresponse impacts on a survey of adults 65 years or older. Face-to-face interviews were conducted with 1155 men and 1942 women concerning their physical, cognitive, and psychological functioning. Don't Know responses increased with a respondent's age, lower health status, lower physical functional status, emotional symptoms, and poor cognitive functioning.

10.2.2 Item Nonresponse Among Refusals and Hard-to-Reach Respondents

There is strong evidence that nonresponse effects differ among the separate components of a sample. Donald (1960) reports that successive waves of a mail questionnaire to a sample of 2768 dues-paying members of the League of Women Voters showed a significant negative relationship between number of contacts (waves) and member involvement. Higher involvement was associated with fewer item missing values. O'Neil (1979) found that converted refusals differed significantly from nonrefusals by age, occupation, education, children in the household, and type of dwelling unit. Differences were even greater for substantive variables, indicating that large nonresponse effects are possible. Stinchcombe et al. (1981), in a survey of South Dakota farmers, report that the principal biasing effect came from survey refusals who, if persuaded to participate, may provide higher frequencies of DK responses. They conclude that bias depends less on the response rate than on the number of refusals that are not converted. Also, the accuracy of estimates of the response bias depends on the number of refusals that are converted. More recently, Chapter 11 of this book reports that there is no evidence that an offer of an incentive or refusal conversion payment increases response rates at the expense of a reduction in data quality. Incentives seem to reduce the frequency of item missing data, particularly among certain demographic groups such as older and nonwhite respondents.

Comparison of item nonresponse between refusal conversion and nonrefusals remains a vital research interest because greater item nonresponse among converted refusals may narrow the difference between these two groups when missing cases are dropped from the analysis. No difference in response means an equal survey error between the two groups would be the expected outcome.

10.2.3 Adjustment for Item Nonresponse

There has been a great deal of literature published on how to deal with the problem of nonresponse so that the most useful information can be obtained from all members of the sample. Lessler and Kalsbeek (1992, pp. 216–233), provide an extensive discussion of nonresponse and particularly discuss the numerous methods to deal with both unit and item nonresponse. Weighting methods are one approach to deal with the nonresponse problem (Little, 1992). Imputation approaches are often used in item nonresponse to fill in the missing values observed in the data collection process (Ford, 1983; Kalton and Kasprzyk, 1986; Lessler and Kalsbeek, 1992, pp. 207–233; and Särndal et al., 1992, pp. 589–594). Two types of imputation methods are used in this chapter to deal with item nonresponse and provide the capability to use complete data analysis procedures. The first approach, a "hot deck" imputation, was selected due to its wide use in survey practice. Another approach that Rubin (1987) proposed is multiple imputation. Rubin (1987, pp. 15–18) discusses a number of advantages of multiple imputations over single imputation that include the ability to account for the sampling variability of the imputed values.

10.2.4 Specification of the Voting Model

Researchers interested in predicting turnout or the division of the vote for specific candidates have generally employed one or more of three distinct theoretical approaches in their work. The first looks at the demographic characteristics of respondents, such as age, education, and income. Higher values on these variables are usually associated with higher levels of registration and voting, as well as a tendency to support Republican candidates (Wolfinger and Rosenstone, 1980). Social psychological variables that predict participation include a higher sense of personal efficacy and caring about the outcome of the election, which combine to give citizens a sense that their votes count (Downs, 1957). Strength of ideology is a predictor of participation, whereas ideology itself is a common predictor of candidate preference, with conservatives more likely to support Republican candidates. Voting also is conceived as a habitual form of behavior, one that requires frequent practice and for which each act increases the likelihood of another. For this reason, past voting behavior is an important predictor of voting in a current election.

A few studies have examined these models and how they are affected by validated and self-reported voting. Katosh and Traugott (1981) used both measures to study defection rates in the 1978 elections among a national sample of respondents. They found that defections among misreporters tended to favor the winning candidate. What has been missing from these studies is attention to the impact of attempts to reduce nonresponse on these estimates and to the relationship between predictor variables and candidate preference.

10.2.5 Effects of Sources of Nonresponse

This investigation will focus on the impact of item nonresponse on survey error among voters. If we consider total nonresponse, the respondent mean for eligible voters can be described as a function of two sources—item and unit nonresponse:

$$\bar{y}_r = \bar{y}_v + (m_1/v)(\bar{y}_r - \bar{y}_{m1}) + (m_2/v)(\bar{y}_r - \bar{y}_{m2}) \qquad (10.2.1)$$

where $\bar{y}_r$ = the response mean of a particular item for eligible voters
$\bar{y}_v$ = mean for the full sample of eligible voters
v = sample size of eligible voters
m_1/v = item nonresponse rate for eligible voters
$\bar{y}_{m1}$ = mean for item nonresponse voters
m_2/v = unit nonresponse rate for eligible voters
$\bar{y}_{m2}$ = mean for unit nonrespondent voters

Imputation is suggested as a method to reduce the m_1/v term to zero.

10.3 METHODS

Data for this study are from results of interviews conducted with adult citizens who resided in telephone households in the state of Oregon. The survey employed a random digit dialing (RDD) design by which telephone numbers were generated from the working exchanges within the state. These numbers were screened for residences, and an eligible adult was randomly selected within each household. A total of 1483 interviews were completed.

10.3.1 Design of the Survey

In the first stage of this design, each county in the state was assigned to one of four strata. The county was selected as a primary sampling unit because of our intent to validate the survey responses concerning registration and voting status in the county clerks' offices, where such records are located. The four strata were comprised of the city of Portland and its suburbs (consisting of seven counties selected with certainty); the other two Standard Metropolitan Areas in the state (consisting of Lane and Jackson counties, also selected with certainty); a random selection of three counties in the eastern section of the state (selected at random and proportional to population size); and four counties selected at random in the Willamette Valley and the western part of the state (also selected proportional to population size).

After contact was made with a residence (secondary sampling unit), a Kish grid (Kish, 1965, pp. 398–401) was used to select a designated respondent at random within the household. If the respondent was not at home, as many calls as possible were made to complete an interview with that person. If the respondent refused the interview, follow-up calls were made to convince that person to complete the interview. A total of 377 respondents, 49% of all refusals, were persuaded to complete the interview. They comprised 25.4% of our sample.

Interviews were conducted between February 1 and March 12, 1996, after the January 1996 election. Overall, the response rate for the survey was 61.4%.

The total cost of a survey involves several components. There are the fixed initial costs, such as senior staff, questionnaire design, and conversion to a CATI application, and the overhead for an interview facility. There are also variable costs that will depend upon the size of the sample and decisions made about the effort expended to contact potential respondents. Questions arise about whether the money spent for increasing response rates might be better spent elsewhere in the project (see Chapter 16, this volume). For instance, a researcher could draw a larger sample or complete more extensive pretests rather than invest more time in interviewer selection and training to reduce refusals and DK responses in the first place.

In our survey, increased costs stemmed from two sources: the higher number of contacts required to complete an interview and the greater cost of interviewers who have special skills to convert refusals. Converted refusals required an average of 7.56 calls to complete an interview while nonrefusers required only 4.01 calls, a statistically significant difference. The final replicate was purchased with names and addresses and, where possible, advance letters were sent to all sampled households. Callbacks and refusal conversion efforts were stopped when they had reached the March 12 date for terminating all interviews. Sample observations were weighted for the number of adults living in the household, the number of separate telephone numbers associated with the household, and a general nonresponse factor associated with the county in which the respondent lived. SUDAAN (Shah et al., 1996) was employed to analyze the weighted data statistically.

Eliminating the fixed initial costs for capitalizing the study to the best of our ability, the marginal cost per completed interview was $36.58 for converted refusals and $22.69 for nonrefusals—a ratio of 1.6 to 1 favoring nonrefusals. Refusal conversion claimed 35% of our interviewing expenses, but contributed 25% to the final data set. Moreover, the field period was extended to March 12, 1996 to make contact with refusals, and this may have raised problems of history or recall effects about participation in the mail vote campaign.

10.3.2 External Criteria for Evaluating Survey Data

One unusual feature of many election studies is that events in the real world associated with the election provide an additional check on the quality of survey data. This approach does not guarantee the quality of responses to any particular question since effects of question wording order, or interviewer quality, can also have an impact on the information that is collected. But calibration of survey results with external measures can increase one's confidence in the quality of the data obtained.

Vote Preferences Among Survey Respondents. In the 1996 survey, respondents were asked for whom they voted in the 1992 presidential election, and 43% said Clinton, 34% said Bush, 16% said Perot, and 7% could not remember their vote or refused to say for whom they had voted. The 1992 Oregon vote for president was 43 percent for Clinton, 33 percent for Bush, and 24 percent for Perot (Barone et al., 1999, p. 1331).

For the 1996 election, exactly the same number of respondents (41.8% of 1010 self-described voters) indicated they had cast ballots for Ron Wyden or Gordon Smith; 12.5% refused to answer the question or indicated that they could not remember their choice. When the data were weighted, each candidate received 41.9% of the vote. A total of 3.1% voted for minor candidates and 13.0% refused or said they could not remember for whom they had voted. This represents a dead heat between the two major candidates. In the election itself, Wyden received 48.4% of the votes cast and Smith received 46.8%.

Balloting Procedures by Survey Respondents. Among the self-reported voters in the survey, 85% reported mailing in their ballots and 15% indicated that they had dropped off their ballots. The secretary of state reported that 86% of the ballots were returned by mail and 14% were dropped off at a site or at the local election office.

Finally, an analysis of turnout in the special December 5, 1995 primary election to determine senate candidates for the general election in January 1996 indicated that 69% of registrants cast ballots, equivalent to 53% of the voting age population in the state. In the survey, 54% of all those interviewed indicated that they had voted in the primary election.

10.3.3 Validity for Self-Reported Registration and Voting

Names and addresses of RDD respondents were sought following the completion of each interview. Commercial reverse directory listings were compared for those who refused to supply their name or address. A total of 1283 names and addresses were obtained, reflecting 87% of the 1483 interviews completed. Eighty-five percent supplied their names and addresses when we offered them a token gift; the remainder came from commercial providers. County election clerks were able to verify 1093 individuals for their registration/voting status, 85% of the names and addresses we had at hand.

A total of 68% reported their voter registration and voting accurately. Another 13% were accurate in telling us they were not registered and had not voted. Nine percent misreported their registration, and six out of 10 in that group (60 individuals) misreported that they had voted. Another 2% said they were not registered when they actually were. The validity of the responses for another 8% of the sample could not be checked. In total, about 81% of the sample was accurate in their turnout responses, 11% was not, and we were unable to verify the remaining 8%.

10.3.4 Impact of Item Nonresponse on the Prediction of Candidate Choice

We fitted a linear logic regression model for binary responses to the data in three stages. The first model deleted all observations with missing values for any of the independent variables and the second and third used "hot deck" and multiple imputation methods, respectively, to adjust the data for missing values. Validity scores for registration, voting, and for the type of response group (converted and nonrefusals) were added to the set of explanatory variables as covariates. Observed and predicted estimates for the response variable are reported for converted and nonrefusals to determine if prediction errors differ significantly between the two groups.

10.3.5 Imputation

For both types of imputation, auxiliary variables were selected from the data set that were highly related to the variables to impute.* The SOLAS (1997) software package was used to create the imputed data sets for both the hot deck and multiple imputation analyses. For the multiple imputation approach, 10 data sets were created. Once the complete data sets were created, the logistic models discussed in Section 10.3.4 were rerun with SUDAAN to obtain the results provided by Rubin (1987, pp. 76–77). In order to obtain an approximate estimate of the mean square error, simulation data sets were created (1000 samples for each data set), and the model was fit with each data set. The bootstrap estimators for the mean square error are also presented in Table 10.4.2 of Davidson and Hinkley (1997) (see also Chapter 25 of that reference).

*Auxiliary variables for the "hot deck" imputation included Level of Education (for Care about the Election Outcome) (Wolfinger and Rosenstone, 1980, pp. 17–28); Strength of Party Identification (for Strength of Ideology) (Kelly and Mirer, 1974); and Years Lived in Oregon (for Past Voting) (Converse and Markus, 1979). Two additional auxiliary variables were included for the multiple imputation treatment: Strength of Party Identification (for Care about the Election Outcome) (Campbell et al., 1960, pp. 141–144); and Political Efficacy (for Past Voting) (Campbell et al., 1960, pp. 63–65).

10.4 RESULTS AND DISCUSSION

Efforts to improve response rates through refusal conversions may well increase the frequency of missing values to substantive questions. This section presents information for the total Oregon sample on, first, the pattern of item nonresponse observed for the dependent variable, for the independent variables or predictors in the voter model, and for the covariate. Second, the proportion of missing items among self-reported voters is summarized. Third, the demographic correlates (age, education, and gender) to the proportion of missing items among voters are reported. Fourth, the results of a logistic regression analysis of the vote model are presented. Finally, the difference in error between refusal conversion and nonrefusal groups is compared for imputation adjustments of missing cases.

10.4.1 The Pattern of Item Nonresponse, Total Sample

The Dependent Variable, Candidate Choice for Senator. Missing values stemmed from two sources—candidate choice and voter turnout. A total of 134 cases represented respondents who would not or could not tell us their choice for a new U.S. senator. Thirty-eight cases were missing from responses to the turnout items. A total of 436 cases represent self-reported nonvoters; 31 cases were voters for minor candidates, and 422 each reported voting for Gordon Smith or Ron Wyden. Since the focus of our analysis is on respondents who voted for one of the two major candidates, the number of missing cases for candidate choice, 134 observations, represents 13% of the available data of 978 cases.

The voting model considers only respondents who said they had voted for either Gordon Smith or Ron Wyden, a total of 844 cases. The difference between the proportion of converted and nonrefusals clearly shows that converted refusals bring in nearly 1.6 more missing voter choice cases compared to nonrefusals. Nineteen percent of the converted refusals did not respond to the vote question, compared to 12% of the nonrefusers ($p = 0.01$), so missing cases from converted refusals are greater for the response variable.

The Independent Variables. The cumulative total of missing values for the independent variables in the model (educational level, efficacy, strength of ideology, care about the outcome of the election, and voting history) was 151 cases, 15% of the available data. Only 31 of these cases had missing values for both the dependent and independent variables. A total of 254 cases, 26% of the available data, represented net missing values for the dependent and independent variables.

Validity of Self-Reported Voting. The largest number of missing data came from this covariate term. Nearly a third of the 978 cases available for analysis had a missing observation on the covariate. A total of 602 were confirmed as valid voters and 58 as valid nonvoters. The remaining 318 could not be verified as either voters or nonvoters from the validation operation.

Adding the net total of 166 missing cases from the independent and dependent variables in the voting model give us a cumulative total of 493 missing observations, 50% of the available data. The magnitude of missing information from the covariate suggests to us that the covariate term should be dropped from the analysis if the coefficient for the term is not statistically significant when the voting model is tested. Dropping the covariate would provide a net gain of 239 cases for analysis.

10.4.2 Nonresponse Sorted by Candidate Choice

The pattern of nonresponse reported in the previous section is based on 978 cases that remained after voters for minor candidates, self-reported nonvoters, and missing turnout cases were deleted. The voting model considered only respondents who said they had voted for either Gordon Smith or Ron Wyden.

Missing observations totaled 359 cases, 42% of the available data (844 cases). This is a slight decrease from the 50% reported for the total sample. Thirty-two percent stems from the covariate and 10% from the independent variables.

Among Smith/Wyden voters, there were twice as many missing observations (24%) among converted refusals than among nonrefusals (11%). The difference is statistically significant at the $p < 0.01$ level. This resulted in a net of 724 cases available for testing the voting model.

10.4.3 Relationship between Demographic Characteristics and Efforts to Convert Refusals

The analysis so far has focused on the variables associated with item nonresponse. We don't want to leave this section without reporting on the effect of education, age, and gender on refusal conversion proportions. There were no significant education or gender effects on refusal conversions. The tabular results suggested a linear trend for age groupings based on the total sample. Higher proportions of refusal conversions are observed among voters in the older age groups. Thirty-one percent of the voters 65 and older were converted refusals, 22% of those 50–64, 20% of those 40–49, 19% of those 30–39, and 18% of those 18–29 ($p < 0.01$). The pattern of refusal conversions, therefore, is not randomly distributed among the electorate but appears to be more successful among older voters.

10.4.4 The Impact of Refusal Conversion on Candidate Choice

The possibility remains that cases dropped from the study due to missing observations from refusal conversion favored one candidate over another. We examined this possibility first through cross-tabulations between refusal groups for voters who said they chose Gordon Smith or Ron Wyden, and we found no difference. A 50/50 split was observed for converted refusals; a 51/49 split among nonrefusals ($p = 0.939$).

Refusal conversion is related to the proportion of missing data, so a log linear analysis was completed to examine the interaction between the two refusal groups, the presence or absence of missing cases, and vote choice. Here, the full sample of 844 cases was employed in the analysis. The expected interaction between the presence of missing information and the refusal conversion group was found, but there was no interaction between the presence of missing cases and vote choice, or between converted refusal groups, presence of missing information, and the choice for senator. There appears to be no empirical evidence to support the notion that attempts to increase response rates by converting refusals favored one candidate over another.

10.4.5 Logistic Regression Analysis of the Vote Model

A model was tested with predictors composed of an individual's educational level, sense of personal efficacy about government, caring about the outcome of the election, strength of liberal or conservative ideology, and voting in past elections. Validation of the person's voting served as a covariate and the individual's refusal/nonrefusal group was added as an indicator variable. A vote for Smith or Wyden was the response category for the dependent variable.

Only three predictors—strength of ideology, caring about the election, and past voting—significantly predicted vote choice. These predictors were employed in a reduced logit model to compare imputation adjustments to the data for the error associated with missing values. The magnitude of error was calculated by summing the proportion of cases in the two off-diagonal "error" cells of a four-fold table formed by crossing predicted outcomes of the model with observed reports of the vote. Error cells (i.e., prediction error) contain those cases in which a Smith vote is predicted but a Wyden vote is observed and vice versa. (See columns four and five of Table 10.1.)

Basically, the results show only slight, nonsignificant difference of proportions in the error cells between converted and nonrefusal groups, in either magnitude or

Table 10.1. Proportion of predicted by observed candidate choices for nonrefusals and converted refusals, by treatment of missing observations

	Predicted × observed cell proportions					
Treatment	Smith/ Smith	Wyden/ Wyden	Smith/ Wyden	Wyden/ Smith	Total	(n)
Converted refusals:						
Missing cases deleted	0.309	0.329	0.168	0.194	1.000	149
"Hot deck" imputation	0.303	0.308	0.197	0.192	1.000	188
Multiple imputation	0.314	0.311	0.186	0.189	1.000	185
Nonrefusals:						
Missing cases deleted	0.308	0.311	0.184	0.197	1.000	588
"Hot deck" imputation	0.308	0.317	0.181	0.194	1.000	656
Multiple imputation	0.316	0.311	0.187	0.186	1.000	653

direction for the three treatments. For the cases in which missing observations were deleted, the difference in error proportions between nonrefusals and converted refusals, 0.0186, is not significant ($p = 0.70$).

10.4.6 Imputation of Missing Values

Three approaches to handle the missing values are compared. The first method is to simply drop any observations that contain missing values from the data analysis. The second approach is to use a "hot deck" imputation, and the third is to use a multiple imputation procedure for the missing cases. Details of these procedures are discussed in Section 10.3.5. The error that was quantified under these approaches and discussed in Section 10.4.5 is shown in Table 10.2.

The results show that the error is largest for the case that had the smallest sample size and where missing values had been deleted. The variance of the difference in error underscores the impact of increased sample sizes for imputed approaches, i.e., the variance decreases as sample size increases (Rubin, 1987; also see Chapters 1 and 22 of this volume). We recognize that the imputed estimates are biased. In order to evaluate the estimates in a way that accounts for the bias introduced by imputation, mean square errors, as discussed in Section 10.3.5, were calculated using a bootstrap approach (Davidson and Hinkley, 1997; Xie and Paik, 1997). The results show that the imputed mean square errors are about half of what was obtained when we simply dropped missing cases from the data and implies an improved accuracy for the imputed estimates.

10.5 SUMMARY AND CONCLUSIONS

The results of the analysis presented here show that the pursuit of respondents through extraordinary attempts to convert refusals increases unit response rates by 15.6 percentage points—from 45.8% to 61.4%. The 377 new cases from converted refusals increased the total sample by 25%. Item nonresponse among the explanatory variables for Smith/Wyden voters occurred in nearly a fourth of the con-

Table 10.2. Comparison of nonsampling error proportions and variances for the difference between nonrefusals and converted refusals, by treatment of missing observations

	Mean square treatment error	Total (n)	Difference in nonrefusal–converted refusal nonsampling error*	Variance of difference
Missing cases deleted	0.0071	737	0.0186	0.0071
"Hot deck" imputation	0.0037	844	–0.0133	0.0016
Multiple imputation	0.0018	838	–0.0028	0.0014

*Differences vary slightly from those reported in Table 10.1 due to rounding error.

verted refusals, compared to 11% for nonrefusers. Through imputation of missing observations, the power of the statistical tests of a vote model based on the additional cases that otherwise are dropped from the analysis is increased, thereby reducing the negative impact of greater item nonresponse that stems from refusal conversion.

In studies of voting behavior that involve turnout and candidate preference, factors that are positively associated with voting are negatively related to the proportion of item nonresponse. That is to say, those with higher levels of education and women are all more likely to vote and less likely to refuse to provide useful answers to survey questions about voting. It may be that difficult-to-reach respondents are less likely to be voters; hence, their absence from a survey does not affect estimates of turnout or candidate preference to any significant degree.

The expenditure of effort and associated costs to boost sample size through refusal conversion would be justified if these additional respondents were contributing useful data for statistical analysis. However, most statistical analysis programs would eliminate these cases because of item nonresponse. Our analysis suggests this problem can be overcome by imputing values for the missing information that stems from item nonresponse.

Interviewing reluctant respondents cost about 1.6 times more per observation than interviewing respondents who require no special persuasion. Converted refusals took 35% of our budget for interviews, but provided only 25% of cases for the sample. Part of the tradeoff is asking oneself if the money could be better spent drawing a larger sample, doing more detailed pretesting of alternative questions, and devoting more effort to the selection and training of interviewers, rather than for chasing reluctant respondents.

The results of our analysis pose an interesting dilemma for survey researchers. The use of additional resources in the data collection phase of a study can generate information for more cases through refusal conversion. At the same time, marginal costs rise and the level of item nonresponse does as well. Such an expenditure of resources to boost the sample may not end up having the desired effect. On the other hand, the use of an imputation technique to convert missing data to useful values will not only increase the analytical sample but may also reduce the errors around estimates for individual values. As imputation algorithms become more sophisticated and more readily and easily available, the decision about how much effort and resources to devote to pursuing respondents during data collection may swing more favorably in that direction.

Our research does not suggest that either persistence in pursuing reluctant respondents is unwarranted or that costs should be transferred from such efforts to simply increasing sample size. In the first place, there is no widely accepted model or general theory to explain why or when such tradeoffs would make a difference. Furthermore, it is conceivable that surveys of voters may represent an unusual set of circumstances whereby the correlates of voting (nonvoting) are also those of survey participation (nonparticipation). But it does open the possibility for a fruitful area of further study of how survey budgets should be allocated.

ACKNOWLEDGMENTS

The data used in the analysis was collected under grants from the Pew Charitable Trust, The Ford Foundation, the Northwest Area Foundation, and the Ralph L. Smith Foundation. The analyses and interpretations are the sole responsibility of the authors. The contribution of David Birkes, Department of Statistics at Oregon State University, is gratefully acknowledged.

CHAPTER 11

The Use of Incentives to Reduce Nonresponse in Household Surveys

Eleanor Singer, *University of Michigan*

11.1 INTRODUCTION

A large number of experiments in recent years have manipulated incentives in telephone and face-to-face surveys, complementing the earlier, and even more voluminous, research literature on the effects of incentives in mail surveys. These experiments are based loosely on various forms of exchange theory (e.g., Adams, 1965; Berger et al., 1972; Homans, 1961, 1974), though many are quite atheoretical. This chapter synthesizes findings about the effects of incentives from both interviewer-mediated and mail surveys in order to increase their usefulness for practicing survey researchers and to stimulate more targeted research by survey methodologists.

The chapter reviews what is known about the intended effects of incentives on response rates in both types of surveys, drawing on two existing metaanalyses (Church, 1993; Singer et al., 1999a) as well as on subsequent work by the same and other authors. It also reviews what is known about such unintended consequences of incentives as effects on response quality and sample composition, concerns about equity, and the development of expectation effects (Groves et al., 1999; Singer et al., 1999b; Singer, 1998, 2000). Finally, it discusses issues of cost effectiveness and suggests avenues for further research. Because of the additional complexities introduced in organizational surveys, they are not included in the present review.

11.2 WHY DO PEOPLE PARTICIPATE IN SURVEYS?

Porst and von Briel (1995) point out that although a great deal is known about survey respondents—their demographic characteristics as well as their answers to thousands of different survey questions—very little is known about why they choose to participate. Based on a content analysis of open-ended responses, their study of 140 participants in five waves of a German methods panel identifies three pure types of participants: those who respond for altruistic reasons (e.g., the survey is useful for some purpose important to the respondent, or the respondent is fulfilling a social obligation—31% of respondents); those who respond for survey-related reasons (e.g., they are interested in the survey topic, or find the interviewer appealing—38%); and those who cite what the authors call personal reasons (e.g., they promised to do it—30%).

More recently, Groves et al. (2000) outlined a theory describing the decision to participate in a survey as the interactive and additive resultant of a series of factors. Some factors are survey-specific, such as topic and sponsorship; others person-specific, such as concerns about privacy; still others are specific to the respondent's social and physical environment; and each is associated with a weight and direction for a given person's decision, moving him or her toward or away from cooperation with a specific survey request.

From this perspective, monetary as well as nonmonetary incentives are an inducement offered by the survey designer to compensate for the absence of factors that might otherwise stimulate cooperation—e.g., interest in the topic of the survey or a sense of civic obligation. Other theoretical frameworks such as social exchange theory (cf. Dillman, 1978), the norm of reciprocity (Gouldner, 1960), and economic exchange (e.g., Biner and Kidd, 1994) can also be used to explain the effectiveness of incentives, and, in particular, the greater effectiveness of prepaid over promised incentives. However, the present perspective is able to account for the differential effects of incentives under different conditions (e.g., for respondents with differing interest in the survey topic or with different degrees of community activism) in a way that other theories cannot easily do.

11.3 INTENDED EFFECTS ON RESPONSE RATES

In an effort to counter increasing tendencies toward noncooperation, survey organizations are offering incentives to respondents with increasing frequency, some at the outset of the survey, as has traditionally been done in mail surveys, and some only after the person has refused, in an attempt to convert the refusal.

The use of incentives has a long history in mail surveys (for reviews, see Armstrong, 1975; Church, 1993; Cox, 1976; Fox et al., 1988; Heberlein and Baumgartner, 1978; Kanuk and Berenson, 1975; Levine and Gordon, 1958; Linsky, 1975; Yu and Cooper, 1983). In such surveys, incentives are one of two factors, the other being number of contacts, which have consistently been found to increase response rates.

A metaanalysis of the experimental literature on the effects of incentives in mail surveys (Church, 1993) classifies incentives along two dimensions: whether the incentive is a monetary or nonmonetary reward, and whether it is offered with the initial mailing or made contingent on the return of the questionnaire. Analyzing 38 studies (yielding 74 comparisons between incentive and control conditions), Church concluded that:

- Prepaid incentives yield significantly higher response rates whereas contingent (promised) incentives do not.
- Prepaid monetary incentives yield higher response rates than gifts offered with the initial mailing.
- Response rates increase with increasing amounts of money.

In Church's (1993) analysis, studies using prepaid monetary incentives yielded an average increase in response rates of 19.1 percentage points, representing a 65% average increase in response. Gifts, on the other hand, yielded an average increase of 7.9 percentage points. The average value of the monetary incentive was $1.38; the average value of the gift could not be computed, given the great diversity of gifts offered and the absence of information on their cost. Results similar to those of Church are reported by Hopkins and Gullikson (1992).

Incentives are also increasingly used in telephone and face-to-face surveys, and the question arises whether their effects differ from those found consistently in mail surveys. A metaanalysis of 39 experiments in interviewer-mediated surveys by Singer et al. (1999a) indicates that they do not, although the percentage point gains per dollar expended are generally smaller than those reported by Church. The main findings from the 1999 study are as follows:

- Incentives improve response rates in telephone and face-to-face surveys, and their effect does not differ by mode of interviewing. As expected, however, effects are much smaller than in mail surveys: on average, each dollar of incentive paid results in about a third of a percentage point difference between the incentive and the zero incentive condition. As in the analyses by Church (1993) and Yu and Cooper (1983), the effects of incentives are linear.
- Prepaid incentives do not differ significantly from promised incentives, although the differences are in the expected direction. However, most studies in the metaanalysis used promised incentives; in the four studies in which prepaid incentives were compared directly with promised ones, prepaid incentives resulted in significantly higher response rates.
- Money is more effective than a gift, even controlling for the value of the incentive.
- Increasing the burden of the interview increases the difference in response rates between an incentive and a zero-incentive condition. However, incentives have a significant effect even in low-burden studies.
- Incentives have significantly greater effects in studies where the response rate without an incentive is low. That is, they are especially useful in compensating for the absence of other motives to participate.

Some studies have found that the effect of incentives diminishes in mail surveys after repeated follow-ups (e.g., Hopkins et al., 1988; Brennan et al., 1991; James and Bolstein, 1992; Shettle and Mooney, 1999). However, Baumgartner and Rathbun (1997) found no reduction in the incentive effect, despite implementation of a procedure modeled on Dillman's (1978) "Total Design Method." Shettle and Mooney's (1999) experiment, involving a mail survey of college graduates that included telephone and personal follow-ups, suggests that incentives work primarily by reducing refusals, rather than by reducing the noncontact rate. Similar findings are reported for the Survey of Consumer Attitudes (SCA), a random-digit-dialed (RDD) national survey of the general population with an N of 500 carried out monthly by the University of Michigan Survey Research Center (Singer et al., 2000).

11.3.1 Lotteries as Incentives

Some researchers, convinced of the value of incentives but reluctant to use prepaid incentives for all respondents, have advocated the use of lotteries as an incentive for stimulating response. The studies reported in the literature—all mail surveys or self-administered questionnaires distributed in person—have yielded inconsistent findings [e.g., positive effects by Balakrishnan et al. (1992), Hubbard and Little (1988), Kim et al. (1996), and McCool (1991); no effects in four studies reviewed by Hubbard and Little (1988) or in the experiment by Warriner et al. (1996)]. A reasonable hypothesis for further testing would seem to be that lotteries function as promised cash incentives with an expected value per respondent (e.g., a $500 prize divided by 10,000 respondents would amount to an incentive of five cents per respondent). Their effect on response rates would be predicted by this value as well as the fact that they are promised rather than prepaid.

11.3.2 Incentives in Panel Studies

Assuring participation is especially important for panel studies, since participation at baseline usually sets a ceiling for the retention rate over the life of the panel. Although some investigators (see, e.g., Presser, 1989) recommend returning to nonrespondents in subsequent waves, this is not often done. Even when it is, cooperation on a subsequent wave is generally predicted by prior cooperation. For this reason, investigators often advocate sizable incentives at the first wave of a panel study.

Analyzing the results of an incentive experiment on Wave 1 of the 1996 Survey of Income and Program Participation (SIPP), a longitudinal survey carried out by the U.S. Census Bureau to provide national estimates of sources, amounts, and determinants of income for households, families, and persons, James (1997) found that the $20 prepaid incentive significantly lowered nonresponse rates in Waves 1–3 compared with both the $10 prepaid and the $0 conditions, but that the $10 incentive showed no effect relative to the zero-incentive group. Mack et al. (1998), reporting on the results through Wave 6 using cumulative response rates, found that an incentive of $20 reduced household, person, and item (gross wages) nonresponse rates in the initial interview and that household nonresponse rates remained significantly lower, with a cumulative 27.6% nonresponse rate in the $0 incentive group, 26.7% in the $10 group, and 24.8% in the $20 group at Wave 6, even though no further incentive payments were made. (SIPP does not attempt to reinterview households that do not respond in Wave 1 or that have two consecutive noninterviews.) Differences between the $10 incentive and the no-incentive group were not statistically significant.

Research on the Health and Retirement Survey (HRS) suggests that respondents who are paid a refusal conversion incentive during one wave do not refuse at a higher rate than other converted refusers when reinterviewed during the next wave (Lengacher et al., 1995).

11.3.3 Summary of Intended Effects

Incentives appear to accomplish their intended effect of increasing response rates to all kinds of surveys, though the size of the effect varies by mode and by other factors affecting willingness to respond. Prepaid incentives are clearly more effective than contingent ones.

There is no evidence that incentives are helpful in making contact with respondents in an RDD survey, nor any theoretical justification for believing that they would be. Thus, if the primary problem is one of finding people at home for such a survey, incentives may not be very useful. However, an experiment by Kerachsky and Mallar (1981) with a sample of economically disadvantaged youths suggests that prepayment may be helpful in locating members of a list sample, especially in later waves of a longitudinal survey. One reason, apparently, is that prepayment (and perhaps promised incentives from a trusted source) may be useful in persuading friends or relatives to forward the survey organization's advance letter or to provide interviewers with a current telephone number for the designated respondent.

11.4 UNINTENDED CONSEQUENCES OF INCENTIVES

11.4.1 Effects on Item Nonresponse

One question often raised about the use of incentives in surveys is whether they bring about an increase in response rate at the expense of response quality. This does not appear to be the case. On the contrary, what evidence there is suggests that the quality of responses given by respondents who receive a prepaid or a refusal conversion incentive does not differ from responses given by those who do not receive an incentive. They may, in fact, have less item-missing data and provide longer open-ended responses (Baumgartner et al., 1998; James and Bolstein, 1990; Singer et al., 2000; Shettle and Mooney, 1999; but cf. Wiese, 1998). Experiments reported by Singer, Van Hoewyk, and Maher (2000) indicate that promised and prepaid incentives reduce the tendency of older people and nonwhites to have more

Don't Know responses and item missing data, resulting in a net reduction in item nonresponse. However, incentives alone explained less than one percent of the variance in item nonresponse in that study.

Findings reported by Mason, Lesser and Traugott in Chapter 10 of this volume suggest that persistent efforts to persuade reluctant respondents to participate may produce more respondents at the price of more missing data. But these authors did not use incentives, and motivational theory suggests that people who are rewarded for their participation would continue to give good information, whereas those who feel harassed into participation may well retaliate by not putting much effort into their answers.

11.4.2 Effects on Response Distributions

Even more troubling, potentially, than an effect on item missing data is the effect of incentives on the distribution of responses. Does offering or paying incentives to people who might otherwise refuse affect their answers to the survey questions?

One reason that incentives might influence response distributions is that they bring into the sample people whose characteristics differ from those who would otherwise be included, and their answers differ because of those differing characteristics. If that is the case, the apparent effect on response distributions is really due to a change in the composition of the sample, and should disappear once the appropriate characteristics are controlled. An example of this process is presented by Berlin et al. (1992), who demonstrate that the apparent effect of a monetary incentive on literacy scores can be accounted for by the disproportionate recruitment of respondents with higher educational levels into the zero-incentive group. There was no significant relationship between incentive level and the proportion of items attempted, indicating that the incentive influenced the decision to participate, but not performance on the test.

A second reason incentives might influence responses is if they have a direct influence on the expression of opinions. A striking example of such influence involving sponsorship is reported by Bischoping and Schuman (1992) in their analysis of discrepancies among Nicaraguan preelection polls in the 1990 election. Bishoping and Schuman speculate that suspicions that preelection polls had partisan aims may have prevented many Nicaraguans from candidly expressing their voting intentions to interviewers, and tested this hypothesis by having interviewers alternate the use of three different pens to record responses: one carried the slogan of the Sandinista party; another, that of the opposition party; the third pen was neutral. The expected distortions of responses were observed in the two conditions that clearly identified the interviewers as partisan. Even in the third, neutral, condition, however, distortion occurred, because polls were apparently not perceived as neutral by many respondents. In the Nicaraguan setting, after a decade of Sandinista rule, a poll lacking partisan identification was evidently regarded as likely to have an FSLN (Sandinista) connection (p. 346); the result was to bias the reporting of vote intentions, and therefore the results of the preelection polls, which predicted an overwhelming Sandinista victory when in fact the opposition candidate won by a large majority.

Still a third way in which incentives might affect responses is suggested by theory and experimental findings about the effects of mood (Schwarz and Clore, 1996). If incentives put respondents in a more optimistic mood, then some of their responses may be influenced as a result. Using 17 key variables included on the SCA, Singer, Van Hoewyk, and Maher (2000) looked at whether the response distributions varied significantly by (a) the initial incentive or (b) refusal conversion payments.

The offer of an initial incentive was associated with significantly different response distributions (at the 0.05 level) on four of the 17 variables; a refusal conversion payment was also associated with significantly different response distributions on four of them. One variable was significantly affected by both types of incentives. In five of these cases, the responses given with an incentive were more optimistic than those given without an incentive. In two cases, they were more pessimistic. In the remaining case, respondents who received an incentive were somewhat more likely to respond "good" and "bad," and somewhat less likely to give an equivocal reply. The effects do not disappear with controls for demographic characteristics; indeed, three additional variables show such effects with such controls. Thus, there is a suggestion that respondents to the Survey of Consumer Attitudes who receive an incentive may give somewhat more optimistic responses than those who do not. Similar findings have been reported by Brehm (1994a). James and Bolstein (1990) reported that respondents receiving an incentive on a mail survey made more favorable comments about the survey sponsor than those who did not, but did not differ in their responses to fixed-alternative survey questions. Nor did Shettle and Mooney (1999) find significant differences in responses in their experimental investigation of incentives in a mail survey of college graduates.

11.4.3 Effects on Sample Composition

Whether to think about the effects of incentives on sample composition as "intended" or "unintended" consequences depends on whether they exacerbate or counteract tendencies to underrepresent certain subgroups of the population. In a 1994 paper presented to a COPAFS workshop, Kulka reported some evidence suggesting that monetary incentives might be especially effective in recruiting into the sample low-income and minority respondents, groups that ordinarily would be underrepresented in a probability sample (Goyder and Warriner, 1999, and the sources cited there; but cf. Groves and Couper, 1998). Reviewing a number of experimental interviewer-mediated studies that provided evidence on the issue of sample composition, including the studies discussed by Kulka, Singer et al. (1999) found that in three such studies there was an indication that paying an incentive might be useful in obtaining higher numbers of respondents in demographic categories that otherwise tend to be underrepresented in sample surveys (e.g., low income or nonwhite race). Five other studies reported no significant effects of incentives on sample composition, and in one study the results were mixed. An early review of the mail literature by Kanuk and Berenson (1975) reported similarly mixed results.

Since then, additional evidence has accumulated suggesting that monetary incentives can be especially effective in recruiting and retaining minority respondents in interviewer-mediated studies. Mack et al. (1998) found that the use of a $20 incentive in the first wave of a SIPP panel was much more effective in recruiting and retaining black households and households in poverty that it was in recruiting and retaining nonblack and nonpoverty households. And a subsequent study by Abreu, Martin, and Winters (1999) found that an incentive of $20 was more effective in recruiting black nonrespondents to a previous wave of a SIPP panel than it was in recruiting white nonrespondents. Both sets of results are in agreement with findings reported by Juster and Suzman (1995). They report that a special Nonresponse Study, in which a sample of people who refused "normal" refusal conversion efforts on the Health and Retirement Survey were offered $100 per individual or $200 per couple to participate, brought into the sample a group of people distinctly different from other participants: they were more likely to be married, in better health, and, particularly, they had about 25% more net worth and a 16% higher income than other refusal conversion households or those who never refused. (In that study, all nonrespondents were sent the incentive offer by FedEx mail; hence, it was not possible to separate the effect of the monetary incentive from the special mailing. In a subsequent small-scale experiment, money had a significant effect on converting refusals, whereas a FedEx mailing did not (Dan Hill, personal communication). Finally, analyses by Singer, Van Hoewyk, and Maher (2000) indicate that a $5 incentive paid in advance to a random half of RDD households for which an address could be located brought a disproportionate number of low-education respondents into the sample; there were no significant differences on other demographic characteristics.

In other words, these studies suggest that, although monetary incentives are effective with all respondents, less money is required to recruit and retain low-income (and minority) groups than those whose income is higher and for whom the tradeoff between the time required for the survey and the incentive offered may be less attractive when the incentive is "small." It should be noted that few, if any, of these studies [Mack (1998) is a notable exception] have explicitly manipulated both the size of the incentive and the income level of the population; the findings reported above are based on ex post facto analyses for different subgroups, or on analyses of the composition of the sample following the use of incentives.

A number of other studies have also reported on the effects of incentives on sample composition. In some of these, it appears that incentives can be used to compensate for lack of salience of, or interest in, the survey by some groups in the sample. For example, the survey reported on by Shettle and Mooney (1999), the National Survey of College Graduates, is believed to be much more salient to scientists and engineers than to other college graduates, and in the 1980s the latter had a much lower response rate. Although this was also true in the 1992 pretest for the 1993 survey, the bias was less in the incentive than in the nonincentive group, though not significantly so. Similar findings are reported by Baumgartner and Rathbun (1997), who found a significant impact of incentives on response rate in the group for which the survey topic had little salience, but virtually no impact in the high-salience group, and by Martinez-Ebers (1997), whose findings suggest that a $5 incentive, enclosed with a mail questionnaire, was successful in motivating less-satisfied parents to continue their participation in a school-sponsored panel survey. Berlin et al. (1992) found that people with higher scores on an assessment of adult literacy, as well as people with higher educational levels, were overrepresented in their zero-incentive group. Groves, Singer, and Corning (1999) also reported a similar result; in their study, the impact of incentives on response rates was significantly greater for people low on a measure of community involvement than for those high on community involvement, who tended to participate at a higher rate even without monetary incentives. In these studies, incentives function by raising the response rate of those with little interest, or low civic involvement; they do not reduce the level of participation of the highly interested or more altruistic groups.

In all of these studies, certain kinds of dependent variables would be seriously mismeasured if incentives had not been used. And the theory of survey participation outlined at the beginning of this paper (Groves et al., 2000) suggests that the representativeness of the sample will be increased by using a variety of motivational techniques, rather than relying on a single one.

11.4.4 Expectation Effects

Effects on Interviewers. Are the consistent effects of incentives in telephone and face-to-face interviews attributable to their effect on respondents, or are they, perhaps, mediated by their effect on interviewers? Clearly, this question does not arise with respect to mail surveys, where incentives have also been consistently effective; but it seemed important to try to answer it with respect to interviewer-mediated surveys. It is possible, for example, that interviewers expect respondents who have received an incentive to be more cooperative, and that they behave in such a way as to fulfill their expectations. (For evidence concerning interviewer expectation effects, see Hyman, 1954; Sudman et al., 1977; Singer and Kohnke-Aguirre, 1979; and Singer et al., 1983; but cf. Lynn, 1999.) Or, they may feel more confident about approaching a household that has received an incentive in the mail, and therefore be more effective in their interaction with the potential respondent.

In order to separate the effects of incentives on interviewers from their effects on respondents, Singer, Van Hoewyk, and Maher (2000) randomly divided all respondents for whom addresses could be obtained into three groups. One third of the group for whom addresses could be obtained were sent an advance letter and $5; interviewers were kept blind to this condition. Another third also received a letter plus $5, and still another third received the letter only. Interviewers were made aware of these last two conditions by information presented on their Computer-Assisted Telephone Interview (CATI) screens.

Large differences were observed between the letter-only and the letter-plus-incentive conditions, but there is no evidence that this is due to the effect of incentives on interviewers. None of the differences between the condition in which interviewers were aware of the incentive and those in which they were not reached statistical significance. Thus, prepayment of a $5 incentive substantially increases cooperation with an RDD survey, and the incentive appears to exert its effect directly on the respondent rather than being mediated through interviewer expectations. This conclusion is in accordance with research by Stanley Presser and Johnny Blair, at the University of Maryland, who also found substantial increases in response rates as a result of small prepayments to respondents to which interviewers were blind (personal communication).

Effects on Respondents. There are concerns that the payment of incentives, especially prepayment, will create expectations for future payment on the part of respondents. The effect of incentives may be direct—that is, arouse expectations on the part of respondents for payment the next time they are asked to participate in a survey—or both direct and indirect, creating a climate that affects even those members of the public who have not themselves been paid for their cooperation. In 1998, Singer, Van Hoewyk, and Maher reported that although people who had received a monetary incentive in the past were significantly more likely to endorse the statement that "people should be paid for doing surveys like this" than those who had not, they were actually more likely to participate in a reinterview 6 months later, in spite of receiving no further payments. Subsequently, the researchers extended this analysis by pooling experiments over 15 months. This permitted them to examine separately the effect of incentives offered at the outset of a survey and of refusal conversion payments, and to control for the demographic characteristics of respondents and for interactions between incentives and age, nonwhite race, Hispanic ethnicity, education, female gender, and income. The effects of both kinds of incentives on response rates 6 months later were negative, but neither effect was significant.

The question remained, however, whether the absence of a significant effect of Time 1 incentives on Time 2 cooperation was because respondents construed the payment as covering both the reinterview (by the same survey organization) and the initial interview, and whether the same results would be obtained if they were approached by a different survey organization. An experiment examining this question, reported by Singer, Groves and Corning (1999), suggests that they would, but respondents may not have construed the second request as coming from a different organization. Thus, this finding is clearly in need of replication.

11.4.5 Summary of Unintended Effects

Incentives do not appear to increase response rates at the expense of response quality, though research is needed about their effects on reliability and validity. However, there are indications that they do sometimes affect the distribution of responses as well as the composition of the sample, and more research is needed on the conditions under which these effects occur.

Incentives appear to increase response rates through their effect on respondents, not interviewers. There is no evidence that incentives create respondent expectations for future payment on the same survey, but whether they create expectations for payment on unrelated surveys remains an open question, and may depend on the size of the incentive and the recency of the experience.

11.5 ISSUES IN THE USE OF DIFFERENTIAL INCENTIVES

Some of the research reported above suggests that it may make economic sense to offer lower incentives to people with lower incomes and higher incentives to those who are economically better off. Another instance of differential incentives is the use of refusal conversion payments, in which respondents who have expressed reluctance, or who have actually refused, are offered payment for their participation but cooperative respondents are not. In both of these situations, the question arises how respondents who received lower, or no, rewards would feel if they learned of this practice, and how this might affect their future participation in this or another survey.

From an economic perspective, the fact that some people refuse to be interviewed may be an indication that the survey is more burdensome for them and that therefore the payment of incentives to such respondents (but not others) is justified. Nevertheless, some researchers are concerned that using incentives in this way will be perceived as inequitable by cooperative respondents, and, that, if they learn of the practice, this will adversely affect their willingness to cooperate in future surveys (Kulka, 1994).

These unintended consequences were the focus of two studies (Groves et al., 1999a; Singer et al., 1999b). The first was done in the laboratory with community volunteers, using self-administered responses to videotaped vignettes. The second was conducted as part of the Detroit Area Study, using face-to-face interviews.

The laboratory experiment resulted in significant negative effects of disclosing differential incentives to subjects on their expressed willingness to participate in the survey. However, this finding was not replicated in the field study. Most respondents in that study (75%) believed survey organizations are currently using incentives to encourage survey participation, and these beliefs were affected by personal experience, so that those who themselves received an incentive in the survey were much more likely to believe this was a common practice. Only half of those who were aware of the use of incentives believed that payments are distributed equally to all respondents; and a large majority of respondents—74%—considered the practice of paying differential incentives unfair.

However, disclosure of differential payments to a random half of the respondents had no significant effect on their expressed willingness to participate in a future survey. About a quarter of each group said they would "definitely" be willing to participate in another survey by the same organization. Even those to whom differential payments were disclosed and who perceived these payments as unfair did not differ significantly in their expressed willingness to participate in another survey by the same organization, although the trend in responses was as predicted: 25.8% versus 32.8% expressed such willingness. Nor were respondents to whom differential incentives had been disclosed significantly less likely to respond to a new survey request, from an ostensibly different organization, a year later, although again the differences were in the hypothesized direction. Given the small sample sizes involved in this experiment, and the fact that rapport with the interviewer may have affected some of the results, this finding of no effects is probably best regarded as tentative.

11.6 ARE PREPAID INCENTIVES COST-EFFECTIVE?

For a variety of reasons, including those discussed in the section immediately above, prepaid incentives to everyone in the sample may be preferable to refusal conversion or other differential payments.

One reason is that interviewers like them. Knowing the household is in receipt of an advance payment, modest though it may be, they feel entitled to ask the respondent to reciprocate with an interview. Furthermore, prepaid incentives are equitable—they reward equally everyone who happens to fall into the sample, and they reward them for the "right" behavior, i.e., for cooperation, rather than refusal. Both of these advantages are likely to make modest prepaid incentives an attractive alternative to refusal conversion payments in many types of surveys. There is also indirect evidence that the use of refusal conversion payments to persuade reluctant respondents leads to increasing reliance on such payments within an organization. Steeh (1999), for example, found that prior to 1995, the linear increase in the percentage of interviews in the SCA resulting from refusal conversion was 0.05% each quarter, compared to 0.287% percent per quarter after 1995. This is also the period in which the SCA substantially increased its use of refusal conversion payments.

Still, the question arises whether prepaid incentives are cost-effective. On the face of it, it would appear that paying a small number of refusal conversion payments to reluctant respondents would be cheaper than paying everyone, even if those initial payments are smaller.

Several studies have concluded that prepaid incentives are cost-effective in mail surveys. For such surveys, the comparison ordinarily has been among incentives varying in amount or in kind, or in comparison with no incentive at all, rather than with refusal conversion payments. Two recent investigations of cost-effectiveness, by James and Bolstein (1992) and by Warriner et al. (1996), have included information on the relative effectiveness of various incentives. James and Bolstein (1992) found that a prepaid incentive of $1 was the most cost-effective, yielding nearly as high a return as larger amounts for about one quarter of the cost. Warriner et al. (1996, p. 9) conclude that for their study, a $5 prepaid incentive was the optimal amount, resulting in a saving of 40 cents per case (because the same response rate could be achieved as in a no-incentive, two-follow-up condition). The $2 incentive resulted in costs per case only a dollar less than the $5 incentive, while yielding a response rate 10 percentage points lower. Similar findings have been reported by Asch, Christakis, and Ubel (1998) in a mail survey of physicians.

For interviewer-mediated studies, as noted above, the comparison is much more likely to be with refusal conversion payments, and the answer is likely to depend on the nature of the study and the importance of a high response rate, on how interesting the study is to respondents (i.e., how many of them are willing to participate even without a prepaid incentive), on whether or not prepaid incentives reduce the effort required, and on a variety of other factors.

Several face-to-face surveys have reported that promised monetary incentives are cost-effective. Berlin et al. (1992), for example, reported on the use of a $20 incentive in a field test experiment with the National Adult Literacy Survey, which entailed completion of a test booklet by the respondent. It resulted in a cost saving per interview over the $0 and $35 incentive conditions when all field costs were taken into account. Similarly, Chromy and Horvitz (1978) reported (in a study of the use of monetary incentives among young adults in the National Assessment of Educational Progress) that when the cost of screening for eligible respondents is high, the use of incentives to increase response rates may actually reduce the cost per unit of data collected.

Singer et al. (2000) investigated this problem in the SCA. They found that a $5 incentive included with an advance letter significantly reduced the number of telephone calls required to close out a case (8.75 calls when an incentive was sent, compared with 10.22 when it was not; $p = 0.05$), and significantly reduced the number of interim refusals (0.282 refusals when an incentive was sent, compared with 0.459 when it was not). As expected, there was no significant difference between the incentive and the no-incentive condition in calls to first contact. The outcome of the first call indicates that compared with the letter only, the addition of a $5 incentive resulted in more interviews, more appointments, and fewer contacts in which resistance was encountered.

Given the size of the incentive and the average cost per call aside from the incentive, sending a prepaid incentive to respondents for whom an address could be obtained was cost effective for the SCA. However, this conclusion depends on the size of the incentive as well as the structure of other costs associated with a study for a given organization, and should not be assumed to be invariant across organizations and incentives.

An argument that can be raised against the use of prepaid incentives is that they may undermine more altruistic motives for participating in surveys. Indeed, prepaid incentives have smaller effects on survey participation for people who score high on a measure of community activism (Groves et al., 2000) than on people low on this characteristic. But this is because groups high in community activism already respond at a high rate. There is no evidence that people high on community activism who are offered a prepaid incentive respond at a lower rate than they would have had they not been offered the incentive. Although there is anecdotal evidence that some people are offended by the offer of an incentive, going so far as to return the incentive to the survey organization, by all accounts such negative reactions are few.

11.7 CONCLUSIONS

Prepaid incentives have been common in mail surveys for many years, although the amounts used are ordinarily quite modest (see Church, 1993). Because of increasing refusal rates in interviewer-mediated surveys and the widely documented success of incentives in counteracting them, we suspect that the use of incentives in such surveys will increase as well. But caution is needed in generalizing the results reported in this chapter, for two reasons. First, most of the experiments were not designed to test theoretically derived hypotheses. Second, many of the findings are based on one or a few experiments, and may not be replicable over time and across survey contexts. For example, most of the experiments agree in reporting smaller effects on response rates for gifts than cash. But there may well be situations in which gifts are more effective because they are particularly well chosen to motivate participation.

Thus, a great deal of specification and replication is needed. Here, I offer only a few modest suggestions for further research, noting that such research ought to be grounded in a theory of survey participation. The number of incentive experiments that could be designed is legion; unless they are guided by theory, they will not contribute to generalizable knowledge.

One question often asked is how large an incentive should be for a given survey. The issue here is the optimum size of an incentive, given other factors affecting survey response. If experiments varying the size of the incentive are designed in the context of a theory of survey participation that allows for changes in motivation over time, some generally useful answers to this question may emerge. In the absence of such theoretically based answers, pretesting is the only safe interim solution.

This chapter has documented effects of incentives on response distributions and sample composition. Research is needed to specify the conditions under which these effects occur. Research is also needed on how paying respondents for survey participation affects both respondent and interviewer expectations for such payments in the long run.

Although a metaanalysis of incentive effects in interviewer-mediated surveys found no differences by mode, this conclusion is in need of further research. For example, promised incentives may play a more useful role in face-to-face surveys, where the presence of the interviewer may engender trust and the delay in payment is relatively brief, than they do in RDD surveys.

Research is needed on the conditions under which incentives not only increase response rates but produce a meaningful reduction in nonresponse bias. Because they complement other motives for participating in surveys—such as interest in the survey topic, deference to the sponsor, or altruism—it is reasonable to hypothesize that incentives would serve to reduce the bias attributable to nonresponse. Whether the use of incentives for this purpose is cost-effective is less easily answered, however, and research is needed on this topic as well.

ACKNOWLEDGMENTS

Many people helped carry out the research on which much of this chapter is based. In alphabetical order, they are: Amy Corning, Nancy Gebler, Bob Groves, Patricia Maher, Kate McGonagle, Trivellore Raghunathan, and John Van Hoewyk. The National Science Foundation supported some of the experiments reported with Grant No. SBR-9513219 to Robert M. Groves and Eleanor Singer. The collegiality of

Richard Curtin, Director of the Surveys of Consumers, and the financial assistance of the Survey Research Center of the Institute for Social Research, University of Michigan, are also gratefully acknowledged. Mick Couper, Don Dillman, Bob Groves, Stanley Presser, and Howard Schuman read various versions of earlier drafts, and I am grateful for their comments even though I haven't always followed their advice.

When studies are included in previous metaanalyses, the present chapter refers only to the metaanalytic conclusions. Incentive experiments carried out after 1991 were retrieved by searching a variety of electronic data bases (Sociofile, WILS, Public Affairs, A Matter of Fact, and Psych Info), as well as the American Statistical Association's *Proceedings of the Section on Survey Research Methods* and the annual programs of the American Association for Public Opinion Research.

CHAPTER 12

The Influence of Alternative Visual Designs on Respondents' Performance with Branching Instructions in Self-Administered Questionnaires

Cleo D. Redline, *U.S. Bureau of the Census*
Don A. Dillman, *Washington State University*

12.1 INTRODUCTION

Chapter 5 examines the cognitive processes that may lead to peoples' decisions not to respond to survey items, independent of the survey's mode of administration. In this chapter, we examine a complexity that is specific to self-administered surveys, stemming from the fact that respondents are often expected to answer certain items, but not others. Research has found that item nonresponse, the failure to answer items that should be answered, is greater in questionnaires that include branching instructions than in questionnaires that do not include them (Turner et al., 1992; Featherston and Moy, 1990; Messmer and Seymour, 1982). However, very little explanation for this is offered, except to say that items with branching instructions cause greater confusion. Consequently, we attempt in this chapter to offer a theoretical framework as to why branching instructions may be confusing, followed by an empirical test of some of the concepts. We end by interpreting the empirical results in light of a proposed model of the question–answer process applied to self-administered questionnaires.

12.2 PAST RESEARCH

A relatively small body of research has attempted to determine what causes item nonresponse. Donald (1960) examined motivational factors for item nonresponse and found that the higher a respondent's involvement in terms of active participation, knowledge and understanding of the study sponsor's organization, and loyalty to it, the lower the incidence of item nonresponse. Bauer and Meissner (1963) looked at the effect of questionnaire length on item nonresponse and found that item nonresponse varied as a result of length, with a two-page questionnaire demonstrating significantly higher item nonresponse than a one-page questionnaire. Ford (1968) examined questionnaire appearance and found that a printed, folded questionnaire did not affect item nonresponse in comparison to a questionnaire composed of copies stapled together. Featherston and Moy (1990) also looked at questionnaire design factors; they report that higher item nonresponse rates were a function of branching instructions, the number of columns, item placement within a column, and question type.

A consistent finding in the research on item nonresponse is that questionnaires, or questions, with branching instructions lead to greater item nonresponse than those without. For example, Messmer and Seymour (1982) found that branching instructions significantly increased the rate of item nonresponse immediately following the instruction. Featherston and Moy (1990) found that the use of branching had a strong negative effect on item response rates. Featherston and Moy (1990, p. 5) suggest that branching instructions actually cause greater rather than less mental burden, to the point where respondents skip items rather than, as they put it, "battle the logic of the item format." While this may be true, a competing explanation could be that respondents accidentally execute branching instructions when they should not.

In addition, other authors have examined alternative branching instruction designs. For example, Turner et al. (1992) concluded that respondents had a greater tendency to see information to the right of an answer category if it was somehow made salient. Jenkins and Dillman (1995) suggested that respondents overlooked branching instructions when the instructions were to the right of an answer category because the instructions were beyond the location where respondents' eyes naturally traveled.

12.3 MENTAL PROCESSES THAT UNDERLIE BRANCHING INSTRUCTION ERRORS

Respondents can make an *error of commission,* which means they fail to branch when instructed to, or they can make an *error of omission,* which means they fail to advance to the next listed question on the page and answer it. We use the terms error of commission and error of omission to convey a relationship between the printed instructions we have given respondents and their answers to questions. In comparison, Beatty and Hermann in Chapter 5 used the terms to describe a relationship between respondents' personal knowledge about themselves and their answers to questions. We propose that errors of commission and omission originate both as a result of the questionnaire and of respondents themselves.

12.3.1 The Questionnaire's Role

To understand the role the questionnaire plays, however, one must first understand that a self-administered questionnaire is composed of more than just verbal language (i.e., words). Jenkins and Dillman (1997) argue that a self-administered questionnaire is composed of nonverbal language in addition to the verbal, with the word "language" used here in the Newell and Simon (1972, p. 65) sense of the word, that is, "anything consciously employed as a sign." In this chapter, we propose that there are two nonverbal languages we must be concerned with: the numeric and symbolic. "Numeric" refers to the use of numbers on a questionnaire and "symbolic" refers to the use of signs, such as arrows. Recent research by Schwarz et al. (1998) lends support to the proposition that these languages may combine to create meaning, for they found that respondents interpreted the verbal label "rarely" as indicating a lower frequency when paired with the numeric value 0 rather than 1.

In addition to language, a self-administered questionnaire also contains "graphic paralanguage." Graphic paralanguage is a term we use to refer to the three fundamental elements of visual perception: brightness and color, shape, and location (Glass and Holyoak, 1986). Generally, "paralanguage" is used to refer to features that accompany speech and contribute to communication, e.g., loudness and tempo and facial expressions and gestures. Visual elements can be thought of similarly; that is, they accompany text and contribute to communication but are not language per se, which is why we have chosen the term graphic paralanguage (or "graphic language" for convenience).

It is important to recognize that the languages of a self-administered questionnaire—the verbal, numeric, and symbolic—never stand alone. By definition, they can only be transmitted through the visual channel via graphic paralanguage. This means that the same verbal language, for instance the words "skip to," can take on an enormous amount of variation when put into print. We can change the size of the words, the font, the color, the background color, and the location.

An evaluation of the 1998 decennial census dress rehearsal questionnaires demonstrates the impact of graphic paralanguage. The relationship question on the short form presents two columns of answer categories, with answer categories referring to relatives on the left and those referring to nonrelatives on the right. A long, white write-in box for other relatives is tucked into the lower right-hand corner of the question space. In comparison, the long form presents one long list of categories, with the long white write-in box inserted into the middle of the list between the relative and nonrelative answer categories. Davis (1999) compared the short form with the long form and found that significantly fewer people selected the nonrelative answer categories on the long form than the short form. Although the verbal, numeric, and symbolic languages are almost identical, the graphic paralanguage is quite a bit different, resulting in respondents not attending to the nonrelative answer categories on the long form because the long, white write-in box visually separates them from the relative answer categories.

Finally, a self-administered questionnaire is more than just language; it is a physical entity that requires physical manipulation by the respondent. In the case of a paper questionnaire, respondents must be able to orient the questionnaire properly in space and turn its pages. Recent cognitive research with questionnaires has suggested that these tasks burden less able readers more than able readers because they require additional work over and above the already taxing work of reading the questions and response options (Dillman et al., 1996). Therefore, it follows that the addition of yet another task—that of executing branching instructions—may only serve to further overload the less able reader.

12.3.2 The Respondent's Role

Perception is a complicated process that relies not only on respondents' seeing an external stimulus (known as bottom-up processing), but also on their expectations (or knowledge) about that stimulus (top-down processing) (Jenkins and Dillman, 1997). Cognitive research has suggested that respondents often think they are supposed to answer every question on a questionnaire (Dillman et al., 1999). Even when respondents begin to answer the questionnaire with the proper realization that they are not supposed to answer every question, if there are long series of questions in which respondents are not required to branch, they can easily fall into the habit of expecting to answer every question.

Another reason respondents may overlook branching instructions is because the questions and response categories absorb their attention at the expense of the navigational aspects of the questionnaire. This might occur with all respondents when they become interested in the content of the questions and response categories. However, it might also occur with less able readers because they may have less ability to handle both the demands of questions and the mechanical aspects of branching through the questionnaire.

The solution to this problem and the previous one regarding respondent expectations may reside with breaking respondents' normal processes of sequentially answering each question. This may be accomplished by introducing variations into the design that will attract respondents' attention. However, it should also be noted that if not properly manipulated, these variations could have the deleterious effect of causing respondents to skip over questions when they should not, leading to increased item nonresponse, or of inducing respondents to answer questions that should be skipped. Thus, isolating the appropriate balance between possible variations is a challenge of particular theoretical and practical interest.

12.4 MANIPULATING THE GRAPHICAL, SYMBOLIC, AND VERBAL LANGUAGES

There are numerous ways in which information can be manipulated to improve the design of a branching instruction.

12.4.1 Manipulating the Graphical Paralanguage of the Branching Instruction

Commonly, designers of self-administered questionnaires locate check boxes to the left of the response options and the verbal branching instruction to the right, as shown here:

1. Which of the following best describes you?
   ☐ I tend to think before I act
   ☐ I tend to act before I think—Skip to 3

Note that the branching instruction is the same *size, shape, color, and brightness* as the rest of the text. Between the response category and the instruction is a dash, which too is of the same color and brightness as the rest of the information. Advancing to the next question (that is, *not* skipping) is signaled by the absence of any information to the right of the nonskip response category.

Cognitive interviews have suggested that respondents frequently fail to see the verbal branching instruction in this location (e.g., Gower and Dibbs, 1989; Jenkins and Ciochetto, 1993; Bogen et al., 1996). A person's vision is sharp only within two degrees, which is equivalent to about 8–10 characters of 12-point type on a questionnaire (Kahneman, 1973). When a respondent is in the process of marking a check box, the branching instruction may be outside the respondent's view. Therefore, we may need to attract respondents' eyes to bring it within view.

Increase the Contrast Ratio. Visually, all of the above information exists in a particular figure–ground format. The contrast ratio between a figure and its ground and between figures is critical to influencing the perception of information (Wallschlaeger and Busic-Snyder, 1992). Up to this point, the figure–ground format we have been working in has been black print (i.e., black figures) on a white background. Black print against a white background is highly visible because of the high level of contrast between the figure (black print) and ground (white paper). Visual search tasks have demonstrated that a target item can be located more rapidly if it is made visually dissimilar from the nontarget items (Foster, 1979). Therefore, one way to attract respondents' eyes to the branching instruction might be to make it look different from its surroundings by increasing the contrast ratio between it and the information surrounding it.

One possibility is to increase the boldness of the branching instruction, as shown below:

2. Which of the following best describes you?
   ☐ I tend to think before I act
   ☐ I tend to act before I think—**Skip to 4**

A second possibility is to increase its size:

☐ I tend to act before I think—Skip to 4

Or, we could manipulate both the figure and ground rather than just the figures. A format developed for questionnaires in the early 1990's and employed by the U.S. Bureau of the Census uses a format that consists of black type on a lightly colored background (i.e., 20% of full color); all answer spaces are highlighted in white to attract respondents' attention.

☐ I tend to act before I think—**Skip to 4**

Black print on a white background contains a higher level of contrast than black print on a 20% colored background (shown here in gray), so we could *change the background of the branching instruction to white* too in hopes of attracting respondents' attention to it. However, the check boxes are also white, so the potential disadvantage of this design is that the more information is placed in white, the less it will look dissimilar from its surroundings.

A second way to change the figure–ground format of the branching instruction is to use *reverse print,* as shown below. The reverse printed branching instruction is white type on a black background, or just as its name implies, the reverse of the main figure–ground composition.

☐ I tend to act before I think— Skip to 4

There are arguments both for and against reverse printing the branching instructions. On the one hand, it is plausible that the high contrast of a reverse-printed branching instruction and the fact it has been made visually dissimilar from the other information on the questionnaire could attract respondents' attention. On the other hand, typographical studies warn against using reverse printing because it is difficult to read (Hartley, 1981; Wallschlaeger and Busic-Snyder, 1992). Also, it could be that if most of what respondents read is printed in black type, they may come to expect the information they are supposed to read to be in black type too and, therefore, not pay attention to the occasional reverse-printed instruction.

Relocate the Branching Instruction. A different way to make branching instructions more visible may be to rearrange information so respondents are more likely to view it. One way to do this is to reverse the locations of the check boxes and response options, as shown here:

3. Which of the following best describes you?
   I tend to think before I act. ☐
   I tend to act before I think. ☐ Skip to 5

Relocate the Verbal Branching Instruction and Increase Its Contrast. The above manipulations can be combined in any number of ways. Since the white answer box is now close to the branching instruction, one logical combination would be to place the branching instruction in a white background too, as a way of emphasizing the connection between the verbal branching instruction and the check box. In the example below, the size and boldness of the verbal branching instruction were increased as well.

I tend to act before I think. ☐ **Skip to 5**

12.4.2 Manipulate Both the Graphical and Symbolic Languages

This section discusses ways of attracting respondents' attention to the branching instruction in its common location by manipulating the symbolic in addition to the graphic language.

Connect the Response Option and Branching Instruction. Symbols, as with verbal language, acquire meaning through use. For instance, an arrow "→" suggests direction. Therefore, replacing the dash in question 1 above with an arrow may convey to respondents that they should go from the response option to the branching instruction:

4. Have you graduated from high school?
   ☐ Yes
   ☐ No → Skip to 6

The intention here is to group information together that the respondent may not otherwise associate as belonging together, in this particular case, the check box, response option, and the branching instruction. In contrast, parentheses around the branching instruction will likely convey to readers that reading the branching instruction is optional rather than mandatory, and in this way act to segregate rather than integrate the information, as can be seen here:

5. Have you graduated from high school?
   ☐ Yes
   ☐ No (Skip to 7)

Connect the Nonskip Response Option with the Next Question. As mentioned earlier, the nonskip situation is implied by the absence of any instructions. Perhaps one way of making respondents more aware of when not to skip is to draw an arrow coming off the left-hand side of the nonskip check box that points to the next question, as demonstrated in the example below:

7. Have you graduated from high school?
   ☐ Yes
   ☐ No → Skip to 9
8. Have you obtained additional schooling beyond the high school level?

12.4.3 Manipulating the Graphic, Symbolic, and Verbal Languages of the Branching Instruction

Prevention Technique. One technique for preventing errors is to educate or train people in their prevention (Wickens, 1992). Training works by altering people's top-down processing about events.

Section 3 makes the point that respondents' top-down processing of the form-filling task is often erroneous. Thus, educating respondents about branching phenomena before they need to execute one and continuing to remind them along the way might help to overcome this. The example below incorporates both the education and reminder prevention techniques:

9. From now on, if a "Skip to" instruction follows a box you mark, skip to the number given. Otherwise, continue with the next question.

   Have you graduated from high school?

   ☐ Yes

   ☐ No—Skip to 14

10. Have you obtained additional schooling beyond the high school level?

    ☐ Yes

    ☐ No

11. Attention: Check for a skip instruction after you answer the question below.

    What is your sex?

    ☐ Male

    ☐ Female—Skip to 16

Detection Technique. Norman (1990) suggests allowing the user to detect and correct errors once they have occurred through the use of feedback. Placing an additional instruction before the subsequent question respondents are supposed to skip after the question requiring respondents to branch might provide effective feedback. The example below illustrates this strategy:

12. Have you graduated from high school?

    ☐ Yes

    ☐ No—Skip to 17

13. (If Yes to 12) Have you obtained additional schooling beyond high school?

12.5 RESULTS FROM AN EXPERIMENTAL TEST

The preceding theoretical framework reveals a large number of individual hypotheses in need of further testing. We reduced this to a manageable number by combining the individual manipulations in two ways that made sense from a theoretical perspective. These were named the "detection" and "prevention" branching instructions. We then tested these instructions against a control, which is the branching instruction used in Census 2000.

The *control branching instruction* is very similar to the common instruction we discussed earlier, with two exceptions. The dash is replaced with an arrow and the verbal branching instruction is italicized rather than normal print, as shown below:

5. Do you have a cellular telephone?
   ☐ Yes
   ☐ No → *Skip to 7*

The *detection branching instruction* is similar to the control with regard to the location of information, but the size and boldness of the verbal branching instruction is increased. Also, a left-hand arrow comes off the nonskip check box and points to a parenthetical instruction meant to provide feedback, as shown below:

5. Do you have a cellular telephone?
   ☐ Yes
   ☐ No → ***Skip to 7***
6. (If yes) Fifteen years from now, do you think the number of adults with cellular telephones will include:
   ☐ Less than one-fourth of the U.S. population?
   ☐ About half of the U.S. population?
   ☐ About three-fourths of the U.S. population?
   ☐ More than three-fourths of the U.S. population?
   ☐ No opinion

The *prevention branching instruction* is quite different from the others. First, the response categories and check boxes are reversed. Second, the verbal instruction was made bold and its background changed to white. Third, an education instruction was placed before question 5, which was the first question to contain a branching instruction on the questionnaire. And last, although this is not depicted below, instructions were placed before every subsequent question that contained a branching instruction to remind respondents to pay attention to the branching instructions.

5. You may be asked to skip over some questions from here on out. It all depends on your answer to questions as you go along. If a skip instruction follows the box you mark below, skip to that number. If a skip instruction does NOT follow the box you mark, then continue with the next question.

   Do you have a cellular telephone?

   Yes ☐
   No ☐ ***Skip to 7***

We developed three versions of a test questionnaire, each using one of the three branching instruction designs described above. Each questionnaire contained the same 50 items, with 24 of the items possessing branching instructions. An important characteristic of the test questionnaire was that respondents could get no clue from the questions themselves about whether they should be answering them.

As described in Redline et al. (1999), a total of 1266 students in classes at Washington State University were randomly asked to complete one of the three forms. Approximately 420 students completed each form, following the general protocol for group administration of questionnaires outlined in Dillman (2000). And in addition, we conducted 48 cognitive interviews both at Washington State University and at the Bureau of the Census with a broad mix of people (Dillman et al., 1999a; Redline and Crowley, 1999).

Both of the experimental forms significantly reduced the percent of commission errors by about half from 20.3% on the control form to 7.4% on the detection form, and 9.0% on the prevention form. However, neither of the experimental forms significantly reduced the proportion of omission errors. Whereas the control form resulted in 1.6% omission errors, the detection and prevention forms produced 3.7% and 3.3% errors, respectively, with statistically significant differences between the control and either of the experimental forms (Redline et al., 1999).

This result seems promising because if college students, who are relatively capable readers to begin with, can be affected in this way, then the chance of affecting the general population, which would contain a greater percentage of less able readers, should be even greater.

12.6 THE INFLUENCE OF QUESTION CHARACTERISTICS ON ERRORS OF COMMISSION AND OMISSION

A substantial amount of variation was observed in the individual error rates across the 24 items. For commission errors, the rates ranged from 0 to 51.7%, and for omission errors they ranged from 0 to 33.9%. Items with higher error rates on one form tended to have higher error rates on the others. Therefore, it seemed likely that question characteristics, as well as the experimental formats, influenced the error rates. Eight ways were identified that described how questions varied from one another in ways that were theorized to affect the ability of respondents to correctly follow the branching instructions. In general, each of these question attributes, which are listed below, are ones that seem likely to place higher cognitive demands on respondents' ability to follow branching instructions, as explained in Dillman et al. (1999b).

1. Being the last question on a page
2. All answer options were directed to branch
3. Write-in answers were requested
4. Answer categories alternated between being directed to branch and continue

5. High number of answer categories
6. High number of words in the question
7. The last answer category contained a branching instruction
8. High distance between check box and branching instruction

The influence of these variables on both errors of commission and omission has been examined in more detail elsewhere. A series of regression analyses was conducted that examined the influence of these eight variables, form type, and potential interactions among them on mean error rates (the dependent variable) for each of the test items (Dillman et al., 1999b). These analyses showed that at the bivariate level (Table 12.1) most of the question attributes were significantly correlated with the number of commission errors (alternative branching instructions was a noteworthy exception), but were not significantly correlated with the making of omission errors.

Initial regression analyses revealed that commission errors and omission errors were significantly influenced by different variables. Commission error rates were primarily influenced by being a write-in question (Table 12.2). In addition, more detailed analyses of commission errors (not shown here) revealed that deleting the powerful effects of the write-in question format did not result in other variables being significant. In contrast, the making of omission errors was primarily influenced by being at the bottom of the page (Table 12.3). However, further partitioning of the data (also not shown here) revealed that being at the bottom of the page led to significant increases in the errors of omission on the detection form only. Also the data revealed that the omission errors were significantly decreased on the detection form when the skip instruction was either associated with the last category or there was a high distance between the skip instruction and the check box. Thus, we concluded that question characteristics, in addition to the visual design of the branching instruction, led to the making of response errors.

Table 12.1. Effect of possessing the specified question characteristic on error rates (Dillman et al., 1999b)

	Commission error			Omission error		
Question characteristics	With (%)	Without (%)	Probability	With (%)	Without (%)	Probability
Bottom of page	15.1%	11.0%	0.02	5.2%	2.3%	0.21
All choices branch	16.2	11.3	0.01	NA	NA	NA
Write-in answer	34.1	11.7	0.00	5.0	2.7	0.50
Alternating branches	14.0	12.3	0.49	1.2	3.3	0.60
Last choice branches	5.4	13.5	0.04	5.8	2.4	0.16
High distance	15.3	9.9	0.00	1.7	3.8	0.39
High number of words	8.7	14.1	0.01	2.3	3.2	0.68

Table 12.2. Effect of question characteristics on commission error rates (Dillman et al., 1999b)

Independent variable	Unstandardized regression coefficient	Standard error	Standardized regression coefficient
Bottom of page	3.69	3.1.3	0.14
All choices branch	5.90	4.31	0.16
Write-in answer	25.16***	4.82	0.56
Alternating branches	1.25	4.19	0.04
High number of choices	0.08	0.60	0.02
Last choice branches	0.78	2.92	0.03
High number of words	–0.13	0.12	–0.12
Intercept	11.81		
R-squared	0.35***		
Model df	7		
Total df	70		

*$p < 0.05$, **$p < 0.01$, ***$p < 0.001$

12.7 DEVELOPMENT OF A MODEL APPLICABLE TO THE QUESTION–ANSWER PROCESS IN SELF-ADMINISTERED QUESTIONNAIRES

It is useful to interpret the above findings in light of the question–answer process. Tourangeau (1984) describes the question–answer process for interviewer-administered surveys as having four steps: (1) comprehending the question, (2) recalling relevant information, (3) making a judgement, and (4) selecting a response. While

Table 12.3. Effect of question characteristics on omission error rates (Dillman et al., 1999b)

Independent variable	Unstandardized regression coefficient	Standard error	Standardized regression coefficient
Bottom of page	5.66**	1.73	0.42
Write-in answer	4.29	2.44	0.22
Alternating branches	0.11	2.14	0.01
High number of choices	–0.60	0.31	–0.28
Last choice branches	–2.23	1.49	–0.19
High number of words	0.05	0.06	0.11
Intercept	4.07		
R-squared	0.25**		
Model df	6		
Total df	61		

*$p < 0.05$, **$p < 0.01$, ***$p < 0.001$

these steps are still at the core of the *self-administered* situation, we propose that the following changes need to be made:

1. Addition of an initial step that takes into account the fact that respondents must perceive and attend to the question stimulus. This is depicted by the addition of Box A in Figure 12.1.
2. Denoting that respondents must perceive and understand the graphic paralanguage and symbolic and numeric languages in addition to the verbal language. This is depicted in Figure 12.1 by inserting the phrase "all languages" in both Boxes A and B.
3. The addition of an entire sequence of activities from F to I to represent the branching instruction phenomena. The column on the left represents the question–answer process for a single question and the column on the right represents the process one must go through to determine the question which should be attended to next.

Thus, Figure 12.1 reveals that the question–answer process in self-administered questionnaires is complicated by the fact that respondents need to go through the

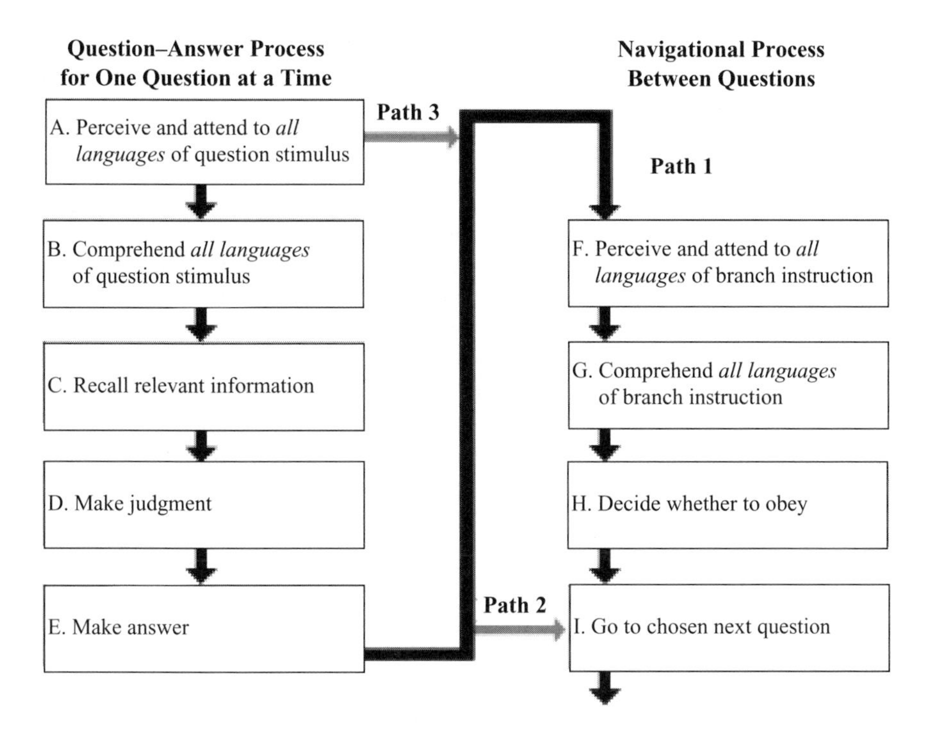

Figure 12.1. Revision of Tourangeau (1984) model for self-administered questionnaires.

same process for branching as for the questions and answers. That is, they need to first perceive the branching instruction in F, then comprehend it in G, etc. Also, this process is made even more complicated by the fact that sometimes respondents are supposed to branch, sometimes not, so we need two paths to represent this, paths marked "1" and "2" in Figure 12.1.

12.7.1 Application of the Model to Errors of Commission

Theoretically, when respondents are instructed to branch they are supposed to travel on the shaded path marked 1, but when they make an error of commission, they accidentally take the unshaded path marked 2. It is clear from the results of the classroom experiment that when respondents come across a question that requires them to write in an answer, they are more likely to erroneously travel Path 2.

The prevention form and the detection forms appear to act upon the commission error process differently. Basically, the prevention form works to induce respondents to stay on Path 1, and given the results of the classroom tests, overall, it was successful at this. However, the data show that it was just as unsuccessful as the other forms when it came to correcting errors of commission associated with write-in responses. In comparison, not only does the detection form attempt to induce respondents to stay on Path 1, but it also provides a mechanism for getting respondents back on this path if they accidentally get off it. If a respondent erroneously takes Path 2, then from I they will go to A. Now at A they should perceive and comprehend the parenthetical instruction, in which case, they will be sent on Path 3 back into Path 1 and on to F, where they will be given a second opportunity to perceive the branching instruction.

12.7.2 Application of the Model to Errors of Omission

Errors of omission appear to occur for different reasons than errors of commission. It seems that, overall, manipulating the information on the experimental forms not only increased the likelihood that respondents would attend to and perceive the branching instruction when we wanted them to, but when we didn't want them to as well. In this case, respondents are supposed to take Path 2, but instead they take Path 1, where they perceive and attend to a branching instruction they are not supposed to. The data show that having to turn the page or go to the top of the next column makes it even more likely on the detection form that respondents will make this mistake.

12.8 CONCLUSION

In this chapter, we attempt to answer the question, "Why do respondents make mistakes when they navigate through a self-administered questionnaire with branching instructions?" We suggest that the answer begins with understanding that respondents must process much more than just the verbal language of a self-administered questionnaire. They must process the graphic paralanguage and numeric and symbolic languages as well as the physical structure. Furthermore, they must interleave the processing of the languages applicable to the questions and responses with those applicable to navigation. Because so much information must be processed, it is not surprising that errors occur. However, an initial experiment has led us to conclude that manipulating the languages as we did reduced errors of commission, but not errors of omission. Consequently, we need to continue to manipulate the languages of the questionnaire and study their effects until we are able to effectively guide respondents through self-administered questionnaires without fail.

ACKNOWLEDGMENTS

Support for this research was provided by the U.S. Bureau of the Census and Washington State University. This chapter reports the results of research and analysis undertaken by Census Bureau and Washington State University staff. It has undergone a more limited review than official Census Bureau publications. This chapter is released to inform interested parties of research and to encourage discussion. We would like to thank Roger Tourangeau for his helpful comments on an initial draft.

PART III

Nonresponse in Diverse Types of Surveys

CHAPTER 13

Evaluating Nonresponse Error in Mail Surveys

Danna L. Moore and John Tarnai, *Washington State University*

13.1 INTRODUCTION

Mail surveys have nonresponse characteristics quite distinctive from those of interviewer-assisted surveys. Refusers and noncontacts are not easily separated, and there are different causes of nonresponse, such as those due to reading problems or not noticing or receiving the questionnaire. Survey researchers have spent much time and effort on improving response rates in mail surveys. That we can be very effective at this is well documented (Dillman, 1978, 1991, 2000; Heberlein and Baumgartner, 1978; Fox et al., 1988; Yammarino et al., 1991; Mangione, 1995). Our objective in this paper is to evaluate whether improving response rates in mail surveys affects nonresponse error.

The survey literature shows that by improving response rates in mail surveys the results match the sample frame variables better on the final wave of follow-up than on the initial wave. Whereas this suggests that improving nonresponse is worthwhile, some of our own work suggests that this may introduce other kinds of error into survey estimates. In this chapter we show that for mail surveys (1) sample frame demographics improve over waves as nonresponse decreases; and (2) outcome variable differences between respondents and nonrespondents sometimes increase over waves as well, which raises the question of whether our methods may be increasing rather than reducing nonresponse error.

13.2 MAIL SURVEYS NONRESPONSE ERROR

Mail surveys have been faulted for their low response rates (Armstrong and Overton, 1977), however, Dillman (1978) and Mangione (1995) have described the kinds of procedures that will ensure high response rates using the mail survey mode.

Groves (1989), among others, has described nonresponse error for linear statistics as consisting of two parts: the nonresponse rate and the difference between survey respondents and nonrespondents. Much of the survey literature has focused on ways of dealing with the nonresponse rate as a way to minimize the error, even though the latter part is probably the more important part of the equation. For instance, if there are no differences between respondents and nonrespondents, then there is no nonresponse error regardless of the response rate. Thus, our focus in this chapter is on the latter issue of the difference between survey respondents and nonrespondents. Evaluating this error requires a data source that is independent of the survey itself. The sample frames for mail surveys often include a variety of demographic and other kinds of administrative information useful for conducting nonresponse analyses. This information can be used to evaluate the extent of differences between respondents and nonrespondents because it is available for all sampled respondents.

There are also differences among mail surveys and other kinds of surveys in the components of unit nonresponse. Groves (1989) describes the three elements that make up nonresponse in household surveys as consisting of (1) refusals, (2) the inability to contact the respondent, and (3) the inability of the respondent to respond to the survey. In most telephone and face-to-face surveys, these three kinds of nonresponse are readily distinguished by interviewers. In telephone surveys for instance, refusals are readily identified and the procedure for dealing with them is to train interviewers in how to avoid them, and conduct refusal conversion attempts. The inability to contact respondents in telephone surveys is usually due to answering machines, people who are using the telephone, respondents who are not at home, and other kinds of no answers. The inability of the respondent to respond to the survey in telephone surveys is generally due to language differences, a hearing or other health-related disability; or the respondent may be incarcerated or otherwise unavailable during the survey period. In mail surveys, these three elements of nonresponse (refusals, noncontact, and inability to respond) are generally indistinguishable from one another, thereby precluding the ability to target them for differential follow-up. The reason for this is that nonresponse to mail surveys is evidenced only by nonreturn of the questionnaire. But nonreturn of the questionnaire provides no information about the reasons for this. Although all three elements of nonresponse are implicated, the majority of nonresponders are most likely refusers or people who intend to respond but then forget to follow through. We base this interpretation on the finding that follow-up efforts to nonrespondents have been shown to be effective at improving response rates (Dillman, 1978; Mangione, 1995).

The remainder of this chapter describes our experiences with two mail surveys and our attempts to both reduce the rate of nonresponse, and to measure the difference between respondents and nonrespondents on characteristics available to us in the entire sample. We then discuss the implications of our findings in relation to the literature on nonresponse bias in mail surveys in general.

13.3 MAIL SURVEY FEATURES REDUCING NONRESPONSE

Past research on evaluating nonresponse to mail surveys has been successful in identifying specific mail survey design features that can be manipulated to improve response rates, such as personalization of mailings, inclusion of stamped return envelope, cover treatments, the number of follow-up contacts, special postage, and use of incentives (Dillman, 1978, 1996; Armstrong, 1975; Heberlein and Baumgartner, 1978; Kanuk and Berensen, 1975; Linsky, 1975; Berry and Kanouse, 1987; Fox et al., 1988; Groves, 1989; Ayidiya and McClendon, 1990; Groves et al., 1992; James and Bolstein, 1992; Church, 1993; Tambor et al., 1993; Biner and Kidd, 1994. Past research has shown that the use of multiple follow-up contacts, special postage, and incentives are the most effective nonresponse reducing factors overall. What this research has not done is to show that improvements in response rates are also successful at reducing the extent to which respondents differ from nonrespondents on characteristics relevant to the study.

13.4 NONRESPONSE ERROR IN MAIL SURVEYS

The primary emphasis of this review is to show which techniques have been used to evaluate nonresponse error, and if any survey design features influence nonresponse error.

Mail surveys have been criticized because of large nonresponse problems. The two main methods for dealing with this problem have been to improve response rates through survey design features (Dillman, 1991; Yu and Cooper, 1983; Woodruff et al., 1998), and through ways of measuring the direction and magnitude of nonresponse error and of using weighting approaches to compensate for this (Armstrong and Overton, 1977; Chen, 1996; Daniel, 1975; Jones and Lang , 1978; Mandell,1975; van Goor and Stuiver, 1998). Others have attempted to describe the characteristics of nonrespondents (Brennan and Hoek, 1992; Goyder, 1987; Gannon et al., 1971), to compare early and late responders to mail surveys (Wellman et al., 1980; Newman, 1962), and to statistically model the nonresponse error in mail survey data (Cameron et al., 1999).

Fillion (1976b) presented a method for correcting nonresponse error that can be used with successive mail questionnaire contacts by using information in successive questionnaire mailings to predict the direction of nonresponse error. Nonrespondents to earlier waves are considered as persons initially resistant to investigation who later consent. Replies to successive waves form a continuum of respondent types from highly motivated to unmotivated individuals. In this study, significant differences between response waves as well as the substantial increases in response confirm the importance of follow-ups as a means of reducing nonresponse error. Fillion computed the percentage error due to nonresponse for variables in the survey and the data revealed an inverse relationship between involvement and response. Involvement was seen as a proxy for saliency of the questionnaire.

Stinchcombe et al. (1981) attribute survey respondents who are never contacted to two factors: availability to be interviewed and refusal. By combining records and using government survey results, they found that factors affecting availability do not vary much for farmers. However, individuals who first refuse to participate in surveys are found to have larger variation in their attitudes, especially about surveys. They found refusers used USDA market information that was derived from the survey data less. Refusers were also more likely to rate the reports as having less accuracy when compared to early responders. The "more difficult to reach" respondents were not significantly different than the responders who were "easy to reach." Nonresponse tendency was found to be strongly related to lack of belief that farmers as a population benefit from government generated agricultural market reports. Refusers and difficult-to-reach respondents believed that they were asked too often to participate in surveys, had concerns over confidentiality, and lacked trust in government assurances of confidentiality when compared to early responders. The authors estimated the size of refusal bias by using a function that takes the difference between the true population proportion estimate and the observed population proportion.

Incentives are often used as a way to motivate respondents and nonrespondents. In their mail survey of college graduates Shettle and Mooney (1999) found that incentives significantly increased survey cooperation and significantly lowered the refusal rate. The cumulative noncontact rate was about the same for both groups, with the incentive group only 2% less. During the final expensive personal visit follow-up to respondents not contacted by phone, the incentive group was found to be significantly less cooperative (15.8% difference), had a higher rate of refusals (4% difference), and had fewer contactable respondents (8.6% difference). While it seems clear that incentives increase overall response, the reasons for this increase are less clear.

13.5 THE EXPERIMENTS

Fundamental to understanding why people decide to participate or to not participate in a mail survey is the idea that the person receiving a survey is motivated by the survey process, the mailing package, the questionnaire, and other contents. The combined aspects of personal and survey-specific factors are assumed to be additive in motivating individuals' to cooperate with a survey request. Singer's (1999) analysis of the use of incentives found that the inclusion of an incentive condition in a survey resulted in significant difference in response rates and that, on average, each dollar of an offered incentive resulted in a third of a percentage point of difference between incentive and the zero-incentive condition. How individual characteristics (gender, age, and location) interact with other types of survey design features in motivating response is of interest because it involves a more comprehensive test of the survey-specific factors.

In the remainder of this paper we present the results of two mail surveys that give us the opportunity to examine the influence of both personal demographic factors as well as survey-specific factors on nonresponse. The first survey mailed a 12-page questionnaire to 6090 new residents of Washington state, identified as those who had surrendered a driver's license in 1996. The purpose of the survey was to determine people's reasons for moving into the state, and their background characteristics. The second survey mailed a 12-page questionnaire about immunization practices to 2472 family practice physicians and pediatric physicians in Washington state in 1998. Both sample frame databases included a variety of demographic variables that allowed us to compare respondents and nonrespondents on these variables, as well as to analyze the disposition of respondents at each wave of contact.

13.5.1 The Washington State Driver's Licensing Study

For this survey, information provided with the sample frame included age, gender, and geographic location; the latter was used to stratify the final sample into 40% metropolitan areas and 60% nonmetropolitan areas. For this chapter, we present only one quarter of the sample frame data for comparison. We compared the sample frame data for all four quarters and found and no significant differences on any demographic characteristics.

The mail survey was designed as a 28-day sequence of four mailings that included a preletter, a questionnaire with a $2 incentive, reminder postcard, and replacement questionnaire. For the whole survey, this procedure yielded an overall 65% response rate. The factors that were manipulated in the experimental design included a color questionnaire cover, $2 incentive, prenotice letter, personalization, and optical scanning. The other demographic variables included in the analysis, available for both respondents and nonrespondents, were age, gender, and metropolitan location. Logistical regression was used to model the effects of the experimental design variables, demographic variables, and their interaction on response.

We use a logistic regression model to evaluate response (a completed returned questionnaire) as a function of sample individuals' characteristics (age, gender, location) and survey design features (questionnaire cover, $2 incentive, prenotice letter, personalization, and optical scanning), and variable interactions. The significant explanatory variables most impacting the probability of response are age, gender, incentive, and age × gender interaction. The only survey design feature that significantly affected response was the use of a $2 incentive. However, when the other factors are removed from the model, the model fit is reduced, which suggests that the other survey design features together have some small positive impact on response, but that it is not significant for any of these other factors alone, when all other factors are held constant. The odds ratios in Table 13.1 indicate the direction and magnitude of the impact on response. For instance, if an individual received a $2 incentive in the questionnaire mailing, the predicted odds of the individual completing and returning the questionnaire increases by 88%. For every additional year of age, the predicted odds of a sampled individual completing and returning a mailed questionnaire increases by 3.3 percentage points, holding all other variables constant.

What the logit model suggests is that when all other variables are held constant,

Table 13.1. Logit analyses of response (0,1) for Washington state driver's licensing study

Variable	B	Wald X^2	Prob > X^2	Odds ratio
Intercept	–1.6759	69.4	0.0001	
Age	0.03211	50.4	0.0001	1.033
Gender (1 = female)	0.8944	21.8	0.0001	2.446
Incentive (1 = $2)	0.6309	31.1	0.0001	1.879
Age × gender	–0.0114	10.9	0.0009	0.989
Gender × incentive	0.0179	0.1	0.8747	1.018
Prenotice (1 = yes)	0.0870	1.4	0.2375	1.091
Topic (1 = retirement)	–0.0084	0.01	0.9320	0.992
Color cover (1 = yes)	0.0697	0.75	0.3864	1.072
Metro location (1 = yes)	0.0316	0.41	0.5241	1.032
Personal (1 = yes)	–0.0903	0.73	0.3927	0.914
Optscan (1 = yes)	–0.0232	0.03	0.8705	0.977

Note: Number of observations = 7483; chi square for covariates: –2; log L = 404.65 with 10 df (p = 0.0001); score = 395.58 with 10 df (p = 0.0001).

age, incentive, and gender all affect response, and the age × gender interaction shouldn't be ignored. Although incentives significantly increase response overall, we can see by evaluating nonrespondents that it is this interaction that distorts the survey results on frame variables and has the largest impact for the study, causing age and gender distributional bias. In other words, it is the age × gender interaction in the presence of the incentive that is the largest contributing factor to bringing in "more of the same" types of respondents, which seems like an imbalance in distribution when it is compared to the no incentive experiment condition (see Table 13.3). This leads us to seek further the impact of increasing response by extra survey efforts that changes the distribution of survey respondents on demographic variables that may further influence the survey outcome variables.

As Fuller (1974) indicates, disproportionate returns due to nonresponse are often treated as equivalent to disproportionate sampling. Adjusting for "underrepresented" or "overrepresented" groups, is often accomplished by evaluating and weighting based on demographic characteristics. Fuller also points out that in the case of disproportionate returns due to nonresponse, inclusion is not random and the probability of an individual's inclusion is not known. For mail surveys, late returns are often weighted to adjust for nonreturns based on the assumption that those responding late to a mail survey are similar to those who do not respond at all. For this population of all people who turned in out of state drivers licenses, we can calculate the full study means for some demographic characteristics [age, gender, or location (metro versus nonmetro)]. These population estimates can be compared to the survey estimates of these same characteristics for respondents and nonrespondents at each stage of follow-up as a measure of the change in response error achieved through additional follow-ups and by incentive contacts.

13.5.2 Demographic Correlates and Incentive Effects in the Driver Licensing Survey

Table 13.2 shows that males tend to be nonrespondent, with greater prevalence among both refusals and total nonrespondents than is true for the total sample. With regard to metropolitan versus nonmetropolitan residential location, Table 13.2 shows that all dispositions are comparable to the sample frame with no significant differences, thus survey recruitment worked equally well for bringing metro and nonmetro licensees into the study. This finding also holds consistent across the waves of contact and the cumulative contact waves. Similarly, the older age persons tend to be nonrespondent. They tend to be underrepresentend among respondents to each successive wave and end the full study with a 14.9% underrepresentation.

For the group receiving the $2 incentive, when completed interviews are compared to the random sample by gender, there is a significant difference in distribution of male and females, with approximately 6% fewer males represented in the completed interviews. However, the nonrespondents are significantly different from the sample frame, with 6.8% more males. For the incentive group, most of the nonresponse error is associated with male respondents who were not reachable and refused.

Evaluation by gender for the completed interviews in the nonincentive group shows another story, with no significant difference in composition (Table 13.3). The refusals are not significantly different than the sample frame for male and female respondents. Unlike the incentive treatment, the gender composition of the nonrespondents for the nonincentive treatment is not significantly different than the sample frame (only 0.7% more males). This finding suggests that when the sample dispositions from the two treatment groups are compared, the non-use of an incentive may actually result in more representative data, whereas the incentive treatment results in accentuating the underrepresentation of males in the final data.

Table 13.3 shows that for the incentive group completed interviews, the percentage of respondents in the various age categories was significantly different than the sample frame. The differences stem from 4.5% fewer completed interviews in the youngest age group (18–29 years), 1.2% fewer in the 30–39 years group, 1.6% more in the 40–49 years group, and 3.8% in the oldest (over 49 years) group. The composition for refusals and nonrespondents were also all significantly different than the sample frame. Refusals were less likely from those under age 39 and more likely from those over age 40. However, for nonrespondents, differences were even larger. There were 7.2% more 18–29 year olds and 7.8% fewer 49 year old nonrespondents. More people under age 40 were nonrespondents than those over age 40.

In summary, considering only gender and age in Table 13.2, when both sides of survey intervention (incentive versus nonincentive) are compared, the effect of the incentive was not to improve the representativeness. Instead, the incentive accentuated the disparity in numbers of males and the numbers of individuals in the oldest age group (those over 49 years) represented in the data. Thinking about nonresponse error, the incentive contributes to adding more error because it is less effective with the underrepresented group, individuals under age 30, and more effective

Table 13.2. 1995 Washington driver's licensing study, tests of contact wave for estimates of population characteristics

	Survey results compared with sample				Completed questionnaire composition for wave of survey contact, each compared to sample			
Measure	Random sample	Completes	Refusals	Nonrespondents	I	II	III	IV
Number	6,090	3,225	108	2013	1055	1016	458	696
Gender (%)								
Male	54.9%	49.0%*	57.4%	61.2%*	49.9%*	45.1%*	49.3%*	53.3%
Female	45.1	51	42.6	38.8	50.1	54.9	50.7	46.7
Location (%)								
Nonmetro	60.1	59.8	65.7	60.5	62.8	60.1	56.1	57.3
Metro	39.9	40.2	34.3	39.5	37.2	39.9	43.9	42.7
Age (%)								
18–29 yrs	36.1	31.5*	21.3*	43.3*	28.6*	29.9*	33.3	36.9
30–39	28.9	27.6	25	31.4	25.3	28.4	29.3	28.9
40–49	18.4	20	22.2	16.2	20.4	20.1	19.7	19.7
Over 49	16.6	20.9	31.5	9.1	25.7	21.6	17.7	14.5
Design factor (%)								
No incentive	6.9	5.0*	5.6	10.0*	3.7	3.6	6.6	8.1*
$2 incentive	93.1	95	94.4	90	96.3	96.4	93.4	91.9

Note: Comparisons of multiple chi square tests of significance using Bonferroni adjustment 4 tests to full sample.

*$p < 0.0025$.

Table 13.3. 1995 Washington driver's licensing study, tests of incentive for estimates of population characteristics

	Incentive group survey results[a] each compared with sample frame					Nonincentive group survey results[a] each compared to sample		
Measure	Random full sample	Completed questionnaires	Refusals	Nonrespondents	Random sample	Completed questionnaires	Refusals	Nonrespondents
Number	5669	3062	102	1813	420	162	6	200
Gender (%)								
Male	55.1	49.1*	56.9	61.9*	53.3	48.2	66.7	54.0
Female	44.9	50.9	43.1	38.1	46.7	51.8	33.3	46.0
Location (%)								
Nonmetro	60.2	62.4	50.0	60.0	60.0	59.7	59.8	60.4
Metro	39.8	37.7	50.0	40.0	40.0	40.3	40.2	39.6
Age (%)								
18–29	35.7	31.2*	19.6*	42.9*	40.5	33.9	50.0	47.0
30–39	28.7	27.5	23.5	31.2	30.5	29.6	50.0	32.0
40–49	18.4	20.0	23.5	16.2	18.8	20.9	0.0	16.0
Over 49	17.4	21.2	33.3	9.6	10.2	15.4	0.0	5.0

Note: Comparisons of multiple Chi square tests of significance using Bonferroni adjustment 4 tests to full sample.

$*p < 0.0025$.

with the over represented group, individuals over age 49. From the data where incentive and nonincentive response effects are compared side by side, we conclude that incentives increased the difference between respondents and nonrespondents and thus added to the nonresponse error by recruiting even more women and older individuals into the data. The net effect is that incentives contributed in two ways to increase response error by adding gender bias and age bias. For the part of the study not using incentives, the sample size is unfortunately considerably smaller, none of the comparisons for gender and age are significantly different. We can still see, however, that we have some differences by age and gender that are in the same direction as the differences resulting from incentives.

To further evaluate whether the significant differences found in response by demographic characteristics predicts nonresponse error, we calculated an outcome score for the survey results. The outcome score was a sum of the "yes" responses to questions about "technologies used in work." Men and women are significantly different on this score, with men reporting using a greater number of technologies at work than women. There is also an age difference, with the youngest age group (18–29-year-olds) significantly different when compared with the next two age groups of 30–39 years and 40–49-year olds. The middle-aged category of individuals reported using a greater number of technologies at work than both the youngest age group and the oldest age group.

13.5.3 Nonresponse Correlates in the Immunization Survey

Previous surveys of physicians suggest that nonresponse error is changed by adding additional contacts or by including or timing an incentive that influences the mix of physicians represented in the final data (Gunn and Rhodes, 1981; Berry and Kanouse, 1987; Guadagnoli and Cunningham, 1989; Tambor et al., 1993; Biner and Kidd, 1994; Del Valle et al., 1997). Berry and Kanouse found physician speciality to be sensitive to incentive payment timing, with early payment having the largest effect on improving response. Only two of the nine specialities were found to be represented significantly differently in the comparison of timing of incentives. Tambor et al. (1993), in an effort to increase response rates, tested whether an incentive increased nonresponse error on a key outcome variable—genetics knowledge—by comparing early respondents with those recruited later. They found no significant difference between treatment groups for genetics knowledge. Gunn and Rhodes found that physician specialties responded quite differently in the presence of an incentive. Comparison of early to late responders by et Del Valle et al. in a randomized trial showed the respondents had different distributions on characteristics of physician speciality, membership types, board certification, and practice settings. They concluded that the third certified mailing was important to the quality of the sample data obtained, not only because it increased the sample size, but because it increased representation of physicians with different characteristics.

In the Washington physician study (Table 13.4), physician speciality as a respondent characteristic is sensitive to the level of survey effort. Family practice physicians increasingly responded at each wave of survey contact, whereas pedi-

Table 13.4. Survey of the population of Washington physicians who immunize study, test of disposition for estimates of population characteristics and survey measures

	Survey results compared with population				Wave of survey contact			
Characteristic	Population	Completes[a]	Refusals	Nonrespondents	I	II	III	IV
Number	2472	1327	88	496				
Physician specialty (%)								
Family practice	68	70.5	87.5	77.8	65.9	74.2	78.2	77.1*
Pediatric	32	29.5	12.5	22.2	34.1	25.8	21.8	22.9
χ^2 (probability)		2.59 (.108)	15.0 (.001)	18.8 (.001)				
Physician practice location								
Rural (%)	15.4	15.7	18.2	15.1	15.3	14.6	14.6	16.4
Urban	84.6	84.3	81.8	84.9	84.7	85.4	85.4	83.6
χ^2 (probability)		0.66 (.81)	0.51 (.47)	0.02 (.89)				
Experimental group[b] (%)								
Priority mail start	50.4	53.9	46.6	47.4	64.3*	79.2*	5.6*	52.8
Telephone start	49.6	46.1	53.4	52.6	35.7	20.8	94.4	47.2
χ^2 (probability)		5.1 (.024)	.40 (.53)	1.17 (.28)				

Note: Comparisons of multiple chi square tests of significance using Bonferroni adjustment 4 tests to full sample.

*$p < 0.0025$.

[a]Completed questionnaires from each contact wave compared on known characteristics using chi square test of proportions where the expected value is that population proportion.

[b]Each observation randomly assigned to experimental group survey sequence. Group 1: priority mail and 1 week postcard reminder is wave I, first class questionnaire mailing is wave II, telephone follow-up with 28 attempts is wave III, and 24 week priority mail follow-ups in wave IV. Group 2: telephone start assignment where 1 to 4 call attempts is wave I, then 5 to 28 call-back attempts is wave II, priority mail follow-up is wave III, 24 week priority mail follow-up is wave IV.

atric physicians responded at a higher level at the first contact wave and response then tapered off at later waves of contact. Priority mail as an experimental treatment was found to be significantly more effective than starting survey contact by telephone mode. The final cumulative effect of extra survey effort in the form of switching modes, using priority mail, and doing a 24 week follow-up by priority mail is the overrepresentation of family practice physicians in the survey results when compared to the sample frame.

Comparisons on a key outcome variable for this study, the Immunization Opportunity score in Table 13.5, shows the consequences of overrepresentation of one type of physician speciality. This score shows family practice physicians taking significantly fewer immunization opportunities than pediatricians at each wave of contact. For family practice physicians, the Immunization Opportunities score is higher at the first wave of contact and decreases through Wave IV, whereas for pediatricians the Immunization Opportunity score is in the opposite direction and increases on average by Wave IV. Because the physician speciality characteristic is sensitive to extra survey effort and the change in Immunization Opportunity score, on average, is in the opposite direction, the overrepresentation of family practice physicians in the final survey data has consequences for the size and direction of the nonresponse error and the resulting data quality for estimates of the population of immunizing physicians. The overrepresentation of family practice physicians in the data tends to mitigate the level of immunization opportunities taken. Using late responders as proxies for nonresponders for each type of physician, we speculate the direction of the nonresponse error for each is different. While we only have information on the frame variable for physician speciality, we can show there is distortion in physician representation and that this is likely to translate into nonresponse errors for survey estimates, but it cannot be assessed whether overall nonresponse error is increased or decreased. To actually determine the size and magnitude of the resulting nonresponse error would require a priori knowledge of the actual immunization opportunities, which we do not have in this study.

13.6 DISCUSSION

We have examined the effects on sample estimates for two groups of mail survey respondents, one a general household sample of drivers license renewers and the other a population of physicians. We show how nonresponse error changes in magnitude and direction when demographic and socioeconomic variables for the sample and population are compared. We then compared outcome variables for each survey for the level and direction of change across the waves with respect to the demographic variables.

What we learned about survey participation and nonrespondents in the physician study is that with extra survey effort there are significant differences between early responders and late responders for both types of physician specialities, and that the differences on one survey outcome variable is in the opposite direction. The imbalance in the percentages of each type of physician represented in the final survey data

Table 13.5. 1997 Washington physician immunization survey, estimates of immunization opportunity; scores by wave of contact for sample characteristics

	Cumulative all completed questionnaires immunization opportunity		Immunization opportunity score (maximum value 17)							
			Wave I		Wave II		Wave III		Wave IV	
	Mean		Mean		Mean		Mean		Mean	
All respondents	11.13		11.20		10.81		11.06		11.20	
	Mean	t	Mean	t	Mean	t	Mean	t	Mean	t
Physician specialty										
Family practice	10.84	–0.364*	10.96	–2.32*	10.56	–1.20	10.80	–1.40	10.73	–2.59*
Pediatric	11.82		11.75		11.49		11.80		13.00	
Location										
Nonmetro	11.34	0.72	11.52	0.82	11.86	1.44	10.53	–0.67	10.82	–0.40
Metro	11.09		11.17		10.60		11.10		11.28	
Survey treatment Group										
Mail start	11.26	1.16	11.50	2.24**	10.79	–0.10	9.93	–0.89	10.86	1.00
Telephone start	10.98		10.74		10.87		11.08		11.60	

*$p < 0.0025$.

has consequences for survey estimates. Because late responders are significantly different than early responders, it demonstrates the importance of pursuing a high level of response. It also demonstrates the importance of selecting survey techniques that work equally well on respondents with differing characteristics important to the study.

What we learned in the driver's licensing survey is that in the presence of an incentive there will be significant differences between males and females and between older versus younger respondents in the final results. For the full study results, the incentive accentuated the representation of women and older individuals. Although it looks like the incentive worked well on all age categories for increasing response, it worked disproportionately better on older members and women. This can best be observed in the regression model of response. Specifically, in the regression results, the variables found significantly associated with response were older age, incentive treatment, female gender, and age by gender interaction. Although the presence of other survey design features improved the model fit, none were significant.

For the driver's licensing study, the outcome survey variable of Technologies Used in Work was significantly different for males versus females, younger versus middle age, and metro versus nonmetro location. We have no frame variable to measure the true score for Technologies Used in Work, therefore, the only way to estimate the change in nonresponse error (direction and magnitude) is to evaluate those demographic variables we have in the frame to see how later responders differ from early responders as they are brought into the study through more survey effort. For the Technologies Used in Work outcome score as we bring in more of the underrepresented males into the survey, the male score slightly declines from Wave I to Wave IV. For females, the score declines after the Wave II. For males, when the differences in means in contact Wave I are compared to Waves II, III, and IV, there is no significant difference. The same is true for females; there is no significant difference in the score comparing Wave I to Waves II, III, and IV. The score is also larger for males than females in Waves I to IV, but not significantly so. We don't have any significant difference when we compare males to females on this score at each wave of contact. The Technologies Used in Work outcome score is only significantly different for men and women cumulatively over the four waves of completed questionnaires. As both more men and women come into the study at later stages of contact, they have a lower mean Technologies Used in Work score and this tends to moderate the extent of the differences between the genders cumulatively. If the remaining nonrespondents after Wave IV are like the late responders in Waves III and IV, then we could speculate that their scores might even be lower and the differences might disappear as more male nonrespondents with lower scores enter the study. But we can't know this for certain, as we have no information on remaining nonrespondents.

13.7 CONCLUSIONS AND IMPLICATIONS

We have conducted two mail surveys for which we have several frame variables that allow us to compare respondents and nonrespondents at several follow-up waves of contact. We conclude, as have many others, that survey design features make a difference in improving response rates. We also conclude that extra survey effort improves survey estimates even if there exists disparity in the frame variables in the end. Survey design features such as incentives, priority mail, and multiple follow-ups bring the more resistive respondents into the study and sooner. For some respondents, this may be the only way to bring them into the study (e.g., younger males come into the study in the greatest proportion in Wave IV). Each wave of contact brings in different respondents based on frame variable comparisons at each contact wave. For the driver's licensing study, we see frame variables improving in representativeness over each wave as nonresponse decreases. However, based on frame variables, outcome variable differences between respondents and nonrespondents also increase over waves of contact. We don't know whether the outcome variable differences (based on frame variables only) reflect improvement in overall nonresponse error or are in fact making this error worse because of bringing in "more of the same" kinds of respondents (for instance, females), but also bringing in more of the underrepresented (males) in the driver's licensing study.

For the driver's licensing study and the physician study, we cannot estimate nonresponse error directly, but we can infer its direction and magnitude by comparing early to late responders on a study variable from the survey that showed significant difference for demographic frame variables and by comparing Wave I results to cumulative survey results (Waves I through IV). For the driver's licensing study, males and younger individuals are underrepresented and females and older individuals are overrepresented in the final survey results for frame variables. By using later responders as proxies for nonrespondents and if less survey contact effort is made by stopping at Wave I, the results show greater differences between Wave I respondents and nonrespondents by overstating the Technologies Used in Work score. In this instance, it seems that the nonresponse error is reduced by increased survey contact in that the overall Technologies Used in Work score decreases with increased contact. The nonresponse error between males and females reduces with increased survey contact. For the physician study, there are differences between family practice and pediatric physicians in response levels and in immunization opportunity score at each wave, and it is hard to evaluate the change in nonresponse error when the direction of change in the Immunization Opportunity Score is in opposite direction. We conclude that it is better to pursue higher levels of response even if the ultimate frame variables are distorted because it is better to represent these late responders in the data than not at all.

CHAPTER 14

Understanding Unit and Item Nonresponse in Business Surveys

Diane K. Willimack and Elizabeth Nichols, *U.S. Census Bureau*
Seymour Sudman, *University of Illinois at Urbana-Champaign*

14.1 INTRODUCTION

Economic data are critical for policy makers, researchers, and business decision makers for monitoring, understanding, managing, and forecasting the economy. The measurement of gross domestic product, production and inventories of farm commodities, energy availability and use, hospital admissions, and school staffing are a few examples of statistics measuring the economy. All of these depend on business surveys.

In this chapter, we highlight problems of unit and item nonresponse for many business surveys in the United States. In Section 14.2, we summarize existing literature on nonresponse in business surveys. In Section 14.3, we describe our research. Through a series of company visits, we investigated unit and item nonresponse problems for large multiunit companies. In Section 14.4, we synthesize our findings with the existing literature and propose a model of factors that affect nonresponse in business surveys. We conclude with suggestions for further research in Section 14.5.

14.2 REVIEW OF THE LITERATURE ON BUSINESS SURVEY NONRESPONSE

14.2.1 The Conceptual Literature

Tomaskovic-Devey et al. (1994), Edwards and Cantor (1991), and Groves et al. (1997) consider aspects of authority, capacity, and motivation relative to respondent selection to have significant bearing on business survey participation. All tend to focus on the critical role of respondent selection relative to both unit and item nonresponse, noting the need for knowledge of the requested data as well as the authority to release data. However, organizational hierarchies often distinguish staff with authority to release data from those having direct knowledge of the data and, thus, the capacity to respond. In addition, divisions of labor and decentralized data require that dispersed knowledge may need to be assembled to satisfy survey requests. This further complicates respondent selection. In addition, businesses' major goal, i.e., to generate profits, may cause them to view some parts of their external environment as "hostile," which frames their motivation for considering outside requests for information, such as surveys (Groves et al., 1997).

14.2.2 Empirical Studies of Unit Nonresponse

Unlike in household surveys, the size of the nonresponding units plays a critical role in the measurement of unit nonresponse in business surveys. The Interagency Group on Establishment Nonresponse (IGEN) (1998) reports common use of two general formulas for response rates in business surveys. The unweighted response rate, calculated as follows, indicates the proportion of eligible units that cooperate in the survey:

$$\text{Unweighted response rate} = \frac{\text{Number of responding eligible reporting units}}{\text{Number of eligible reporting units in survey}}$$

Also calculated is a weighted response rate, which considers the importance assigned to reporting units in terms of their sampling weights (i.e., inverse of the probability of selection):

$$\text{Weighted response rate} = \frac{\begin{array}{c}\text{Total weighted quantity for}\\ \text{responding reporting units}\end{array}}{\begin{array}{c}\text{Total estimated quantity for}\\ \text{all eligible reporting units}\end{array}}$$

The weighted rate measures the proportion of some estimated population total that is contributed by respondents. Since it is common in business surveys for a small number of large businesses to account for a major proportion of the population total, the weighted response rate is considered a better indicator of the quality of the estimate.

There appears to be no clear declining trend over time for business survey response rates, unlike in household surveys (see Chapter 3). Studies of domestic U.S. and international government sponsored censuses and surveys show mixed trends in nonresponse rates over time, with nearly the same number of surveys showing increasing as decreasing nonresponse (IGEN, 1998; Christianson and Tortora, 1995).

An authoritative sponsor and a legal mandate clearly produce higher response rates among businesses, just as in households. Internationally, Christianson and

Tortora (1995) report unit nonresponse rates ranging from 0–25% for censuses and from 0–60% in surveys, with medians of 3 and 13%, respectively. Table 14.1 shows that unweighted unit response rates for selected censuses and surveys sponsored by U.S. government statistical agencies range from approximately 60% to over 90% (IGEN, 1998). The median nonresponse rate from Table 14.1 is 21%, somewhat higher than the international rates. On the other hand, university survey researchers appear to suffer from substantially higher nonresponse rates in business surveys than do government agencies. Various studies report nonresponse rates ranging from 35% (Osterman, 1994a, 1994b; Spaeth and O'Rourke, 1994) to nearly 50% (Paxton et al., 1995; Tomaskovic-Devey et al., 1994).

Evidence for mandatory authority can be seen in Table 14.1, where the simple average response rate is 86% for mandatory surveys compared to 81% for voluntary reporting, not accounting for other design differences. Controlled experiments show mandatory survey response rates 13–23% higher than those for voluntary reporting (Tulp and Kusch, 1993; Cole et al., 1993; Groves et al., 1997).

Some evidence suggests response rate improvement using targeted respondent selection methods to identify the respondent with specific knowledge appropriate to the survey topic, rather than addressing the survey generically to the office of the chief executive (Ramirez, 1996). Shatos et al. (1998) and Christianson and Tortora (1995) acknowledge that finding the appropriate respondent often requires getting past a gatekeeper, which is a formal permanent role in businesses (e.g., receptionist, secretary) not found in households. Pietsch (1995) stresses that "profiling" businesses—"procedures to verify the proper respondent and address, as well as the status of the business and its establishments"—is particularly cost effective for large multiunit companies.

However, Moore and Baxter (1993) did not find an overall response rate difference using a specific contact name when other Total Design Method (Dillman, 1978) techniques were used, suggesting the utility of follow-up activities. Experimental studies associated with the U.S. Census Bureau's 1990 Pollution Abatement Costs and Expenditures study showed that phone calls following initial mail-out or using a series of mail follow-ups improved response to the voluntary survey, while use of certified mail was most successful, increasing response rates by 32 percentage points (Tulp and Kusch, 1993). A nonexperimental study by Shatos et al. (1998) found a higher response for a nongovernment business survey where a telephone contact was the primary mode with a follow-up mail survey for nonrespondents.

In addition, common business survey practice offers multiple response modes concurrently, including electronic reporting options such as disk by mail, Internet, electronic data interchange, and touchtone data entry. Although studies have shown that companies appear to prefer electronic modes, this has not translated into higher response rates in controlled experiments (Ramos et al., 1998; Bond et al., 1993).

There is some indication that the salience of the survey topic may improve business response rates. Fecso and Tortora (1981) and Jones et al. (1979) found that farm operators participating in the Crop and Livestock Survey (sponsored by the National Agricultural Statistics Service) reported using more of the survey results.

Table 14.1. Response rates for selected censuses and surveys sponsored by U.S. government agencies

Sponsoring agency	Survey name (frequency[a]) (legal authority[b])	Unweighted unit response rate
National Agricultural Statistics Service	Agricultural Survey (Q) (V)	82.4% (1996); 80.5% (1997)
	Agriculture Resource Management Study (A) (V)	59% (1996); 72% (1997)
	Census of Agriculture (5 years) (M)	85% (1992); 86% (1997)
Census Bureau	Annual Survey of Manufactures (A) (M)	90% (1996)
	Census of Manufactures (5 years) (M)	85%–90% (1992)
	Company Organization Survey (A) (M)	86% (1996)
	Construction Value Put-In-Place Surveys (M) (V)	54%–90% (1996)
Energy Information Administration	Commercial Building Energy Consumption (3–4 years) (V)	87%
	Consumption & Expenditures Survey (3–4 years) (V)	89%
Bureau of Labor Statistics	Current Employment Statistics Program (M) (M and V)	70% active reporters (1997)
	Hours at Work Survey (A) (V)	70% (1994)
	International Price Program (A) (V)	79% (1997)
	Occupational Safety and Health Survey (A) (M)	92% (generally)
	Occupational Employment Statistics Survey (A) (M and V)	70% (1996)
National Center for Health Statistics	National Ambulatory Medical Care Survey (Continuous) (V)	70% (1996)
	National Hospital Ambulatory Care Survey (Continuous) (V)	95% (1995)
	National Hospital Discharge Survey (Continuous) (V)	94.7% (1996)
National Center for Education Statistics	Schools and Staffing Survey (3 years) (V)	80%–90%
Bureau of Transportation Statistics	Commodity Flow Survey (5 years) (M)	76% (1992/1993)

Source: IGEN, 1998.

[a]Frequency: A = annual, Q = quarterly, M = monthly.

[b]Legal authority: V = voluntary, M = mandatory.

They were also more aware of the different data users, and were generally more positive toward government data collection activities than were nonrespondents.

14.2.3 Empirical Studies of Item Nonresponse

Measurement of item nonresponse for business surveys is complicated by difficulty defining eligibility for some data items and by including data obtained during call-backs for initially missing, misreported, or questionable responses (Wakim, 1987). Missingness may not be substantive when a questionnaire section containing components and their sum is incomplete, but the missing values can easily be calculated from the items present. Editing and imputation rates overstate item nonresponse because corrections for erroneous data are often intermingled with imputation for missing items.

The difficulty in monitoring item nonresponse in business surveys is suggested by Christianson and Tortora's (1995) survey of international government statistics offices. Only 32 of 104 government-sponsored surveys reported item nonresponse rates. They ranged from 0–39% for major variables, with a median rate of 10%. Similar rates were reported at the U.S. Census Bureau, where weighted item nonresponse rates for 11 economic censuses ranged from 1–28%, with a mean of 12% and a median of 10%. For 29 economic surveys, the range was from 0–50%, with a mean of 14% and a median of 10% (King and Kornbau, 1994).

As with unit response rates, the mandatory/voluntary requirement influences item nonresponse. In the U.S. Census Bureau's 1990 Survey of Industrial Research and Development, the response rate for four critical questions was 20 percentage points lower in an experimental voluntary panel than in the traditional mandatory panel (Cole et al., 1993; Worden and Hamilton, 1989).

14.3 THE RESPONSE PROCESS IN LARGE MULTIUNIT COMPANIES

In this section, we describe qualitative exploratory research conducted at the U.S. Census Bureau during 1998–1999 to study the statistical reporting process in large multiunit companies, defined primarily in terms of payroll and employment. Typical of business surveys, the largest firms account for a very large portion of census statistics and survey estimates. We also believed these large companies face more complex reporting issues than smaller companies. Thus, these results cannot be assumed to apply to medium- and smaller-sized firms, or to reporters not located at corporate headquarters. Although statistical analyses and tests are not appropriate, our investigation provides insights into the behaviors of business survey respondents, suggesting hypotheses about influences on those behaviors.

14.3.1 Methodology

Unstructured interviews were conducted during group meetings with company headquarters staff responsible for government reporting during site visits to thirty large multiunit companies. We attempted to arrange meetings with 37 companies, but seven did not participate for a variety of reasons. The companies were selected judgmentally to represent a variety of industry types, with many companies diversified into multiple industries. Both public and privately owned companies were selected, as well as companies with varying degrees of foreign involvement. The selected companies also exhibited differing cooperation rates on various Census Bureau surveys and censuses.

The Census Bureau's business register containing company contact names was used to arrange the meetings. Typically, the company was represented by two to six members of the headquarters financial reporting staff responsible for completing government surveys. Census Bureau participants included a project researcher, one or two subject area specialists, and the project sponsor, a Census Bureau senior executive.

Meetings typically lasted at least three hours and were audiotape recorded with permission from the company. Using a protocol tailored for each company based on background research, a variety of topics were discussed, including company organization and information system structure, availability of data, respondent selection and response strategies, and perceptions of confidentiality and burden. We did not focus on a single Census Bureau census or survey report form; instead, we typically reviewed one or two of the 1997 Economic Census forms the company received, and two of the current survey forms completed at corporate headquarters. This chapter describes findings that relate to unit and item nonresponse. Other papers (Willimack et al., 1999a, 1999b; Nichols et al., 1999a, 1999b) provide results about other topics addressed in this research.

14.3.2 Findings

Organizational Changes. One of our most striking findings was the frequency with which large companies change their organizational structures due to mergers, acquisitions, divestitures, closings of divisions, or internal restructuring. These frequent changes appeared to affect the timeliness of their response to surveys. Organizational changes added workload for company reporters, affecting their ability to respond to surveys. These changes also had significant implications for communicating and maintaining the desired unit for reporting data, particularly when that reporting unit differed for different surveys.

Respondent Selection/Identification. We learned that a major consequence of frequent organizational changes is frequent staff turnover among those assigned to government reporting. Professional mobility was also a reason for turnover. It was rare to find the same reporter for consecutive economic censuses, which occur every five years.

Large companies typically staff a unit with primary responsibility for most, but not all, of the external government reporting for both regulatory and statistical purposes (Nichols et al., 1999a). In many of our interviews, corporate reporting staffs were unfamiliar with reporting requirements fulfilled by other staffs, such as human resources or tax departments.

The respondent is typically selected depending on access to the necessary data and other competing job responsibilities. We discovered use of documentation and previously completed reports for ongoing data requests, even if there is a new respondent. This suggests that any item nonresponse will, at minimum, be consistent with prior reports, and not necessarily increase with a change in the respondent. Staff turnover is more likely to affect unit nonresponse when there is no historical knowledge of who completed the form in previous periods and when the mailing label does not specify this person.

Timing and Priorities. When considering the availability of staff, survey due dates are considered relative to other tasks. Staff workloads are heaviest during the first quarter of the company's fiscal year, when annual reports, Securities and Exchange Commission (SEC) filings, and tax returns must be prepared. Survey due dates falling within this period are problematic, since statistical data requests receive lower priority.

Data may not be available for statistical reporting until figures are summarized for internal reports prepared on a corporate accounting timetable. Most companies indicated that data would not be released to the Census Bureau until it had been released to stockholders or reviewed by upper management. If corporate timing cannot be accommodated by the data collector, unit nonresponse may occur.

Data Availability and Retrieval. Our company visits revealed that the availability of data lies along a continuum. Some data are directly available from company information systems and some data are compiled or calculated based on data that can be directly retrieved. Other data require some degree of effort to retrieve, involving multiple data sources and/or providers. Some data do not exist at all in company records. Examples from the economic census illustrate these problems:

- *Establishment revenue:* This is very problematic for service industries that have a networked structure, such as communications, transportation, financial, or consulting firms, where the term "establishment" is not meaningful and geographic breakouts of revenue do not exist.
- *Detailed operating costs:* Detailed data on operating costs are not available to corporate reporters (e.g., utilities costs, which are frequently included with facility rental fees, or the cost of paper isolated from general office supplies).
- *Detailed merchandise/product lines:* Companies generally keep track of revenues by merchandise lines, which may not match those requested in the economic census. The more detailed the census definitions of merchandise lines, the less likely the companies are to keep data that way.
- *Employment:* Although most companies are able to report employment information by establishment, financial reporting staffs are often unable to meet the requirement that these data be reported for the pay period including March 12, the reference date for quarterly tax forms (941s).

- *Employer Identification Numbers (EINs):* EINs, which identify legal entities, are maintained by legal and/or tax departments and are generally not available to corporate financial offices.

For data that are not available or are available with difficulty, company reporters resort to estimation strategies. The most common and simplest estimation method reported during our company visits was to allocate revenues relative to establishment payroll figures. In general, company reporters preferred to use a reasonable estimation scheme rather than leave an item blank.

Once procedures were developed to retrieve data for one establishment in a company, these could be applied to any number of them. In addition, survey response verification was often performed by financial staff managers, and selected reported data items were summed across all establishments for comparison with readily available corporate aggregates.

Authority of the Survey Sponsor. Concerns about data confidentiality did not appear to be a reason for unit or item nonresponse on Census Bureau surveys for the large companies we visited. Company reporters expressed a general trust that the Census Bureau keeps data confidential. They also noted that most of the data requested were already public from SEC 10K filings and annual reports.

Companies reported making a distinction between statistical requests and requests from taxing and regulatory agencies, such as the Internal Revenue Service (IRS), the SEC, and the Federal Trade Commission, which are required to keep their businesses open and growing. Companies also distinguished between mandatory and voluntary surveys, though legal mandates were not considered a threat to the viability of their businesses. Response, therefore, may not be timely. Nonresponse to voluntary monthly surveys was attributed to management decisions about resource use, since additional work is required to prepare unaudited or unsummarized data for survey response.

Topic Salience. Respondents in our study did not see a direct benefit to their companies for completing statistical reports. Instead, they noted the cost in terms of resources used to fulfill these requests. Data reporters were not familiar with the published data series since they were not direct users, and they were generally uninterested in the publications we gave them.

Survey Procedures. Callbacks were met with mixed reviews. Some companies clearly felt callbacks associated with questionable or missing data or late responses presented extra reporting burden and were something to be avoided. Others relied heavily on callback reminders.

These large companies saw benefit in prenotification for all upcoming requests. They thought annual notification, including brief descriptions of needed data, the names of previous reporters for recurring surveys, and survey due dates, could help them during their resource planning stage. They implied this type of information would help response rates.

In addition, all respondents were enthusiastic about electronic reporting. Not only were these staff computer savvy, using computers and electronic spreadsheets daily, but the data necessary to complete requests were often stored on computers. Some companies had translated our paper questionnaires into in-house electronic versions, especially those that acquired data from several sources across the company. Automating the questionnaire appears to be a logical step in terms of reducing burden and, hopefully, increasing response.

14.4 A PROPOSED CONCEPTUAL FRAMEWORK FOR BUSINESS SURVEY PARTICIPATION

Based on findings from our company visits, along with key points from the literature cited earlier, we present a conceptual framework for business survey participation in Figure 14.1. We adopt the dichotomy used in the Groves and Couper (1998) framework for household survey participation of factors that are either under or out of the researcher's control, because this is useful for considering practical survey design as well as empirical research. We modify their major dimensions of the household model to fit the business environment. Out of the researcher's control are factors relating to the external environment, the business, and the respondent. Under the researcher's control are survey design features. In making the survey participation decision, we conceive that businesses weigh response burden against business goals, which are both functions of these four major dimensions.

14.4.1 External Environment

Since businesses can assign a direct economic cost to completing surveys, general economic conditions may influence business survey participation. A weak economy may cause businesses to be more protective and reluctant to disclose information to outsiders. It may also cause staff layoffs, resulting in fewer employees available to fulfill survey requests.

The survey-taking climate impacts businesses even more than households. Our company visits highlighted the large number of surveys that businesses receive, particularly from multiple government agencies. Also noted was the volume of routine legal and/or regulatory requirements imposed upon businesses, mainly by government agencies. Our findings showed that these reporting requirements receive priority over survey participation. Further, since businesses do not distinguish among the various government agencies, seeing instead a single government, they add up the burden of the multiple requests and make resource allocations accordingly.

14.4.2 The Business

A key factor related to both unit and item nonresponse is the availability of data for the business. Business characteristics, such as size, type, industry participation,

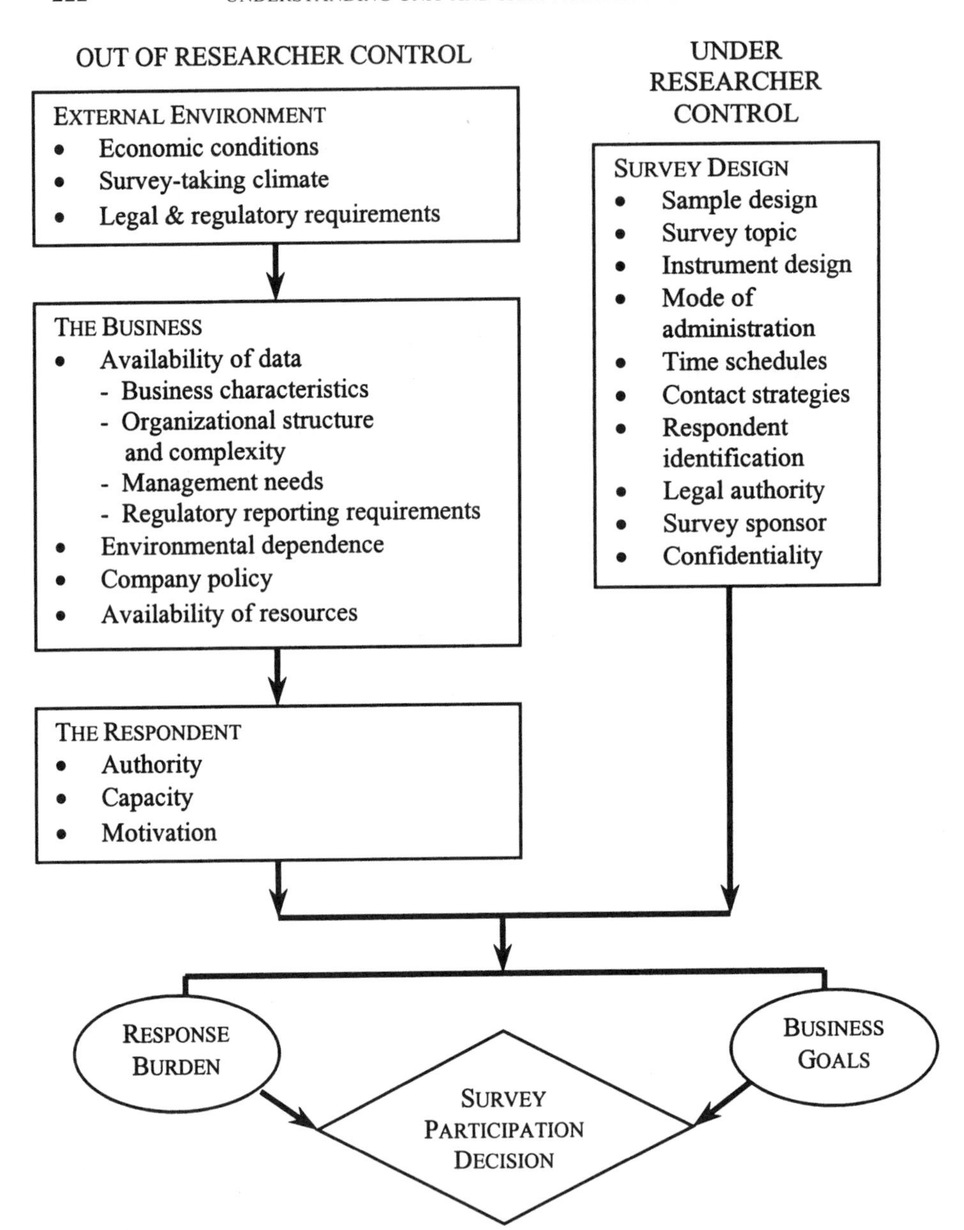

Figure 14.1. A conceptual framework for business survey participation.

public versus private ownership, and foreign involvement, influence the types and levels of data kept in business records. In addition, our study corroborates Tomaskovic-Devey et al. (1994), showing that organizational structure and complexity drive the structure and location of business records, as well as which data are available to whom. Management needs, along with legal and regulatory reporting requirements, influence several dimensions of record formation and data availability: the type of data and their level of detail or aggregation, the structural units for which data are kept, the timing of updates and summarization, the location of the data, and the staff positions with data access. The more closely survey requests match recorded information, particularly data that are routinely retrieved for various purposes, unit and item response are facilitated (Edwards and Cantor, 1991).

According to Tomaskovic-Devey et al. (1994), environmental dependence affects both response capacity and motive. "Boundary-spanning units" manage business interactions with outsiders while protecting the technical core, where more intimate knowledge of and access to data exist. The corporate reporters we visited acted as boundary-spanners with survey-takers, interfacing with local data providers in the technical core to retrieve data. Businesses that are more dependent on their environments, such as publicly traded firms, have higher motivation to disclose information, while those that are insulated or in unregulated environments are more protective of information. Among companies we visited, those in more volatile industries were more protective of their data, because releasing information could result in loss of competitive advantage (Groves et al., 1997).

Environmentally influenced motivation may be institutionalized with a company policy on survey participation. Businesses may have policies that they will only respond to surveys that are legally mandated. We found, however, that informal policies against reporting on voluntary surveys were primarily driven by burden and resource issues.

The availability of resources is a major factor affecting business ability to respond to surveys. The availability of staff, the timing of data requests relative to other priorities and workload, and the amount of effort needed to compile data from multiple sources all have bearing. All companies in our study spoke of these resources in the context of burden—the burden of multiple government data requests, regardless of their purpose; the burden imposed by the timing of the requests relative to more urgent priorities; the burden of callbacks requesting clarification or additional data—and thus sought efficiencies in responding to statistical data collections. Nonresponse reduction will be enhanced by efficient use of business resources in the survey response task.

14.4.3 The Respondent

We purport that selection of the respondent ultimately belongs to the business. Thus, attributes of the respondent are out of the control of the researcher.

Both authority and capacity are required to provide business data, but these need not reside in the same person. "Authority" has traditionally referred to the ability of business staff to take responsibility for releasing data to the outside (Edwards and

Cantor, 1991; Tomaskovic-Devey et al., 1994). In our study, this authority appeared to reside with mid-level financial managers who subsequently assigned the response task to their staff. Thus, authority includes the ability to delegate or assign the survey completion task itself to subordinate staff, who have the capacity to retrieve the data. "Capacity" refers to knowledge of data sources and the ability to retrieve and compile data from multiple data sources and providers.

Once the response task is assigned, it becomes a job responsibility of the assigned staff person. Their motivation, or strength of attention to the response task, is determined by competing priorities and job performance evaluation criteria, both set by their supervisors. In addition, our research suggests that business respondents, who are typically accountants, are also motivated by a professional standard of consistency—that financial information about the company released in various formats and outlets appears consistent. This likely contributes to more complete response and better data quality.

14.4.4 Survey Design

Under the researcher's control are aspects of the survey design, which impact the potential for both unit and item nonresponse in business surveys. Sample designs in business surveys tend to include the largest businesses with certainty, because they contribute disproportionately to most economic statistics, particularly estimates of population totals. This causes the same businesses to be included in many surveys, and may result in nonresponse as burden accumulates. However, companies we visited saw little merit in subsampling individual establishments to limit survey burden, because of data retrieval routines and company initiated data verification requiring aggregation across all establishments.

In household surveys, the salience of the survey topic is believed to increase response rates. However, our company visits revealed that company reporters tend to be unfamiliar with and unconcerned about the published data series generated from statistical collections. To them, the survey topic manifests itself in the specific data items requested, and the more critical issue is the availability of these data. Unit and item response will be higher for survey topics for which data are available and easily retrieved.

Instrument design, such as the length and format of the questionnaire, can affect both unit and item nonresponse (Paxton et al., 1995). However, our company visits suggest that design of the questionnaire and instructions likely have a stronger effect on the accuracy of completed data items than on item nonresponse.

A number of survey design elements may impact unit or item nonresponse by facilitating or hindering business efficiency. The mode of administration affects the resources required to communicate survey response. Offering multiple alternative modes enables businesses to choose the one that presents efficiency in reporting. Our study respondents embraced electronic reporting, perceiving it to reduce reporting burden and facilitate survey response, since their data were stored in automated information systems.

In addition, our research indicates that survey time schedules are considered relative to business schedules for data availability and other reporting priorities. During periods when business workloads are defined by various internal and external reporting requirements, such as the first quarter of the fiscal year, unit response is likely to suffer. Contact strategies, such as prenotification, callbacks, or follow-ups, can be used to encourage cooperation during data collection or to acquire missing data. These can also be used for respondent identification to locate the appropriate respondent prior to survey distribution. Our company respondents suggested that prenotification, an advance survey schedule, and identification of previous reporters would help them plan for upcoming work, facilitating efficient resource use.

Legal authority, the survey sponsor, and confidentiality affect unit response rates. Empirical evidence presented in Section 14.2.2 shows that mandatory surveys achieve higher response rates than do voluntary ones. Government agencies achieve higher response rates, even on voluntary surveys, than do academic researchers. We speculate that academic researchers likely get better response than commercial market researchers, although perhaps not better than business trade associations. Finally, even businesses that routinely exchange data with the outside, such as publicly held companies, nonprofit organizations, and government entities, consider some types of data to be strategically sensitive at particular times. Thus, confidentiality pledges are necessary to assure response.

14.4.5 The Decision to Participate

We conceive that factors of the external environment, the business, the respondent, and the survey design manifest themselves, from the perspective of the business respondent, in terms of response burden and business goals. Business respondents weigh these when deciding to participate in a survey, creating a cost–benefit model.

Response burden appears to be a function of many factors—frequency and timing of data requests, the similarity of requested data from multiple data collectors, the availability of data from business information systems, the retrievability of the data, the distribution of data across multiple organizational units, follow-up procedures and callbacks, and the data collection mode. These factors represent tangible and intangible costs of survey participation to businesses.

In making survey participation decisions, business respondents weigh the burden of response against business goals. The primary goal of a business is to generate a profit for its owners. Although businesses depend on information to operate effectively, often making extensive use of government data, our research shows that data providers are usually not well aware of data use by others in their own companies. Thus, survey participation is considered a nonproductive activity, resulting in a cost to the business that does not generate profit. As a result, in order to reduce costs, businesses seek efficiencies in the response task.

Businesses also strive to be good corporate citizens and to have a positive public image. In the long run, this yields a better climate for their activities and a more supportive business environment that ultimately leads to increased profits. Thus, responding to government data requests, particularly mandatory surveys, is considered a corporate responsibility, even though businesses acknowledge that lack of compliance will not put them out of business.

14.5 IMPLICATIONS FOR RESEARCH ON BUSINESS SURVEY NONRESPONSE

Research is needed to understand the left side of model and its effect on survey response. This will help survey researchers identify and test survey procedures that will facilitate, rather than hinder, business processes.

Further research should study the survey response process in medium- and smaller-sized firms, and for data reporters at other business levels or locations, since our research did not address these differences. Findings should be evaluated relative to our model to identify pertinent components. For example, we hypothesize that survey response among smaller firms is more strongly associated with availability of resources, which impacts respondent selection and identification.

In addition, an empirical study of business record-keeping practices is needed to provide a description of factors related to data availability relative to various business characteristics. Research is also needed to determine the appropriate level of respondent authority for survey participation, along with the relationship between survey response and the respondent roles of authority versus capacity. All of these will provide insights around which survey procedures can be designed and tested.

Our model also has implications for survey procedures. Since survey participation represents a cost with no associated profit, burden reduction lies in survey organizations designing procedures that reduce the costs of response through facilitating efficiencies in business reporting. This explains requests from the companies we visited for advance notice of surveys to help them plan and integrate survey requests into existing processes, as well as their enthusiasm for electronic reporting to ease the reporting activity. A research agenda should include measures of the effectiveness of these procedures in reducing burden and increasing response.

Burden reduction may also be achieved through flexibility in survey procedures, such as providing multiple alternative response modes, offering concessions in timing, relaxing data needs, varying follow-up procedures, and making efforts to identify appropriate respondents. Customizing survey procedures to business respondents may improve response. However, survey organizations must contend with broader needs for customization and a wider variety of tailoring methods for business surveys than are needed or used in household surveys. Nevertheless, we conclude by suggesting that this "designed inconsistency," or "controlled variation," in survey design features will likely prove superior to using standardized survey procedures in business surveys.

Research on these and other related topics will improve our understanding of business survey response processes from the perspective of the businesses and lead toward improved survey designs that reduce respondent burden and improve survey participation.

ACKNOWLEDGMENTS

Research reported in this paper was conducted while the late Professor Sudman was at the U.S. Census Bureau under an Intergovernmental Personnel Agreement. This chapter is dedicated to his memory. The authors thank Thomas L. Mesenbourg, Assistant Director for Economic Programs at the U.S. Census Bureau, for his significant contribution to and support of this research. They also gratefully acknowledge helpful review comments from Robert Groves, Joseph Garrett, W. Sherman Edwards, Charles P. Pautler, and Eileen O'Brien. The views expressed are the authors' and do not necessarily reflect those of the U.S. Census Bureau.

CHAPTER 15

Nonresponse in Web Surveys

Vasja Vehovar, Zenel Batagelj, Katja Lozar Manfreda, and Metka Zaletel, *University of Ljubljana, Slovenia*

15.1 INTRODUCTION

Web questionnaires have been placed on thousands of Web sites and Internet survey panels now include millions of Internet users. In addition, technology continues to improve all aspects of doing Web surveys: standardized software, user-friendly interfaces, attractive multimedia, merger of technologies (Web, TV, phone, VCR), high speed of transmission, and low access costs. During the next years, increased Internet penetration, massive Web usage, and technological improvements will further expand the application of Web surveys. However, nonresponse to such surveys is a serious problem. Our purpose in this chapter is to discuss the nonresponse process, factors that contribute to its occurrence, and its consequences.

The Web survey mode is based on computer-assisted self-administered questionnaires answered without the presence of the interviewer. The questionnaires are based on HTML forms usually presented in standard Web browsers, and the responses are immediately transferred through electronic networks, usually the Internet. We further limit our discussion to the basic Web survey mode, where respondents record their answers manually (with a keyboard or a mouse), written questions are the core layout on the screen, there is no on-line interaction (help) with the interviewer, and multimedia are only used to illustrate the survey questions.

Our discussion of Web surveys draws on an extensive on-line literature base we have compiled (http://websm.org). In addition, data are reported from the Research on Internet in Slovenia (RIS) national project (RIS, 1996–2001), conducted at the Faculty of Social Sciences, University of Ljubljana since 1996. In particular, we refer to RIS Web surveys in which participants were invited via e-mail solicitation (with two follow-ups) through addresses from the public directory. In 1998, one tenth (n = 6,500) of active Slovenian Internet users participated in the RIS Web survey. Due to a small population (2 million) and moderate Internet penetration (15%) a large post-Web telephone survey (n = 10,000 households) enabled a study to be done of the units that were aware of the Web survey but did not participate.

15.2 NONRESPONSE PROCESS IN WEB SURVEYS

Nonresponse occurs across several stages of the Web survey process. In Figure 15.1 we outline the key steps of involvement and the corresponding groups of participants that also form the basis for calculations of nonresponse indicators.

In Web surveys, the target population often differs dramatically from the operational population. In the general population, those who actually access the Web usually represent only a minor part of the target population. Many of those who can connect to the Web do not actually use it. In the United States, for example, out of around 100 million Internet users only 40% actually accessed the Web from home in the month of July 1999 (http://www.nielsen-netratings.com).

The frame population refers to the units that can actually be reached. Typically, when e-mail is used, the frame is further restricted to e-mail users. In Slovenia, for example, as of January 2000, only 60% of monthly Internet users possessed a personal e-mail address. In addition, frames of e-mail addresses are rare and incomplete and a quarter of Internet users explicitly rejected being included in these frames (Batagelj and Vehovar, 1998). This percentage is similar to the customers of e-commerce Web sites who will not reveal their e-mail address (Enander and Sajti, 1999).

When e-mail is used, incorrect spellings—which usually survive postal delivery—are fatal. Together with changed, multiple, or fictitious addresses (Comley, 1996), the percentage of messages that are immediately returned is usually high, particularly when the e-mail address list has not been edited. Even when users (re-

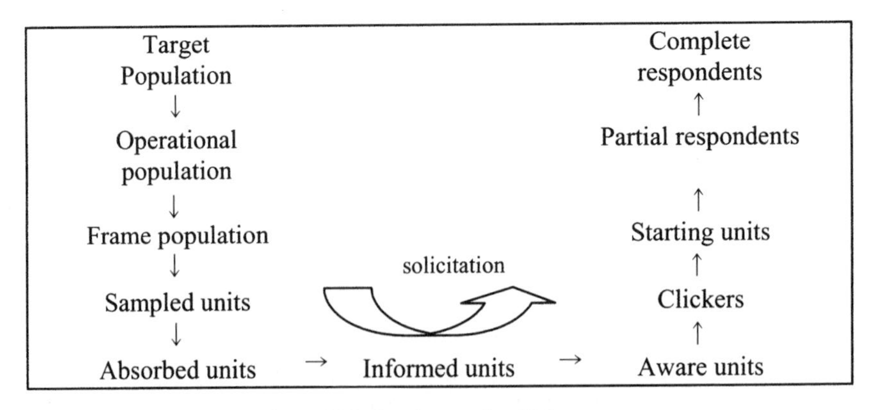

Figure 15.1. Involvement in a Web survey.

spondents) type the e-mail address (i.e., e-commerce customers), at least 10% of the e-mails are returned (Flemming and Sonner, 1999, Enander and Sajti, 1999), but this figure can be as high as 35% (Comley, 1996) or even higher when less accurate frames are applied.

However, not all wrong addresses are returned to the sender; some are absorbed by the network together with the correct ones. The informed units are those that actually receive the e-mail message. The exact proportion of absorbed but lost e-mails is extremely difficult to calculate. In a highly controlled e-mail solicited Web survey (Enander and Sajti, 1999) the post-Web telephone survey revealed that 7% of the sampled units claimed they had not received the invitation and an additional 28% were not sure about receiving it.

The informed units may remain unaware of the invitation for a variety of reasons and only the aware units can decide whether to participate in the survey or not. This group—although difficult to measure—is the base for the calculation of cooperation rates, the percentage of respondents among eligible and contacted sample units (Groves and Couper, 1998). In the above-mentioned study of e-commerce customers, this rate was between 51% and 76% (Enander and Sajti, 1999), while in the RIS 98 Web survey this percentage was estimated at 39%.

Among the aware units, only the clickers actually try to locate the Web survey page. In the RIS 98 Web survey, this rate was 44% (among aware units), which is relatively high. The highest reported click-through rate among contacted units was 76% (Dillman, 2000, p. 374) in a study of customers where the telephone was used for the initial contact, e-mail for reminders and a two dollar incentive was sent by mail.

However, not all clickers proceed to the survey questions. Only the starting units do; others may not be persuaded by the invitation, or may have gained access with no intention of responding. In addition, some potential respondents observe the survey questions without answering, or, they start answering but then stop. To distinguish partial respondents from nonrespondents, the completion of an initial block of questions is usually required. Of course, the complete respondents must, in addition to finishing the questionnaire, successfully perform the submission procedure.

Partial response refers to the percentage of respondents who complete only a portion of the questionnaire among all respondents (partial and complete). It is alternatively expressed as attrition rate (Kehoe and Pitkow, 1996), failure rate (Jeavons and Bayer, 1997), or drop-off rate (Kottler, 1997a,b). Partial nonresponse depends on a variety of factors and can vary from 5% for a survey of users who previously (in a screening questionnaire) agreed to participate (Kottler, 1997a) to 37% (Batagelj et al., 1998) for a long and complex questionnaire.

We can conclude that the stages of the nonresponse process in Web surveys are relatively complex. The corresponding indicators are undeveloped and are lacking standardization. The same holds true for optimal design strategies; therefore, we do not really know whether the proper approach was undertaken. As a consequence, the available research exhibits low values and extreme variation in response rates, e.g., the percentage of completed questionnaires among all eligible units included in the sample.

The reported response rates in e-mail solicited Web surveys are below 50% in the vast majority of available studies. In addition, in telephone-recruited and e-mail-solicited Web surveys the overall response rate hardly reaches 30% (Comley, 1997, 1998; Flemming and Sonner, 1999; Hollis, 1999), regardless of the context. Even in the RIS 98 Web survey—with three e-mail invitations applied in a standard TDM pattern (but no incentives)—the response rate was 39%, which can be compared to a related RIS telephone survey response rate of 70%. Similar results were observed in a survey of e-commerce customers (Enander and Sajti, 1999).

15.3 PARTICIPATION IN WEB SURVEYS

In general, the basic features determining survey cooperation (characteristics of respondents, social/technical environment, and survey design) play a role in Web surveys as well. However, some differences exist (Figure 15.2) compared to conventional (face-to-face, telephone, mail) household and establishment surveys.

15.3.1 Social and Technological Environment

The social environment affects participation in Web surveys indirectly, through general economic development, telecommunication policy, educational system, technological tradition, etc. In some countries (e.g., Slovenia and the United Kingdom) school children are provided free e-mail addresses, which may impact their participation. Similarly, advanced Internet adoption, as in the United States or Scandinavia, increases the potential for cooperation in Web surveys. A general survey climate (Groves and Couper, 1989), perception of direct marketing, legiti-

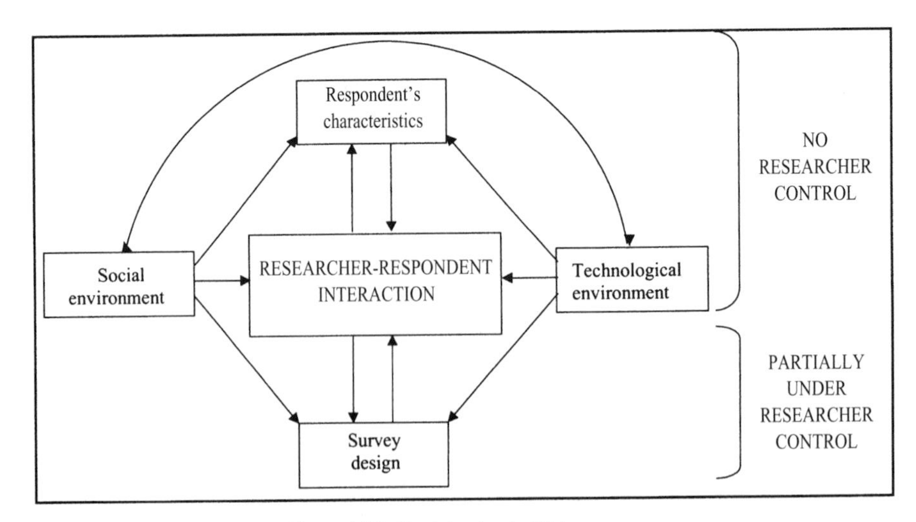

Figure 15.2. Participation in Web surveys.

macy of surveys and their sponsors, data protection scandals, and opinion leaders, also influence participation. In addition, an Internet-specific social exchange climate may exist. For example, in the early years of Internet adoption, a sense of comradery existed among Internet users, which could help participation in Web surveys.

Legal regulations are also of extreme importance. In some countries strict laws exist regarding incentives (Eichman, 1999), or, privacy regulations require encryption technologies (Clayton and Werking, 1998). Internet users are also increasingly reluctant to reveal personal information if conditions of its use are not clearly specified (Pitkow and Kehoe, 1997). However, evidence exists that respondents to computer-assisted modes of surveying are less critical about privacy compared to paper-and-pen interviews (Beckenbach, 1995; de Leeuw and Nicholls, 1996; Weisband and Kiesler, 1996). With Web surveys, these attitudes may change, because of the lack of Internet privacy laws (Clayton and Werking, 1998).

The attitude towards spamming, e.g., unsolicited commercial e-mail messages, is also extremely important in e-mail-solicited Web surveys. Though an invitation to participate in a Web survey is not considered spamming, it is unclear how respondents may perceive it. Currently, spam regulations in the European Union (Banks, 1998) and in the US Congress (Everett-Church, 1999a, 1999b) have no direct restrictions on e-mail survey solicitation, other than the explicit right to opt out of the list. Professional standards (ESOMAR, 1997, 1998; ARF, 1999) also allow for unsolicited e-mail invitations, but encourage that they be minimized. Nevertheless, antispam activities on the Web (Everett-Church, 1999a, 1999b; Rickard, 1999) are extremely inconvenient. The same is true also for the ISP's restrictions on the number of e-mails delivered from a single address (Banks, 1998; Sheehan and Hoy, 1999). Due to these difficulties, market researchers recognize that for a Web survey one needs a panel of Internet users who previously agreed to participate (Nadilo, 1999).

Of course, attitudes about unsolicited e-mail vary considerably from country to country. In the early stages of Internet penetration and limited commercial activities on the Web, as in Slovenia in 1998, only a minority of Web users regularly received spammed messages. Similarly, in the RIS 1998 Web survey, out of 19,000 units from the public e-mail directory that received an invitation, only 1% clicked the option to be removed from the list. However, the RIS telephone survey in 1999 showed that only a half of the Internet users regarded an e-mail invitation as an appropriate tool in Web surveys.

The social environment interacts with the technological environment. General telecommunication and information infrastructure is especially critical for successful implementation of Web surveys. Across countries, considerable differences can be observed with respect to Internet penetration, which has, by the year 2000, surpassed 50% of the active population only in a few developed countries such as the United States, Canada, and the Scandinavian countries. In general, the developed world had reached 20–30% penetration rate by the year 2000, but in some of the largest countries (China, Russia, India, etc.) as well as in the remaining developing countries, penetration was well below 10% (http://www.nua.ie).

Low-quality Internet networks may be an important limitation. For example, a joint Slovenian–Russian research effort (RINE, 1998) had to adapt a Web questionnaire to an e-mail version due to the limitations of Web usage in Russia. High costs of Internet access in countries without a flat Internet or telephone access fee can interfere with participation. The local telephone charges per hour (typically, between one and two U.S. dollars) that are in addition to the ISP monthly fee are not negligible.

15.3.2 Respondent's Characteristics

Response rates to surveys usually vary across social–demographic categories, survey experience, interest in the survey topic, and other attitudes. In addition, with Web surveys, computer literacy and one's orientation towards computer use also become extremely important.

Social-Demographic Characteristics. The intensity of computer and Internet usage is the most important predictor of cooperation in a Web survey, even when observed within the social–demographic categories defined by age, gender, education, and income (Kehoe and Pitkow, 1996; Batagelj and Vehovar, 1999; Vehovar et al., 1999). Of course, when computer orientation is not controlled, it appears that the usual characteristics of Internet usage also determine the participation in Web surveys: respondents are younger, educated, richer, and male (Batagelj and Vehovar, 1998; Flemming and Sonner, 1999; GVU, 1994–1999).

As a typical illustration, we can observe the RIS 98 Web survey where characteristics of e-mail-solicited nonrespondents were analyzed in the post-Web telephone survey. The features determining computer and Internet usage were almost linearly related to involvement in the survey (Table 15.1). The Pew Research (Flemming and Sonner, 1999) study confirms these findings. Of course, with increased Internet penetration, the above differences are becoming less radical.

Attitudes and Other Psychological Predispositions. A positive attitude toward survey participation in general (measured by participation in previous surveys) increases the chances for completing a Web survey (Gonier, 1999; PR Newswire, 1998). In addition, persons not willing to participate in traditional surveys may still participate in Web surveys (PR Newswire, 1998). In particular, younger males, a traditionally difficult group from which to obtain response, have positive attitudes towards the participation in these surveys (Balden, 1999).

In general, respondents like computer-assisted self-interviewing (Beckenbach, 1995; Witt and Bernstein, 1992; Zandan and Frost, 1989) and Web survey designers have enormous resources for increasing the fun and satisfaction of some respondents. Research has already shown that specific design features significantly impact the satisfaction (Batagelj et al., 1998). However, a novelty effect (de Leeuw and Nicholls, 1996; Pilon and Craig, 1988) may also be present that will diminish with time. Some researchers (Onyshekvych and McIndoe, 1999; Venter and Prinsloo,

Table 15.1. Characteristics of participants in the RIS 98 Web survey (data source in parentheses)

Data source	General population (Statistical Office)	Internet users (RIS telephone survey)	Nonrespondents (RIS telephone survey)	Partial respondents (RIS Web survey)	Complete respondents (RIS Web survey)
Women (%)	53	40	40	27	18
Average age	42	32	32	30	30
Having university education (%)	11	38	39	47	49
Speaking English well or fluently (%)	56	63	57	85	90
Weekly users (%)	—	69	70	94	98
Average number of years of Internet use	—	1.6	1.6	2.1	2.7
Users reading computer magazines (%)	—	25	23	40	48
Cooperated in previous RIS Web surveys (%)	—	—	12	18	26

1999) are already warning that the willingness to participate in Web surveys is declining because of respondents being oversurveyed.

In general, survey participation depends strongly on the survey topic, its salience, and the respondent's involvement (Clausen and Ford, 1947; Franzen and Lazarsfeld, 1945; Groves and Couper, 1998; Jansen, 1985; Kojetin et al., 1993; Martin, 1994; Pearl and Fairley, 1985; Roeher, 1963). For Web surveys in particular, intensive users with strong attitudes toward Internet issues are more likely to participate in Internet use surveys (Findlater and Kottler, 1998). Similarly, satisfied customers are more likely to participate in customer satisfaction Web surveys (Enander and Sajti, 1999) and more involved e-commerce users are more likely to participate in Web surveys on e-commerce (Elder, 1999). On the other hand, a strong aversion to certain topics (e.g., drugs, politics) unrelated to the Internet has been observed (Batagelj and Vehovar, 1999).

Respondent's Technical Equipment. The respondent's technical equipment affects Web survey participation. For certain Web surveys, only the technologically advanced users with the latest version of browser, higher speed of Internet access, and better PC platform and monitor can participate. Similarly, for e-mail-solicited surveys, it has been reported that the users with e-mail software enabling "clickable" URL addresses are more likely to participate (Chisholm, 1998). Inadequate equipment also makes the Web survey longer, unpleasant, difficult or even impossible (Batagelj and Vehovar, 1999; Batagelj et al., 1998; Dillman, 1998, 2000; Dillman et al., 1998; Kehoe and Pitkow, 1996; Nichols and Sedivi, 1998; Sheehan and Hoy, 1999).

15.3.3 Design of a Web Survey

Invitation to a Web Survey. An e-mail invitation to participate in a survey clearly has cost and time advantages over the telephone, mail, fax, or personal invitations. In addition, it provides easy access to Web questionnaires. On the other hand, an e-mail invitation is less noticeable than other types and it is also extremely sensitive to typing errors and changes in addresses. In addition, it is often perceived as commercial spam, particularly when the information in the head of the message (lines "From," "To," "Subject") is unclear. The e-mail headline is therefore extremely important in order to catch the attention and encourage participation (Coomber, 1997; Tuten, 1997). A prenotice e-mail has been suggested (Comley, 1996; Sheehan and Hoy, 1999; Venter and Prinsloo, 1999), but may generate high presurvey refusal rates (Sheehan and Hoy, 1999). Some other features of e-mail invitations that have been shown to have a positive impact on participation are the sponsoring organization (Woodall, 1998) and personalization (Nadler and Henning, 1998).

RDD telephone surveys are sometimes used to solicit e-mail addresses for web surveys (Farmer, 1998; Flemming and Sonner, 1999; Hollis, 1999). However, in RIS telephone surveys only 65% of the respondents with e-mail address were willing to reveal it. In the United States, the corresponding percentages in the Pew Research telephone recruiting survey (Flemming and Sonner, 1999) was much lower (36%). The same was found to be true for commercial surveys in the United States in which incentives were applied (Hollis, 1999).

Follow-ups. Follow-ups improve response to Web surveys, just as they also do for personal, telephone, and mail surveys (Dillman, 1978, 1991; Dillman et al., 1974; Goyder, 1985a, 1987; Heberlein and Baumgartner, 1978; Kanuk and Berenson, 1975; Scott, 1961), as well as e-mail surveys (Mavis and Brocato, 1998; Mehta and Sivadas, 1995; Schaefer and Dillman, 1998; Sheehan and Hoy, 1999). Evidence exists that e-mail reminders contribute up to one third of the final sample size (Batagelj and Vehovar, 1998; Enander and Sajti, 1999; Flemming and Sonner, 1999). In addition, follow-ups contribute to a more representative sample since late respondents often differ from early respondents (Batagelj and Vehovar, 1998; Willke et al., 1999).

Usually, the majority of responses in Web surveys are received within the first few days of the data collection period (Balden, 1999; Batagelj and Vehovar, 1998; Comley, 1996, 1998; Enander and Sajti, 1999; Flemming and Sonner, 1999; Kottler, 1998; Venter and Prinsloo, 1999; Willke et al., 1999). This suggests that, in comparison to traditional mail surveys, the time intervals between follow-up strategies should be shortened (Schaefer and Dillman, 1998).

Incentives. Incentives have proved to be an efficient way of increasing response rates in mail, telephone, and personal surveys (Dillman, 1991; Lankford et al., 1995; Shettle and Mooney, 1999; see also Chapter 11, this volume). For Web surveys, the research findings vary from low impact of incentives (Enander and Sajti, 1999) to a prevailing positive effect on the response rate (Venter and Prinsloo, 1999; Woodall, 1998).

Of course, the respondents attracted by incentives may differ in their characteristics (Dillman, 1978), and this may contribute to more effective follow-up (Enander and Sajti, 1999). Incentives can cause people to ponder the respondent burden with the final consequence being nonresponse (Groves and Couper, 1998). They can also stimulate respondents to answer more than once when access is not controlled (Batagelj and Vehovar, 1998) or produce biased responses (Chisholm, 1998).

Research Organization. Mentioning the name of the sponsoring organization is becoming extremely important to participation. This trend is additionally reinforced with the increased concern for privacy. In the RIS Web surveys, loyalty measured by participation in previous RIS surveys is one of the most important determinants of participation (Batagelj and Vehovar, 1999).

Length of Data Collection Period. A prolonged data collection period has only a limited impact on the number of responses, even in unsolicited Web surveys. For example, in the self-recruited (unsolicited) part of the RIS 97 survey only 13% of responses were received in the last 10 (of the 40) days of the data collection period (Batagelj and Vehovar, 1998). However, more women, older users, and new users responded in this period, suggesting that an extended data collection period brings respondent characteristics closer to those of the target population.

Questionnaire Design. The design of a Web questionnaire has a limited impact on the initial decision to participate in a Web survey but is strongly related to partial nonresponse, item nonresponse, and data quality. In general, computer-based surveys have a positive effect on item nonresponse and data quality (Saris, 1998). On the other hand, the role of advanced technological features, such as intensive use of graphics, images, animations, cookies, and links to other web pages, is much more ambiguous. In addition, even some crucial features of Web questionnaire design are still waiting for a comprehensive evaluation of their effects on response.

One-Page Versus Multiple Pages Design. Both alternatives, the whole questionnaire on one scrolling HTML page or a questionnaire divided to several HTML pages, have certain advantages and disadvantages (Clayton and Werking, 1998; Dillman, 2000; Farmer, 1998; Kottler, 1997b; Spain, 1998; Vehovar and Batagelj, 1996). Item nonresponse may be greater when one page is used, unless advanced Java applets and Java Scripts are applied. However, increased download times and lack of browser support may prevent their application. On the other hand, the responding time and the danger of abandoning the questionnaire are potentially higher with multiple page designs. There is also evidence of no difference in the partial nonresponse rates for the two designs in a relatively short questionnaire (7 minutes), although the completion time is 30% longer for the multiple-page design (Vehovar and Batagelj, 1996). On the other hand, use of an extreme one-question-per-page design in the RIS 98 Web survey resulted in strong complaints (Batagelj et al., 1998). Different results were reported by Zukerberg et al. (1999); no differences were observed in either the questionnaire completion time or the respondents' satisfaction. That may have been due in part to the laboratory conditions of this experiment.

Advanced Graphics. Advanced graphics may improve the respondents' motivation, and generate a precious feeling of having "fun" while answering a Web questionnaire. However, the potential for negative effects from technological limitations, distraction of respondents and biased answers must be taken into account. An experimental study of two questionnaire designs—a fancy and plain one—(Dillman, 2000) provides an important warning about the extensive use of graphics. Respondents to the plain version, which required only one-third the computer memory of the fancy version, completed more pages and write-in boxes, were less likely to drop out, spent less time, and were less likely to have to return to the questionnaire in order to complete it. Similarly, the graphical aids (logotypes) used in measuring Web page visitation in the RIS 98 Web survey increased the percentage of respondents abandoning the survey in that block of questions from 0.5% when no logotypes were used to 4% when logotypes were used (Lozar Manfreda, 1999). The graphical version was especially a problem for respondents with older versions of browsers and those answering from dial-up access.

Progress Indicator. Graphical symbols, for example a "progress bar" that shows the proportion of questionnaire so far completed, conveys an important sense of orientation in the questionnaire completion process. When missing, this was one of the most frequent complaints (Batagelj et al., 1998). In addition, its application may slightly increase the response rate (Couper et al., 1999) without any other effects on data quality and item nonresponse. On the other hand, in lengthy surveys, the progress indicator may remind people of the length and cause them to abandon the survey prematurely. In certain circumstances, it may also increase the download times of each page (Couper et al., 1999).

Quality-Check Reminders. The extent of edit control is one of the crucial design issues for which a definitive research answer has not yet been obtained. In principle, forcing respondents to answer questions properly can prevent any item nonresponse or inconsistent response. However, besides known (but disappearing) technical difficulties with Java Script applications, the respondent's frustration associated with these requirements is likely to lead to premature terminations (Dillman, 2000; Dillman et al., 1998; Zukerberg et al., 1999). Soft reminders that allow the respondent to proceed, even when the error is not corrected, seem a reasonable alternative to hard (forced) corrections (Zukerberg et al., 1999).

Questionnaire Length. In general, shorter questionnaires achieve higher response (Dillman et al., 1994; Groves and Couper, 1998) and we can expect similar results for Web surveys. In the RIS Web surveys, respondents had no problems in answering a 12 minute questionnaire, but a 20 minute questionnaire was a problem (Batagelj and Vehovar, 1999). The research organization InfoTek found that questionnaires that take more than 15 minutes to complete have a very high probability of some questions not being answered or the questionnaire being abandoned prematurely (Farmer, 1998). On the other hand, questionnaire lasting more than 40 minutes have been reported as effective (Wydra, 1999).

Mixed-Mode Environment. The solicitation procedures in Web surveys often introduce certain aspects of a mixed-mode approach. More and more often sample units are being offered the option of answering questions by a different mode. The Web survey option helps target segments that are difficult to reach with other modes and increases response rates through offering an alternative survey option.

The mixing of modes used for solicitation and response makes the mixed-mode environment extremely complex. This is particularly true for the context in which the respondent selects the preferred survey option. In some circumstances, they prefer the Web to the paper questionnaire (Chisholm, 1998), and in others the opposite is true (Vehovar et al., 2000). Nevertheless, evidence exists that a Web option increases the response rate in mail/Web surveys of the general population (Comley, 1996), in establishment surveys (Vehovar et al., 2000), and also in Web/telephone surveys (Onyshekvych, 1999).

15.4 VALIDATION PROBLEM

Web surveys have been used extensively only since the mid-1990s (GVU, 1994). However, conducting them has already became a profitable industry, with a code of ethics (ESOMAR, 1997, 1998; ARF, 1999), tens of software packages and hundreds of research papers (RIS, 1999–2001). Despite this, an impression exists that Web surveys are still a questionable survey mode. This is often stated as a "validation problem" (Bruzzone, 1999; Hollis, 1999; Kottler, 1998; Nadilo, 1999) and is based on the discrepancies arising from comparisons of the results of Web surveys with parallel (control) surveys performed by traditional modes.

However, in probability Web surveys, or in controlled experiments, the available research shows no significant differences between Web surveys and e-mail (Chisholm, 1998), telephone (Gonier, 1999; Terhanian and Black, 1999), mail (Gonier, 1999), or mall intercept surveys (Gonier, 1999; Nadilo, 1999; Willke et al., 1999). On occasion, minor differences are observed, such as the inclination of Web respondents towards more extreme responses (Findlater and Kottler, 1998; Gonier, 1999; Wydra, 1999), more "don't know" answers, and a higher recognition of advertisements (Hollis, 1999). The Web survey mode itself is therefore a valid survey mode, which has already been demonstrated for other types of self-administered computer surveys (Saris, 1998). The threat to validity thus arises only from nonresponse and noncoverage problems, particularly in nonprobability Web surveys, which are, unfortunately, quite common. As a consequence, Web surveys are often automatically associated with the validation issues in nonprobability Web surveys, although this is nothing but the usual problem of statistical inference without scientific/probability sampling (Hollis, 1999). What makes the issue of nonprobability samples so fresh and unique with Web surveys is only the temptation to ignore inferential limitations, because fast, reliable (at the respondents' level), and cheap data can be collected so easily.

In this context, Web surveys are often mentioned as a "replacement technology" (Black, 1998; Hollis, 1999) in the sense that they replace telephone surveys as telephone surveys replaced face-to-face surveys in the 1970s. However, such replacement should be restricted only to the probability Web surveys replacing other probability surveys, or nonprobability Web surveys replacing other nonprobability surveys. Nevertheless, an implicit claim exists that the nonprobability (panel) Web surveys could also replace the probability telephone surveys of the general population (Black, 1998; Comley 1996; Kottler, 1997a, 1997b; Nadilo, 1999). In principle, this is extremely questionable as the actual Web coverage rates are much lower than telephone coverage rates were at the time of replacing the face-to-face surveys 20 years ago. In addition, the reported nonresponse rates are usually twice as high in Web surveys compared to telephone surveys, not to mention that Web surveys have no sampling frame comparable to the RDD frame used for the telephone surveys .

On the other hand, there is an evidence that nonprobability (panel) Web surveys perform relatively well even in the case of election predictions for the general population (Comley, 1997; Terhanian and Black, 1999). This success can be partially explained by the robustness of the variables and with the advanced selection and modeling/adjustment techniques. Nevertheless, this poses the question of why pay for expensive probability samples when high response rates, adequate sampling frames, and probability samples may not be needed. There is no doubt that further success of nonprobability (panel) Web surveys would seriously alter professional standards as well as the role of probability samples and nonresponse issues in survey research.

Of course, there is also much evidence that points to the fact that, even after adjustment (i.e., weighting), considerable discrepancies among estimates from Web and telephone surveys remain (Batagelj and Vehovar, 1999; Elder, 1999; Flemming and Sonner, 1999; Gates and Helton, 1998; Vehovar et al., 1999). Kish's (1998) comment that nonprobability samples "can be correct for the majority of variables, but sometimes they are painfully wrong," provides another context for understanding the relative success of certain nonprobability Web survey panels.

To summarize, the so-called validation problem of Web surveys relates only to the eternal quest for replacing inconvenient probability surveys with convenient nonprobability ones. Nonprobability Web surveys bring nothing new to this problem except the massive usage arising from convenience, speed, and costs. On the other hand, however, it seems that with sophisticated Web survey panels, the above quest might come a small step closer to its aim.

15.5 CONCLUSIONS

The nonresponse process in Web surveys is much more complex than for other survey modes. First, technology emerges as an additional factor that interacts with other features at the respondent, at the society, and at the survey organization levels.

Second, technological changes are occurring extremely fast, permanently complicating a whole array of issues. Third, the very nature of the Web survey mode often introduces components of mixed-mode surveys at the solicitation as well as at the responding stage.

We can conclude that at the beginning of the millennium, the available evidence shows relatively low response rates to Web surveys. This may be due to limited Internet penetration, to the technology supporting only the basic Web survey mode or, more likely, to relatively undeveloped Web survey solicitation techniques. In addition, the probability Web surveys are currently limited only to special populations and to certain mixed-mode studies.

With respect to nonprobability Web surveys and Web surveys with extremely low response rates, we can observe some promising results arising from advanced selection and adjustment/modeling techniques. An alternative approach that gives legitimacy to these surveys is the context of qualitative research, which is a legitimate instrument not conditioned with statistical inference issues and nonresponse problems. Web surveys can be thus applied as a preliminary qualitative method in a quantitative study or a follow-up qualitative method in quantitative research (Morgan, 1998).

The future expansion of the Web surveys can be observed in the light of transforming all computer-assisted survey modes to the Web (Comley, 1998; Onyshekvych and McIndoe, 1999). The self-administered Web option now represents the third stage—after computer-assisted telephone and personal interviews—in the evolution of the computerized survey industry (Baker, 1998). In addition, data input and survey management of other survey modes also converge to automated computer processes performed on the Web. This perhaps presents the final integration step in the computer-assisted collection of survey information. However, interactive and multimedia help from the interviewer may also appear in the near future and radically change the prevailing self-administered trends.

In future nonresponse research in Web surveys, the following areas need special attention:

- Solicitation strategies. The existing comparisons of response rates with other survey modes may be unfair to the Web surveys, as we do not know the proper strategy to attract cooperation. The mode of the contacts, their combination, frequency, and intervals of follow-ups thus need to be studied.
- Incentives. Research on incentives (type, elasticity, combinations) is extremely important, as they seem to play a crucial role in providing sufficient cooperation in Web surveys.
- Questionnaire design. Some key features still await an answer: one scrolling page versus multiple-page design, forced reminders and checking controls, the role of graphics, and the design/implementation of instructions. All these aspects have to be observed in a broader context of data quality.
- Respondents' satisfaction. The factors that increase satisfaction are extremely important because of the growing role of the research organization's brand in conducting Web surveys.

- Confidentiality issues. The identification procedures (password, cookies, automatic identifiers) are critical for cooperation, particularly in the context of anonymity and confidentiality.
- Survey participation. Understanding participation in different stages of the Web survey process may lead to efficient postsurvey adjustment.
- Costs and errors in the mixed-mode environment. The low costs and high nonobservation errors of the Web surveys force us to introduce mixed-mode surveys, so explicit models for the costs and errors are needed for these complex settings.

CHAPTER 16

Nonresponse in Exit Polls: A Comprehensive Analysis

Daniel M. Merkle, *ABC News*
Murray Edelman, *Voter News Service*

16.1 INTRODUCTION

Survey methodologists have devoted a considerable amount of time to studying nonresponse in sample surveys. One of the main focuses of this research has been on factors that influence response rates, with the goal of improving them under the assumption that a higher response rate will lead to lower survey error (for a review see Groves, 1989). Although this assumption makes sense theoretically, it can rarely be tested because the population values needed to compute survey error measures are often not known. This chapter uses data from Voter News Service's (VNS) Election Day exit polls to further study factors that influence response rates, and it takes the important next step of looking at the impact of response rates on survey error.

Data from exit polls can make a unique contribution to the study of nonresponse. Unlike most surveys that employ relatively homogenous interviewer pools, VNS employs a larger, more heterogeneous group of interviewers for each election. As with other types of in-person surveys, it is possible to obtain, by observation, some basic demographic information on nonrespondents. Finally, and most important, a measure of error can be computed for the key variable of interest in exit polls—the vote. This error measure, along with the response rate, can be computed by precinct for each interviewer.

After a discussion of the methodology, we begin by exploring the influence of a variety of factors on response rates. First we look at the voter characteristics of age, race, and gender. Next, we explore which interviewer characteristics and Election Day factors predict response rates. Then we look at the extent to which the interviewer characteristics of age, race, and gender interact with these same voter characteristics in terms of nonresponse. Finally, we attempt to model the relationship between response rates and survey error.

16.2 METHODOLOGY

The data used in this chapter come from the VNS exit polls conducted from 1992 through 1998. VNS, a consortium of ABC, the Associated Press, CBS, CNN, FOX, and NBC, is the main source of exit poll data in the United States, providing data to its six member organizations and approximately 100 television stations, newspapers, magazines, and radio stations. This section provides details about how these exit polls are conducted and a description of how the nonresponse and survey error measures are operationalized in this study.

16.2.1 How the Exit Polls Are Conducted

The VNS exit polls are conducted using a two-stage sampling design. In the first stage, a stratified, systematic sample of precincts is selected in each state, proportionate to the number of votes cast in a previous election. In the second stage, interviewers systematically select voters exiting polling places on Election Day, using a sampling interval based on the expected turnout at that precinct. The interval is computed so that approximately 100 interviews are completed in each precinct. In the presidential years of 1992 and 1996, there was an exit poll in every state and the District of Columbia. In the nonpresidential election years of 1994 and 1998, there were exit polls in states with gubernatorial or senate races (36 in 1994 and 45 in 1998). For the elections from 1994 through 1998, all interviewers were hired and trained by VNS. For the 1992 exit polls, the interviewers were hired and trained by the former Chilton Research Services, now TNS Intersearch. This is the only substantive difference in methodology across these four election years.

One interviewer is assigned to each precinct in the sample. Interviewers are instructed to work with a polling place official to determine the best place to stand. Ideally, the interviewers are located inside the polling place where all voters must pass them. Unfortunately, sometimes interviewers must stand outside, some distance from the door. As voters exit the polling place, the interviewer approaches the sampled voter, shows the questionnaire and asks him or her to fill it out. The self-administered questionnaires are one or two sides of a piece of paper, depending on the newsworthiness of the races in the state. After the voter fills out the questionnaire, he or she places it in a "ballot box." As discussed in more detail below, interviewers also keep track of nonrespondents throughout the day.

Interviewers take a 10-minute break from interviewing each hour to tally the responses to the vote questions and their observations of the nonrespondents. Interviewers call VNS three times during the day to report their data. In local time, the first call is around 9:00 AM, the second around 3:00 PM, and the last call shortly before poll closing. During each call, interviewers report their vote and nonresponse tallies and read in the question-by-question responses from a subsample of the questionnaires.

All interviewers are asked to fill out a questionnaire after they complete their assignments on Election Day. The interviewer questionnaire includes a number of background questions, such as basic demographics and prior interviewing experience, as well as questions related to the interviewing experience on Election Day.

16.2.2 Operationalization of Measures

Response Rates. On Election Day, VNS has the interviewers keep track of nonrespondents on a worksheet. On this sheet, interviewers code the nonrespondent's age (18–29, 30–59, 60 and over), race (White/other or Black/Hispanic) and gender (from observation).

In 1992, the second author did an evaluation of exit poll interviewers' ability to estimate age. Interviewers made correct estimates 88% of the time. Seven percent of the time, interviewers overestimated age, and 6% were underestimates.

There are two types of nonrespondents in exit polls: refusals and misses. A refusal occurs when a sampled voter is asked to fill out the questionnaire and declines. A miss is when the interviewer is unable to ask a sampled voter to fill out the questionnaire. This can occur when the interviewer is too busy to approach the selected voter or when the voter does not pass the interviewer. About three-fourths of nonresponse in exit polls is attributable to refusals and about a quarter to misses.

Due to the way the nonresponse data are tabulated on Election Day, it is not always possible to compute separate refusal and miss rates by precinct or by state. For this reason, the overall response rate [i.e., completed questionnaires/(completed questionnaires + refusals + misses)] is used in many of the analyses reported here. However, whenever possible, refusal and miss rates are considered separately. Because most of the nonresponse is due to refusals, overall response rates are highly correlated with refusal rates. For example, in 1996 the correlation between overall response rates and refusal rates at the precinct level was –0.83 ($p < 0.01$, $n = 1372$).

The precinct-level response rates are normally distributed in each of the four election years. They range from around 10% on the low end to around 90% on the high end, with most falling in the 45–75% range.

In the analyses reported in this paper, we weight the precinct-level response rates by the number of interview attempts (i.e., the denominator of the response rate calculation which is the total number of sampled voters the interviewer approached or should have approached based on their sampling interval) (cf., Groves and Couper, 1998, p. 73). This is done to give relatively more weight to precincts with more attempts because their response rate estimates are more stable.

Survey Error. On Election Day, VNS gathers the actual vote in each exit poll precinct for the vote projections. These data are also used to measure the accuracy of the exit poll in each precinct. Exit poll error is operationalized in two ways using these data. First, the "signed error" is computed by subtracting the Dem–Rep difference (i.e., the Democratic vote percentage minus the Republican vote percentage) in the exit poll in each precinct from the actual Dem–Rep difference for the main race in that state. The distribution of the signed error for each of the four election years studied here closely approximates a normal distribution. Second, the "absolute error" is the absolute value of the signed error and, as a result, is positively skewed. The analyses below use the square root of the absolute error in order to better satisfy the assumptions of the statistical tests. It is important to note that these variables measure total survey error rather than just nonresponse error.

16.3 INFLUENCE OF SAMPLE VOTER CHARACTERISTICS

Groves and Couper (1998) hypothesize that "those who are alienated or isolated from the broader society/polity would be less likely to cooperate with survey requests that represent such interests" (p. 131). Groves and Couper discussed this social isolation hypothesis in relation to household survey requests. It should be generally applicable to exit polling as well because of the in-person nature of the survey request, except that we would expect the hypothesized effects to be attenuated because of the restricted range of the social isolation variable. Exit polls only interview people who made the effort to go to their polling location on Election Day. If the social isolation hypothesis is correct, those who feel relatively isolated would be less likely to go and vote in the first place.

Of the three demographic characteristics studied here, age is the most strongly related to response rates. Consistent with previous research, older voters (60 and above) have much lower response rates across all four years (Table 16.1). Although this finding is consistent with the social isolation hypothesis, we believe that it is a partial explanation at best. Another plausible reason for less participation by older voters is fear and suspicion of strangers given perceived or actual physical vulnera-

Table 16.1. Response rates by voter age, race, and gender

	1992 (n = 88,230)	1994 (n = 103,188)	1996 (n = 133,018)	1998 (n = 101,302)
Age				
18–29	63.4%	55.6%	**57.5%**	51.0%
30–59	65.5%	58.1%	**58.1%**	55.7%
60+	49.0%	47.5%	43.9%	46.2%
Race				
White	**61.7%**	55.5%	54.2%	52.2%
Black/Hispanic	**61.0%**	51.1%	57.4%	54.8%
Gender				
Male	60.1%	**54.6%**	53.8%	52.0%
Female	62.9%	**55.2%**	55.6%	53.1%

Note: The table entries in **bold** are *not* significantly different from each other at $p < 0.05$; the rest are significantly different. In computing the standard errors for the statistical tests, a design effect of 1.7 was used for age and gender, and 3.5 was used for race.

bility. As we shall see, the interaction between interviewer and voter age provides support for this explanation. Other possible reasons include poor health, lower education and literacy, lower levels of exposure to survey requests, and possibly less of an appreciation of the value of surveys. From a practical standpoint, the lower response rates among the elderly, and the fact that this is a consistent finding in VNS' exit polls, highlight the importance of performing a nonresponse adjustment for age in exit polls.

Similar to Groves and Couper's findings for household surveys, the results for race do not support the social isolation hypothesis (Table 16.1). Contrary to the hypothesis, in two of the years, 1996 and 1998, minorities actually had slightly higher response rates than whites. In 1994, whites had a higher response rate, and in 1992 there was no difference.

Consistent with some previous research, women's response rates are slightly higher than men's (Table 16.1). Groves and Couper suggest that the tendency for men to be less cooperative may be related to the differences in social roles played by males and females. They note that women are more likely to answer the telephone and typically take more responsibility for establishing and maintaining social relationships.

16.4 INFLUENCE OF ELECTION DAY FACTORS, VOTER CHARACTERISTICS, AND INTERVIEWER CHARACTERISTICS

This section, using data from the 1992 and 1996 exit polls, employs multiple regression to look at the influence of three sets of variables on precinct-level response rates: Election Day factors, voter characteristics, and interviewer characteristics (see Appendix for a description of the variables). This makes it possible to determine the contribution that each variable makes in explaining response rates while controlling for the others.

16.4.1 Election Day Factors

The interviewing environment varies by polling location. Various Election Day factors, largely out of the interviewer's control, can have an impact on the ability to get people to fill out the questionnaire. The Election Day factors from the interviewer questionnaire consist of the following: where the interviewer stood in relation to the precinct, the number of exits worked, whether or not they had a problem with election officials or police, and their ability to keep up with the interviewing rate. We also include two measures of questionnaire length (i.e., the number of questions on the front of the questionnaire and whether or not the questionnaire had a back side) as a burden indicator.

16.4.2 Voter Characteristics

The voter characteristics used in the regressions are precinct-level data on voters' age, race, and gender. As already seen in Table 16.1, some types of voters are more likely to comply with a request to participate in an exit poll. Therefore, we would expect precincts with proportionally more of these voters to have higher response rates. Therefore, we include in the analysis information about the precint's demographic makeup, specifically the proportion of men, minorities, and older people voting at the precinct on Election Day. These variables are computed by combining data from the exit poll respondents and the interviewers' estimates of the age, race, and gender of the nonrespondents.

16.4.3 Interviewer Characteristics

There is more variability in the characteristics of VNS' interviewers than is typically the case with other types of in-person surveys, and we have data from a very large number of interviewers each election year. Although the VNS interviewing pool was not selected to be representative of all potential exit poll interviewers, it does allow us to make response rate comparisons among a large and diverse group of individuals. Also, it is operationally impossible to randomly assign interviewers to precincts in the type of large-scale national exit polls VNS conducts. We address this problem by using multiple control variables to account for many of the differences between the precincts.

In addition to demographic variables, the VNS interviewer questionnaire also asked about prior telephone, face-to-face, and exit poll interviewing experience. As Groves and Couper (1998) point out: "The prevailing belief in the survey industry is that interviewer experience is a critical factor in gaining cooperation from sample persons" (p. 200). Their analysis found that experience, measured in years worked, was positively related to response rates.

Unlike most survey organizations that have relatively permanent interviewing staffs, VNS interviewers are independent contractors who work once every two years. As such, VNS interviewers tend to have less survey experience than those employed at more traditional survey organizations. Therefore, VNS measures experience in terms of number of surveys the interviewer has worked on rather than years of experience. The three interviewer experience variables were combined into a composite measure using factor analysis. The internal consistency of the three items (i.e., Cronbach's alpha) is 0.77 in each of the two years.

16.4.4 Results

Four multiple regressions were run using the same three sets of independent variables in each. Overall precinct-level response rates were used as the dependent variable in the first two regressions (one for 1992 and one for 1996). This was done to permit comparisons between the two years, because it was only possible to separate out refusals and misses in the 1996 data. However, as mentioned above, refusals and misses are qualitatively different types of nonresponse and likely have different causes. Therefore, two more regressions were run on the 1996 data using refusal rates and miss rates as separate dependent variables.

Election Day Factors. VNS has long recognized the importance of the interviewer's position at the polling place and has emphasized this in interviewer training and Election Day support. Interviewers are instructed to stand as close to the voting area as possible where they can see each voter and where the voters must pass them before exiting the building.

Several factors may hinder an interviewer's ability to work at this position. Some states have laws which specifically keep interviewers some distance from the exit door (e.g., Arizona, Minnesota, and South Dakota). More problematic is the Election Day official or police officer who attempts to enforce "electioneering" or "loitering" statutes that do not pertain to exit polling.

The importance of interviewing position is demonstrated in the regression analyses (Table 16.2). Of all the Election Day factors (not including the questionnaire length control variables), interviewing position is the strongest predictor of re-

Table 16.2. Multiple regressions predicting response rate

	1992 Response rate	1996 Response rate	1996 Refusal rate	1996 Miss rate
Election Day factors				
Number of questions on front of questionnaire	–0.21**	–0.25**	0.29**	0.03
Does questionnaire have a back side?	–0.19**	–0.11**	0.12**	0.05
Interviewing position	–0.17**	–0.22**	0.17**	0.16**
Problems with officials	–0.13**	–0.08**	0.06	0.10**
Trouble keeping up with the interviewing rate	–0.12**	–0.17**	–0.01	0.28**
Number of exits	–0.02	–0.08**	0.03	0.11**
Voter characteristics				
Percent male voters	–0.09**	–0.06*	0.04	0.04
Percent minority voters	–0.05	0.04	–0.08*	0.05
Percent older voters	–0.20**	–0.17**	0.18**	0.09**
Interviewer characteristics				
Age	0.19**	0.25**	–0.24**	–0.08*
Education	0.08*	0.02	0.00	–0.03
Race	0.03	0.05	–0.04	–0.00
Gender	0.01	–0.05	0.03	0.02
Interviewing experience	0.01	0.04	–0.05	–0.00
Multiple R	0.52	0.51	0.45	0.42
Adjusted R-squared	0.26	0.25	0.19	0.16
n	846	1076	1028	1028

Note: Table entries are standardized beta coefficients. See Appendix for a description of how the variables are coded.

*$p < 0.05$.

**$p < 0.01$.

sponse rates in each year. As one would expect, response rates decline as the interviewer moves further away from the voting room. Those who have a problem with election officials or police also have lower response rates in both years, even after controlling for interviewer location. Most often, these problems consist of situations where an official makes the interviewer work at the electioneering distance. Therefore, this effect likely represents the impact of standing extremely far from the polling place. This is because the highest category in the interviewing position variable is "over 50 feet," whereas the electioneering distance is most often 100 feet or more from the precinct.

It is noteworthy that interviewing position is a significant predictor of both refusal and miss rates (Table 16.2). As interviewers move further away from the polling place, more voters are missed, and it is more difficult to get voters to cooperate. It stands to reason that miss rates would increase when interviewers must stand far away because a greater number of voters will not pass them. The increase in refusal rates as interviewers move further away may be due to the shift in voters' attention away from the quiet act of voting and toward their next order of business. In addition, when interviewers must stand some distance from the polling place, there may also be some confusion on the part of voters between the interviewers and the partisans electioneering for a candidate. It is also possible that those who stand inside the polling place obtain more cooperation because it may appear that the exit poll is sanctioned by the election officials and is thus part of the voting process.

In both years, "trouble keeping up with the interviewing rate" is a significant predictor of response rates. The more trouble interviewers have keeping up with the interviewing rate, the lower the response rate. This makes sense. Interviewers will be more likely to miss sampled voters if they are overworked. The 1996 data confirm that trouble keeping up with the interviewing rate affects miss rates but not refusal rates. One other Election Day factor, number of exits worked, is negatively related to response rates in 1996 but not 1992. The 1996 data show that the number of exits is related to miss rates but not refusal rates.

The results for the length of questionnaire control variables are consistent with previous experimental research. Longer questionnaires have lower response rates. In both 1992 and 1996, the more questions on the front side of the questionnaire, the lower the response rate. In addition, questionnaires with a back side had lower response rates than those without. As would be expected, breaking out the response rate into its two components shows that the questionnaire length variables are related to refusal rates but not miss rates.

Voter Characteristics. Consistent with the individual-level data reported above, precincts with more older voters have lower overall response rates. The proportion of older voters in the precinct is related to refusal rates and, to a lesser extent, miss rates. Also consistent with the individual-level data, precincts with more men have lower response rates in both years, though the relationship disappears when refusal and miss rates are considered separately. The proportion of minorities is unrelated to response rates in 1992 and 1996. There is a slight negative relationship between minorities and refusal rates in 1996, suggesting that precincts with more minorities have slightly lower refusal rates.

Interviewer Characteristics. After controlling for the Election Day factors and precinct-level voter characteristics, the only interviewer characteristic that stands out as a significant predictor in both years is age. There is a relatively strong, positive linear relationship between age and response rates in both years. In fact, the effect for age is similar in magnitude or larger than each of the Election Day factors. Looking at the 1996 data, much of the relationship between age and response rates is due to the impact of age on refusal rates. Older interviewers experience fewer refusals.

One possible explanation for this is that older interviewers may appear more professional than their younger counterparts and, therefore, may be taken more seriously. The VNS training manual instructs interviewers to "wear clean, conservative, neatly pressed clothing" and not to wear jeans or T-shirts. It was thought that younger interviewers might be more likely to depart from these instructions. In 1996, VNS added a question on how the interviewer was dressed on Election Day to see if this accounted for some of the impact of age. However, the hypothesis that interviewer dress would be related to response rates was not supported.

Another noteworthy finding is the lack of a relationship between interviewer experience and any of the response rate measures. Because exit polls are a unique kind of survey research, it may be less important to get experienced survey interviewers than it is to get intelligent, motivated, and somewhat outgoing individuals who are able to follow sometimes complicated instructions.

16.5 INTERACTION BETWEEN INTERVIEWER AND RESPONDENT CHARACTERISTICS

According to Groves and Couper (1998) "one should be more willing to comply with the requests of liked others" (p. 34). One factor that has been shown to increase liking is similarity of background. Therefore, we hypothesize that response rates will be higher when the interviewer and sampled voter have similar backgrounds. Given the research showing that interviewer characteristics can interact with respondent characteristics and affect answers to survey questions (e.g., Kane and Macaulay, 1993; Anderson et al., 1988; Groves and Fultz, 1985; Schaeffer, 1980), it is possible that similar effects operate as potential respondents decide whether or not to comply with a survey request. As Groves and Couper (1998) note: "The use of race or gender matching by survey organizations may be an attempt to invoke liking through similarity, as well as reducing the potential threat to the householder" (p. 35).

For each interviewer characteristic, we computed the mean response rate for each of the same voter characteristics using the 1992 and 1996 data. This was done as a separate analysis because these data come from a different data file than the one used above.

The similarity hypothesis is not supported for race as there was no interaction between interviewer race and response rates by race (data not shown). The nature of interview assignments could be one reason that no interaction was found. Interviewers tend to be assigned to precincts that match their racial backgrounds, either on purpose or by default. It is possible that an experimental design that randomly assigned interviewers to precincts or voters would produce different results. The test of the interaction hypothesis is more robust for gender and age because these characteristics are less clustered by precinct (i.e., interviewer attempts by gender and age group are less variable across precincts). As with race, there was no interaction between interviewer gender and response rates by gender (data not shown). This finding may not generalize very well to household surveys because there may be more hesitancy on the part of some women to invite a male interviewer into their home.

Figure 16.1 provides evidence of an interaction between interviewer and voter age. In 1992 and 1996, the response rates of older voters vary quite a bit by interviewer age, with older voters responding much less often to younger interviewers. Middle aged voters show a similar, but less pronounced trend. Response rates for younger voters, however, are less influenced by interviewer age; they are almost as likely to respond to interviewers of any age. Therefore, it appears that much of the main effect for voter age on response rates observed above is due to some type of dynamic whereby older voters are not as comfortable interacting with younger interviewers, whereas the opposite does not appear to be the case. As noted above, older voters may have more of a fear or suspicion of strangers and feel physically vulnerable. The perceived threat may be diminished when they are approached by an older interviewer.

Although there is evidence of an interaction for age, the pattern of the findings does not offer strong support for the similarity hypothesis. Only one of the three in-

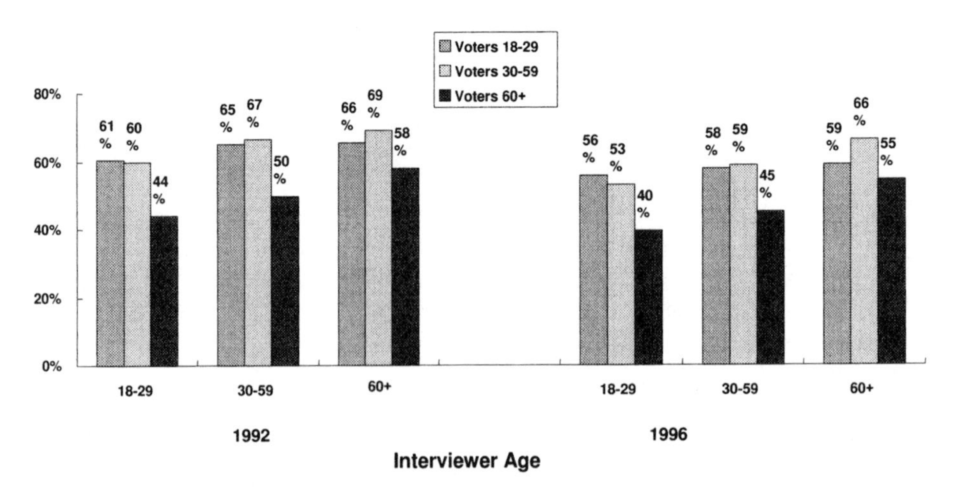

Figure 16.1. Response rates by age of interviewers and voters.

terviewer age groups, those 60 and older, has a higher response rate among voters of that same age group. Older voters react much more favorably to older interviewers, whereas younger and middle-age voters do not react more favorably to interviewers in their respective age groups.

16.6 RESPONSE RATES AND SURVEY ERROR

The presence of nonresponse in a sample survey does not necessarily mean there is nonresponse error. The magnitude of nonresponse error is a function of the proportion of nonrespondents and how different they are from respondents (cf. Groves and Couper, 1998). As mentioned above, one unique aspect of the VNS exit polls is that they make it possible to look at the relationship between survey response rates and error using a very large number of cases.

It is commonly assumed that surveys with lower response rates will have a greater amount of survey error than those with higher response rates. But, contrary to conventional wisdom, there is either very little or no correlation between response rates and the two exit poll error measures (Table 16.3). In the three election years from 1994 through 1998, there is no correlation between response rates and the signed error, and in 1992 there is only a small correlation of 0.10. Separating out refusal and miss rates using the 1996 data produces similar results. There is no relationship between refusal or miss rates and the signed error.

The analysis for absolute error uses partial correlations controlling for sample size. This is done to control for the impact of sample size on absolute error. The partial correlations for 1994 through 1998 are between –0.04 and –0.07, and for

Table 16.3. Relationship between response rates and survey error

	Response rates				Refusal rate	Miss rate
	1992	1994	1996	1998	1996	1996
Correlations						
n	1005	885	1205	894	1148	1148
Signed error	0.10**	0.00	–0.01	0.01	–0.01	0.01
Absolute error	–0.13**	–0.07*	–0.04	–0.07*	0.02	0.04
Multiple regressions						
n	740	na	1076	na	1028	1028
Signed error	0.10*	na	–0.05	na	–0.00	0.06
Absolute error	–0.11*	na	–0.03	na	0.03	0.01

Note: Correlations for the signed error are Pearson product–moment correlations, and for the absolute error they are partial corrections controlling for sample size. Table entries for the multiple regressions are beta coefficients controlling for the Election Day factors, voter characteristics and interviewer characteristics that were also used in Table 16.2.

*$p < 0.05$.

**$p < 0.01$.

1992 it is slightly larger at –0.13. Though these correlations are in the expected direction, they are each very small. Knowing the response rate has almost no predictive power when it comes to explaining exit poll error. Looking at the 1996 refusal and miss rates separately also yields no correlation with the absolute error.

The next step in modeling response rates and error is to determine whether there are any nonlinear relationships in these data. Figure 16.2 shows a scatterplot of the signed error and response rates for 1996. The scatterplots for the other three years are stikingly similar to the one shown here. These plots provide little encouragement for the existence of nonlinear relationships in these data. If anything, the error measures appear to be randomly distributed across the various response rate levels. Even so, we ran a variety of nonlinear regressions (i.e., logarithmic, inverse, quadratic, cubic, power, compound, S, logistic, growth and exponential) using response rates to predict each error measure within each of the four years. None of these models explained any more variance in the error variables than the more traditional linear models presented in Table 16.3.

As a final step, we fit some multivariate models predicting error using response rates while controlling for all of the Election Day factors and voter and interviewer characteristics discussed above. As seen in Table 16.3, the relationship between response rates and error is virtually unchanged when these multiple controls are included. When the 1996 refusal and miss rates are substituted (simultaneously in the same equation) in place of overall response rates, the results are similar. Neither refusal rates nor miss rate are significant predictors of error in the multivariate context.

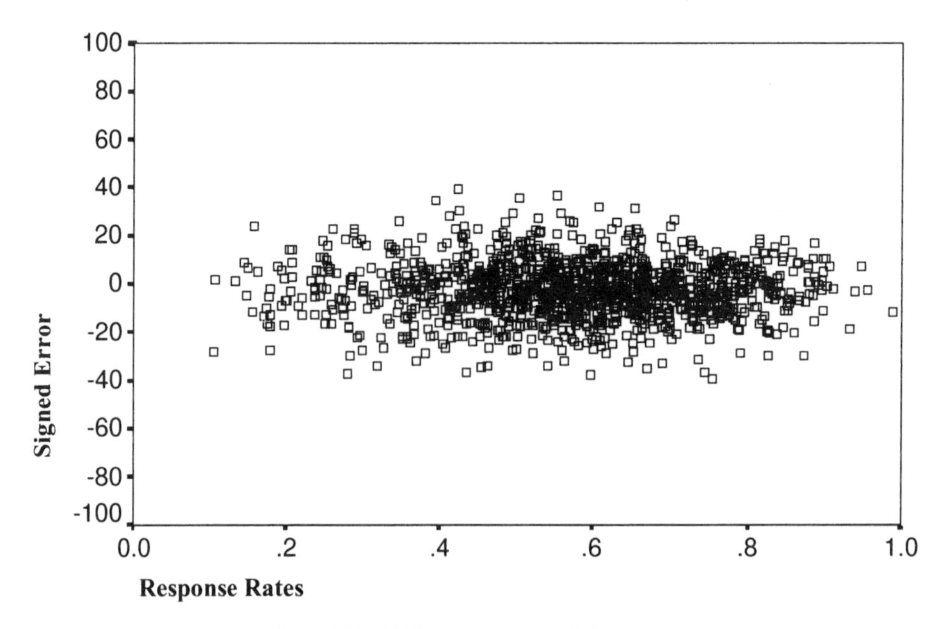

Figure 16.2. 1996 response rates and signed error.

One possible limitation of these findings is that response rates were not randomly assigned to precincts. However, we feel this problem was effectively limited by using multiple control variables that accounted for many of the important differences between the precincts. It is also important to keep in mind that the error variables employed here are measures of total survey error rather than just nonresponse error. Therefore, another possible explanation for these findings is that there is, in fact, consistent nonresponse error that it is being canceled out in the aggregate by another consistent source of survey error. However, this explanation is not very compelling. It is quite implausible that an additional source of exit poll error is also related to response rates and consistently operating in the opposite direction of nonresponse error across all four of these elections.

The results in this section buttress Krosnick's (1999) conclusion that "the prevailing wisdom that higher response rates are necessary for sample representativeness is being challenged" and that "it is no longer sensible to presume that lower response rates necessarily signal lower representativeness." However, this is not to say that response rates are never related to survey error. In a recent experimental study we conducted to explore ways to improve exit poll response rates, the manipulation we employed slightly increased response rates but, at the same time, significantly increased the bias in the vote estimates (Merkle et al., 1998). This is an example where response rates were related to error, but in the unexpected direction—the bias in the exit poll was higher in the experimental condition with the higher response rate.

16.7 CONCLUSION

Unlike most research on survey nonresponse, the strengths of this study are the extremely large, heterogeneous nature of VNS' interviewing pool, the wide range of response rates and our ability to actually compute a measure of error. This study identifies a number of factors that impact exit poll response rates and takes the next step of looking at the relationship between response rates and survey error.

One of the main findings concerning factors that influence response rates is the importance of age. Both interviewer and voter age influence response rates. Older voters have lower response rates, whereas older interviewers produce higher response rates. There is also an interaction between these two variables, indicating that the influence of interviewer age on response rates is driven largely by middle-age and older voters who are less likely to respond to younger interviewers. This suggests that a strategy of employing older interviewers would increase response from older and middle-aged voters, without negatively impacting the response rates of younger voters.

This study provides little support for the similarity hypothesis, which predicts an interaction between voter and interviewer characteristics. It should be noted that the data for age and sex provide a much more robust test of this than the data for race. There are two reasons for this. First, race tends to be more highly clustered by precinct. Second, there is a tendency for interviewer race to correspond to the racial composition of the precinct. There was evidence of an interaction for age, but the pattern of results offered only partial support for the similarity hypothesis. Older voters reacted much more favorably to older interviewers, whereas younger and middle-age voters did not react more favorably to interviewers in their respective age groups.

Of all the Election Day factors, interviewing position is the strongest predictor of response rates. This confirms the importance of the resources VNS puts into ensuring that interviewers can stand as close to the polling place as possible. Another important Election Day factor is the interviewing rate and making sure interviewers are not overworked. Not being able to keep up with the interviewing rate needlessly lowers the response rate by increasing the number of voters who are missed. Interviewing rates should be computed prior to Election Day, taking into account the expected turnout at each location. It is also necessary to monitor these rates when interviewers call in on Election Day and to make adjustments for interviewers who are being overworked because of a higher than expected turnout.

After identifying a number of factors that influence exit poll response rates, we turned our attention to the more critical issue of the impact of response rates on survey error. This is the most extensive study of the relationship between response rates and survey error reported to date, using two error measures across four elections with results from 4000 individual precinct-level exit polls. The most intriguing and surprising finding is the lack of any substantial linear or nonlinear relationship between response rates and error.

This should serve as an eye-opener to those who pursue higher response rates for their own sake without considering the impact on survey error. Devoting limited resources to increasing response rates with little or no impact on survey error is not money well spent, especially when that money might be better spent reducing other sources of error (see Groves, 1989). Even worse, if one ignores the implications for survey error, it is certainly possible that increases in response rates can actually increase survey error (cf. Merkle et al., 1998).

APPENDIX

Variables Used in Regression Analyses

1. Election Day Factors

Number of exits worked.

Interviewing position (1 = Inside the room where people vote, 2 = Outside the room, but inside the building where voters could be seen exiting the voting room, 3 = Outside the building, but at the door, 4 = Outside the building, between 10 and 50 feet from the door, 5 = Outside the building, more than 50 feet from the door).

Problems with election officials or police (1 = yes, 0 = no).

Trouble keeping up with the interviewing rate (3 = yes, all day, 2 = only during busy times, 1 = no).

Number of questions on the front of the questionnaire.

Did the questionnaire have a back side (1 = yes, 0 = no).

2. Voter Characteristics

Percent male voters—The proportion of voters that day who were men.

Percent minority voters—The proportion of voters that day who were Black or Hispanic.

Percent older voters—The proportion of voters that day who were 60 or older.

3. Interviewer Characteristics

Gender (1 = female, 0 = male).

Race (1 = White, 0 = nonwhite).

Education (4 = college graduate, 3 = some college, 2 = High school graduate, 1 = some high school).

Age (in years).

Interviewer experience—A scale of three variables: number of telephone surveys worked on, number of face-to-face surveys worked on, number of exit polls worked on.

CHAPTER 17

Nonresponse in the Second Wave of Longitudinal Household Surveys

James M. Lepkowski and Mick P. Couper, *University of Michigan and Joint Program in Survey Methodology*

17.1 INTRODUCTION

Longitudinal or panel surveys collect data from the same household or person repeatedly over several rounds or waves of data collection. Panel surveys are subject to nonresponse and present additional complications of missing waves of data. Since there are treatments elsewhere in this volume of item and initial unit nonresponse in longitudinal surveys (e.g., Pfefferman and Nathan, Chapter 26), the focus here is on wave nonresponse for a unit that has already participated in a baseline interview.

Groves and Couper (1998) advocate that survey researchers should "design for nonresponse" as a component of total survey design. We extend this notion to panel surveys in which the first wave can be used to collect information to inform efforts to locate sample units and maximize the likelihood of cooperation at subsequent waves. An initial investment in suitable data collection at the first, or prior, wave can yield estimated response propensities and improve survey efficiency at later waves.

The survey literature is full of good ideas for locating respondents, making initial contact, and obtaining cooperation in longitudinal surveys (see for example Groves and Hansen, 1996; Laurie et al., 1999; Ribisl et al., 1996). Comprehensive theories that place these ideas in a broader framework to understand the nonresponse process and foster development of methods to reduce and compensate for nonresponse are more rare. The survey literature does contain rich information about potential predictors of nonresponse in longitudinal surveys. Unfortunately, the information often is limited to demographic or geographic measures. Furthermore, although the literature contains many examples of modeling attrition in panel surveys (e.g., Fitzgerald et al., 1998; Lillard and Panis, 1998; Zabel, 1998), few of these models examine components of the process separately (for exceptions, see Couper et al., 1996; Iannachione, 1998).

In this chapter, we examine how social, psychological, economic, and situational measures may enhance the likelihood of locating subjects and improve the chances of gaining cooperation in longitudinal surveys. We examine location and cooperation processes separately to determine if, as expected theoretically, different predictors are needed for each process. If so, data can be collected to inform approaches tailored to the location or cooperation process. We present a theoretical framework for understanding nonresponse in panel surveys, and present findings from empirical models of location and cooperation propensity from two social surveys.

17.2 PANEL SURVEY DESIGN

Panel survey design varies along dimensions that have implications for panel nonresponse: the frequency of data collection, the time between data collections, the number of waves, and the type of unit being followed. Variation in length of time between waves has an effect on longitudinal nonresponse. Information on location will be more recent for shorter times between interviews. The number of panel waves also affects nonresponse, since nonresponse is typically cumulative, increasing with successive waves. Repeated interview requests increase the perceived burden of participation and decrease the likelihood of cooperation.

Other operational features also influence nonresponse. Many longitudinal surveys of individuals do not attempt to contact those who were not interviewed at an initial wave, and others do not attempt to follow those who do not respond at later waves. Surveys may contact units between waves to update information on location or to motivate units to cooperate with a future interview request.

The content of a survey, the sponsoring organization, the data collection organization, and mode of data collection also influence nonresponse, although the role of these elements in panel surveys is less well understood. We suspect that these effects may well diminish with time as units in a panel survey become familiar with the survey questions and develop a degree of trust that the survey is legitimate and will be used for the public benefit. Mode differences in response rates may also diminish with time as the degree of trust in the survey process increases among respondents, or as more compliant respondents are self-selected.

One advantage of dealing with nonresponse in later waves of panel surveys is that in the first wave of the survey one can obtain a substantial vector of possible causes and correlates of later wave cooperation. On the other hand, in panel surveys of relatively long duration, the cumulative effects of nonresponse over several waves can increasingly threaten the representativeness of the remaining sample cases.

17.3 TOWARD A THEORY OF NONRESPONSE IN PANEL SURVEYS

Once the first wave of data collection is completed, a survey organization has much more information about the process than basic design elements. The organization knows a good deal about each unit with each successive wave. At the same time, the sample unit also has more information about the nature of the request being made. The availability of this information to both parties in the survey process can have a substantial effect on longitudinal survey nonresponse. For this reason, we believe the nonresponse process in later waves of a panel surveys differs in important ways from cross-sectional surveys or the initial wave.

We advocate a framework for understanding the nonresponse process at a second (or later) wave of a panel survey that divides the process into three conditional processes: location, contact given location, and cooperation given contact. For each process, models can be posited that use information from prior waves of data collection on the sample unit to predict the outcome of the process. We examine potential predictors for the outcome of each of these processes in turn.

Location. Locating a sample unit in a cross-sectional study is largely considered to be a straightforward field operation. Locating sample households or persons in subsequent waves of a panel survey is also straightforward, provided the unit has not changed location from one wave to the next. The difficulty arises when units change addresses. The propensity of locating units in later panel waves is determined by factors related to whether or not the unit moves.

If the unit is a person, follow-up efforts can be concentrated on that individual. But in some longitudinal surveys, the unit is a household or family. Whole-household moves pose different location problems than partial household moves, particularly if all household members have moved but have not moved to the same new location. If some household members remain at the original address, the likelihood of finding sample persons who have moved out from the household is much greater than if the entire household has moved.

Survey design features that affect location propensity include the length of the study (the longer the panel, the greater the likelihood of moving), the interval between waves (the longer the interval between waves the greater the likelihood of moving), and the nature of contacts between waves. Tracking rules for cost reasons may limit the geographic area for follow-up. The amount and quality of information collected by the survey organization specifically for tracking movers is also driven by cost considerations. Survey resources can be used to increase the number of attempts made to follow individuals, the types of tracking resources that can be used, and the variety of resources that are employed in any given longitudinal survey.

Household characteristics affect the likelihood of moving, and thus the location propensity. Geographic mobility is related to the household or individual life stage, as well as cohort effects. For example, younger people and those with higher education levels are more mobile. The number of years at a residence, household tenure, and community attachments through family and friends also determine the likelihood of moving.

Life events are related to moving likelihood. Births or deaths, marriage, job changes, and crime victimization affect the likelihood of moving. Life events may increase the difficulty of locating individuals: a name change in marriage or following divorce makes it more difficult to track and locate someone who has moved since the last panel wave.

The social aspects of community attachment may also affect the likelihood of moving, or provide more data on units that do move. Individuals engaged in the civic aspects of their community are posited to be more stable and less likely to move. Those linked to their current community life are likely to leave many traces to their new address; they are likely to be politically, socially, and economically engaged in their new community. Their lives are accessible through public databases such as telephone directories, credit records, voter and library registration, membership in religious or cultural organizations, or children in schools.

Socially isolated individuals will be more difficult to track following a move, as will those who withdraw from community attachments. Unlisted numbers and other means to assure privacy may accompany life changes such as divorce. Evidence of unwillingness to be found may be obtained through various indicators of reluctance to participate in the first wave, and the amount and quality of information provided at that time.

Contact. Contact difficulty is a significant nonresponse factor in first waves, but it is relatively small in later waves, given successful location. Attempts to verify the location of a sample household or member often result in contact with that sample unit, making it hard to separate the contact effort from the location effort. For the most part, once the location of the sample unit has been identified, the contact task is similar to that for a cross-sectional survey. However, interviewers have prior knowledge about the likely at-home patterns of the household and information on best times to call. Given this, the number of call attempts to contact should be lower in later waves of a panel survey.

Cooperation. The sample person's willingness to cooperate with subsequent waves of data collection will be shaped by a variety of situational factors at the time of the request (as in cross-sectional surveys) as well as their recollection of the prior wave experience. The strength of the effect of prior wave experience on later cooperation may depend on several factors: the salience of the experience, the time between waves, whether there is interwave contact from the survey organization, and whether the same or a different interviewer is used (e.g., Hill and Willis, 1998; Lurie et al., 1999; Zabel, 1998).

Assessing prior wave experiences may be an important component in understanding the cooperation process. Several studies have shown that interviewer assessment of respondent enjoyment with and understanding of a prior wave interview is strongly related to cooperation (e.g., Kalton et al., 1990). Any nonresponse model for longitudinal surveys should control for both proximate, current wave factors as well as distant, prior wave indicators of survey experience. It may be possible that an interviewer armed with such knowledge and properly trained may be able to "immunize" a respondent at a prior wave against subsequent wave nonresponse by reinforcing a positive experience.

Groves and Couper (1998) concluded that much of the decision to cooperate with a one-time survey request is shaped at the time of the request, rather than being a well-formed predisposition on the part of the sample household. It is not clear whether this emphasis on current situational factors is also true in later waves of a longitudinal survey.

A Theoretical Model. In the spirit of the Groves and Couper model, we present a theoretical framework that identifies the various influences on nonresponse in longitudinal surveys and provides a structure for linking the effect of those influences.

The propensity of response R_t at time t (Little, 1986) is the product of three conditional propensities. Let $\mathbf{x_L}$, $\mathbf{x_C}$, and $\mathbf{x_I}$ denote vectors of covariates for the ith subject which are expected to be related to the location, contact, and cooperation probability. These vectors may be overlapping, containing many if not all of the same predictors. Then the response propensity is given by $\Pr\{R_t|\mathbf{x_L}, \mathbf{x_C}, \mathbf{x_I}\} = \Pr\{I_t|C_t, L_t, \mathbf{x_I}\} \times \Pr\{C_t|L_t, \mathbf{x_C}\} \times \Pr\{L_t|\mathbf{x_L}\}$. Each conditional propensity represents a step in the nonresponse process that may be modeled separately for each wave.

The covariates may need to contain contemporaneous and lagged factors. For the location propensity model at time $t = 2$, the vector $\mathbf{x_L}$ may be comprised of contemporaneous covariates U at time $t = 2$ and a separate set of lagged factors V at time $t - 1 = 1$, the same or different factors as those in U. A logit model for location propensity at time t may be given as $\text{logit}(\Pr\{L_t|\mathbf{x_L}\}) = U\lambda_t + V\zeta_{t-1}$, where λ_t and ζ_{t-1} are vectors of parameters for time t factors and for lagged effects from time $t - 1$, respectively. Similarly, for factors Y at the current wave and W at the prior wave, the logit model for contact given location may be expressed as, $\text{logit}(\Pr\{C_t|L_t, \mathbf{x_C}\}) = Y\varphi_t + W\theta_{t-1}$ for parameter vectors φ_t and θ_{t-1}, respectively. For cooperation at wave t, conditional on a subject having been located and contacted, a logit model linking factors X at time t and Z at time $t - 1$ is $\text{logit}(\Pr\{I_t|C_t,L_t, \mathbf{x_I}\}) = X\beta_t + Z\gamma_{t-1}$, where β_t and γ_{t-1} are vectors of parameters.

The model is illustrated in Figure 17.1. For each process there are survey design as well as household and person characteristics that affect the final status. Some of these characteristics may be related to more than one process. There is also a process cycle repeated at each wave of a longitudinal survey. Lagged unit characteristics and survey design features from prior waves may influence the outcome for each process.

Not made explicit in this general formulation is whether the different processes are related. For example, those who are hard to contact may be less cooperative once found. People who do not want to be found, or who did not provide detailed tracking information because of their reaction may need an estimated propensity from a prior process as part of the current process factors. For example, an estimated location propensity may be a predictor of cooperation propensity.

Noncontact once a subject has been located is typically a relatively rare status in longitudinal surveys. The two surveys examined subsequently have very few noncontact events among those who have been located. Noncontact propensity will therefore be dropped from further consideration empirically, even though it is a part of the overall response process.

The application of the model to the study of nonresponse on the second wave of a panel should contain covariates that are related to location or cooperation propensity. The following factors appear to be important for location propensity: *survey*

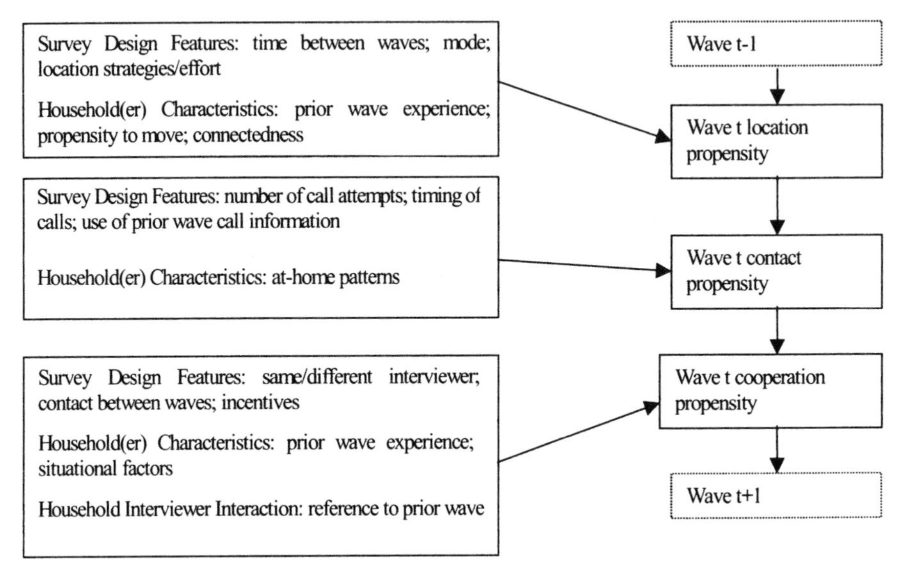

Figure 17.1. Model for nonresponse in longitudinal surveys.

design features such as the number of panel waves, length of time between panel waves, and topic; *organizational efforts* ranging from between-wave contact efforts, use of tracking specialists during data collection, and varied resources used to track respondents who have moved; *respondent characteristics* such as sociodemographic characteristics, economic status, or geographic location at last interview; respondent *community attachment factors* include indicators of contacts within the community, participation in community events, strength and nature of relationships with relatives and friends, tenure (i.e., rent or own), and social integration; *willingness to be found* indicators include income, problems paying bills, serious money problems, recent moves, and previous marriage. Finally, *survey experience,* such as the extent to which the subject understood questions, enjoyed the interview, or had difficulty remembering questions, may be weakly related to location propensity.

For cooperation propensity, several of the same factors that influence location propensity may also be important: *survey experience* is time lagged and may be more closely related to cooperation than to location; *respondent characteristics* such as sociodemographic, economic, and geographic factors, most of which do not vary over time; *situational factors,* including a respondent's physical or mental status or economic circumstances such as unemployment or chronic financial stress, may increase or decrease willingness to cooperate, probably assessed at a prior wave; *social integration* at a prior wave such as marital status or summary measures of contacts with friends, relatives, and others; and *organizational efforts* such as repeated call backs, incentives, or refusal conversion efforts.

Models of the nonresponse process for panel surveys depend on the availability of these kinds of variables at prior waves of measurement. In the next section, we examine data from two longitudinal surveys that collected a number of these factors at the initial wave of data collection.

17.4 EMPIRICAL EXAMPLES OF PANEL NONRESPONSE

17.4.1 Location and Cooperation in the Americans' Changing Lives (ACL) Survey

The ACL is a panel survey of a national stratified multistage probability sample of the U.S. civilian noninstitutionalized population ages 25 years and older selected in 1986. African Americans were chosen at twice the rate of non-African Americans, and older persons at twice the rate of those under age 60. Wave 1 interviews collected information from 3617 respondents (67% response rate). Wave 2 interviews were conducted approximately $2\frac{1}{2}$ years later in 1989, following up all 3451 Wave 1 respondents who were alive at the Wave 2 interview. A third wave was conducted in 1994, but we focus only on the first two waves here.

The location and cooperation status of subjects at Wave 2 are outcomes of interest. Wave 1 respondents who died prior to Wave 2 are excluded. Wave 1 respondents who completed, refused, or did not complete a Wave 2 interview due to health or other reasons are coded as located, whereas those not found at Wave 2 are coded as not located. Wave 1 respondents who completed an interview at Wave 2 are coded as cooperating, leaving those who were located but refused or not interviewed for other reasons coded as not cooperating. Table 17.1 shows the distribution of location and cooperation status at Wave 2 for Wave 1 respondents.

The large number of potential Wave 2 location and cooperation predictors available in Wave 1 serve as lagged predictors in the model. Eight sociodemographic and geographic variables were selected: age, gender, race, ethnicity, education, im-

Table 17.1. Wave 2 location and cooperation status for ACL and NES Wave 1 respondents

	ACL		NES	
Status	*N*	%	*N*	%
Location				
Total	3617		1769	
Located	3315	96.1	1587	94.0
Not located	136	3.9	101	6.0
Died	166	—	81	—
Cooperation, given location				
Total	3617		1769	
Cooperating	2867	86.5	1359	82.5
Not cooperating	448	13.5	228	17.5
Died or not located	302	—	182	—

puted annual family income, region, and urban, suburban, or rural residence. Seven community attachment variables were used: frequency of talking on the telephone; visiting with friends or relatives; attending meetings in the community; satisfaction with the home; whether living in an intimate relationship; whether caring for a relative or friend; and whether renting or owning their home.

Eight measures of social integration were chosen for the analysis: marital status, summary scales of informal and formal social integration, friend and relative total support, and participation in voluntary activities. Four measures of situational circumstances were chosen: satisfaction with health and employment status, and summary scales of chronic financial stress and depression. Six interviewer assessments of the respondent's survey experience during the Wave 1 included interviewer assessments of respondent ability to express themselves and to understand the questions, cooperation with and enjoyment of the interview, difficulty remembering things asked about in the interview, and the extent to which the respondent required frequent repetition of questions. Finally, four indicators of the extent to which respondents were accessible to survey efforts to locate them are included: number of marriages, and whether the respondent had moved to a new address in the last three years, had difficulty paying bills, and had serious money problems.

A total of 37 Wave 1 predictors were examined for Wave 2 location and cooperation status. Logistic models were estimated using the SUDAAN software system (Shah et al., 1997), accounting for the stratified multistage sample design and weights to compensate for unequal probabilities of selection as well as Wave 1 nonresponse.

Three sets of models were examined for Wave 2 location and cooperation status, separately. Bivariate associations for the 37 Wave 1 predictors were assessed through Wald F statistics for the null hypothesis that the predictor had no association with the outcome. A full model then regressed the subset of the 34 predictors that had significant ($p < 0.05$) results from the bivariate and full regression models are presented in Table 17.2. The three predictors that did not reach significance in the bivariate models are excluded from the table. Finally, predictors that had statistically significant associations ($p < 0.05$) with location or cooperation in the full model were then regressed on location or cooperation, respectively, in reduced models. The results of the reduced models closely parallel those of the full model and are thus not shown here. The direction of the association is noted for each predictor by the presence or absence of a negative sign before each bivariate Wald F statistic. (No association changed direction from the bivariate to the full or reduced models.) Negative associations decrease location or cooperation propensity.

The bivariate associations in Table 17.2 suggest that the effects of several variables differ on location versus cooperation propensity. For example, those who rent their home have lower location and cooperation propensities, but the effect appears larger for location. Survey experience variables have little effect on location, but several are significant predictors of cooperation. All the community attachment variables are significantly related to location, as are the situational measures and the willingness to be found indicators. As might be expected, prior survey experience

Table 17.2. Bivariate and full models for ACL Wave 1 predictors of Wave 2 location and cooperation

		Location		Cooperation	
Wave 1 predictors	*df*	Bivariate	Full model	Bivariate	Full model
Sociodemographic/geographic					
Age (6 groups)	5	5.89***	2.32 +	(–)6.34***	1.48
Female	1	2.79	5.08*	6.69***	3.15 +
African American	1	(–)[a]17.27***	27.37***	0.08	1.77
Hispanic	1	9.12**	0.49	2.32	0.09
Education (6 groups)	5	4.09**	3.23*	6.25***	0.66
Family income (10 groups)	9	9.52***	1.25	2.68*	1.75
Region (4 groups)	3	7.30***	9.27***	0.46	0.90
Urban/rural	2	(–)4.69*	1.69	0.54	2.45 +
Community attachment					
Frequently talk on phone	3	6.95***	1.14	4.65**	2.79 +
Frequently visit with friends	3	7.77***	1.92	1.06	1.68
Frequently attend meetings	3	6.34***	2.88*	7.97***	2.31 +
Satisfaction with home	2	6.31**	0.02	0.29	0.30
Care for relative/friend	1	6.43*	4.67*	3.86 +	2.78
Rent home	1	(–)80.48***	8.09**	(–)11.79**	9.00**
Social integration					
Marital status	4	8.63***	0.71	0.44	1.63
Informal social integration	1	2.63	3.19 +	6.79*	6.45
Formal social integration	1	7.43**	1.77	2.00	2.12*
Support from friend/relatives	1	13.67***	0.31	0.56	0.02
Volunteers at education group	1	0.32	1.69	16.85***	4.65
Volunteers for political group	1	0.27	0.31	6.14*	0.52*
Volunteers for senior citizens group	1	0.56	0.37	6.22*	0.48*
Other volunteer activities	1	1.05	0.11	13.96***	1.84
Situational circumstances					
Satisfaction with health	4	2.95*	3.60*	1.38	0.46
Employment status	2	17.19***	6.19**	2.48 +	1.03
Chronic financial stress	1	(–)73.01***	0.88	0.11	2.03
Survey experience					
Ability to express self	3	2.38 +	0.63	22.15***	3.93
Understanding of questions	3	2.36 +	0.59	12.22***	0.19
Cooperation with interview	3	1.79	2.41 +	18.63***	4.00
Enjoyed interview	3	4.57**	3.33*	6.37**	0.14*
Difficulty remembering	3	1.76	0.41	(–)15.82***	1.69
Frequency of repetition	1	(–)5.53*	6.31*	(–)13.36***	0.86*
Willingness to be found					
Moved in last 3 years	1	(–)35.23***	2.37	0.43	0.22
Difficulty paying bills	4	(–)9.64***	1.57	1.41	1.51
Serious money problems	1	(–)12.79***	0.75	0.94	0.00

Note: [a]Indicates a negative association between predictor and location/cooperation status; + $0.10 > p \geqq 0.05$; * $0.05 > p \geqq 0.01$; ** $0.01 > p \geqq 0.001$; *** $0.001 > p$.

has little relationship with location propensity, but willingness to be found factors are strongly associated with Wave 2 location propensity.

ACL full models show weakened or eliminated associations. Notable was a lack of association for the social integration and willingness to be found measures. On the whole, location propensity appears to be related moderately to sociodemographic and geographic measures, with weak association for community attachment and situational factors.

For cooperation, mild association for a community attachment measure (renting a home) occurs, and as expected in the theoretical model, situational circumstances and willingness to be found are not associated with cooperation. But other expected associations do not appear.

The ACL reduced models (not shown) indicate that sociodemographic and geographic measures are associated with location but not with cooperation. Community attachment, particularly renting a home, is related to location propensity, and mildly associated with cooperation propensity. Survey experience is associated with both location and cooperation, with no clear pattern among those measures.

17.4.2 Location and Cooperation in the NES Panel

The biennial National Election Studies (NES) are based on stratified multistage area probability samples of households of the U.S. civilian noninstitutionalized population ages 18 years and older (Miller et al., 1992). The present study is limited to the 1990 NES participants followed in 1992. These two years are especially valuable because in 1990 a set of observational measures on nonresponse (see Couper, 1997) were added. Interviewers in 1992 recorded information about any contact events with the sample household prior to the completion of an interview. These measures can thus serve as current wave (Wave 2) predictors of location and cooperation propensity, with stronger associations for cooperation than location propensity expected.

A total of 2000 persons were interviewed in the 1990 NES, and 1769 were released for follow-up in 1992. Of the Wave 2 interviews, 146 were partials. Those who died between the Wave 1 and 2 interviews are excluded from both propensity models (see Table 17.1). The location and cooperation rates are slightly lower than for ACL.

The NES location and cooperation status were each regressed on the available predictors in a three-step process, as for the ACL analysis, again accounting for the stratified multistage sample design. Weights were not employed in the NES analysis because the NES was essentially self-weighting at the household level.

Predictors in the ACL location and cooperation models were also available for the NES models, including sociodemographic or geographic, survey experience, and willingness to be found. The NES had available, though, several particularly valuable measures as predictors that were not available in the ACL. Contact measures collected prior to the Wave 2 interview indicating reluctance to be interviewed were available.

With the exception of income, the same sociodemographic and geographic measures as in the ACL are used. The NES has a limited set of indicators of community attachment: whether renting or owning a home, the number of young children in the household, and whether the respondent attends church services. NES has only a single variable that could be construed as a measure of social integration: marital status. A single item on whether religion is important in the respondent's life and three measures of political integration (whether the respondent participates in nonvoting political activities, discusses politics with relative and friends, and voted in the 1990 Congressional election) were included as political integration predictors. The NES has only a single measure of situational circumstances at Wave 1—employment status. The NES has a large number of measures of political knowledge and interest that may serve as indicators of social alienation from the community. Eight such measures were selected for the present analysis: the frequency with which the respondent follows government and public affairs, party identification, interest in politics, knowledge about political figures, whether any of named figures were not known, and three items on political alienation.

Eleven measures of survey experience are used in the NES analysis. Seven of these are either based on comments made during contact prior to completion of the interview or based on interviewer assessments of the respondent during the Wave 1 interview: whether the respondent or someone else at the household said they were too busy to be interviewed, whether they were not interested in the interview, whether they had no knowledge about political matters, whether they exhibited behavior or made comments indicating reluctance to be interviewed, whether the respondent refused to answer the Wave 1 income item, the Wave 1 item missing data rate, and interviewer assessments of the respondent's cooperation with and interest in the interview. Also used were interviewer assessments of whether the respondent was suspicious about the purposes of the interview and if the respondent gave indications that the interview was too long. Finally, two measures are used of the willingness of the respondent to be found: whether the respondent had moved to a new address in the last three years and whether a telephone number was provided.

The same three sets of models as in the ACL analysis—bivariate, full, and reduced—are used to predict location and cooperation propensity. The results of fitting the first two types of models (bivariate and full) on the NES are given in Table 17.3. The reduced models are not shown.

For the NES bivariate models, few of the sociodemographic measures appear strongly associated with either location or cooperation at Wave 2. Renting a home and having young children in the home are associated with lower location propensities, demonstrating the effect again of community attachment measures for location propensity. The effect of these variables on cooperation is more modest. The effect of social and political integration measures differs for location and cooperation propensity. Notable are the several political knowledge and interest variables related to location propensity, suggesting that community alienation or isolation may contribute to location. As with the ACL model, all the survey experience measures have significant associations with cooperation, but fewer appear to affect location propensity. In contrast, as we would expect, the accessibility measures are related to

Table 17.3. Bivariate and full model for NES Wave 1 predictors of Wave 2 location and cooperation

		Location		Cooperation	
Wave 1 predictors	*df*	Bivariate	Full model	Bivariate	Full model
Sociodemographic					
Age	5	3.77**	2.57*	1.79	3.19*
Female	1	0.95	3.45 +	5.09*	1.14
African American	1	(–)[a]6.52*	0.89	3.09 +	3.88 +
Hispanic	1	3.39 +	0.36	3.91 +	2.18
Region	3	0.44	1.91	0.94	2.06
Urban/rural	2	1.63	0.88	3.19 +	4.44*
Education	3	2.6 +	1.71	5.11**	6.89*
Community Attachment					
Rent home	1	(–)46.08**	0.29	0.71*	0.96
Child under 6 years	1	(–)5.44*	0.47	5.2*	2.54
Social/political integration					
Marital status	3	3.7*	1.23	0.46	1.41
Active politically	1	1.6	0.58	10.15**	2.95 +
Discuss politics with friend/relatives	1	4.41*	1.00	5.65*	0.44
Voted 1990	1	30.59**	3.26 +	0.35	1.49
Situational circumstances					
Employment status	2	2.83 +	0.84	0.56	0.70
Political knowledge/interest					
Follow government/public affairs	3	1.62	1.54	3.8*	0.95
Party identification	3	4.62**	1.90	0.6	0.99
Interested in politics	2	3.82*	0.18	3.3 +	0.05
Knowledge of political figures	1	15.51**	2.47	10.81**	5.32*
Any political names unrecognized	1	5.84*	0.08	0.17	2.39
Public officials don't care	4	3.87*	1.05	5.4**	3.05*
Survey experience					
Comments: too busy	1	0.13	0.35	51.81**	10.99**
Comments: not interested	1	0.05	0.85	5.59*	0.14
Reluctant behavior	1	0.01	0.65	24.16**	15.78***
Refused income question	1	5.49*	4.45*	6.68*	1.67
Cooperation with interview	2	5.16*	2.15	7.27**	0.07
Interested in interview	2	4.9*	1.79	9.53**	2.77 +
Suspicious of interview	1	2.44	0.54	13.43**	3.93 +
Interview too long	1	0	5.43*	5.89*	0.01
No political knowledge	1	0.27	0.29	8.65**	1.76
Missing data rate	1	(–)24.22**	6.03*	3.12 +	0.09
Accessibility					
Moved in last 3 years	1	(–)49.27**	5.60*	1.52	8.17**
Telephone number provided	1	27.35**	12.67**	0.64	0.83

Note: [a]Indicates a negative association between predictor and location/cooperation status; + $0.10 > p \geqq 0.05$; * $0.05 > p \geqq 0.01$; ** $0.01 > p \geqq 0.001$; *** $0.001 > p$.

location propensity, but less so to cooperation. Again, it is clear that the factors that affect these two types of nonresponse are not the same.

The NES location full model for 32 measures shows that only six measures have associations significant at the $\alpha = 0.05$ level. These six are dominated by the survey experience and willingness to be found measures. For the cooperation full model, eight measures showed an association at the $\alpha = 0.05$ level. The strongest associations appear to be with respect to the two survey experience measures and the mobility indicator.

Unlike the ACL, the NES full models indicate that the principal predictors are survey experience and willingness to be found measures. Sociodemographic and geographic measures play a role in location and cooperation, but modest bivariate associations for these variables are largely explained by the theoretically more interesting survey experience and willingness measures.

Few of the associations remain significant in the NES reduced model (not shown). However, the cooperation model shows notable associations with sociodemographic and geographic measures, social and political integration, political knowledge and interests, and survey experience. Unlike the ACL reduced models, sociodemographic and geographic factors play a role in the NES cooperation reduced model, but not in the location model. Community attachment is not as well measured in the NES as the ACL, and the associations are not statistically significant. But survey experience and various forms of integration show associations with both location and cooperation in both NES and ACL. However, the NES reduced models notably indicate that willingness to be found is a factor in both location and cooperation propensity, a finding that is absent in the ACL.

17.5 CONCLUSION

The empirical findings in the present study are subject to a number of limitations. Fitting models with large numbers of predictors when the degrees of freedom are limited in survey variance estimation is problematic. The bivariate and reduced models have, except for the NES reduced cooperation model, sufficient degrees of freedom to be confident that the Wald statistics are adequate for model generation purposes. The full model Wald statistics are based on limited degrees of freedom, and thus used only as rough guides for selecting predictors to include in the final model.

The model selection strategy is limited for purposes of exposition. A more complete model fitting strategy would examine combinations of predictors beyond those examined in the reduced models to be sure that collinearity in the full models was not masking important data.

The empirical models employ at times weak proxy measures for various constructs that should be related theoretically to location or cooperation propensity. Further, several useful constructs are not represented at all. For example, a variety of measures could be added to give greater insight into the role of interviewers in location and cooperation propensity.

This chapter was meant to illustrate the kind of investigations that we believe should be conducted to examine the nature of location and cooperation propensity in longitudinal surveys. The greatest barrier to developing a more complete understanding of the location and cooperation process is the lack of suitable predictors in large data sets. Designers of longitudinal surveys should seriously consider adding variables of the types examined here to predict location and cooperation more reliably. This could lead to an improved understanding of the processes that produce nonresponse in later waves of panel surveys. In turn, this may lead to more effective methods for reducing nonresponse rates and to more effective adjustment procedures based on richer information.

PART IV

Statistical Inference Accounting for Nonresponse

CHAPTER 18

Weighting Nonresponse Adjustments Based on Auxiliary Information

Jelke G. Bethlehem, *Statistics Netherlands*

18.1 INTRODUCTION

This chapter considers weighting adjustment for unit nonresponse. The remainder of this section introduces the general theoretical framework necessary to describe various weighting techniques. Section 18.2 gives a general introduction to adjustment weighting, and describes poststratification as the traditional and most frequently used weighting technique. Section 18.3 is about a weighting technique based on the general regression estimator. Section 18.4 describes a weighting technique often called raking or iterative proportional fitting. Section 18.5 considers linear and multiplicative weighting as special cases of a more general method called calibration estimation. Section 18.6 describes a practical example of a survey in which a number of different weighting techniques were applied. The chapter concludes with some discussion in Section 18.7.

18.1.1 Population and Sample

Let the finite survey population U consist of a set of N elements. The values of target variable Y for these elements are denoted by $Y_1, Y_2, \ldots, Y_N$. Objective of the sample survey is assumed to be estimation of the population mean

$$\bar{Y} = \frac{1}{N} \sum_{k=1}^{N} Y_k \tag{1.1.1}$$

However, note that the theory can also be applied to estimators of other population characteristics. A sample of size n is selected from the population. It is assumed throughout this chapter that samples are simple random without replacement. Let $y_1, y_2, \ldots, y_n$ denote the values of the target variable for the n selected elements.

18.1.2 A Model for Nonresponse

To analyze the impact of nonresponse on the characteristics of estimators, the phenomenon of nonresponse must be incorporated in the theory of sampling. Here, the random response model is assumed; see e.g., Bethlehem and Kersten (1985) and Oh and Scheuren (1983). This model assigns to each element k in the population a certain, unknown probability ρ_k of response when contacted in the sample. Let

$$\bar{y}^* = \frac{1}{m}\sum_{i=1}^{m} y_i \tag{1.2.1}$$

denote the mean of the m ($m < n$) available observations. The obvious approach would be to use this estimator to estimate the population mean. Unfortunately, this is not an unbiased estimator. Bethlehem (1988) shows that the expected value of this estimator is approximately equal to

$$E(\bar{y}^*) \approx \frac{1}{N}\sum_{k=1}^{N} \frac{\rho_k Y_k}{\bar{\rho}} \tag{1.2.2}$$

Therefore, the bias of this estimator is approximately equal to

$$B(\bar{y}^*) \equiv E(\bar{y}^*) - \bar{Y} \approx \bar{Y}^* - \bar{Y} = \frac{C(\rho, Y)}{\bar{\rho}} \tag{1.2.3}$$

in which

$$C(\rho, Y) = \frac{1}{N}\sum_{k=1}^{N} (\rho_k - \bar{\rho})(Y_k - \bar{Y}) \tag{1.2.4}$$

can be seen as the population covariance between response probabilities and the values of the target variable. A similar expression for the bias is given by Tremblay (1986). From expression (1.2.3) it follows that the estimator is approximately unbiased if there is no correlation between target variable and response behavior. The stronger the relationship between target variable and response behavior, the larger the bias. The size of the bias also depends on the amount of nonresponse. The more people are inclined to cooperate in a survey, the higher the average response probability will be, resulting in a smaller bias.

18.2 WEIGHTING TECHNIQUES

18.2.1 Weighting Sample Surveys

To correct for a nonresponse bias, adjustment weighting is often carried out. Each observed element i is assigned an adjustment weight w_i. So, the simple sample mean (1.2.1) is replaced by a new estimator

$$\bar{y}_w = \frac{1}{n} \sum_{i=1}^{n} w_i y_i \tag{2.1.1}$$

Weighting adjustments are based on the use of auxiliary information. Auxiliary information is defined as a set of variables that not only have been measured in the survey, but for which information on the sample or population distribution is also available. Often, in case of unit nonresponse, there are no auxiliary variables for which the sample distribution is available. An exception is the situation in which the sampling frame contains auxiliary variables such as sex and age in a population register. If there is no sample distribution, weights can also be based on the population distribution of auxiliary variables. Often, statistical agencies have this type of information with respect to a limited number of auxiliary variables. In the reminder of this chapter, it is assumed that auxiliary information is only available at the population level.

18.2.2 Poststratification

Poststratification is a well known and often used weighting method. An early paper by Holt and Smith (1979) describes poststratification as a robust technique that can improve the precision of estimates in case of full response. Thomsen (1973) shows how poststratification reduces a possible bias due to nonresponse. More recent considerations of poststratification include papers by Little (1986, 1993a) and Chapter 19 in this volume by Gelman and Carlin.

To carry out poststratification, one or more qualitative auxiliary variables are needed. Here, only one such variable is considered. The situation for more variables is not essentially different. Suppose there is an auxiliary variable X having L categories, so it divides the population into L strata. The poststratification estimator is defined by

$$\bar{y}_{ps} = \frac{1}{N} \sum_{h=1}^{L} N_h \bar{y}_h \tag{2.2.1}$$

in which N_h is the number of population elements in stratum h, and $\bar{y}_h$ is the mean of the n_h available observations in stratum h.

Poststratification assigns identical adjustment weights to all elements in the same poststratum. The weight w_i for an element i in stratum h is equal to

$$w_i = \frac{N_h/N}{n_h/n} \tag{2.2.2}$$

In case of full response, the poststratification estimator is unbiased. If nonresponse occurs, the estimator may be biased. It can be shown the bias of this estimator is equal to

$$B(\bar{y}_{ps}) = \frac{1}{N} \sum_{h=1}^{L} N_h B(\bar{y}_h) = \frac{1}{N} \sum_{h=1}^{L} N_h \frac{C_h(\rho, Y)}{\bar{\rho}_h} \tag{2.2.3}$$

in which $C_h(\rho, Y)$ is the covariance between the response probabilities and the values of the target variable within stratum h, and $\bar{\rho}_h$ is the mean of the response probabilities in stratum h. If there is no relationship between response probabilities and values of the target variable within each stratum, all stratum covariances are zero, and thus the bias of estimator (2.2.3) vanishes. Two situations can be distinguished in which this is the case:

1. The strata are homogeneous with respect to the target variable. Then this variable shows little variation within strata and, consequently, the stratum covariances will be small.
2. The strata are homogeneous with respect to the response probabilities. Then these probabilities show little variation within strata and, consequently, the stratum covariances will be small.

Holt and Smith (1979) showed that in the first situation the variance of the estimator is also reduced. So, poststratifications based on strata that are homogeneous with respect to the target variable both reduces variance and bias. Poststratification based on strata that are homogeneous with respect to the response probabilities reduces the bias but not necessarily the variance. Similar observations were made by Little (1986) and Tremblay (1986).

As more variables are used in a weighting scheme, there will be more strata. Therefore the risk of empty strata or strata with too few observations will be larger. There are two solutions for this problem. One is to use less auxiliary variables, but then a lot of auxiliary information is thrown away. Another is to collapse strata, as discussed by Kalton and Maligalig (1991), Little (1993), and Chapter 19 in this volume by Gelman and Karlin.

Another problem in the use of several auxiliary variables is the lack of a sufficient amount of population information. For example, if the population distributions of the two variables age and sex are known separately but the distribution in the cross-classification is not known, then poststratification by age and sex cannot be carried out because weights cannot be computed for the strata in the cross-classification.

One way to solve this problem is to use only one variable, but that would mean ignoring all information with respect to the other variable. What is needed is a weighting technique that uses both marginal frequency distributions simultaneously. There are two weighting techniques that can do that: linear weighting and multiplicative weighting. These two techniques are described in the next sections.

18.3 LINEAR WEIGHTING

18.3.1 The General Regression Estimator

In the case of full response, the precision of the simple sample mean can be improved if suitable auxiliary information is available. Suppose there are p auxiliary variables. The values of these variables for element k are denoted by the vector X_{k1}, $X_{k2}, \ldots, X_{kp}$. The vector of population means is denoted by $\overline{X}$. If the auxiliary variables are correlated with the target variable, then for a suitably chosen vector $B = (B_1, B_2, \ldots, B_p)'$ of regression coefficients for a best fit of Y on X, the residuals $E_k = Y_k - X_k B$ vary less then the values of target variable itself. The ordinary least squares solution B can, in the case of full response, be estimated asymptotically design unbiased by

$$b = \left(\sum_{i=1}^{n} x_i x_i'\right)^{-1} \left(\sum_{i=1}^{n} x_i y_i\right) \tag{3.1.1}$$

where y_i and x_i denote sample values. Using (3.1.1), the general regression estimator for the full response can be written as

$$\bar{y}_{\text{REG}} = \overline{X}' b \tag{3.1.2}$$

Under nonresponse, a modified version of general regression is:

$$\bar{y}^*_{\text{REG}} = \overline{X}' b^* \tag{3.1.3}$$

in which $\bar{x}^*$ is the analogue of $\bar{y}^*$, as defined in (1.3.1), and b^* is the analogue of b based on the available data. Bethlehem (1988) shows that the bias of this estimator is approximately equal to

$$B(\bar{y}^*_{\text{REG}}) = \overline{X} B^* - \overline{Y} \tag{3.1.4}$$

where B^* is defined by

$$B^* = \left(\sum_{k=1}^{N} \rho_k X_k X_k'\right)^{-1} \left(\sum_{k=1}^{N} \rho_k X_k Y_k\right). \tag{3.1.5}$$

The bias of this estimator vanishes if $B^* = B$. Thus, the regression estimator will be unbiased if nonresponse does not affect the regression coefficients. Practical experience (at least in The Netherlands) shows that nonresponse often seriously affects estimators like means and totals, but less often causes estimates of relationships to be biased. Particularly if relationships are strong (the regression line fits the data well), the risk of finding wrong relationships is small.

By writing

$$B^* = B + \left(\frac{1}{N} \sum_{k=1}^{N} \frac{\rho_k X_k X_k'}{\bar{\rho}}\right) \overline{E}^* \tag{3.1.6}$$

where

$$\overline{E}^* = \frac{1}{N} \sum_{k=1}^{N} \frac{\rho_k E_k}{\bar{\rho}} \tag{3.1.7}$$

two conclusions can be drawn. First, B^* and B will be approximately equal if quantity (3.1.7) is small. So a good fit of the regression model will result in small residuals, and thus will reduce the bias. Second, B^* and B will be approximately equal if (3.1.7) is close to or equal to 0, and this will be the case if there is little or no correlation between the residuals of the regression model and the response probabilities. This condition is satisfied if the data are missing at random (MAR) and the auxiliary variables concerned are included in the regression model.

The theory presented in this section shows that use of the general regression estimator has the potential of improving precision and reducing the bias in the case of ignorable nonresponse. Therefore, it forms the basis for linear weighting adjustment techniques.

18.3.2 Linear Weighting with Qualitative Variables

Bethlehem and Keller (1987) have shown that the general regression estimator (3.1.2) can be rewritten in the form of the weighted estimator (2.1.1), where the adjustment weight w_i for observed element i is equal to $w_i = v'X_i$, and v is a vector of weight coefficients which is equal to

$$v = n\left(\sum_{i=1}^{n} x_i x'\right)^{-1} \tag{3.2.1}$$

Poststratification is a special case of linear weighting. To show this, qualitative auxiliary variables are replaced by sets of dummy variables. Suppose there is one auxiliary variable with L categories. Then L dummy variables $X_1, X_2, \ldots, X_L$ are introduced. For an observation in a certain stratum, the corresponding dummy variable is assigned the value 1 and all other dummy variables are set to 0. Consequently, the vector of population means of these dummy variables turns out to be equal to $\underline{X} = (N_1/N, N_2/N, \ldots, N_L/N)'$, and v is equal to $v = (n/N)(N_1/n_1, N_2/n_2, \ldots, N_L/n_L)'$.

18.4 MULTIPLICATIVE WEIGHTING

If linear weighting is applied, correction weights are obtained that are computed as the sum of a number of weight coefficients. It is also possible to compute correction weights in a different way, namely as the product of a number of weight factors. This weighting technique is usually called raking or iterative proportional fitting. Here it is denoted by multiplicative weighting, because weights are obtained as the product of a number of factors contributed by the various auxiliary variables.

Multiplicative weighting can be applied in the same situations as linear weighting as long as only qualitative variables are used. It computes correction weights by means of an iterative procedure. The resulting weights are the product of factors from all cross-classifications.

The iterative proportional fitting technique was described by Deming and Stephan (1940). Skinner (1991) discusses application of this technique in multiple frame surveys. Little and Wu (1991) describe the theoretical framework and show that this technique comes down to fitting a loglinear model for the probabilities of getting observations in strata of the complete cross-classification, given the probabilities for marginal distributions. To compute the weight factors, the following scheme must be carried out:

1. Introduce a weight factor for each stratum in each cross-classification term. Set the initial values of all factors to 1.
2. Adjust the weight factors for the first cross-classification term so that the weighted sample becomes representative with respect to the auxiliary variables included in this cross-classification.
3. Adjust the weight factors for the next cross-classification term so that the weighted sample is representative for the variables involved. Generally, this will disturb representativeness with respect to the other cross-classification terms in the model.
4. Repeat this adjustment process until all cross-classification terms have been dealt with.
5. Repeat steps 2, 3, and 4 until the factors do not change any more.

18.5 OTHER ASPECTS OF WEIGHTING

18.5.1 Calibration Estimation

Deville and Särndal (1992) and Deville et al (1993) have created a general framework for weighting of which linear and multiplicative weighting are special cases. Assuming simple random sampling, their starting point is that adjustment weights have to satisfy two conditions:

1. The adjustment weights w_i have to be as close as possible to 1.
2. The weighted sample distribution of the auxiliary variables has to match the population distribution, i.e.

$$\bar{x}_W = \frac{1}{n} \sum_{i=1}^{n} w_i x_i = \bar{X} \tag{5.1.1}$$

The first condition sees to it that resulting estimators are unbiased, or almost unbiased, and the second condition guarantees that the weighted sample is representative with respect to the auxiliary variables used.

To reach their goal, Deville and Särndal (1992) introduce a distance measure $D(w_i, 1)$ that measures the difference between w_i and 1 in some way. The problem is now to minimize

$$\sum_{i=1}^{n} D(w_i, 1) \tag{5.1.2}$$

under condition (5.1.1). This problem can be solved by using the method of Lagrange. By choosing the proper distance function, both linear and multiplicative weighting can be obtained as special cases of this general approach. For linear weighting, the distance function D is defined by $D(w_i, 1) = (w_i - 1)^2$, and for multiplicative weighting the distance $D(w_i, 1) = w_i \log(w_i) - w_i + 1$ must be used.

Deville and Särndal (1992) and Deville et al (1993) only consider the full response situation. They show that then estimators based on weights computed within their framework have asymptotically the same properties. This means that for large samples, it does not matter whether linear or multiplicative weighting is applied. Estimators based on both weighting techniques will behave approximately the same. Note that although the estimators behave in the same way, the individual weights computed by linear or multiplicative weighting may differ substantially.

Under nonresponse, the situation is different. Generally, the asymptotic properties of linear and multiplicative weighting will not be equal under nonresponse. The extent to which the chosen weighting technique is able to reduce the nonresponse bias depends on how well the corresponding underlying model can be estimated using the observed data. Linear weighting assumes a linear model to hold with the target variable as dependent variable and the auxiliary variables as explanatory variables. Multiplicative weighting assumes a loglinear model for the cell frequencies. An attempt to use a correction technique for which the underlying model does not hold will not help to reduce the bias.

18.5.2 Limited Weights

There are several reasons why a statistician may want to have some control over the values of the adjustment weights. One reason is that extremely large weights are generally considered undesirable. Another reason to have some control over the values of the adjustment weights is that application of linear weighting might produce negative weights. For some discussion of these issues see, e.g., Deville et al. (1993), Huang and Fuller (1978), and references cited therein.

18.5.3 Variance Estimation

Computation of adjustment weights allows for publication of better quality population statistics. However, to be able to judge the usefulness of statistics, it is important to be able to compute variances of estimates.

The theory of linear weighting provides formulae for the computation of the variances of estimators of population characteristics. However, computation of these variances requires the second-order inclusion probabilities to be available. Particularly for large samples, this is a considerable computational effort. For multiplicative weighting, there are no straightforward variance expressions.

A different approach to a variance estimation is offered by replication methods. Detailed discussions of this approach to variance estimation are in the chapters by Shao (Chapter 20) and by Lee et al. (Chapter 21) in this volume.

18.6 A PRACTICAL EXAMPLE

18.6.1 The Survey

Since 1995, Statistics Netherlands has developed an integrated system of social surveys. It is know by its Dutch acronym POLS (Permanent Onderzoek Leefsituatie). More details about POLS can be found in Statistics Netherlands (1998).

POLS is a continuous survey. Every month, a sample is selected. The target population consists of people of age 12 and older. The samples are stratified two-stage samples. In the first stage, municipalities are selected within regional strata with probabilities proportional to the number of inhabitants. In the second stage, an equal probability sample is drawn in each selected municipality. Sampling frames are the population registers of municipalities. The samples are self-weighted.

In this example, the effect of weighting on one social participation variable is studied. This is the variable recording whether or not a person is doing any volunteer work. It is to be expected that there is a relationship between social participation and response behavior: people participating more in social activities will also be more inclined to respond. In 1997, the sample size of the thematic module on social participation was 6672, with a response percentage of 56.6%.

18.6.2 The Auxiliary Information

For this example, the only population information used was taken from the *Statistical Yearbook of Statistics Netherlands* (Statistics Netherlands, 1998). It contains frequency distributions with respect to five variables: sex, age, marital status, province of residence, and degree of urbanization of the area of residence.

The ideal situation would be to have the complete crossing of these five variables. However, the *Statistical Yearbook* contains only information with respect to partial crossings. Table 18.1 contains counts on the population distribution of sex by age by marital status.

Table 18.1. Population distribution of age × sex × marital status (× 1000)

	Male				Female			
Age	Unmarried	Married	Widowed	Divorced	Unmarried	Married	Widowed	Divorced
12-19	752.4	0.4	0.0	0.0	716.5	3.5	0.0	0.0
20–29	981.5	185.7	0.2	10.2	785.0	330.6	0.7	22.7
30–39	445.4	795.1	1.9	72.1	283.5	879.3	5.7	93.8
40–49	164.7	899.0	6.9	113.9	103.1	882.9	21.5	138.4
50–59	67.3	732.9	15.8	86.3	44.4	675.9	56.1	98.8
60–69	42.0	519.2	31.7	42.6	41.4	458.9	140.0	51.5
70–79	21.4	308.4	52.5	16.6	43.0	239.9	254.3	27.9
80+	8.0	84.0	50.4	4.0	35.0	49.6	243.9	12.4

Table 18.2. Population distribution of province × age (× 1000)

	Age				
Province	12–19	20–44	45–64	65–79	80+
Groningen	49.3	222.1	127.8	48.4	20.8
Friesland	61.5	225.7	144.0	65.7	21.6
Drenthe	43.7	165.9	114.3	53.9	15.3
Overijssel	106.2	404.2	240.2	110.8	31.9
Flevoland	33.8	115.7	53.0	21.8	4.2
Gelderland	183.4	720.6	443.4	195.7	56.9
Utrecht	103.7	439.4	238.6	102.2	32.5
Noord-Holland	220.5	999.9	574.4	254.3	82.1
Zuid-Holland	316.2	1307.9	759.5	347.1	117.7
Zeeland	35.0	130.1	89.2	44.1	15.6
Noord-Brabant	218.7	898.7	562.4	225.2	57.9
Limburg	100.8	429.5	289.8	127.1	30.8

With respect to the two variables "province" and "degree of urbanization," only the crossing with age is available, as displayed in Tables 18.2 and 18.3, respectively.

Note that in Tables 18.2 and 18.3, the age variable only has five categories, whereas in Table 18.1 it has eight categories. In a linear weighting model, this causes no problems. It is possible to use both age variables simultaneously.

Poststratification can only be used if one of these tables is used. Also, Table 18.1 cannot be used as is, because it contains four empty cells. To address this, one may collapse the strata with other strata; see e.g., Tremblay (1986), Kalton and Maligalig (1991), Little (1993a), and Chapter 19 in this volume by Gelman and Carlin. For example, the age categories 12–19 and 20–29 could be merged into a new age category, 12–29.

18.6.3 Analysis of the Auxiliary Variables

To develop a weighting method, Table 18.4 compares population and response distributions for each auxiliary variable. For the variable "age," nonresponse is

Table 18.3. Population distribution of degree of urbanization × age (× 1000)

	Age				
Province	12–19	20–44	45–64	65–79	80+
Very strong	223.2	1196.6	565.3	293.4	113.0
Strong	336.5	1468.4	839.0	389.8	114.6
Moderate	317.1	1223.7	766.3	321.7	90.5
Little	333.7	1226.3	820.6	328.8	93.4
None	262.3	944.7	645.4	262.6	75.8

Table 18.4. Comparing population and response distributions of the auxiliary variables (%)

Variable	Response	Population	Difference
Age			
12–19	12.8	11.1	1.7
20–29	15.9	17.5	–1.6
30–39	20.5	19.4	1.1
40–49	17.9	17.6	0.3
50–59	14.0	13.4	0.6
60–69	10.0	10.0	0.0
70–79	6.5	7.3	–0.8
80+	2.5	3.7	–1.2
Marriage status			
Unmarried	32.7	34.2	–1.5
Married	57.2	53.2	4.0
Widowed	5.2	6.0	–0.8
Divorced	4.9	6.7	–2.8
Province			
Groningen	2.7	3.5	–0.8
Friesland	4.3	3.9	0.4
Drenthe	2.3	3.0	–0.7
Overijssel	6.8	6.7	0.1
Flevoland	1.8	1.7	0.1
Gelderland	15.4	12.1	3.3
Utrecht	5.4	6.9	–1.5
Noord-Holland	14.0	16.1	–2.1
Zuid-Holland	18.0	21.5	–3.5
Zeeland	2.7	2.4	0.3
Noord-Brabant	17.6	14.8	2.8
Limburg	9.1	7.4	1.7
Sex			
Male	48.6	49.1	–0.5
Female	51.4	50.9	0.5
Urbanization			
Very strong	11.8	18.0	–6.2
Strong	24.0	23.8	0.2
Moderate	23.2	20.5	2.7
Little	23.3	21.1	2.2
None	17.7	16.5	1.2

highest for people between 20 and 30 years old (mainly not at home) and elderly people (mainly refusal). Marital status shows a relatively high response for married people.

Divorced people tend to respond less than average. Response is particularly high in the provinces Gelderland and Noord-Brabant. There is a lot of nonresponse in the more densely populated and more industrialized provinces of Noord-Holland and

Zuid-Holland. This phenomenon is also reflected in the variable degree of urbanization.

This analysis indicates that at least the variables "marital status," "province," and "urbanization" should be included in the weighting model. Note that the last two variables are partially, but not completely, confounded.

18.6.4 Weighting Models

Table 18.5 displays results of a number of different weighting models applied to this example using the software package Bascula, developed by Statistics Netherlands; see, e.g., Bethlehem (1996).

A clear pattern can be distinguished in Table 18.5: the more auxiliary information is used, the lower the estimate of the percentage of people doing volunteer work. Of course, the effectiveness of a model cannot be judged by just looking at the deviation from the unadjusted estimate. However, use of more information also leads to a decrease in standard error, and this is an indication of better fitting models. Standard errors were computed using the method of balanced half samples; see Renssen et al. (1997).

Poststratification by sex, marital status and age is not possible due to empty cells. But even merging the age categories 12–19 and 20–29 would produce cells with less than five observations, which could produce unstable weights. Therefore, instead of attempting to carry out the poststratification Sex × MarStat × Age8, the linear weighting model (Sex × Age8) + (Sex × MarStat) was used. Application of this model produces a decrease in the estimate of 1.1%.

Model 11 in Table 18.5 contains the maximum possible weighting model. The population information required for the term (Sex × Age8) + (Sex × MarStat) is taken from Table 18.1, and that for the term Age5 × Urban from Table 18.2. Note that

Table 18.5. Estimates of the percentage of people doing volunteer work, based on various weighting models

	Weighting model	Number of parameters	Estimate	Standard error
1	No weighting	0	43.4	1.2
2	Sex	2	43.4	1.2
3	Province	12	43.3	1.2
4	MarStat	4	42.9	1.2
5	Urban	5	42.9	1.0
6	Age8	8	42.8	1.2
7	Age5 × Province	60	42.9	1.2
8	(Sex × Age8) + (Sex × MarStat)	22	42.3	1.1
9	Age5 × Urban	25	42.5	1.0
10	Sex + Age8 + MarStat + Urban + Province	23	42.1	1.0
11	(Sex × Age8) + (Sex × MarStat) + (Age5 × Urban) + Province	53	42.0	0.9

only the term Province is used and not Age5 × Province, because the response table contains cells with too few observations. Population counts for Province are taken from Table 18.3. Application of this maximum weighting model shows the greatest decrease in the estimate: from 43.3 to 42.0. In addition, the simpler model 10 performs almost as well as the maximum model 11. This is an indication that in this example the main effects of the auxiliary variables play an important role in reducing the bias of the estimates, whereas all kinds of interaction effects are less important.

18.7 DISCUSSION

18.7.1 How to Get Auxiliary Information

In practice, it is very difficult to assess the possible negative effects of nonresponse. And even if such effects can be detected, it is no simple matter to correct for them. Vital for useful detection and correction techniques is the availability of at least some information about the nonrespondents. There are two ways of obtaining more information. First, one may get information from nonrespondents themselves; see, e.g., the method of Hansen and Hurvitz (1946) and the Basic Question Approach of Kersten and Bethlehem (1984). Second, one may obtain more information about nonrespondents without contacting them, using the sampling frame; population registers, which contain variables like sex, date of birth, marital status, and household composition; observation by interviewers of the type of neighborhood, type of house, age of house, type of town, etc.; or statistical agency records.

18.7.2 How to Use Auxiliary Information

The available auxiliary information may provide insight about a possible bias due to nonresponse. In principle, this only applies to estimates for auxiliary variables and not for the target variables of the survey. However, if there are sufficiently strong relationships between auxiliary variables and target variables, so that models can be constructed to predict target variables from auxiliary variables, conclusions with respect to a possible bias carry over to the target variables.

To be able to assess and remove a possible nonresponse bias, it is important to collect as much auxiliary information as possible. However, use of too many auxiliary variables may lead to unstable weights, which in turn may cause problems with point estimation and variance estimation; see, e.g., Little (1993a) and the chapter by Gelman and Carlin in this volume (Chapter19).

CHAPTER 19

Poststratification and Weighting Adjustments

Andrew Gelman, *Columbia University, New York*
John B. Carlin, *Royal Children's Hospital and University of Melbourne, Australia*

19.1 INTRODUCTION

Poststratification (PS) is a form of weighting adjustment that reduces discrepancies between a sample survey and the population. We examine the method from a model-based, Bayesian perspective. We clarify the difference between PS and inverse-probability (IP) weights, and describe difficulties with PS weighting when the number of poststrata become large. These problems motivate further development of the model-based PS approach (Holt and Smith, 1979; Little, 1991, 1993), which is linked to design-based approaches via a basic PS identity.

We focus on estimation of the population mean of a single survey response in a one-stage sampling design, but also sketch extensions to ratio and regression estimation (Section 19.4.1) and to cluster samples (Section 19.4.2). Section 19.5 illustrates the potential advantages of a model-based PS approach with two examples. We do not attempt a thorough literature review; more comprehensive references on weighting and sample survey analysis appear in books such as Lohr (1999).

19.2 BASIC WEIGHTING AND POSTSTRATIFICATION

The essential idea of weights in sample survey estimation is to provide weighted estimators that are (approximately) unbiased for the target population quantity under repeated sampling under the same sampling plan. A more fundamental notion is to correct for known differences between sample and population, whether these discrepancies arise from sampling fluctuation, nonresponse, frame errors, or other sources.

Most mainstream statistics packages do not provide for inferences (that is, standard errors as well as point estimates) using so-called sampling or "probability" weights. One exception is Stata (Stata Corporation, 2001) which distinguishes between sampling weights and so-called "analytic" weights in regression, where the data values are weighted for differences in residual variance. Some similar capabilities are available in the latest release of SAS (An and Watts, 1998).

Survey organizations differ in their usage of weights to correct for discrepancies between sample and population and it is hard to define just how much weighting is "enough." Voss et al. (1995) report on weights for national political polls by news organizations in the 1988 U.S. general election campaign. At one extreme, some of the ABC News/Washington Post polls used only small weighting adjustments for sex (for example, 1.04 for men and 0.96 for women). In contrast, the CBS News/New York Times polls included weights proportional to number of adults in household divided by the number of telephone lines, then used ratio weights to match the sample to the population for sex × ethnicity and age × education. Despite these differences, Voss et al. (1995) found the weighted average estimates for population quantities of interest to be similar for the different survey organizations.

Two types of weights—inverse-probability (IP) and poststratification (PS)—need to be carefully distinguished. The basic difference is that the former are known at the time the survey is designed whereas the latter are calculated after the data are collected, with the weight (multiplier) for each poststratum proportional to the number of units in the stratum in the population, divided by the number of units in the sample in this stratum. Both kinds of weights have the same intuitive interpretation, but they have a different statistical standing, with potential implications for the estimation of standard errors; for more discussion see Section 19.3.3.

How should the weights be used? There is general agreement that for estimating population means and ratios, weighted averages are appropriate. That is, the weighted estimate of a mean $\overline{Y}$ is $\Sigma_{i=1}^{n} w_i y_i / \Sigma_{i=1}^{n} w_i$, and the weighted estimate of a ratio $\overline{Y}/\overline{X}$ is $\Sigma_{i=1}^{n} w_i y_i / \Sigma_{i=1}^{n} w_i x_i$. If more complicated quantities such as regression coefficients are estimated, or if standard errors are required, current practice and textbook recommendations vary. For estimating the coefficients of a regression, most survey statisticians include the survey weights in "estimating equations" or pseudoscore statistics, which also take the form of a (weighted) average over the sample (Binder, 1983; Carlin et al., 1999). Modelers might ignore the weights, reasoning that regression relationships should be validly estimable even from a nonequally weighted sample as long as the model is adequate. This approach is vulnerable to model misspecification. A better approach is to include all the information used in the survey weighting as additional covariates in the regression, and perform an unweighted regression of y on the augmented X matrix (DuMouchel and Duncan, 1983, Pfeffermann, 1993). This approach is discussed in Section 19.4.1.

Poststratification. Poststratification (PS) may be defined simply as the use of stratified sample estimators for unstratified designs. For example, if the distribution of demographic variables in the sample differs from the distribution in the population based on a Census, survey estimates for each demographic category or poststratum are combined with population poststatum totals from the Census. Considerable gains in precision can be made by poststratifying on variables that are predictive of survey outcomes. However, a more important reason for poststratification is to correct for differential nonresponse between PS cells. For example, the well-educated are much more likely than the poorly educated to respond to national telephone opinion polls (e.g., Little, 1996).

An extension of PS to more than one variable is called raking, which applies iterative proportional fitting (Deming and Stephan, 1940) to obtain weighted sample counts that match the population on a set of margins. Other approaches to the use of auxiliary information to improve survey estimates include the regression and ratio estimates, and extensions such as those considered in Deville and Särndal (1992) and Deville et al. (1993). These methods are discussed in Chapter 18 of this book and are not considered further here.

19.3 A UNIFYING FRAMEWORK

19.3.1 Notation

We adopt a unified notation, derived from Little (1991, 1993), for weighting and poststratification of sample surveys. We follow standard practice and focus on a single survey response at a time, labeling the values on units i in the population as Y_i, $i = 1, \ldots, N$, and in the sample as y_i, $i = 1, \ldots, n$. To start with, we assume the goal is to estimate the population mean $\theta = \overline{Y} = \Sigma_{i=1}^{N} Y_i/N$. We suppose a population is divided into J PS cells, with population N_j and sample size n_j in each cell $j = 1, \ldots, J$, with $N = \Sigma_{j=1}^{J} N_j$, and $n = \Sigma_{j=1}^{J} n_j$. For example, if the population of U.S. adults is classified by sex, ethnicity (white or nonwhite), four categories of education, four categories of age, and 50 states, then $J = 2 \times 2 \times 4 \times 4 \times 50 = 3200$, and the cell populations N_j would be (approximately) known from the public-use subset of the long form of the U.S. Census.

We define π_j as the probability that a unit in cell j in the population will be included in the sample. For some designs, π_j is known but, in general, when nonresponse is present, it can only be estimated. The estimate n_j/N_j is natural for equal probability designs; smoothed estimates can perform better if cell sample sizes are small. We label the population mean within cell j as $\theta_j = \overline{Y}_j$ and the sample mean within cell j as $\overline{y}_j$. The overall mean in the population is then

$$\theta = \overline{Y} = N^{-1} \sum_{j=1}^{J} N_j \theta_j \tag{19.1}$$

which we refer to as the basic PS identity. We focus on weighted estimates of the form

$$\hat{\theta} = \sum_{j=1}^{J} W_j \hat{\theta}_j \tag{19.2}$$

where the cell weights W_j sum to 1. A variety of procedures yield estimates of the form (19.2). Classical weighting methods generally avoid any modeling of the responses and restrict themselves to unsmoothed estimates $\hat{\theta}_j = y_j$ and weights W_j that depend only on the n_j's and N_j's (as well as IP weights, where present), but not on the y_j's, thus yielding population estimates of the form,

$$\hat{\theta}_W = \sum W_j \bar{y}_j = \left(\sum_{i=1}^{n} w_i \right)^{-1} \sum_{i=1}^{n} w_i y_i \tag{19.3}$$

where $w_i = W_{j(i)}/n_{j(i)}$ is the *unit weight* of the items i in cell j, $j(i)$ denoting the cell in which unit i belongs. The challenge in design-based methods is for the unit weights w_i to capture the unequal sampling fractions in the different cells without being so variable as to lead to an unstable estimate of θ.

An alternative approach sets the cell weights to the population proportions—that is, $W_j = N_j/N$ for each j—and using a probability model or smoothing procedure to construct the estimates $\hat{\theta}_j$ from the sample means $\bar{y}_j$ and sample sizes n_j. Thus,

$$\hat{\theta}_m = N^{-1} \sum_{j=1}^{J} N_j \hat{\theta}_j \tag{19.4}$$

All these procedures assume a constant probability of inclusion within cells, where inclusion encompasses both selection and response. In the presence of nonresponse, it is desirable to poststratify as finely as possible, so that the implicit assumption of equal probability of inclusion is reasonable within each poststratification cell (with these probabilities being allowed to vary between poststrata).

19.3.2 Poststratification Inference

An essential part of survey estimation is going beyond point estimates to provide credible measures of uncertainty. We shall follow standard practice here and assume sample sizes are large enough that normal theory inferences are acceptable, so that we can base inferences on point estimates and standard errors.

For standard survey problems, classical design-based and Bayesian model-based calculations tend to give similar inferences as long (a) sample sizes are large enough that sampling distributions of estimands of interest are approximately normal, (b) the inferences take into account design features such as stratification and clustering, and (c) the model uses noninformative prior distributions (see, e.g., Gelman et al., 1995, chapters 4 and 7). This similarity allows one to use model-based calculations to get reasonable repeated-sampling inference or, conversely, to use design-based standard errors to make probability statements about unknown population quantities.

With complex weighting schemes, the design-based perspective can be used to derive variances of weighted estimates that account for the design factors by using the form (19.2) and recognizing the sampling variability of the unit weights. In PS weighting, the weights $W_j = N_j/N$ are fixed and the cell estimates are simply $\hat{\theta}_j = \bar{y}_j$.

Under simple random sampling, we can use the variance formula for a stratified mean:

$$\text{var}(\hat{\theta}) = \sum_{j=1}^{J} W_j^2 \sigma_j^2/n_j \qquad (19.5)$$

where σ_j^2 can be estimated from the within-stratum sample variance. [For expression (19.5), we ignore the generally very minor correction arising from the randomness of the $1/n_j$ factors.]

With IP weights, raking, or iterative proportional fitting, the cell weights W_j depend on the vector of data sample sizes $n = (n_1, \ldots, n_J)$. As a result, the sampling variance can be decomposed as

$$\text{var}(\hat{\theta}) = E[\text{var}(\hat{\theta})|n] + \text{var}[E(\hat{\theta}|n)] \qquad (19.6)$$

the first term of which is for random sampling within cells, and the second term of which accounts for the randomness in the cell weights. For a complex survey design, (19.6) can be estimated using linearization or jackknife-type methods (Binder, 1983, Lu and Gelman, 2000).

Design- and model-based inferences begin to differ when sample sizes become small or models become more complicated, both of which happen when poststratification is applied with many cells. In this case, the sample size in each cell becomes small and design-based approaches may suffer problems of excess variance. Model-based inferences may provide an attractive alternative for such problems, as discussed in Sections 19.3.4 and 19.3.5 and illustrated in Section 19.5. In a Bayesian analysis, standard errors for estimates such as $\hat{\theta}_m$ in (19.4) can be computed directly via Monte Carlo simulation of the posterior variance of the parameter vector θ (e.g., Gelman et al., 1995).

19.3.3 The Difference Between PS and IP Weighting

The formula (19.3) can be applied with IP weights $w_i \propto 1/\pi_{j(i)}$ or with PS weights $w_i \propto N_{j(i)}/n_{j(i)}$, so these two forms of weights are often confused. We illustrate the difference with two simple examples. Consider first a simple random sample of $n = 200$ adults from a population divided into male and female poststrata, with population counts $N_1 = N_2 = 500{,}000$, and sample counts $n_1 = 90$, $n_2 = 110$. The IP weights are then equal for all the units, and the corresponding IP-weighted mean $\hat{\theta}^{\text{IP}} = \bar{y}$ (see, e.g., Lohr, 1999). The PS weights are proportional to 1/90 for the men and 1/110 for the women, and the PS mean is $\hat{\theta}^{\text{PS}} = \bar{y}_1/2 + \bar{y}_2/2$. The two estimates also differ in their standard errors. The estimate $\hat{\theta}^{\text{IP}}$ has standard error $\sigma/\sqrt{n}$, where σ is the population standard deviation of the Y_i values, and $\hat{\theta}^{\text{PS}}$ has approximate standard error $\sqrt{(N_1/N)^2\sigma_1^2/n_1 + (N_2/N)^2\sigma_2^2/n_2}$, where σ_1 and σ_2 are the within-stratum standard deviations in the population, and are typically smaller than σ since we tend to poststratify on variables that are relevant for the survey responses of interest. For example, if $\sigma = 30$ and $\sigma_1 = \sigma_2 = 20$, then $\hat{\theta}^{\text{IP}}$ has standard error 2.1 and $\hat{\theta}^{\text{PS}}$ has standard error of approximately 1.4. (The approximation arises from treating the poststratum sample sizes as fixed, and is correct to two significant figures.) Thus PS and IP weighting give different estimates and standard errors in this example.

Consider the same sample data with $n_1 = 90$ and $n_2 = 110$, but with $N_1 = 450{,}000$ and $N_2 = 550{,}000$. Then $\hat{\theta}^{\mathrm{PS}} = \hat{\theta}^{\mathrm{IP}} = \bar{y}$, since the PS weights are now equal ($450{,}000/90 = 550{,}000/110$). Even though the two estimates are the same for this sample, $\hat{\theta}^{\mathrm{PS}}$ still has a lower standard error in repeated samples, reflecting the fact that it generally provides a better "fit" to the population.

Our second example (Gelman and Little, 1998) concerns a household survey in which households are sampled with equal probability, and then a single individual is selected from each sampled household. The probability of an individual being selected is thus inversely proportional to the size of the household. However, composition of the sample is also affected by nonresponse. One source of nonresponse is nonavailability—no one answers the phone, or no one receives the message on the answering machine. It seems reasonable that the likelihood of a call being answered increases with the size of household. Another source of nonresponse is refusal to participate in the survey.

Gelman and Little (1998) compared the distribution of household sizes in the U.S. Census to the distribution in three series of national opinion polls (two sets of preelection telephone polls conducted by CBS News and the in-person National Election Study conducted by the University of Michigan), and computed weights for adults poststratified by household size. Table 19.1 compares these to the IP weights, which are simply proportional to the number of adults in the household. (The IP weight in the last row is not exactly 4 because that category includes households with 4 or more adults.) The PS weights for the household sizes are lower than the IP weights, presumably because adults in smaller households are harder to reach. The discrepancy is smallest with the in-person NES poll, which makes sense since it had the highest response rate of all these surveys. Thus it appears that IP weights in this example overly weight the adults in larger households. Interestingly, Table 19.1 shows that the PS weights are also less variable than the IP weights.

A key assumption of the PS weighting is that the Census counts represent the target population of the survey. In general, the Census includes people who are apathetic about politics, and one might argue that a political poll should represent the

Table 19.1. IP and PS weights for three surveys, all scaled so that the weight is 1 for respondents from households with 1 adult

		PS weights for 3 surveys		
N of adults in household	IP weights	Early CBS	Late CBS	NES
1	1	1.00	1.00	1.00
2	2	1.32	1.38	2.00
3	3	1.35	1.53	2.30
4+	4.25	0.95	1.20	2.55

people who plan to vote in the election. For this particular example, however, there is no substantial correlation between number of adults in a household and the likelihood of voting. The discrepancy between the sample and population is more plausibly explained by nonresponse, caused primarily by the difficulty of reaching anyone in a small household. This differential nonresponse remains after adjusting for the demographic variables (Gelman and Little, 1998). Thus PS weighting on household size seems to us more reasonable than IP weighting here.

19.3.4 Difficulties with Classical Weighted Estimates

The estimate (19.3) with weights $w_i \propto N_{j(i)}/n_{j(i)}$ is unbiased if probabilities of inclusion within the poststrata are constant. If the weights are too variable, as occurs if the n_j's are small, then $\hat{\theta}_W$ will itself have unacceptably high variance. The extreme case is cells for which $n_j = 0$ but $N_j \neq 0$. For example if $J = 3200$ as in Section 19.3.1 and n is 1500, as is typical in national polls, then most of the cells will be empty. A common solution is to pool weighting cells, or to adjust for fewer variables (for example, adjusting by margins of the demographic variables but not by the 50 states). The choice of which cells to pool is somewhat arbitrary and pooling contradicts the goal of including in the analysis all variables that affect the probability of inclusion, which is a basic principle in both classical and Bayesian sampling inference.

An alternative to pooling is to keep all the weighting cells but "smooth the weights," that is, modify the w_i so that they are less variable. In practice, this approach reduces the weights of cells with small sample sizes. Elliott and Little (2000) note that estimates of θ of the form (19.2) can be stabilized in two ways: by smoothing the weights W_j, or by estimating the cell means θ_j using a hierarchical model. Weight smoothing is usually done without reference to the responses y_i, whereas the amount of smoothing in a fitted hierarchical model depends on the variance of y_i between and within strata.

19.3.5 Difficulties with Model-Based Estimates

Standard design-based methods avoid overt dependence on models, so there is an understandable resistance among survey sampling practitioners to the use of models for survey response. Modeling assumptions can appear arbitrary and there are legitimate concerns about sensitivity of inferences to model misspecification. At a minimum, model-based methods should be able to give answers similar to classical methods in settings where the classical methods make sense. This means that models for survey outcomes must at least include all the information used in design-based weighting estimates. This is possible in principle (see Section 19.5 for two examples), although issues remain about how information in unequal probability sampling schemes should be handled.

If we resolve to include all the observable variables affecting the probability of selection in the model, the model may be quite complicated and require many assumptions. In the notation of Section 19.3.1, we must model the θ_j's conditional on all crossclassification variables—for example age, sex, ethnicity, education, age, and state—and all their interactions. The challenge is to construct models for this problem that yield resonable estimates (19.4) and associated inferences.

19.4 MORE COMPLICATED SETTINGS

19.4.1 Estimating Ratios and Regression Coefficients

So far we have focused on estimating the population mean (19.1) or subgroup means. One may also be interested in more complex estimands, most notably ratios and regression estimates. Ratios arise in various ways, for example as means of subgroups with unknown population proportions. For example, suppose we are interested in θ, the average income of supporters of the Republican candidate for President. If we let Y_i be the income response in the population and U_i be the indicator for supporting the Republican, then

$$\theta = \left(\sum_{i=1}^{N} U_i Y_i\right)\left(\sum_{i=1}^{N} U_i\right)^{-1} = \overline{V}/\overline{U}$$

where $V_i = U_i Y_i$. The design-based bias and standard error of a ratio estimate are estimated using a Taylor-series expansion; in the Bayesian context, one would need to model Y and U jointly [perhaps by modeling U given X, then Y given (U, X), where X represents the variables that determine the PS categories].

Regression estimates commonly arise in analytical studies of survey responses that attempt to understand what variables U are predictive of an outcome of interest Y. This can be directly incorporated into our framework by including the variables in U as predictors, thus regressing Y on (U, X), where X represents the variables used in any weighting and poststratification scheme. If the probability of inclusion is constant or units in the sample are conditional on these X variables, any analysis that conditions on X [in this case, a regression of Y on (U, X)] would yield valid inferences without any need for weighting in the estimates (DuMouchel and Duncan, 1983). For example, Gelman and King (1993) model vote preferences as a function of party identification and political ideology (the variables U in our notation) as well as demographics (the variables X used in the weighting) using an unweighted regression. See also the rejoinder in Gelman et al. (1998).

What if one is ultimately interested in the regression of Y on U without conditioning on X? We can derive this "marginal regression" from our joint regression of Y on (U, X) by averaging over X, which means averaging over the distribution of X in the population. The weights then reenter to adjust for differences between sample and population. The marginal regression can be written as

$$E(Y|U) = E[E(Y|U, X)] = \sum_{X} p(X|U)E(Y|U, X)$$

and the summation on the right side requires additional modeling of the joint distribution of (U, X) in order to estimate the conditional distribution $p(X|U)$. [With no U's, this reduces to $p(X)$, which corresponds to the cell population N_j used in poststratification.] Modeling and estimating $p(X|U)$ seems like a reasonable task, but we are not aware of any examples in the literature.

19.4.2 Cluster Sampling and Unequal Sampling Probabilities

In cluster sampling, the population or subset of the population is partitioned into clusters, only some of which are sampled. This fits naturally into a hierarchical model that includes a parameter for each cluster. The key difference from stratification or poststratification is the need to generalize to the unsampled clusters. This requires a hierarchical model in which the cluster parameters are assumed to be drawn from a common distribution. In addition, depending on the design, it may also be necessary to estimate the population sizes of the clusters. This problem becomes more elaborate with unequal probability sampling designs such as probability proportional to size, where it is possible for sampling probabilities π_j to be known even though population sizes N_j are not. There are various reasons why one might want to define poststratification cells for which the population totals N_j are unknown, and one example is described in the next section.

Challenges arise when clustering is combined with weighting or poststratification. For example, consider a national survey of personal interviews that is clustered geographically (for example, with 20 persons interviewed in each of 50 counties selected at random with probability proportional to size) and then poststratified by demographics (for examply, age, sex, ethnicity, etc.). A full model analysis with estimate (19.4) requires estimates $\hat{\theta}_j$ for cells j defined by all the counties in the U.S. cross-classified by all the demographic categories used in the weighting/poststratification. This is possible in principle, but experience with this sort of modeling is limited, and further research is needed to see what sort of relatively simple models could work reliably here.

Classical weighting methods based on (19.3) can incorporate cluster sampling. Initially, the units in sampled cluster k are assigned the weight $N_k/(n_k p_k)$, where N_k is the number of units in the cluster, n_k is the number of units in the cluster included in the sample, and p_k is the probability of selection of cluster k (see, e.g., Lohr, 1999). This expression is general enough to include sampling with probability proportional to size, or an approximate measure of size, and the n_k in the denominator automatically corrects for unequal nonresponse between clusters (implicitly assuming, as is standard with these weighting or modeling procedures, that nonresponse is uncorrelated with the outcome under study). These weights can then be ratio-adjusted to poststratify on demographics, as discussed in Section 19.2.1.

As always, the standard error of the weighted estimate (19.3) should reflect the sampling design. A generally reasonable approximation is to compute standard errors from variation in the observed cluster means (Kish, 1965). That is, if clusters $k = 1, \ldots, K$ have been sampled, (19.3) is expressed as a weighted average of cluster means:

$$\hat{\theta}_w = \left(\sum_{k=1}^{K} \omega_k z_k\right)\left(\sum_{k=1}^{K} \omega_k\right)^{-1} \tag{19.7}$$

where $\omega_k = \Sigma_{i\in k} w_i$ and $z_k = \Sigma_{i\in k} w_i y_i / \Sigma_{i\in k} w_i$. The variance of $\hat{\theta}_w$ in (19.7) can be computed using standard ratio estimation formulas or the jackknife (see e.g., Lohr, 1999).

19.4.3 Item Nonresponse

Weighting and poststratification are designed to correct for sampling and unit nonresponse. Within a modeling framework with parameters θ_j for mean response within poststrata, it is natural to model responses to the set of survey questions as a multivariate outcome. Such a multivariate model can be used to impute missing items (e.g., Rubin, 1996; Schafer, 1997). A full multivariate analysis of survey responses could in principle be used to impute both unit and item nonresponse, but unit nonresponse is still usually handled by weighting methods.

19.5 EXAMPLES OF POSTSTRATIFICATION MODELS

The model-based approach is also useful when there is partial but not complete information on the PS cell totals N_j. For example, Reilly and Gelman (1999) analyze a series of national opinion polls, focusing on the question of how strongly the respondent approves of the President's job performance. A highly effective predictor of Presidential approval is the respondent's party identification, defined as Democrat, Republican, or neither. Using party identification as a poststratifier would seem to be hopeless since its distribution in the population is not known. However, this distribution is estimated in a closely spaced time series of polls, and party identification is known (and observed) to change only slowly over time. It is thus possible to fit a time-series model to party identification, estimate the PS counts N_j, and use these to poststratify and get more precise estimates of average Presidential approval at each time point. In this example, weekly snapshots of public opinion were analyzed, with sample sizes of about 40 to 60 in each survey. The model-based PS estimates reduced the estimation variances by factors of about 1.3 relative to the simple unadjusted estimates. This approach to modeling the distributions of poststratifiers warrants further development.

The model-based approach to poststratification can be combined with models for small-area estimation (e.g., Fay and Herriott, 1979, Dempster and Raghunathan, 1987). Gelman and Little (1997) used this approach to derive state estimates from a series of national preelection polls. Two natural approaches to this problem are (1) the classical method of assigning unit-level design-based weights and then computing weighted means for each state, and (2) the naive Bayesian method of shrinking the mean of each state toward the national average, with the amount of shrinkage determined by the variance of the sample mean in each state. However, the classical method yields highly variable estimates for all but the largest states, and the naive Bayesian method ignores design information and thus fails to correct for known sampling biases (e.g., that women and more educated persons are more likely than men and less educated persons to respond to a telephone survey).

Define the binary response y_i equal to 1 if the respondent supported or leaned toward supporting the Republican candidate for President and 0 if the respondent supported or leaned toward the Democrat. (Respondents who supported other candidates or had no opinion were excluded from our analysis.) Gelman and Little fit a model of the form $n_j \bar{y}_j \sim \text{Bin}(n_j, \theta_j)$ and $\text{logit}(\theta_j) = (X\beta)_j$, with X including indicators for each state and for all the demographic variables recognized as relevant by the survey organization in its weighting scheme: sex × ethnicity, age × education, and region of the country. State indicators were modeled as random effects, so that the Bayesian inference shrinks the state differences toward zero after adjusting for the variables that affect nonresponse.

Once the model has been fitted and inferences obtained for all θ_j's, poststratified estimates of the population mean within each state k are computed as $\theta_k = \Sigma_{j \in k} N_j \theta_j / \Sigma_{j \in k} N_j$, where the summations are over all poststratification cells within state k, and the cell sizes N_j are from the Census. Our Bayesian computation yields 1000 posterior simulation draws of the vector β. The vector of cell means θ_j is computed for each draw of β, and then summed to yield the vector of state means θ_k. For each θ_k, we can take the 1000 simulation draws and compute a point estimate as the median of the draws and 50% or 95% intervals from the appropriate quantiles (see Gelman et al., 1995).

This modeling approach performs quite well, as can be seen by comparing our inferences, which were based on polls immediately preceding the presidential election, to the state-by-state outcomes of the election itself. Figure 19.1 displays result versus prediction, by state, for four estimation methods: classical weighting ("raking"), the Bayesian estimate with hierarchical variance set to infinity ("unsmoothed," which contains no shrinkage and is thus very similar to the classical estimate), the Bayesian estimate with hierarchical variance set to 0 ("var = 0," which overshrinks by assuming that states are identical after demographic adjustments), and finally hierarchical Bayes, which has the lowest prediction errors. This is a fair test of the model, since the actual election results were *not* used in any way in the estimation procedure. Figure 19.1d suggests that the hierarchical model does not shrink the data enough toward the nationwide mean. As discussed by Gelman and Little (1998) and Little and Gelman (1996), this extra variation in the predictions could be caused by a pattern of nonignorable nonresponse that varies between states; see also Krieger and Pfeffermann (1992).

Figure 19.2 shows that the poststratified hierarchical estimates had lower errors than the classical weighting in 41 out of the 48 states (Alaska and Hawaii were not included in these surveys). In this example, the Bayes estimates work well because the model used all the information in classical weighting in a reasonable way. A useful feature of the model-based approach is its direct computation of posterior uncertainties. The average width of the 50% intervals for the 48 states estimates is 0.57, and 20 out of the 48 intervals contain the actual result for that state. (By comparison, the model-based 50% intervals for the raking estimates

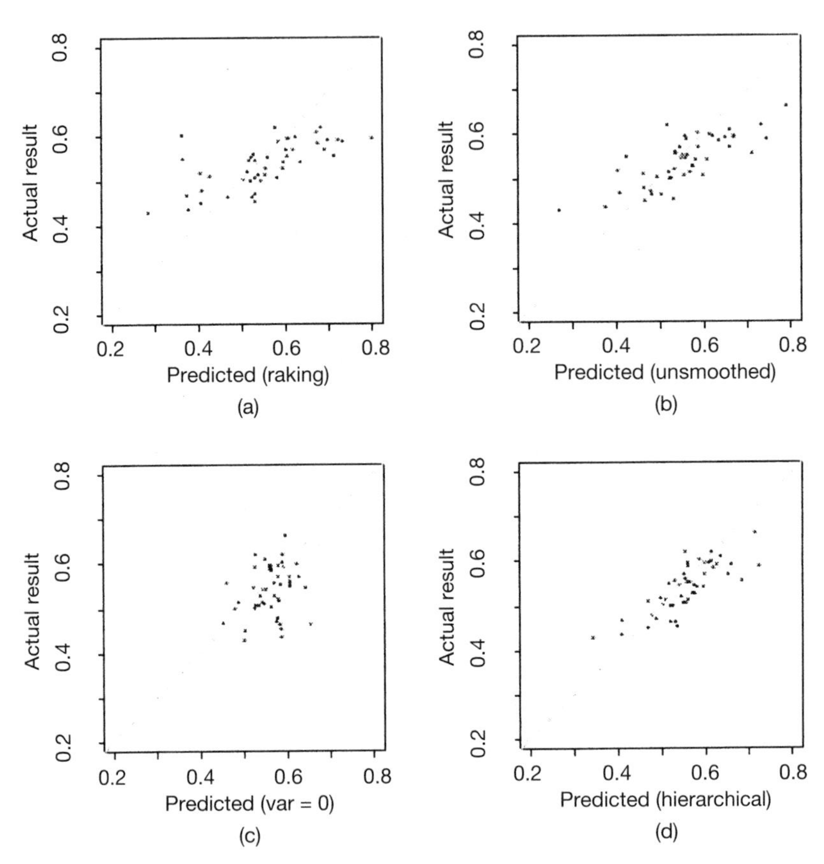

Figure 19.1. Election result by state versus posterior median estimate based on four methods. (a) Raking on demographics, (b) regression modeling including state indicators with no hierarchical model, (c) regression model setting state effects to zero, and (d) regression model with hierarchical model for state effects.

have an average width of 0.69, and only 18 of these intervals contain the actual results.)

19.6 CONCLUSION

Many sample survey texts focus on design unbiasedness as the primary motivation for weighting, but we agree with Holt and Smith (1979) that ". . . it is the structure of the population, rather than the sample design, which an estimator should reflect." We think it may be helpful to regard weights as a tool for ensuring that inferences reflect as well as possible the structure of the target population. Extending this notion suggests that other efforts to capture population structure as part of the survey

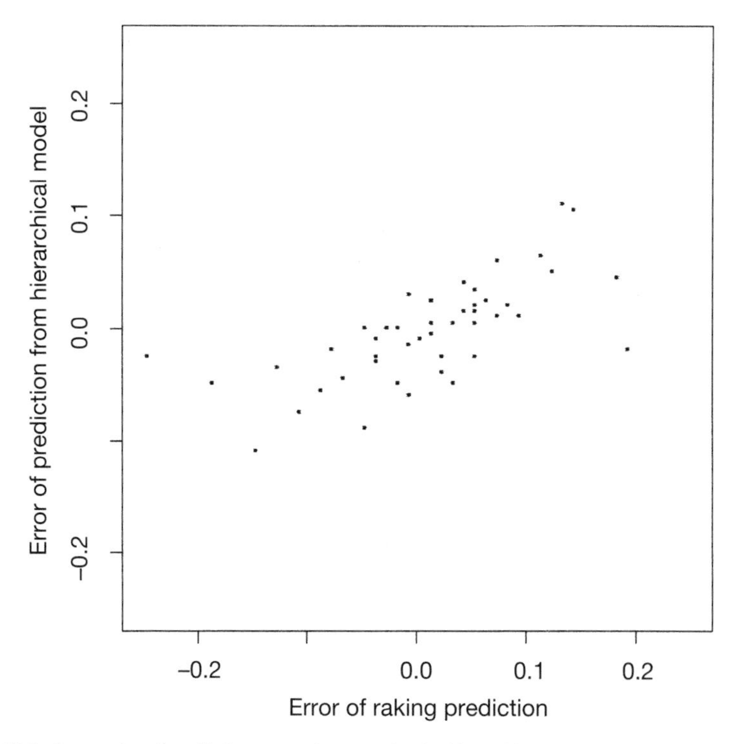

Figure 19.2. Scatterplot of prediction errors, by state, for the hierarchical model versus the classical raking estimate. (The errors of the hierarchical model are lower for most states.)

analysis task will be fruitful, and we have described examples where this was achieved through appropriate modeling.

The preelection polls example in Section 19.5 illustrates an approach to the problem of large numbers of poststrata, which challenges traditional design-based methods. This example also shows how a successful model-based approach works by conditioning on variables relevant in the design and nonresponse and then using population information on these variables to estimate population averages of interest. The first example in Section 5 illustrates a model-based PS approach when the population PS sizes are estimated.

We have attempted to clarify where existing methods may be open to improvement by greater investment in modeling technology. In particular, the goal of conditioning on all variables that might affect nonresponse leads to a large number of potential poststratification cells and thus many parameters θ_j in (19.1); Section 19.5 illustrates how hierarchical models can be used to estimate all these parameters simultaneously. Further work is needed, however, to define models that successfully incorporate adjustments currently made with operationally straightforward techniques such as IP weighting and raking (see Little and Wu, 1991). Our goal is a unified approach to survey estimation that combines the benefits of modeling population structure while remaining "backwards compatible" with the more traditional design-based adjustment techniques.

ACKNOWLEDGMENTS

We thank the National Science Foundation for support through grants SBR9708424, SES-9987748, and Young Investigator Award DMS-9796129. The second author is grateful to Dr. C. Hendricks Brown and the Department of Epidemiology and Biostatistics, University of South Florida, for sabbatical support during this work. We also thank the Department of Statistics, Columbia University, New York, USA, Clinical Epidemiology and Biostatistics Unit, Royal Children's Hospital, and University of Melbourne, Australia.

CHAPTER 20

Replication Methods for Variance Estimation in Complex Surveys with Imputed Data

Jun Shao, *University of Wisconsin*

20.1 INTRODUCTION

Variance estimation is a key element in sample surveys. When there are imputed nonrespondents, treating them as observed data and applying variance estimation formulas appropriate for the case of no nonresponse produces substantial underestimation when the proportion of nonrespondents is appreciable. We consider the situation where a single imputation is performed and identification flags are added to imputed values. There are two types of methods of obtaining unbiased or nearly unbiased variance estimators. The first type is to derive theoretically an approximate variance formula (under some models/assumptions) for the given estimator of interest, and then obtain a variance estimator by replacing unknown quantities in the approximate variance formula with their estimators. Our main focus is the second type of methods, the replication methods. A replication method creates a number of replicated datasets (called pseudoreplicates) and estimates the variance of a given estimator by the sample variance of replicate estimators, where each replicate estimator is the same as the original estimator but is computed based on a pseudoreplicate. Compared with the first type of method, a replication method requires longer computing time to calculate variance estimators but has the advantages of (1) requiring no separate theoretical derivations of a variance formula for each problem, which can be difficult or messy; (2) programming ease in complex situations; (3) using a unified recipe for various problems; and (4) to some degree, robustness against violations of models/assumptions.

A crucial part for the success of a replication method is how the pseudoreplicates are constructed. In the presence of imputed values, pseudoreplicates should be created to take nonresponse and imputation into account. Several popular replication methods are discussed in the subsequent sections.

We assume that a sample S is drawn from a population P according to the following stratified multistage sampling plan. The population P is stratified into H strata with N_h clusters in the hth stratum. In the first stage of sampling, $n_h \geq 2$ clusters are selected without replacement from stratum h according to some sampling plan, and the clusters are selected independently across the strata. The first stage sampling fraction, $\Sigma_h n_h / \Sigma_h N_h$, is assumed negligible, although n_h/N_h may be nonnegligible for some h's. A second stage sample, a third stage sample, and so on may be taken from each sampled cluster, using some sampling plan independently across the sampled clusters. According to the sampling plan, survey weights w_i, $i \in S$, are constructed so that for any set of values $\{z_i : i \in P\}$,

$$E_S\left(\sum_{i \in S} w_i z_i\right) = \sum_{i \in P} z_i \tag{1}$$

where E_S is the expectation with respect to S. Let y be a variable (item) of interest. If there is no nonresponse, the well known Horvitz–Thompson estimator of the population total $Y = \Sigma_{i \in P} y_i$ is

$$\hat{Y} = \sum_{i \in S} w_i y_i \tag{2}$$

which is unbiased because of (1). Let $\tilde{y}_i$ denote the imputed value when y_i is a nonrespondent and $\tilde{y}_i = y_i$ when y_i is a respondent. An analog of the Horvitz–Thompson estimator in (2) based on imputed data is

$$\tilde{Y} = \sum_{i \in S} w_i \tilde{y}_i \tag{3}$$

To ensure that $\tilde{Y}$ is unbiased or nearly unbiased, imputation is often done by first dividing P into several imputation classes and then imputing nonrespondents in an imputation class using respondents within the same imputation class, independently across the imputation classes. Imputation classes are formed so that one of the following assumptions holds.

Assumption D. Within the kth imputation class, the response probability for a given variable is a constant (but may be different for different variables and for different imputation classes) and the response statuses for different units are independent.

Assumption M. Within the kth imputation class, whether or not a unit responds does not depend on the variable being imputed but may depend on a covariate vector $\boldsymbol{x}$ without nonresponse, and for the variable being imputed (say y),

$$y_i = \psi_k(\boldsymbol{x}_i) + \sqrt{v_{ki}}\, \varepsilon_i \tag{4}$$

where $\psi_k(\cdot)$ is an unknown regression function, $v_{ki} = v_k(\boldsymbol{x}_i)$ for a known function $v_k(\cdot) > 0$, ε_i's are independent random errors with mean 0 and unknown variance $\sigma_k^2 > 0$.

Assumption D is called the uniform response assumption and is adopted when one uses the customary design-based approach. Assumption M is for the model-assisted approach. The response mechanism under assumption M is called "unconfounded" by Lee et al. (1994), and "missing at random" by Rubin (1976).

We mainly consider the following two types of regression imputation methods that include many popular imputation methods (e.g., mean imputation, ratio imputation, and random hot deck imputation) as special cases.

1. Deterministic regression imputation. Assume that $\psi_k(\boldsymbol{x}_i)$ in (4) is a parametric linear regression function $\beta_k \boldsymbol{x}_i$, where β_k is a row vector of unknown parameters. Within the kth imputation class, a nonrespondent y_i is imputed by $\tilde{y}_i = \hat{\beta}_k \boldsymbol{x}_i$, where

$$\hat{\beta}_k' = \left(\sum_{i \in R_k} w_i v_{ki}^{-1} \boldsymbol{x}_i \boldsymbol{x}_i' \right)^{-1} \sum_{i \in R_k} w_i v_{ki}^{-1} \boldsymbol{x}_i y_i \tag{5}$$

and R_k is the set of y-respondents within imputation class k. Two important special cases are (1) $\boldsymbol{x}_i \equiv 1$ (no covariates) and $v_{ki} \equiv 1$, in which case $\hat{\beta}_k$ is the weighted mean of respondents in imputation class k and the imputation method is called mean imputation; and (2) $\boldsymbol{x} = x$ is univariate, $x > 0$, and $v_k(x) = x$, in which case $\hat{\beta}_k$ is a ratio estimator and the imputation method is called ratio imputation.

2. Random regression imputation. Random regression imputation adds a random term to a regression imputed value, i.e., within the kth imputation class, a nonrespondent y_i is imputed by

$$\tilde{y}_i = \hat{\beta}_k \boldsymbol{x}_i + \sqrt{v_{ki}}\, \tilde{\varepsilon}_i$$

where, given the observed data, $\tilde{\varepsilon}_i$'s are independent with mean 0 and variance

$$\sigma_{\varepsilon,k}^2 = \sum_{i \in R_k} w_i (r_{ki} - \bar{r}_k)^2 / \sum_{i \in R_k} w_i$$

$r_{ki} = v_{ki}^{-1/2}(y_i - \hat{\beta}_k \boldsymbol{x}_i)$ and $\bar{r}_k = \Sigma_{i \in R_k} w_i r_{ki} / \Sigma_{i \in R_k} w_i$. In the special case of $\boldsymbol{x}_i \equiv 1$ and $v_{ki} \equiv 1$, this random regression imputation with $\tilde{\varepsilon}_i$ drawn from normalized residuals is the same as imputing nonrespondents by an iid sample from $\{y_i : i \in R_k\}$, with each y_i being selected with probability $w_i / \Sigma_{i \in R_k} w_i$, which is the weighted hot deck imputation method in Rao and Shao (1992).

20.2 THE JACKKNIFE

One replication method that appropriately takes nonresponse and imputation into account is the adjusted jackknife method proposed in Rao and Shao (1992). To introduce Rao and Shao's adjusted jackknife, let us first describe how the jackknife (Quenouille, 1949) works in the case of no nonresponse. We consider the jackknife method that deletes one first-stage sampled cluster at a time. Let $\boldsymbol{Z} = \{y_i, \boldsymbol{x}_i, w_i : i \in S\}$ denote the dataset including covariates and survey weights and let $\boldsymbol{Z}_r = \{y_i, \boldsymbol{x}_i, w_i^{(r)} : i \in S\}$ be the rth pseudoreplicate after deleting cluster r, $r = 1, \ldots, R$, $R = \Sigma_h n_h$. Note that each $\boldsymbol{Z}_r$ is the same as $\mathbf{Z}$ except that the survey weight w_i is changed to 0 if i is in the cluster r and $n_h w_i/(n_h - 1)$ if i is not in the cluster r and both i and r is in stratum h. The jackknife variance estimator for the Horvitz–Thompson estimator $\hat{Y} = \hat{Y}(\boldsymbol{Z})$ in (2) is defined by

$$v_{\text{Jack}} = \sum_{r=1}^{R} c_r \left[\hat{Y}(\boldsymbol{Z}_r) - \frac{1}{R} \sum_{t=1}^{R} \hat{Y}(\boldsymbol{Z}_t) \right]^2 \tag{6}$$

where $c_r = (n_h - 1)/n_h$ when cluster r is in stratum h and $\hat{Y}(\boldsymbol{Z}_r)$, called the rth replicate estimator, is the same as $\hat{Y}(\boldsymbol{Z})$ except that $\boldsymbol{Z}$ is replaced by $\boldsymbol{Z}_r$. Note that the computation of $\hat{Y}(\boldsymbol{Z}_r)$, $r = 1, \ldots, R$, is routine. It can be done repeatedly using a program similar to that for computing $\hat{Y}$. When variance estimation for an estimator other than $\hat{Y}$, say $\hat{\theta} = \hat{\theta}(\boldsymbol{Z})$, is concerned, formula (6) can still be used with the simple change of replacing $\hat{Y}(\boldsymbol{Z}_r)$ by $\hat{\theta}(\boldsymbol{Z}_r)$.

In the presence of imputed nonrespondents, treating imputed values as observed data and using formula (6) leads to the naive jackknife variance estimator that equals the right-hand side of (6) with $\boldsymbol{Z}_r$ replaced by $\tilde{\boldsymbol{Z}}_r = \{\tilde{y}_i, \boldsymbol{x}_i, w_i^{(r)} : i \in S\}$, where $\tilde{y}_i$ is an imputed value if y_i is missing and is y_i if y_i is observed. Because $\tilde{Y}$ in (3) usually has a larger variance than $\hat{Y}$ due to nonresponse and imputation, and because formula (6) does not take nonresponse and imputation into account (i.e., for the same sample unit, the imputed values in $\tilde{\boldsymbol{Z}}_r$ and $\tilde{\boldsymbol{Z}}_t$, $r \neq t$, are the same), the naive jackknife variance estimator usually underestimates the variance of $\tilde{Y}$.

One approach to obtain a nearly unbiased jackknife variance estimator is to modify pseudoreplicates $\tilde{\boldsymbol{Z}}_1, \ldots, \tilde{\boldsymbol{Z}}_R$ so that the imputed values in $\tilde{\boldsymbol{Z}}_r$ and $\tilde{\boldsymbol{Z}}_t$ for the same unit are different and this difference recovers the variance component due to nonresponse and imputation. Rao and Shao's adjusted jackknife is the first attempt at using this idea. For the weighted hot deck imputation method, the random regression imputation (Section 20.1) in the special case of $\boldsymbol{x}_i \equiv 1$ and $v_{ki} \equiv 1$, Rao and Shao (1992) proposed to adjust each imputed value $\tilde{y}_i$ in $\tilde{\boldsymbol{Z}}_r$ to

$$\tilde{y}_{i,r}^{\text{Adj}} = \tilde{y}_i + \frac{\sum_{i \in R_k} w_i^{(r)} y_i}{\sum_{i \in R_k} w_i^{(r)}} - \frac{\sum_{i \in R_k} w_i y_i}{\sum_{i \in R_k} w_i} \tag{7}$$

Let $\tilde{\boldsymbol{Z}}_r^{\text{Adj}}$ be the adjusted rth pseudoreplicate. Rao and Shao's adjusted jackknife variance estimator is then obtained by using formula (6) with $\boldsymbol{Z}_r$ replaced by $\tilde{\boldsymbol{Z}}_r^{\text{Adj}}$.

Note that the last term on the right hand side of (7) is the imputation expectation of $\tilde{y}_i$ and the second to last term on the right hand side of (7) is the imputation expectation of $\tilde{y}_i^{(r)}$, the imputed value obtained by using weighted hot deck imputation and the respondents in the rth pseudoreplicate $\boldsymbol{Z}_r$. Thus, Rao and Shao's adjusted jackknife can be extended to any imputation method as follows. Let I denote a particular imputation method. For a nonrespondent y_i, $\tilde{y}_i = I_i(\boldsymbol{Z})$, which indicates that imputation uses the respondents in $\boldsymbol{Z}$ and the imputation method I. Let $\tilde{y}_i^{(r)} = I_i(\boldsymbol{Z}_r)$, the imputed value by using imputation method I and the respondents in the rth pseudoreplicate. Rao and Shao's adjusted jackknife can be applied by modifying (7) to

$$\tilde{\boldsymbol{y}}_{i,r}^{\mathrm{Adj}} = \tilde{y}_i + E_I(\tilde{y}_i^{(r)}) - E_I(\tilde{y}_i) \tag{8}$$

where E_I is the imputation expectation. When the random regression imputation method (Section 20.1) is used, for example, $E_I(\tilde{y}_i) = \hat{\beta}_k \boldsymbol{x}_i$ and $E_I(\tilde{y}_i) = \hat{\beta}_k^{(r)} \boldsymbol{x}_i$, where $\hat{\beta}_k^{(r)}$ is the same as $\hat{\beta}_k$ in (5) but with w_i replaced by $w_i^{(r)}$.

Why does adjustment (8) provide a nearly unbiased variance estimator? Let us consider deterministic imputation (given the data). When the imputation method I is deterministic, $E_I(\tilde{y}) = \tilde{y}$ and, therefore, adjustment (8) reduces to $\tilde{y}_{i,r}^{\mathrm{Adj}} = \tilde{y}_i^{(r)} = I_i(\boldsymbol{Z}_r)$, i.e., adjustment (8) amounts to reimputing nonrespondents in $\tilde{\boldsymbol{Z}}_r$ using the respondents in $\boldsymbol{Z}_r$ and using the same method I that is used to impute nonrespondents in $\boldsymbol{Z}$. This discovery leads to the second approach for deriving jackknife variance estimators (or, more generally, replication variance estimators) that take nonresponse and imputation into account. That is, we reimpute each pseudoreplicate $\tilde{\boldsymbol{Z}}_r$ using respondents in $\boldsymbol{Z}_r$ and using the same imputation method I. Let $\tilde{\boldsymbol{Z}}_r^{\mathrm{Imp}}$, $r = 1, \ldots, R$, be reimputed pseudoreplicates. The reimputed jackknife variance estimator is then defined by (6) with $\boldsymbol{Z}_r$ replaced by $\tilde{\boldsymbol{Z}}_r^{\mathrm{Imp}}$.

Although reimputation is the same as adjustment (8) for deterministic imputation, reimputation is easy to interpret. Consider, for example, the special case of a single imputation class and ratio imputation, where $\tilde{Y}$ is given by

$$\hat{Y}(\tilde{\boldsymbol{Z}}) = \sum_{i \in R} w_i y_i + \sum_{i \notin R} w_i x_i \frac{\sum_{i \in R} w_i y_i}{\sum_{i \in R} w_i x_i} \tag{9}$$

Without reimputation, we use (6) with

$$\hat{Y}(\tilde{\boldsymbol{Z}}_r) = \sum_{i \in R} w_i^{(r)} y_i + \sum_{i \notin R} w_i^{(r)} x_i \frac{\sum_{i \in R} w_i y_i}{\sum_{i \in R} w_i x_i}$$

whereas with reimputation, we use (6) with

$$\hat{Y}(\tilde{\boldsymbol{Z}}_r^{\mathrm{Imp}}) = \sum_{i \in R} w_i^{(r)} y_i + \sum_{i \notin R} w_i^{(r)} x_i \frac{\sum_{i \in R} w_i^{(r)} y_i}{\sum_{i \in R} w_i^{(r)} x_i}$$

If $\hat{Y}(\tilde{\boldsymbol{Z}})$ in (9) is viewed as a function of three weighted averages, then clearly $\hat{Y}(\tilde{\boldsymbol{Z}}_r^{\text{Imp}})$ not $\hat{Y}(\tilde{\boldsymbol{Z}}_r)$ is the right jackknife replicate estimator to use in applying (6).

When random imputation is used, a crucial factor for reimputation is how the random terms are generated in the process of reimputing pseudoreplicates. There are two simple ways. The first one is to use the same random terms in the original imputation, in which case reimputation is again the same as adjustment (8). The second one is to use independent random terms across R pseudoreplicates, which produces variance estimators different from those using adjustments. Although independent reimputation has the attractive feature of programming ease, it sometimes leads to overestimation of the variance of $\tilde{Y}$, since excessively large variation may be created using independent random terms. For the jackknife, reimputing pseudoreplicates independently does produce overestimation (Burns, 1990, Rao and Shao, 1992).

20.3 THE BALANCED HALF SAMPLE AND RELATED METHODS

The idea of adjustment or reimputation can be used to obtain other replication variance estimators. In this section we consider the balanced half sample (McCarthy 1969) and related replication methods. Although the balanced half sample and related replication methods are computationally more complicated than the jackknife, they can be applied to variance estimation for nonsmooth estimators, such as sample quantiles.

Consider the special case of $n_h = 2$ for all h. In the case of no nonresponse, the balanced half sample variance estimator for the Horvitz–Thompson estimator $\hat{Y}$ in (2) is

$$v_{\text{BHS}} = \frac{1}{R}\sum_{r=1}^{R}\left[\hat{Y}(\boldsymbol{Z}_r) - \frac{1}{R}\sum_{t=1}^{R}\hat{Y}(\boldsymbol{Z}_t)\right]^2 \tag{10}$$

where $\boldsymbol{Z}_r = \{y_i, \boldsymbol{x}_i, w_i^{(r)} : i \in S\}$ is the rth pseudoreplicate with $w_i^{(r)} = 2w_i$ if $e_{rh} = 1$ and 0 if $e_{rh} = -1$ for i in the hth stratum, and e_{rh} is the (r, h)th element of a matrix constructed from an $R \times R$ Hadamard matrix by choosing any H columns excluding the column of all +1's and $H + 1 \leq R \leq H + 4$. Note that assigning 0 weight to unit i is equivalent to deleting unit i and, therefore, $\boldsymbol{Z}_r$ contains half of $\boldsymbol{Z}$ and is called a half sample.

When there are imputed nonrespondents, we again use $\tilde{\boldsymbol{Z}}_r = \{\tilde{y}_i, \boldsymbol{x}_i, w_i^{(r)} : i \in S\}$ to denote the rth pseudoreplicate with imputed values. The naive balanced half sample variance estimator for $\tilde{Y}$ in (3) is given by (10) with $\boldsymbol{Z}_r$ replaced by $\tilde{\boldsymbol{Z}}_r$. The adjusted balanced half sample variance estimator can be obtained by using (10) with $\boldsymbol{Z}_r$ replaced by $\tilde{\boldsymbol{Z}}_r^{\text{Adj}}$ which is obtained by adjusting the imputed values in $\tilde{\boldsymbol{Z}}_r$ according to (8). The reimputed balanced half sample variance estimator can be obtained by using (10) with $\boldsymbol{Z}_r$ replaced by $\tilde{\boldsymbol{Z}}_r^{\text{Imp}}$, which is obtained by reimputing the imputed values in $\tilde{\boldsymbol{Z}}_r$.

However, independent reimputation leads to overestimation when a random imputation method is adopted. This problem can be fixed if we independently reimpute every nonrespondent in $\tilde{\boldsymbol{Z}}_r$ twice. Equivalently, we can use the following procedure. Instead of creating a half sample $\boldsymbol{Z}_r$ by deleting half of clusters and doubling the weights for the other half clusters, we can obtain a "half sample" $\boldsymbol{Z}_r'$ by replacing the values of units in $\boldsymbol{Z}$ corresponding to $e_{rh} = -1$ by those corresponding to $e_{rh} = +1$ and without changing their weights. Let $\tilde{\boldsymbol{Z}}_r'$ be the same as $\boldsymbol{Z}_r'$ except that nonrespondents are replaced by $\tilde{y}_i$'s. Obviously, $\hat{Y}(\boldsymbol{Z}_r) = \hat{Y}(\boldsymbol{Z}_r')$ and $\hat{Y}(\tilde{\boldsymbol{Z}}_r) = \hat{Y}(\tilde{\boldsymbol{Z}}_r')$. But the imputation variance of $\hat{Y}(\tilde{\boldsymbol{Z}}_r'^{\text{Imp}})$ is the same as the imputation variance of $\hat{Y}(\tilde{\boldsymbol{Z}}_r)$, where $\tilde{\boldsymbol{Z}}_r'^{\text{Imp}}$ is obtained by reimputing $\tilde{\boldsymbol{Z}}_r'$. That is, independent reimputation based on $\boldsymbol{Z}_r'$, $r = 1, \ldots, R$, produces a nearly unbiased balanced half sample variance estimator.

The balanced half sample can be extended to the case where some n_h's are not 2. The key is how to construct pseudoreplicates $\boldsymbol{Z}_1, \ldots, \boldsymbol{Z}_R$. Once a suitable set of pseudoreplicates are constructed, adjustment or reimputation discussed previously can be applied to take the nonresponse and imputation into account. The construction of balanced or approximately balanced pseudoreplicates is discussed, for example, in Kish and Frankel (1970), Gurney and Jewett (1975), Gupta and Nigam (1987), Wu (1991), Sitter (1993), Rao and Shao (1996), and Shao and Chen (1999).

20.4 THE BOOTSTRAP

The bootstrap (Efron, 1979) can be used not only for variance estimation, but also for other inference purposes such as setting confidence sets.

We first consider the case where all n_h are large. Let S_r^* be the bootstrap sample generated by drawing a simple random sample of size n_h with replacement from n_h first stage clusters in stratum h, and independently repeating the process across H strata. The rth bootstrap pseudoreplicate is $\boldsymbol{Z}_r^* = \{y_i, \boldsymbol{x}_i, w_i : i \in S_r^*\}$, where S_r^*, $r = 1, \ldots, R$, are independent bootstrap samples. In the case of no nonresponse, the bootstrap variance estimator for $\hat{Y}$ in (2) is defined by the right hand side of (10) with $\boldsymbol{Z}_r = \boldsymbol{Z}_r^*$ and a large integer R. It follows from the law of large numbers that conditional on the observed data, as $R \to \infty$, this bootstrap variance estimator tends to $V^*[Y(\boldsymbol{Z}^*)]$, where $\boldsymbol{Z}^* = \boldsymbol{Z}_r^*$ for any fixed r and V^* is the variance with respect to bootstrap sampling, given the observed data.

In the presence of imputed nonrespondents, the ideas described for the balanced half sample method in Section 20.3 can still be used. Using an adjustment similar to that in (8), Shao and Sitter (1996) proposed an adjusted bootstrap method. In the case of random imputation, bootstrap with independent reimputation is also straightforward: each bootstrap pseudoreplicate $\boldsymbol{Z}_r^*$ is reimputed using the same imputation method, independently across the bootstrap pseudoreplicates (Efron, 1994; Shao and Sitter, 1996). Because the bootstrap sample has the same size as the original sample, independent reimputation does not cause overestimation.

For the bootstrap, reimputation can be naturally motivated as follows. In the case of no nonresponse, the spirit of the bootstrap is to mimic the statistical behavior of

$\{\boldsymbol{Z}, \hat{Y}(\boldsymbol{Z})\}$ by using that of $\{\boldsymbol{Z}^*, \hat{Y}(\boldsymbol{Z}^*)\}$, conditional on $\boldsymbol{Z}$, where $\{\boldsymbol{Z}^*, \hat{Y}(\boldsymbol{Z}^*)\}$ is the bootstrap analog of $\{\boldsymbol{Z}, I(\boldsymbol{Z})\}$. When imputation is applied with method I, the bootstrap analog of $\{\boldsymbol{Z}, I(\boldsymbol{Z}^*), \hat{Y}(I(\boldsymbol{Z}))\}$ is $\{\boldsymbol{Z}^*, I(\boldsymbol{Z}^*), \hat{Y}(I(\boldsymbol{Z}^*))\}$, where $I(\boldsymbol{Z}^*)$ indicates that the bootstrap dataset is reimputed using respondents in $\boldsymbol{Z}^*$ and using the same imputation method I. On the other hand, the bootstrap without reimputation uses the conditional statistical behavior of $\{\tilde{\boldsymbol{Z}}^*, \hat{Y}(\tilde{\boldsymbol{Z}}^*)\}$ to mimic that of $\{\boldsymbol{Z}, I(\boldsymbol{Z}), \hat{Y}(I(\boldsymbol{Z}))\}$, treating $\tilde{\boldsymbol{Z}} = I(\boldsymbol{Z})$ as the observed dataset, which clearly misses the important component of nonresponse and imputation.

For stratified sampling with some small n_h's, the previously described bootstrap procedure has to be modified, even in the case of no nonresponse (Rao and Wu, 1988, Sitter, 1992). The simplest modified bootstrap is the one generates S_r^* by drawing a simple random sample of size $n_h - 1$, instead of n_h, with replacement from n_h first stage clusters in stratum h, independently across H strata. The use of bootstrap sample size $n_h - 1$ is to adjust the loss of one degree of freedom in stratum h due to variance estimation without knowing the stratum mean, which is important when some n_h's are small. The rth bootstrap pseudoreplicate should be modified to $\boldsymbol{Z}_r^* = \{y_i, \boldsymbol{x}_i, w_i^{(r)} : i \in S_r^*\}$ with $w_i^{(r)} = w_i n_h/(n_h - 1)$ for i in stratum h. In the presence of imputed nonrespondents, the adjusted bootstrap can be applied without any modification. In the case where a random imputation method is adopted, independent reimputation has to be applied similarly to the case of balanced half sample (Section 20.3) to avoid overestimation, since the sizes of the bootstrap sample S_r^* and S are not close when some n_h's are small.

20.5 EVALUATIONS OF REPLICATION METHODS

In what sense do we say a variance estimator is nearly unbiased or asymptotically unbiased? This section considers this issue for the replication variance estimators discussed in the previous sections.

Like any other statistical procedures, a replication method for variance estimation should be evaluated both theoretically (asymptotically) and empirically. We first discuss asymptotic properties of replication variance estimators introduced in Sections 20.2–20.4. An asymptotic framework for survey problems can be described as follows. The population P is assumed to be a member of a sequence of populations indexed by v. Let M be the total number of ultimate units in P, $n = \Sigma_h n_h$ and $N = \Sigma_h N_h$, which depend on v but v is suppressed for simplicity of notation. As $v \to \infty$,

1. $n \to \infty$ and $n/N \to 0$
2. $nM^{-1} \max_{h,j} \Sigma_{i \in S(h,j)} w_i$ is bounded, where $S(h, j)$ is the set of sampled units in stratum h and cluster j
3. $n\text{Var}(\hat{Y}/M)$ is bounded away from 0
4. $n^{1+\delta} \Sigma_{h,j} E\|z_{hj} - E(z_{hj})\|^{2+\delta}$ is bounded for some $\delta > 0$, where $\|\ \|$ is the usual vector norm and $z_{hj} = \Sigma_{i \in S(h,j)} w_i z_i/M$ with $z_i = y_i$ or $\boldsymbol{x}_i$.

Under this asymptotic setting, $\{v/V[\hat{Y}(\widetilde{\boldsymbol{Z}})]\} \to 1$ in probobability, where v is any of the replication variance estimators discussed in Sections 20.2–20.4 expect for the jackknife variance estimator with independent reimputation when random imputation is used. Result (14) also holds when $\hat{Y}(\widetilde{\boldsymbol{Z}})$ is replaced by other types of estimators, for example, functions of several weighted averages (totals) or sample quantiles. More details can be found in Rao and Shao (1992, 1999), Shao et al. (1998), Shao and Chen (1999), and Shao and Sitter (1996).

Next, we present some empirical results from a simulation study. A stratified one-stage cluster sample from an artificial population was considered in Shao et al. (1998). For each of 32 strata, two clusters were selected with equal probability and with replacement. The sampling fraction was $n/N = 64/1000 = 0.064$. Each cluster contains 20 units with within-cluster correlation coefficient 0.1. For each sampled unit y_{hij}, a uniform random variable u_{hij} was generated. If $u_{hij} \leq p$, where p is the response rate assumed the same for all units, then y_{hij} is a respondent; otherwise y_{hij} is a nonrespondent. The values of p considered in the simuation are 90%, 80%, 70%, 60%, and 50%. The nonrespondents were imputed using the weighted hot deck mothod described in Section 20.1. Estimation of the population average and the population median were considered. The relative bias (RB) and the coefficient of variation (CV) of the naive (unadjusted) and the adjusted balanced half sample (BHS) variance estimators, and the coverage probability (CP) of asymptotic 95% confidence intervals based on these variance estimators are reported in Table 20.1. It is clear that the naive BHS variance estimator seriously underestimates the true variance, whereas the adjusted BHS variance estimator performs well.

Finally, we present results from a real data analysis in the National Immunization Program Record Check Study (Nixon et al. 1996). Immunization

Table 20.1. Relative bias (RB), coefficient of variation (CV), and coverage probability (CV) based on 2000 simulations

Point estimator	Response rate p (%)	Naive BHS estimator (%)			Adjusted BHS estimator (%)		
		RB	CV	CP	RB	CV	CP
Sample mean	90	−12.31	27.82	91.85	7.57	31.55	95.25
	80	−28.39	36.04	89.25	8.75	32.70	94.70
	70	−41.79	45.05	85.55	10.09	33.37	94.75
	60	−56.10	57.51	79.20	5.16	30.85	94.05
	50	−64.11	65.06	73.80	9.49	34.31	94.70
Sample median	90	−15.05	33.33	89.80	3.20	35.19	94.00
	80	−27.06	38.43	88.30	7.85	38.52	94.55
	70	−36.47	44.14	84.85	14.83	43.62	94.70
	60	−52.16	55.87	78.00	6.89	40.17	93.00
	50	−59.66	62.57	74.05	10.70	45.11	93.80

data for 19–35-month-old children were obtained from households in a supplement to the National Health Interview Survey which is a three-stage probability sample of households. Two clusters per stratum were selected from 62 strata (metropolises or counties). Within each sampled cluster, area segments and permit segments were selected, and within those segments households were selected. Data were collected on the number of doses each eligible child received for five antigens—DTP, Polio, MMR, Hib, and Hep-B—and three combinations, 431 (at least 4 doses of DTP, 3 of Polio, and 1 of MMR), 4313 (at least 4 doses of DTP, 3 of Polio, 1 of MMR, and 3 of Hib), and 43133 (at least 4 doses of DTP, 3 of Polio, 1 of MMR, 3 of Hib, and Hep-B). The items of interest are the best values of immunization status for the five antigens and three combinations, where the best value for a child's immunization status was determined by comparing and reconciling the responses from two surveys. The best values were missing for 378 of 1230 sampled children. Random hot deck imputation was done independently in 20 imputation classes formed by cross-classifying five control variables. Based on the best values (observed or imputed), estimated proportions of children who received the recommended doses for five antigens and three combinations are listed in the third column of Table 20.2 ($\hat{\theta}$). For comparison, estimated proportions based on 852 respondents are listed in the second column of Table 20.2 ($\hat{\theta}_R$). $\hat{\theta}_R$ is actually biased, since the response rates in different imputation classes are different. In this example, however, the difference between $\hat{\theta}_R$ and $\hat{\theta}$ is small compared to the estimated variances of $\hat{\theta}_R$ and $\hat{\theta}$ given in Table 20.2. The fourth column of Table 20.2 contains variance estimates v_R for $\hat{\theta}_R$, computed using the standard balanced half sample (BHS) formula with no imputed data. The balanced half samples were constructed using a 64×64 Hadamard matrix. The fifth and the sixth columns of Table 20.2 contain the naive (unadjusted) BHS variance estimate v_{BHS} and the adjusted BHS variance estimate v_{ABSH} for $\hat{\theta}$, respectively. It can be seen from Table 20.2 that the effect of using the wrong variance estimator is quite substantial: first, v_{BHS} is even smaller than v_R, although the variance of $\hat{\theta}$ is larger that that of $\hat{\theta}_R$; second, the ratio $v_{\text{ABHS}}/v_{\text{BHS}}$ (listed in column 7 of Table 20.2) can be as large as 3.34 and has an average value 2.24.

TABLE 20.2. Immunization data example: statistics by antigen

Antigen	$\hat{\theta}_R$	$\hat{\theta}$	v	v_{BHS}	v_{ABHS}	$v_{\text{ABHS}}/v_{\text{BHS}}$
DTP	77.3	76.3	2.02	1.30	2.66	2.04
Polio	84.8	83.6	1.61	1.04	2.43	2.34
MMR	89.4	88.6	1.51	0.66	2.19	3.34
Hib	91.1	90.4	1.51	0.96	1.88	1.95
Hep-B	19.2	18.1	2.25	1.61	3.03	1.88
431	75.7	74.8	1.99	1.21	2.62	2.17
4313	74.2	73.2	2.13	1.21	2.69	2.22
43133	15.8	14.8	2.34	1.37	2.66	1.94

20.6 SUMMARY AND DISCUSSION

20.6.1 Summary

Replication methods such as the jackknife, the balanced half sample, and the bootstrap provide nearly unbiased and consistent variance estimators when adjustment or reimputation is applied to take imputation into account. The jackknife does not produce good variance estimators for point estimators that are not smooth enough, although it is the simplest replication method in terms of computation. When a random imputation method is used, applying a replication method with independent random reimputation may lead to overestimation of variances. Some modifications have to be applied (Section 20.3). Although independent random reimputation requires more computations than adjustment (8), the former can be easily applied to cases where different estimators such as sample mean and sample quantiles are considered, whereas the latter has to be applied separately for each estimator.

20.6.2 Relationship Between Replication with Reimputation and Multiple Imputation

Rubin (1978) proposed a multiple imputation method for variance estimation discussed in depth in other chapters in this book. A common feature of the multiple imputation method and the replication methods with reimputation (or adjustment) is that repeated imputations are required. However, a major difference is that a replication method reimputes using respondents in pseudoreplicate $\boldsymbol{Z}_r$, whereas in multiple imputation, repeated imputation are obtained using the respondents in the original dataset $\boldsymbol{Z}$. Because of this, multiple imputation is not applicable to any deterministic imputation method.

Nonnegligible First Stage Sampling Fraction. If the first stage sampling fraction $\Sigma_h n_h / \Sigma_h N_h$ is not negligible (see, for example, Shao and Steel, 1999), then the application of replication methods in the presence of imputed values is not easy. Some discussions can be found in Shao and Steel (1999). This problem is still under investigation.

Multivariate Imputation. We considered only the case of univariate y. Results in previous sections can be easily extended to multivariate $\boldsymbol{y}$ with vector imputation for unit nonresponse or with marginal item imputation for item nonresponse. Although marginal item imputation produces asymptotically unbiased estimators of population marginal totals, it does not produce valid estimators for population parameters measuring relationship among items, e.g., the correlation coefficient between two components of $\boldsymbol{y}$. Some adjustment has to be made to obtain asymptotically unbiased estimators of correlation coefficients based on marginal imputation (see Skinner and Rao, 1999). For a suitable joint imputation method, replication methods with reimputation or adjustment discussed previously can be applied for variance estimation. Results for correlation coefficients can be found in Shao and Wang (1999).

Nonignorable Nonresponse. The assumed response mechanism in assumption D or assumption M is called the ignorable response mechanism by Rubin (1976). For nonignorable response, i.e., the case in which response probabilities depend on the values of variables being imputed, it is hard to obtain an imputation method that produces a nearly unbiased estimator based on imputed data. Once such an imputation method is found, replication methods with adjustment or reimputation can be applied for variance estimation, but theoretical and empirical studies should be performed to examine properties of replication variance estimators.

CHAPTER 21

Variance Estimation from Survey Data under Single Imputation

Hyunshik Lee, *Westat, USA*
Eric Rancourt and Carl E. Särndal, *Statistics Canada*

21.1 INTRODUCTION

Imputation is commonly used to fill in substitutes for missing survey data. When this step has been completed, it is also common to treat imputed data as true observations and to use standard variance estimators. However, this approach may lead to severe underestimation of the true variance. This problem was recognized early on and multiple imputation was proposed as a solution in a Bayesian framework (Rubin, 1978). Nonetheless, single imputation has been and is still widely used, particularly by statistical agencies, because of its operational convenience. Therefore, there is a need to provide valid variance estimation techniques for survey data with singly imputed values. In this chapter, we focus on the problem of variance estimation of a point estimator of a finite population parameter in the presence of single imputation. Merits and demerits of single and multiple imputation are debated elsewhere (e.g., Fay, 1996; Rao, 1996; Rubin, 1996; Chapter 22, this volume, for single imputation; and Chapter 23, this volume, for multiple imputation).

To address the problem, various approaches have been proposed in the literature. After laying out theoretical foundations, we review these approaches in Section 21.2, focusing on their basic principles. This is a relatively young and emerging area of research and some of very recent developments are not covered by the review. The approaches included in the review are first theoretically compared, then empirically evaluated via a simulation study in Section 21.3. Based on the results of Section 21.3, we give recommendations and concluding remarks in Section 21.4.

21.2 APPROACHES TO VARIANCE ESTIMATION UNDER SINGLE IMPUTATION

Since Särndal (1990), many approaches have been proposed to address the problem of variance estimation for singly imputed data. Earlier, Ford (1983) suggested reimputation for the replication variance estimators under hot deck imputation, which Burns (1990) used unsuccessfully to address the problem with the jackknife technique. In this section, we review all major approaches found in the literature after 1990. For simplicity, we focus on the case where the target parameter is a population total, so that $\theta = Y = \Sigma_U y_k$, where Σ_U is shorthand for summation over the finite population U. The estimation of many other parameters can be handled by the fact that they can be expressed as a function of population totals.

The estimator $\hat{\theta}$, designed to estimate θ for the ideal situation of 100% response, is called the prototype estimator. For example, we may use the generalized regression (GREG) estimator (see Särndal et al., 1992, pp. 225–234), given by

$$\hat{Y}_{\text{GR}} = \Sigma_s a_k g_k y_k \tag{21.1}$$

where s denotes the sample, a_k is the sampling weight, and g_k (called the g-factor) is a modifier to a_k. The g-factor is computed using an auxiliary vector $\mathbf{x}_k$, available at the population level to improve the estimator by exploiting a regression relationship between y_k and $\mathbf{x}_k$. The GREG estimator includes many well-known special cases, such as the Horvitz–Thompson (HT) estimator, obtained when $g_k = 1$ for all k, and the ratio estimator, obtained when $g_k = X/\hat{X}_{\text{HT}}$ for all k with a scalar $\mathbf{x}_k = x_k$, where $X = \Sigma_U x_k$ is the population total of auxiliary x-variable and $\hat{X}_{\text{HT}} = \Sigma_s a_k x_k$ is the HT estimator. (Chapter 18 discusses a similar special case of the GREG estimator).

The usual procedure after imputation is to take the formula for the prototype estimator and apply it to the imputed data. The result is called the imputed estimator, denoted by $\hat{\theta}_{\text{I}}$. Thus, the imputed GREG estimator is $\hat{\theta}_{\text{I}} = \hat{Y}_{\text{I,GR}} = \Sigma_s a_k g_k y_k^*$ and the imputed HT estimator is $\hat{\theta}_{\text{I}} = \hat{Y}_{\text{I,HT}} = \Sigma_s a_k y_k^*$, where $y_k^* = y_k$ (observed y-value) if $k \in r$ (response set) and $y_k^* = \hat{y}_k$ (imputed value) if $k \in o = s - r$ (nonresponse set). The total error of the imputed estimator is $\hat{\theta}_{\text{I}} - \theta = (\hat{\theta} - \theta) + (\hat{\theta}_{\text{I}} - \hat{\theta})$. The first term on the right-hand side is the sampling error and the second is the imputation error, which is due to variability associated with the response mechanism and the imputation procedure. The goal of variance estimation must be to estimate the total variance. A naive approach is to take the variance estimator appropriate for the prototype $\hat{\theta}$ under full response (which we call the standard variance estimator) and to apply it to the imputed data set $\{y_k^* : k \in s\}$. This will usually yield a large underestimation of the total variance.

The approaches proposed for a more correct estimation of the total variance fall into two broad classes, which differ in the probabilistic set-up governing the sampling and response processes. As in Fay (1991), the two set-ups are depicted in the following diagram:

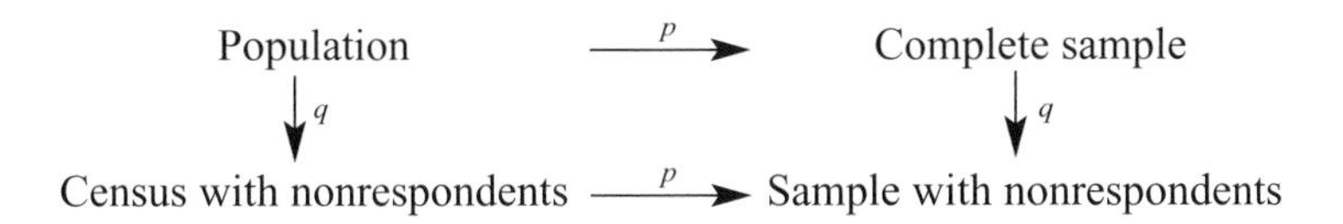

where p is a known sampling design and q is an unknown response mechanism (RM). In order to proceed, one has to make an assumption about the unknown RM.

The upper (pq) path is more natural, since nonresponse occurs after the sample s is selected (Dalenius, 1983). Under this path, an RM is defined as a conditional probability distribution given s, denoted as $q(r \mid s)$. In the model-based framework, a frequently used RM is missing at random (MAR), as discussed in Chapter 1. In the survey sampling context used in this chapter, different RM's are also used. As in Lee et al. (1994), an RM is said to be unconfounded when $q(r \mid s, \mathbf{z}_s, y_s) = q(r \mid s, \mathbf{z}_s)$ and $\text{Prob}(k \in r \mid s) > 0$ for all $k \in s$, where $\mathbf{z}_s = \{\mathbf{z}_k : k \in s\}$ is an auxiliary data set used for imputation and $y_s = \{y_k : k \in s\}$. The auxiliary data set is assumed to be available from an external source for the full sample. This definition is closely related to that given by Rubin (1987, p. 39). A stronger assumption than the unconfounded RM is the uniform RM, which implies that the sample units respond independently of each other with the same probability. Often, the uniform RM is assumed to hold within groups (imputation classes) and imputation is then carried out group by group. Such RM's are suitable for design-based or model-assisted arguments. Some authors argue that the lower (qp) path has advantage over the upper path, as discussed later in this section.

The RM plays an important role in the theory for nonresponse compensation. If imputation is carried out under a wrong assumption about the RM, then a point estimator obtained from the imputed data can be seriously biased. Even under "good imputation," $\hat{\theta}_\text{I}$ is not free of bias. We can hope that the bias is small, but since it is nonzero, the mean square error (MSE) is a more relevant indicator of the quality of $\hat{\theta}_\text{I}$ than its variance. Using the pq-probabilistic path, the MSE is

$$\text{MSE}_{pq}(\hat{\theta}_\text{I}) = E_{pq}(\hat{\theta}_\text{I} - \theta)^2 = V_p(\hat{\theta}) + E_p V_{\text{cIMP}} + E_p(B_c^2) + 2\text{Cov}_p(\hat{\theta}, B_\text{c}) \quad (21.2)$$

Here $V_p(\hat{\theta})$ is the variance of the prototype $\hat{\theta}$. The sum of the last three terms of (21.2) measures the increase in MSE caused by nonresponse followed by imputation. The first of these involves the conditional variance $V_{\text{cIMP}} = V_q(\hat{\theta}_\text{I} \mid s)$; the last two terms contain the conditional bias, $B_\text{c} = E_q(\hat{\theta}_\text{I} \mid s) - \hat{\theta}$. The covariance term may be numerically small, but $E_p(B_\text{c}^2)$ can represent a large addition to the MSE. If $B_\text{c} = 0$ for all s (which never holds exactly in practice), then (21.2) becomes

$$V_{\text{TOT}} = V_{\text{SAM}} + V_{\text{IMP}} \quad (21.3)$$

where $V_{\text{TOT}} = V_{pq}(\hat{\theta}_\text{I}) = \text{MSE}_{pq}(\hat{\theta}_\text{I})$, $V_{\text{SAM}} = V_p(\hat{\theta})$, and $V_{\text{IMP}} = E_p V_{\text{cIMP}} = E_p V_q(\hat{\theta}_\text{I} \mid s)$ are, respectively, the total variance, the sampling variance, and the imputation variance of $\hat{\theta}_\text{I}$. In the simulation study (Section 21.3.2), we evaluate the average performance of the different approaches to variance estimation (presented in Sections 21.2.1 to 21.2.5) in relation to value of the MSE. Two strong reasons to compare with the MSE rather than with the variance are: (1) it gives a reminder that an assumption of zero bias in $\hat{\theta}_{\mathrm{I}}$ (although usually made implicitly by users) is unjustified; (2) the MSE is the appropriate indicator of accuracy, but since the bias cannot be estimated, there is no choice but to estimate the variance (21.3), as in the approaches that we consider. (The result is often an underestimation of the MSE, as the simulation shows; by contrast, these approaches estimate the variance quite well, even if $\hat{\theta}_{\mathrm{I}}$ is considerably biased.) Formula (21.3) represents the total variance as a sum of two components. As explained later, to provide separate estimates of the two is important for survey management.

Another important factor in variance estimation is the imputation method. Many imputation methods used in practice assume implicitly or explicitly an imputation model (denoted by ξ) of the form

$$y_k = \mathbf{z}_k' \beta + \sqrt{c_k}\varepsilon_k \tag{21.4}$$

where β is the column vector of regression coefficients and the c_k are suitably defined constants. The error terms ε_k are usually assumed to be independently indentically distributed (iid) with $E_\xi(\varepsilon_k) = 0$, and $E_\xi(\varepsilon_k^2) = \sigma^2$, where σ^2 is unknown, and $E_\xi(\varepsilon_l \varepsilon_k) = 0$, if $l \neq k$. After fitting (21.4) using the respondent data, we obtain the residuals $e_{kr} = (y_k - \mathbf{z}_k' \hat{\mathbf{B}}_r)/\sqrt{c_k}$ for $k \in r$, with $\hat{\mathbf{B}}_r = (\Sigma_r \omega_k \mathbf{z}_k \mathbf{z}_k' / c_k)^{-1} \Sigma_r \omega_k \mathbf{z}_k y_k / c_k$, where the ω_k are specified weights, and the presence of c_k is justified by the variance structure of (21.4). Some authors advocate the choice $\omega_k = a_k$ (e.g., Shao and Sitter, 1996). (For more discussion on single imputation, see Chapter 22.)

21.2.1 The Two-Phase Approach

The probabilistic set-up under the pq-path resembles the usual set-up for a two-phase sampling design, where $p(s)$ and $q(r \mid s)$ correspond to the first and the second phase sampling procedures, respectively. The unknown RM, $q(r \mid s)$, is often assumed to be uniform or uniform within strata. This set-up, also called "quasirandomization," has been used to handle the nonresponse problem by weighting adjustment (e.g., Oh and Scheuren, 1983; Särndal and Swensson, 1987).

The total variance of $\hat{\theta}_{\mathrm{I}}$ is as given in (21.3). The objective of this approach is to find variance component estimators, $\hat{V}_{\mathrm{SAM}}$ and $\hat{V}_{\mathrm{cIMP}}$, such that (i) $E_pE_q(\hat{V}_{\mathrm{SAM}}) = V_{\mathrm{SAM}}$, and (ii) $E_q(\hat{V}_{\mathrm{cIMP}}) = V_{\mathrm{cIMP}}$ for every fixed s. A pq-unbiased variance estimator of the total variance V_{TOT} is then obtained by taking $\hat{V}_{\mathrm{TOT}} = \hat{V}_{\mathrm{SAM}} + \hat{V}_{\mathrm{cIMP}}$, since $E_pE_q(\hat{V}_{\mathrm{cIMP}}) = V_{\mathrm{IMP}}$. Note that $\hat{V}_{\mathrm{TOT}}$ is sensitive to the assumption made about $q(r \mid s)$.

Rao and Sitter (1995) studied the two-phase approach for the HT prototype estimator under SRS and the uniform RM. Lee et al. (2000) provide a variance estimation formula for a more general case where the prototype is the GREG and the sampling design is arbitrary.

21.2.2 The Model-Assisted Approach

The model-assisted approach was first formulated by Särndal (1990, 1992). It has been studied by Deville and Särndal (1994) for the regression imputed HT estimator, by Rancourt et al. (1994) for the nearest neighbor imputed HT estimator, and by Gagnon et al. (1996) for the imputed GREG.

The probabilistic background for the model-assisted approach consists of the sampling design $p(s)$, the RM $q(r \mid s)$, and a third distribution, the imputation model ξ given in (21.4). Since the imputation model is an explicit element of this approach, the appropriate variance concept is the anticipated pq-variance, or the ξpq-variance, of the imputed estimator $\hat{\theta}_{\text{I}}$, denoted by $E_{\xi}V_{pq}(\hat{\theta}_{\text{I}})$. Taking the model expected value of both sides of (21.3), we obtain

$$E_{\xi}V_{pq}(\hat{\theta}_{\text{I}}) = E_{\xi}V_{\text{TOT}} = E_{\xi}V_{\text{SAM}} + E_{\xi}V_{\text{IMP}} \tag{21.5}$$

As in the two-phase approach, we seek estimators of the two variance components, V_{SAM} and $V_{\text{IMP}} = E_p(V_{\text{cIMP}})$. The model serves as an instrument in deriving the component estimators, $\hat{V}_{\text{SAM}}$ and $\hat{V}_{\text{cIMP}}$, such that $E_{\xi}\{E_pE_q(\hat{V}_{\text{SAM}}) - V_{\text{SAM}}\} = 0$ and $E_{\xi}\{E_q(\hat{V}_{\text{cIMP}}) - V_{\text{cIMP}}\} = 0$ for every fixed s. Then a ξpq-unbiased estimator of the total variance is obtained by taking $\hat{V}_{\text{TOT}} = \hat{V}_{\text{SAM}} + \hat{V}_{\text{cIMP}}$. About the RM, it is sufficient to assume that it is unconfounded. This assumption allows changing the order of the operators, $E_{\xi}E_pE_q$ into $E_pE_qE_{\xi}$, and back, without affecting the value of the expectation. It is then easy to verify that $E_{\xi}\{E_pE_q(\hat{V}_{\text{TOT}}) - V_{\text{TOT}}\} = 0$, which shows that $\hat{V}_{\text{TOT}}$ is ξpq-unbiased. To construct $\hat{V}_{\text{cIMP}}$, we need a model unbiased estimator $\hat{\sigma}^2$ of the unknown σ^2 in (21.4). The resulting $\hat{V}_{\text{SAM}}$ and $\hat{V}_{\text{cIMP}}$ are slightly different from those obtained in the two-phase approach.

Stochastic imputation methods usually provide enough variability to estimate the sampling variance component by the standard formula. The same holds for nearest neighbor imputation, since it behaves roughly like a stochastic imputation, at least for modest nonresponse rates. For deterministic regression imputation, $\hat{V}_{\text{SAM}}$ requires a correction term. An alternative to the approach with a corrective term is to amend the imputed data so that they contain the variability required to yield a "correct" level when the standard variance formula is computed on the amended imputed data set. For regression imputation, this can be achieved by adding a randomly selected residual e_k^* to the regression prediction $\mathbf{z}_k'\hat{\mathbf{B}}_r$. This is only for the purpose of variance estimation and can be carried out within imputation classes.

21.2.3 The All Cases Imputation Approach

Montaquila and Jernigan (1997) proposed the all cases imputation (ACI) approach. Applying the imputation method also to the respondents, the imputation variance component is estimated using both imputed and observed y-values. Specifically, the imputation variance is estimated using the residuals $\hat{y}_k - y_k$, $k \in r$. The sampling variance component is estimated by the standard variance estimator applied to the imputed data set $\{y_k^*: k \in s\}$, as defined earlier.

Montaquila and Jernigan (1997) provided a variance estimation formula for the HT estimator under the stratified simple random sampling design when imputation classes coincide with the strata. Krenzke et al. (1998) studied the ACI approach for regression imputation with a randomly added residual.

The approach implicitly assumes that the standard variance estimator is approximately unbiased for the sampling variance component. This is often true for stochastic imputation methods (e.g., hot deck) under an appropriate RM. It is empirically shown to work well also for nearest neighbor imputation since this imputation method behaves much like a stochastic one (see Section 21.3.2). On the other hand, for deterministic imputation methods such as mean, ratio, and regression, it underestimates the variance.

21.2.4 Replication Approaches

Replication/resampling variance estimation techniques for prototype estimators are computer-intensive but are becoming more popular as computers continue to become more powerful. They include the jackknife, the balanced repeated replication (BRR), and the bootstrap. The main advantage of these techniques is that the variance of complicated nonlinear estimators can easily be estimated without theoretical derivation of the variance formula. However, when applied naively to imputed data, they underestimate the variance. To correct the underestimation problem, two approaches have been proposed; adjustment and reimputation. The reimputation approach suggested by Ford (1983) and tried by Burns (1990) for hot deck imputation causes an overestimation for the jackknife technique, as shown in Rao and Shao (1992). However, the adjustment approach works well for the jackknife and BRR techniques. For the bootstrap, either an adjustment or reimputation is used, depending on the sample size. The *pq*-probabilistic path was implicitly employed for the adjusted jackknife and BRR, while the opposite *qp*-path was assumed by Shao and Sitter (1996) for the bootstrap. Chapter 20 gives detailed discussion of these approaches.

Rancourt (1999) proposed a way to apply the jackknife approach to nearest neighbor imputation, which has been shown empirically to work well, at least for a simple case. Chapter 20 discusses another approach that uses a partial adjustment and partial reimputation.

One of the drawbacks of the replication approaches is the difficulty in incorporating the finite population correction (fpc). The jackknife and the BRR approaches generally overestimate the variance for without-replacement sampling and appreciably so when the fpc is nonneglibile. Lee et al. (1995) provided an fpc-corrected adjusted jackknife variance estimator for a simple case. Steel and Fay (1996) also studied an fpc-corrected adjusted jackknife variance estimator for stratified simple random sampling. However, for more complex sample designs, the problem remains unresolved. The bootstrap approach has an opposite problem, where it generally underestimates the variance.

Sitter and Rao (1997) proposed a linearized form of the adjusted jackknife variance estimator for ratio imputation under simple random sampling. It does not require replication and is approximately model and design unbiased under the ratio model unless the RM is confounded. The fpc can easily be incorporated in contrast to the adjusted jackknife variance estimator.

21.2.5 Other Approaches

Tollefson and Fuller (1992) studied the variance estimation of the imputed HT estimator with hot deck imputation under a superpopulation structure. They assumed that the population is divided into imputation classes in which the MAR mechanism holds within imputation class.

Shao and Steel (1999) used the *qp*-probabilistic set-up for composite imputation wherein more than one imputation method is used and/or imputed values are again used for imputation of other variables

Assuming that the imputed estimator is unbiased, the total variance under the *qp*-path is given by

$$V(\hat{\theta}_I - \theta) = E_q V_p(\hat{\theta}_I) + V_q E_p(\hat{\theta}_I - \theta) \tag{21.6}$$

Note that E_q and V_q are defined at the population level and no longer conditional on the sample s. Shao and Steel (1999) considered particularly the imputed HT estimator under stratified sampling.

If imputation is composite but the imputed estimator can be expressed as a smooth function of estimated totals of the y-variable and the variables involved in imputation, then the variance components in (21.6) can be estimated using the usual technique (Taylor, Jackknife, etc.). The validity of the resulting variance estimator can be proved under the uniform within groups RM.

After conditioning on a particular response set at the population level, the response status of a population unit can be treated as a population characteristic and therefore, the fpc can easily be incorporated in the first component [i.e., in $V_p(\hat{\theta}_I)$]. Another advantage of this method (later referred to as the POP method) is that an estimate of the second variance component is robust since it does not depend on the RM or the imputation model. Moreover, the ratio of the second component to the first is $O(f)$, where f is the overall sampling fraction. Therefore, if f is small, the second term, which is more difficult to compute, can be ignored. Note, however, that the variance components in (21.6) do not correspond to the two variance components (sampling and imputation variances) given in (21.3) and, thus, they are less useful from the survey management point of view.

21.3 COMPARATIVE DISCUSSION OF THE VARIANCE ESTIMATION APPROACHES

21.3.1 Theoretical Comparisons

Separate Variance Components. In (21.3), the total variance is decomposed into two main components: sampling variance and imputation variance. Not all approaches provide separate estimates for these two components. The model-assisted approach (MOD) provides the components rather naturally. However, resampling and replication approaches, jackknife (JKNF), bootstrap (BOOT), and BRR, do not provide them. All others—two-phase (2PH), all-cases-imputation (ACI), linearized jackknife (LINJ), except POP—can provide estimates of the two components. POP gives separate variance components but their interpretation is different.

Minimization of the sampling variance is one of the key goals in designing an efficient survey. However, a more appropriate goal should be to minimize the total variance (including the imputation variance). This important point deserves more attention (see Gagnon et al., 1997). From the survey management point of view, it becomes necessary to estimate the two variance components separately, so that the survey resources can be allocated in a cost efficient manner.

Computational Burden. Even though computers are becoming more powerful, computational burden remains a concern when the data set is large. Approaches based on explicit formulae (MOD, 2PH, LINJ, and POP) are much less burdensome in computation. ACI requires more computation since imputation has to be done for the respondents. Replication approaches are even more computer-intensive, especially BOOT if reimputation is used. Ordering the approaches by level of computational burden, we have (2PH, MOD, LINJ, POP) → ACI → (JKNF, BRR) → BOOT.

A parallel can be made between the replication approaches and the multiple imputation approach. The replication approaches do not have much computational advantage over multiple imputation unless the number of multiple imputations is large, but they have the important advantage that they can be applied to situations where imputation is not proper (see Rubin, 1987 for a definition of proper imputation). In fact, any single imputation variance estimator can be used for multiply imputed data sets regardless of whether the imputation method is proper or not. This was pointed out by Shao et al. (1998) for BRR but applies for any other valid single imputation variance estimator.

Information Needs for Variance Estimation. Besides the usual information needed for the standard variance estimator, all approaches need additional information on the imputation status of each record, the imputation method, the imputation classes, and the auxiliary variables. In the case of nearest neighbor imputation, identification of donors is also needed, except for the ACI approach.

Adaptability. In general, the replication approaches can relatively easily handle variance estimation for various imputed estimators. They can also handle complex situations in which more than one imputation method is used for a given variable. However, there is some difficulty in incorporating the fpc, because these approaches do not separate the sampling variance and imputation variance, and the fpc must be applied to the sampling variance component only. For other approaches, a variance formula must be worked out for each imputed estimator according to the sample design and the imputation method. For complex situations, this derivation may be nontrivial but incorporation of the fpc is easy. POP appears to be the only approach that addresses the case in which imputed values are used for imputing other variables but BOOT can also handle this case if the imputation procedure can be replicated.

Response Mechanism. All approaches described in Section 21.2 have been designed to work at least for some form of the uniform within-groups RM. However, ACI and 2PH use a more restrictive form in which imputation classes cannot cross stratum boundaries. MOD, JKNF, and LINJ are applicable under the unconfounded RM, which is a weaker assumption than uniform within-groups RM. POP can be used under the unconfounded RM if the variance formula is developed using the model-assisted principle. Under the confounded RM, all approaches are sensitive to misspecification of the RM. Also, imputed estimators are biased and correction of the bias becomes a more pressing issue; cf. Rancourt et al. (1994).

Imputation Model. Although all imputation methods assume a model, either explicitly or implicitly, MOD uses the model assumption explicitly. As a result, it is more sensitive to deviations from the assumptions. On the other hand, if the model is correct, the approach provides a variance estimator with a smaller variance. All other approaches are based on quasirandomization or can be applied under the design-based framework and are robust to misspecification of the imputation model. JKNF is approximately design and model unbiased (ξp-unbiased) under a linear imputation model and uniform within-groups RM.

21.3.2 Empirical Comparisons

We briefly describe the Monte Carlo simulation study we conducted to compare empirically the reviewed variance estimators, and then present the results with discussion. For detailed description and discussion, see Lee et al. (2000).

As described in Lee et al. (1994), we generated two artificial populations of size 100—one following the ratio model (21.4 with $c_k = z_k$) and the other following the nonlinear regression model (of one auxiliary variable z) with a mild positive second-degree term (so the shape of the curve is concave). Simple random samples of size 30 were drawn and nonresponse was simulated with an expected nonresponse rate of 30% using three different RM's so that the response probability is (i) constant (uniform RM), (ii) decreasing with z (unconfounded RM), and (iii) decreasing with y (confounded RM). The imputed HT estimator is used. Two types of imputation methods were studied, namely, ratio (RAT) and nearest neighbor (NN).

The variance estimation formulae used in the simulation are given in Lee et al. (2000). For JKNF, we used the fpc-corrected formula proposed by Lee et al. (1995). We created two pseudoclusters to implement BRR by randomly dividing the sample into two equal-sized groups. Then we applied BRR (assuming that the number of strata is one and stratum sample size is two) with $L = 2$ balanced replicates. We repeated the procedure 50 times to stabilize the variance of BRR and applied the fpc-correction as for JKNF. We did not implement ACI for ratio imputation since it is not applicable. We also included, for comparison, the standard (ordinary) variance estimator (ORD) and the multiple imputation (MI) variance estimator with $M = 5$ and $M = 50$ for ratio imputation, where M is the number of imputations for each missing value.

For NN imputation, MOD and ACI have explicit formulae. For JKNF, we used the adjustment proposed by Rancourt (1999). We also used the same adjustment for BRR and BOOT. For other approaches we used the same formulae used for ratio imputation and so the resulting variance estimators are negatively biased, even under the ideal situation (ratio population and uniform RM). We suppressed this latter part of the results in the presentation to save space.

The sampling experiment was repeated 50,000 times. Three measures were used to evaluate the performance of the variance estimators: (i) the relative bias (RB), which is the average of the 50,000 variance estimates compared on a relative basis to the Monte Carlo MSE of the point estimator; (ii) the root mean squared error (RMSE) of the 50,000 variance estimates; and (iii) the coverage rate (COVR) of the 50,000 confidence intervals computed at a nominal 95% level using the normal score 1.96. With this number of iterations, the Monte Carlo error is expected to be about 1.5% for RB.

The purpose of the simulation was to evaluate the approaches under the ideal situation and to see how sensitive they are to violations of the imputation model and/or RM assumptions. In the following, we discuss the results presented in Table 21.1 for each combination of imputation method and RM under each of the two populations. The results for multiple imputation are discussed separately. As expected, ORD greatly underestimated the variance (usually by over 30%). For this reason, we exclude ORD from the following discussion.

Under the Ratio Population. *RAT-Uniform.* All variance estimators worked well; their RB's lie within ±10%. For JKNF and BRR, the small positive RB's (4.6 and 5.2, respectively) could have been much larger if the fpc had not been used. BOOT underestimated slightly (–6% RB), as expected since the bootstrap procedure incorporated the fpc, which was also automatically applied to the imputation variance unnecessarily. In other words, the variance was overcorrected. RMSE's are fairly close to each other for all approaches except BRR, perhaps due to the way it was implemented. BRR is meant for stratified sampling, but we created pseudoclusters for a single stratum design. All approaches have fairly good COVR's, achieving close to the nominal 95%.

RAT-Unconfounded. In terms of RB, all approaches performed well, having RB within ±10%. They also have very good COVR's. JNKF and BRR have a somewhat larger RMSE. The RB of 2PH would have been much larger (RB = –14.8%) if we had not used the improved formula derived in Lee et al. (2000).

RAT-Confounded. All variance estimators are severely biased in estimating the MSE, with RB ranging from –34% to –22% because the point estimator is (negatively) biased (–7.7% of the population total). This confirms the well-known fact

Table 21.1. Simulation results for two populations (ratio and concave) and two imputation methods (ratio and NN)

	Uniform			Unconfounded			Confounded		
Approach	RB	RMSE[a]	COVR	RB	RMSE[a]	COVR	RB	RMSE[a]	COVR
	Ratio Population – Ratio Imputation								
ORD	–30.9	25.2	88.6	–39.8	35.4	86.7	–56.7	54.7	76.4
2PH	2.0	22.5	94.2	–8.6	25.7	93.4	–33.7	38.1	84.5
MOD	7.0	23.5	94.9	4.9	29.4	95.0	–24.3	33.8	86.6
JKNF	4.6	25.9	94.4	8.8	35.5	95.1	–22.7	36.0	86.6
LINJ	2.4	23.2	94.2	–0.6	28.9	94.3	–28.3	36.2	85.7
BRR	5.2	31.7	93.5	9.4	41.7	94.3	–22.2	39.9	86.0
BOOT	–6.2	25.3	92.5	–8.1	31.5	92.9	–33.9	41.1	83.8
POP	1.3	22.7	94.1	–1.8	28.3	94.2	–29.2	36.5	85.5
MI (5)	12.1	37.6	94.8	12.4	53.2	94.8	–13.0	47.1	87.7
MI (50)	12.3	26.9	95.4	13.4	36.2	95.9	–13.9	32.7	88.5
	Ratio Population – NN Imputation								
ORD	–35.0	34.3	87.3	–44.8	48.8	84.3	–59.5	68.5	73.8
MOD	6.6	33.4	94.7	12.3	53.6	95.2	–20.6	47.0	87.3
JKNF	–2.5	33.4	93.4	–3.7	44.8	93.3	–30.1	49.4	84.7
BRR	–2.4	38.9	92.5	–3.3	49.8	92.7	–29.7	52.7	84.0
BOOT	–14.3	32.4	91.3	–19.2	41.6	90.6	–40.8	54.8	81.5
ACI	–4.4	29.5	93.2	–21.4	37.6	90.6	–42.4	53.7	81.3
	Concave Population – Ratio Imputation								
ORD	–22.0	32.2	90.7	–33.7	54.6	89.5	–30.0	41.5	87.7
2PH	–1.4	30.3	93.8	–15.2	43.2	93.5	–9.9	34.5	91.2
MOD	12.8	37.7	95.2	1.9	47.3	95.8	8.9	42.1	93.6
JKNF	1.0	36.9	93.7	–2.0	53.6	95.1	–0.1	44.4	92.0
LINJ	–1.2	31.2	93.8	–11.6	44.8	94.0	–7.1	36.4	91.5
BRR	1.2	46.6	92.8	–1.3	63.8	94.2	0.4	53.2	91.4
BOOT	–8.7	36.8	92.2	–17.4	52.2	92.4	–13.7	42.2	90.0
POP	–1.7	30.8	93.7	–12.2	44.5	94.0	–7.7	36.0	91.5
MI (5)	11.3	52.0	95.1	–2.8	70.4	95.2	12.4	62.3	93.9
MI (50)	11.7	39.5	95.6	–3.0	50.8	95.8	12.9	45.7	94.4.
	Concave Population – NN Imputation								
ORD	–31.9	44.0	87.7	–40.9	60.7	85.1	–47.6	69.8	81.1
MOD	13.15	47.5	95.1	28.8	95.0	95.7	4.3	63.9	92.2
JKNF	–12.9	39.5	91.4	–14.3	50.4	90.9	–26.5	55.2	86.7
BRR	–12.8	47.8	90.5	–14.1	57.5	90.1	–26.4	60.7	85.8
BOOT	–19.9	42.2	89.9	–24.9	53.1	88.5	–34.6	60.4	84.6
ACI	–8.6	33.8	92.4	–22.8	45.9	89.6	–31.3	53.4	86.0

[a]The RMSE's are in hundreds.

that treatment of nonresponse is very sensitive to the difference between unconfounded and confounded RM's.

NN-Uniform. MOD, JKNF, BRR, and BOOT designed for NN imputation explicitly or implicitly worked well with a negligible negative bias, except for BOOT with RB of –14% due to over-correction of the fpc. The replication approaches tend to have a larger RMSE. COVR's are all reasonably good.

NN-Unconfounded. MOD, JKNF, and BRR have a small RB (within ±5%). MOD has a moderate positive bias, whereas others (BOOT and ACI) have substantial negative biases. COVR's are slightly on the low side, but still acceptable. Theoretically, BOOT suffered from the same problem mentioned in NN-Uniform.

NN-Confounded. All approaches are negatively and heavily biased in estimating the MSE of the biased point estimator, as for the case of RAT-Confounded. The overestimating tendency of MOD helps reduce the bias, but it is still severe at –21% RB.

Under the Concave Population. Note that this population represents a mild violation of the assumed imputation model.

RAT-Uniform. Approaches that are less dependent on the imputation model (2PH, JNKF, LINJ, BRR, BOOT, and POP) performed well, except BOOT. BOOT did not fare well because of the problem noted above. MOD has a moderate positive bias but an excellent COVR. The other approaches also have fairly good COVR's with over 92%. RMSE of BRR is again strikingly large.

RAT-Unconfounded. MOD, JNKF, and BRR are nearly unbiased, while others are negatively and moderately biased. COVR's are still good for all the approaches but RMSE's are generally higher than other cases.

RAT-Confounded. All approaches except BOOT worked well, in striking contrast with the results for the same RM under the ratio population. Two incorrect assumptions (RM and imputation model) combined to create an extreme case. COVR's remained over 90%.

NN-Uniform. The bias is more visible and mostly negative, except that of MOD. All COVR's are around or over 90% but fall somewhat short of the nominal value of 95%, except for the MOD, which has a COVR of 95%.

NN-Unconfounded. The bias deteriorates further in the same direction as under the uniform RM. MOD still has an excellent COVR and others have around or slightly lower than 90% COVR.

NN-Confounded. The bias is negative and unacceptably large for all approaches except for MOD, which performed surprisingly well with only 4% RB and 92% COVR.

Multiple Imputation (MI). Being model-based, the simulation results for the multiple imputation variance estimator are somewhat similar to those of MOD, in terms of RB. As the multiple imputation theory suggests, COVR is good under uniform RM and unconfounded RM. Its RMSE is substantially higher than those of the single imputation estimators when $M = 5$ but becomes more or less the same when $M = 50$.

21.4 RECOMMENDATIONS AND CONCLUSIONS

Because of its large bias, the ordinary variance estimator (ORD) must be ruled out, unless the nonresponse rate is very small. All the approaches to variance estimation studied in the simulation performed well under the conditions for which they are designed. However, the approaches reacted in different ways to deviations from these conditions. BOOT underestimates the variance because automatic incorporation of the fpc in the procedure causes overcorrection.

All approaches seem to be robust to a mild violation of the particular type of imputation model assumption we studied. MOD showed some sensitivity to this assumption, but the results under the Concave population are acceptable. However, the approaches may behave very differently under different type of violations of the imputation model assumption.

When the assumption about the RM is violated, all approaches can fail, which is especially noticeable when the true RM is confounded. 2PH and ACI, developed for the uniform RM, are particularly sensitive to the truth of this assumption. Thus, the user should be vigilant in choosing an RM.

It should be noted that the bias criterion we used is with respect to the MSE of a point estimator. In fact, all approaches worked fairly well with small to moderate biases in estimating the variance of the point estimator, even under the confounded RM.

If computational burden is an issue, then JKNF, BRR, and BOOT should be avoided. However, if the imputed estimator is a complicated statistic, then these approaches have an advantage of not requiring explicit derivation of variance formula. For NN imputation, the approaches studied appear to work well. In particular, the adjustment used for JKNF and BRR worked well.

All approaches examined in the simulation yielded fairly good coverage rates, except when the bias of the point estimator is severe, as it is in most cases for the confounded RM. However, the coverage rate was on the low side in most cases.

All approaches almost always require flagging imputed values on the data file as well as specification of the imputation method(s), the imputation auxiliary variables, and the imputation classes. For NN imputation, all approaches except ACI require identification of the donors. It is a good practice to always make donor information available along with the other necessary information.

It is generally of great interest to survey managers, for continuing surveys in particular, to obtain separate estimates of the two variance components, the sampling variance, and the imputation variance. This will facilitate an efficient allocation of resources among the different survey processes. Therefore, the survey manager should seriously consider this aspect in choosing an approach to variance estimation. The approaches in Sections 21.2.1 and 21.2.2 are particularly suitable from this perspective. When such a choice is not feasible (e.g., dictated by the available software), one should still strive to assess the relative size of both variance components. For example, one can get the sampling variance component separately (e.g., using the standard variance estimator with stochastic imputation or NN imputation); the imputation variance component follows easily by subtraction from the total variance estimated by the selected approach.

ACKNOWLEDGMENTS

This paper is an abridged version of Lee et al. (2000) presented at the International Conference for Survey Nonresponse, Portland, 1999, which provides much more detailed discussion on the subject. We are very much indebted to Jean-François Beaumont for his active participation in producing all simulation results and for his comments. We also thank John Eltinge, Thomas Belin, and Rod Little for their useful comments.

CHAPTER 22

Large-Scale Imputation for Complex Surveys

David A. Marker, David R. Judkins, and Marianne Winglee,
Westat, Rockville, Maryland

22.1 INTRODUCTION

There are two main challenges in the field of imputation. One is to maximize the use of available data to both minimize mean square error for univariate statistics and to preserve covariance structures in multivariate datasets despite complex patterns of missing data. The other is to reflect the uncertainty caused by item nonresponse in variance estimates from imputed datasets. This paper discusses work toward meeting the first challenge.

Hot deck imputation methods have long been used at large-scale data publishers (Kalton and Kasprzyk, 1986; Brick and Kalton, 1996). A simple hot deck involves defining imputation cells based on a cross-classification of variables (referred to as a partition on the dataset), randomly matching donors and beggars (cases with and without values for the target variable being imputed), and transferring values from donors to beggars. Some of the advantages of simple hot decks are reasonably low cost, reduced nonresponse bias on univariate statistics, reduced variance due to use of partially complete records, univariate plausibility, ease of use by secondary analysts, and cross-user consistency.

Cost is low because there is no model fitting of the variable to be imputed. The partitions used in hot decks are often based on a priori intuition by the user. Nonresponse bias is reduced to the extent that variables used in the partitions are associated both with the propensity not to respond and with the characteristic itself. When using imputed datasets, variances are reduced (compared to complete-case analysis methods) because all cases are used in each analysis. This reduces the variance for X in particular if it is seldom missing while Y is often missing. By univariate plausibility, we mean that every imputed value is within the range of values that were observed to have actually occurred for that variable. By cross-user consistency, we mean that when user A cross-tabulates X with Y and user B cross-tabulates X by Z, both users get the same marginal mean for X. This cross-user consistency does not hold when users apply available-case methods of analysis. These advantages of cost, reduced bias, reduced variance, univariate plausibility, ease of analysis, and cross-user consistency have been powerful motivators to keep using simple hot deck methods.

Despite these advantages, hot deck users have been dissatisfied with simple hot decks for three reasons. First, simple hot decks are frequently not optimized for each target variable. It is not uncommon for a single partition to be set up in terms of simple frame and demographic variables and then for this single partition to be used to impute a large number of target variables. This keeps the cost of imputation low but neglects the fact that the best set of predictors will vary by target variable and might include some partially reported variables. Second, although the set of values imputed for each variable may be plausible when considered in isolation, the imputed values may be nonsensical when used in cross-tabulations. This latter phenomenon is referred to as attenuation of association (Kalton and Kasprzyk, 1986). Third, naive variance estimates on imputed datasets are biased downward, leading to confidence intervals that are unduly narrow.

In the last decade, considerable progress has been made in developing new hot deck procedures to improve univariate mean square error and avoid attenuation of association. Naturally, the cost of imputation has risen as these new procedures have been developed and applied, but this extra cost has led to major benefits, as will be illustrated in the examples that follow. Independently, other design-based statisticians have been working on postimputation variance estimation methods, but the two efforts have not been synthesized.

Other imputation methods have also seen active development work. In particular, nearest-neighbor (NN) techniques and Bayesian methods have seen active development work. NN techniques are closely related to hot decks but they more fully utilize continuous covariates and leave little or no randomness in the imputation process. NN techniques involve defining a distance metric on the covariate space and then transferring the value of the target variable to a beggar from the nearest donor. When the metric is based on a parametric model for the target variable, it is often referred to as predictive mean matching. Hot deck methods and NN techniques can be referred to collectively as semiparametric techniques. In this chapter, we contrast semiparametric methods with Bayesian methods. Although the Bayesian methods do have considerable advantages for some applications, we think that hot deck or NN methods are a better solution for most imputation problems. In Section 22.2, we develop the arguments about the limitations of Bayesian imputation methods and why semiparametric techniques are more useful for many applications. In Section 22.3, we review some of the hot deck techniques that have been developed for multivariate imputation. In Section 22.4, we review three applications. In Section 22.5, we offer our views for future research and development.

22.2 BAYESIAN METHODS VERSUS HOT DECKS AND NEAREST NEIGHBOR TECHNIQUES

Bayesian methods of imputation involve Monte Carlo Markov chain (MCMC) methods to simulate the posterior distribution of vector parameters given the data, a model, and a prior distribution. They can be used on all parametric distributions. It is possible to use them to estimate the joint posterior for a mixed vector of categorical and continuous variables. Also, it is possible for the model to reflect not just fixed population parameters, but random parameters, thereby allowing the preservation of both fixed structure and components of variance. Bayesian methods allow for the preservation of a large number of low-order associations. There is no need to categorize continuous covariates. Bayesian methods may be compactly and transparently described to sophisticated statistical audiences. Furthermore, Bayesian methods can be used to estimate variances for statistics based on partially imputed data in a way that at least roughly reflects the uncertainty due to imputation and having to estimate the components of variance. Clearly, this is an attractive set of properties.

In contrast to Bayesian methods, hot decks do not allow random effects and cannot preserve low-order associations without also trying to preserve associations of all orders (thereby reducing the number of low-order associations that can be effectively preserved). Continuous predictors must generally be categorized (although one can always use one continuous predictor within cells to define distance between donors and cases with missing values) and it is sometimes difficult to transparently describe the procedures used to deal with cells with small sample sizes. Nearest-neighbor (NN) techniques are more like Bayesian methods in being able to use multiple continuous predictors and having more transparent descriptions. However, the methods of variance estimation that have been developed for hot decks and NN techniques can only be used on univariate statistics.

However, there are also weaknesses in the Bayesian methods. Most fundamentally, Bayesian methods require that an explicit parametric model be formulated and fit for each target variable. Some common variables in surveys such as personal income do not appear to follow any known distributions. Transformations can help with skewness, but they are useless for discontinuities in the distribution. These discontinuities can be caused by measurement error, but they can also represent real phenomena. Semiparametric methods handle discontinuities beautifully. If there is a point mass in the distribution (e.g., number of hours worked per week), hot decks and NN techniques will impute a mass to the point. If there are natural gaps and limits, hot decks and NN techniques will never impute values in the gaps or outside the limits. These are critically important properties in which semiparametric methods perform better than Bayesian methods.

Second, Bayesian methods require the specification of prior distributions on parameters. Under certain circumstances, improper priors may be assigned to some parameters. Recent research has shown that it is necessary to assign proper priors to the parameters for the components of variance in a mixed model for a binary variable; otherwise, the posterior distribution may not be proper (Natarajan and McCulloch, 1995 and Ghosh et al., 1998). Furthermore, even if it is permissible to assign improper priors to the fixed effects in a mixed model, there is no such thing as an uninformative prior. A prior that is flat for β will not result in a flat prior for β^2. We think it best in official government statistics to minimize the use of priors. From our point of view, although the parametric models and prior distributions on parameters make the imputation process transparent to professional statisticians, these same features are baffling to most consumers of government statistics. For the majority of users, the simple procedures of grouping or matching similar cases and transferring values from donors to beggars are far easier to describe.

Third, Bayesian methods are difficult to apply unless one assumes that all the variables to be imputed have normal distributions. For that very special case, Schafer (1999) distributes software called NORM. For all other parametric families, custom software must be developed. The advent of BUGS (Spiegelhalter, et al. 1996) makes this task less difficult than it used to be, but it is still the case that programming Bayesian methods is more difficult than programming hot decks. This can lead not just to higher costs for the imputation task but can also lengthen the time needed to produce imputed datasets. NORM can lead to quick imputations for multivariate normal problems, but the effort to develop parametric models that capture the discontinuities in real distributions is daunting. That is not to imply that multivariate semiparametric methods that preserve covariate structures are quick and inexpensive to create. They can also be expensive and require months of lag between data collection and data publication. Still, it is our judgment that semiparametric methods are less expensive to implement and once created require less maintenance and can be carried out by less-skilled personnel in repeated surveys.

Fourth, Bayesian methods can require significant computing capabilities. While computing power continues to grow less expensive, it is still the case that MCMC methods on long vectors require massive amounts of computation. The multivariate semiparametric methods discussed here are also computer intensive, but less so by orders of magnitude.

Fifth, the variance estimates produced by Bayesian methods do not meet the claims that are made about them except in simple situations. Rubin (as clarified in 1996) laid out the conditions for "proper imputation." If a proper imputation method is used, then multiple imputation may be used to construct confidence intervals that are frequency valid. Work by Binder and Sun (1996) suggests that finding proper methods may be extremely difficult when complex designs are used unless the imputer's model for the variable being imputed is correct. That appears to be a strong additional assumption. Schafer (1997, p. 145) has abandoned Rubin's concept of proper imputation entirely, substituting the concept of Bayesian proper imputation. As noted by Judkins (1998), while Schafer's concept appears to be easy to satisfy, it has not been shown to lead to any desirable frequentist properties such as consistency.

Further complicating the question of the existence of proper imputation methods, Fay (1992) has demonstrated that a method that is proper for some analyses is not proper for others. As emphasized in Judkins (1996), Rubin's concept of proper imputation must be reevaluated for each specific analysis. Meng (1994) lays out conditions under which multiple imputation will lead to valid inference for unplanned analyses, but these conditions appear too complex for almost any secondary analyst to assess.

The only situations in which multiple imputation has been clearly demonstrated to lead to valid inference are planned analyses of stratified simple random samples. From these simple, albeit encouraging, demonstrations, hopes are raised that Bayesian imputation methods can lead to valid inference for a broad range of unplanned analyses on surveys with complex designs. For example, in work by Schafer, et al., (1996), an artificial population was built out of several years of survey reports. Stratified simple random samples (without clustering) and MAR (missing at random) structures were then simulated using the conjoined sample as a test population. Multiple imputation performed well for planned analyses on this simple structure. This, however, leaves open the question of how it performs on clustered datasets (the rule for all national demographic studies) and how it performs for unplanned analyses.

We conclude that the problem of variance estimation for unplanned analyses on surveys with complex designs is not solved whether one uses Bayesian methods or semiparametric methods. In both cases, more research is required to develop good variance estimation procedures. In Section 22.5, we discuss the directions that we feel are most promising for postimputation variance estimation. Even though such methods have not been developed yet, we believe that it is still important to provide general-purpose imputations that will preserve covariate structures.

22.3 MULTIVARIATE HOT DECKS

Along with our colleagues, we have developed a number of multivariate hot deck techniques. These include the n-partition hot deck, the common-donor hot deck, the cyclic n-partition hot deck, and the full-information common-donor hot deck. These are all built around a simple hot deck method that can be programmed in macro form. For this method, donors are randomly matched with beggars within the cells of a user-defined partition, and the reported value of the target variable is copied from donor to beggar. If a particular cell has no donors, then an automated procedure is specified to "reach" across cell boundaries to find a donor. (Reaching is also sometimes used for a thin donor pool, where the there are some donors, but more beggars than donors.) If variables $Z_1, \ldots, Z_k$ define the partition, and no donor is available within a cell, then the algorithm will first weaken the match on Z_k to find a donor. (Nearest neighbor imputation results if Z_k is continuous.) If weakening the match on Z_k does not lead to a donor, then the macro also relaxes the stringency of the match requirement on Z_{k-1}, and so on.

This reaching procedure forces each run to successfully identify donors for all beggars, thereby making it ideal for a production environment. By ordering both the $\mathbf{Z}$ vector and the values of each component of the $\mathbf{Z}$ vector, the user has considerable control over the reaching procedure. The choice of the partition is often driven by the analytic objectives of the study rather than being driven by an exploration of the data. While it is possible to conduct exploratory analyses of the database to inform choices about the partition, such analyses can add greatly to the cost and time required to complete the imputation project.

The n-partition hot deck uses a separate hot deck for every variable. The cells of each partition consist of completely reported variables and previously imputed variables. Partially reported variables can be used to define the partitions but it should be noted that this can cause difficulties if a missing value for one of the predictor variables is associated with a high rate of missing data for the target variable. This can easily occur if many interviews are broken off part way through. This method is excellent for reducing the mean-square error on marginal means while preserving the shape of marginal distributions, but weak on preserving covariance structures beyond those between the target variables and the partition variables.

The common-donor hot deck uses a single hot deck for a group of variables. Within the cells of the partition, donors are randomly matched with beggars and the entire target vector is copied from the donor record to the beggar record. If the beggar record is partially reported, the user may choose to either impute over the reported elements of the vector or to just copy over the missing portions. If the vector is imputed over, then reported data are lost, thereby increasing variances and the risk of nonresponse bias. If only the missing portions of the vector are filled in, then the imputed and reported portions of the vector may be quite inconsistent with each other.

As discussed in Judkins (1997), the cyclic n-partition hot deck (CNP hot deck) involves an iterative cycling through n-partition hot decks. On the first iteration, simple methods are used to fill in starting values for the imputation vector. On second and subsequent iterations, any variable may be used to help define the partition for any other variable, since all variables do have valid values. If any of the variables being imputed have continuous distributions, then rules for categorizing them must be worked out in advance. (An obvious alternative would be to develop cyclic nearest-neighbor procedures based on predicted values from parametric models for the target variables.) This method was inspired by Bayesian methods but retains the semiparametric features of the hot deck. No strong assumptions are required about distribution shapes or about prior distributions for parameters. Instead, deliberate choices are made about which features of the covariance structure deserve the best preservation efforts.

The full-information, common-donor hot deck (FICD hot deck) was first described in Judkins et al. (1993), but the name suggested here is new. This method uses a different common-donor hot deck for every distinct pattern of missingness in the target vector. Each common-donor hot deck uses a partition that is defined by all the reported elements of the vector and possibly additional covariates. This type of approach had been hinted at in Fahimi et al. (1993) and in Winglee et al. (1993). However, in the Fahimi paper, average values were substituted into variables with values that had not yet been imputed, and in the Winglee paper, missing values were treated as legitimate match categories. The former approach did not prevent anomalies and the second approach led to thin donor pools.

The FICD hot deck has the nice feature of not requiring any cycling and thus avoiding questions of convergence. However, if the target vector is very long, the number of distinct patterns of missingness can be very large, thereby requiring a large number of hot decks to be run. The CNP hot deck requires far fewer hot decks, with just one per element of the target variable per iteration. With fewer distinct hot decks, it is feasible to put more thought into the best partition for each element of the target vector.

22.4 THREE LARGE-SCALE IMPUTATION EXAMPLES

Examples are provided from three complex surveys demonstrating some of the complex issues faced by large-scale imputation. The three surveys are the 1994 National Employer Health Insurance Survey, the 1992–1996 Medicare Current Beneficiaries Survey, and the IEA (International Association for the Evaluation of Educational Achievement) Reading Literacy Study. In all three applications, it was assumed that nonresponse was ignorable (Little and Rubin, 1987), as do most Bayesian MCMC methods.

22.4.1 National Employer Health Insurance Survey (NEHIS)

The 1994 National Employer Health Insurance Survey (NEHIS) collected information on the health insurance plans offered by 40,000 private-sector establishments and governments, and 50,000 health insurance plans. More than 100 variables were collected for each establishment and each health plan. Fifty of these variables were selected for imputation. Restrictions placed on borrowing strength across different types of plans resulted in the need to set up almost 150 separate imputation procedures. There were dozens of potential covariates that could be used and there was the desire to support a broad range of secondary analysis. It was therefore decided to conduct a fairly thorough exploratory analysis of each of the target variables to detect the most important covariate structures and then use this knowledge to shape the partitions accordingly.

The item response rates for the150 variables to be modeled varied from 99% to 25%; but in almost all cases the response rates were above 70% (see Table 22.1). Even though the government did not plan to publish estimates for the few variables

Table 22.1. Item response rates for imputation variables

Response rate	Percent of variables
95–100%	37%
90–95%	21
85–90%	13
80–85%	8
75–80%	9
Below 75%	12

with low response rates, it planned to use them in a variety of modeling efforts, since no other source exists for this information.

As with many complex surveys, variables were measured at different levels. For NEHIS, data were collected at the firm (corporation) level, establishment level, the health insurance plan level, and the plan within establishment level. It is important to retain this structure in the imputed data. Further complicating the imputation, the data were subject to numerous logical consistency requirements. Also, many of the target variables had limited allowable ranges.

Methods. The basic approach taken to imputing the 150 variables was the *n*-partition hot deck. Generally, variables with low rates of missing data were imputed prior to other variables. Also, variables that were seen as causally prior to others were imputed in assumed causal order. Variables that had been previously imputed became eligible to be used as predictors for subsequent hot decks.

Examination of bivariate correlations and patterns of missingness were used to identify best covariates for each hot deck. Highly correlated covariates with low missing rates were chosen to define the hot deck cells. Continuous covariates were split into three or five categories. For most of the hot decks, the variables defining the partition were all categorical. However, for some variables, a single continuous predictor was admitted as the final variable in the string of predictor variables defining the partition. The automatic reaching feature of the macro described above then resulted in a nearest neighbor imputation within the cells defined by the rest of the predictor variables. This procedure has the advantage of easy implementation and its results are comparable to those obtained from regression imputation with randomly added empirical residuals (Aigner et al., 1975).

Groups of variables were set up that could be imputed based on the same vector of covariates. Some groups were imputed using the common-donor hot deck. Variables missing only one or two percent of the time were deterministically imputed by mean or modal imputation within cells. Given the low rate of missing data, the resulting distortion of the univariate marginal distributions for these variables was thought to be acceptable. Of the approximately 150 imputation models implemented in the NEHIS imputation task, 60% used hot decks with all categorical predictors, 30% used hot decks with several categorical variables and one continuous predictor, and 10% used deterministic imputation. For full details see Yansaneh, et al. (1998).

The various consistency requirements discussed above necessitated both an edit–impute cycle and an edit–construct cycle. The dataset before imputation was edited, missing values were imputed, and then the imputed data were edited again. Any values that failed edits and were set to missing were then reimputed. Similarly, during the course of imputation, an impute–construct cycle was implemented for sets of variables with algebraic relationships that needed to be maintained; once a value was imputed, others could be logically constructed using that value. These constructed variables were then subject to their own edits.

Discussion. The imputation procedure resulted in a database with 50 variables that can be analyzed by secondary analysts using complete-case methods. This significantly increases the utility of the resulting dataset for multivariate analyses by providing consistency across tabular analyses. We believe that the risk of nonresponse bias on marginal means has been sharply reduced. Attenuation of key relationships has been minimized. For example, after imputing enrollment in family coverage for over 10% of health plans, its correlation with overall enrollment (single plus family coverage) remained at 0.98.

The final database contains flags for imputed values and an indication of how the imputed value was created (from a donor versus resulting from an edit constraint). The imputation procedures are quite straightforward and easily described in documentation in terms of the partition used for each target variable, so users can decide on their appropriateness for their analyses. This approach, however, just like Bayesian methods, was labor-intensive and time-consuming. The large effort was a result of the number of imputation variables combined with the very complex logical and edit constraints imposed on the resulting data.

22.4.2 Medicare Current Beneficiaries Survey (MCBS)

The main activity of the MCBS is to collect data on about 660,000 health care events per year on about 12,000 beneficiaries. These events can be doctor visits, inpatient hospitalizations, dental visits, purchases of prescription drugs, purchases of adult diapers, and so on. For each event, interviewers attempt to collect the cost and a complete record of payments by the patient and all third parties.

Table 22.2 shows the level of partial data that is being preserved by imputation. It is important to have an imputation system that can either impute payments given a known cost or impute cost given partial payment data. By imputing costs and payments, it was possible to preserve the partial data present on 51% of events.

It is interesting to note that among the 66% of noninstitutional respondents with medical events who should have some knowledge of their medical expenses (that is, they were alive and eligible at the end of the year, not eligible for Medicaid at any time during the year, and not covered by capitated payment plans during the year), just 3.8% were able to give a complete cost and payment record for every event

Table 22.2. Frequency of edited and imputed cost and positive payment data level of missingness

All events	Events
Complete data	34%
Cost reported but payments incomplete	22%
Cost missing and payments incomplete	29%
All cost and payment data missing	15%
Event count[a]	561,655

[a]Excludes events of persons in facilities entire year and placeholder records for persons eligible for Medicare in 1996 but not drawn into the sample in time for collection of event-level data. Also, multiple purchases of same prescription medicine within 4 month period not counted separately.

they reported during the year (or was reported to HCFA by a provider). (This is not a typographical error: at least some data were missing on at least one event for 96% of the best reporters in the sample. Other persons with some events during the year had prefect reporting rates below 1%.) Clearly, a pure weighting strategy for handling missing data would not work on this survey. Without imputation, there would be no point in conducting medical expenditure surveys.

Methods. The imputation procedure has two major steps. The first step uses FICD hot decks to impute the vector of parties making positive payments related to a specific health care event. The second step uses cyclic n-partition hot decks and other procedures to impute the cost of the event and the actual payments, consistent with the vector of known and imputed payers. Consideration was given to applying the FICD hot decks to the payment amount and cost vectors, but the number of hot decks required for distinct payer patterns and missingness in the payment vector was considered to be too large. Instead, three systems were developed. For events with no cost or payment data, a simple n-partition hot deck was first run to impute cost and then a common-donor hot deck was run to impute the cost payment vector conditioned on cost. For events with partial payment data, a cyclic n-partition hot deck was run with some special features to keep payments additive to cost. Finally, a separate system based on expert knowledge was used for dual Medicaid–Medicare beneficiaries, since as a group they were mostly unable to provide any information on costs and payments. The details are in Judkins et al., (1993) and England et al. (1994). However, a large team at Westat and HCFA contributed to the effort. (Some of the team members were Frank Eppig, Kim Skellan, Dave Gibson, John Poisal, Gary Olin, Mary Laschober, Ian Whitlock, and Diane Robinson.)

Discussion. The algorithm has a number of desirable properties. First, all payments were imputed to be positive and to sum to the cost of the event. This would not have been easy to achieve with parametric model-based methods because of the strongly nonsmooth nature of the distribution of payment amounts. The most common payment is $0 for most payers. Second, all imputed payments were consistent with the set of known and potential payers. This means that if a person did not have a certain type of coverage, then a strictly positive payment by that source was never imputed. Also, if a person mentioned that a particular source had paid but they did not know the amount of the payment, then a strictly positive payment amount was imputed. Payments were never imputed for insurance or programs for which the beneficiary had no coverage at the time of the event. Third, no postedit partial cost or payment data were discarded. As shown in Table 22.2, this resulted in a considerable saving of partial data.

Fourth, the method was computationally feasible on the available computer hardware. Fifth, the method has required very little manual review and retraining. There was a large expenditure of professional labor to create and test the system in 1993 through 1995, but since that time there has been very little professional labor. Sixth, similar events belonging to the same person were imputed to have similar payer participation patterns. Seventh, the marginal distribution of costs by nature of event seemed reasonable. Eighth, the cost sharing arrangements that were imputed mostly echoed observed patterns.

22.4.3 IEA Reading Literacy Study

The aim of the IEA Reading Literacy Study was to assess school children's reading proficiency. The U.S. component of the IEA Reading Literacy Study was conducted in the 1990–1991 school year. The three-stage study (school districts, schools, and classes) involved national samples of over 6,000 grade 4 and 3,000 grade 9 students. Sample students were given tests to evaluate their reading levels and comprehension. In addition, the students, their teachers, and their school principals completed questionnaires about background factors related to the students' reading achievement. The amount of missing data for each item was usually small. About 80% of questions on the fourth grade student questionnaire (over 90% for ninth graders) had item nonresponse rates of 10% or less. Teacher and principal item nonresponse rates were even lower.

Methods. In this study, a combination of methods similar to the *n*-partition hot deck and the common-donor hot deck were used. Most of the variables used in the partitions were naturally categorical. The items in the questionnaires were imputed sequentially, following roughly the logical sequence of the questionnaires. The imputed values of some variables were used in the subsequent imputation of other variables. For a few particularly tricky relationships such as maternal and paternal education levels, FICD hot decks were used.

Discussion. As part of methods research on imputation for the IEA, an analysis was conducted with several different methodologies. For full details, see Winglee et al. (1994, and in press). Regression models, estimated from the data set completed by imputation, were compared with the corresponding models estimated using three other methods of handling the missing data that do not involve imputation: the complete case analysis (CC), in which cases with missing values for any of the variables involved in the analysis are discarded; the available case (AC) analysis, in which all the reported data are used to derive the sample means and variance–covariance matrix employed in the regression analysis; and the EM algorithm, in which a maximum likelihood method of estimation is used (Dempster et al., 1977).

The CC analysis is easy to implement but clearly inefficient in its use of data. In the regression models examined in this paper, almost a third of the sampled students were discarded. The CC approach assumes that the complete cases are completely missing at random (Little and Rubin, 1987), always a strong assumption. For this data set, there is clear evidence that the students with one or more missing values of the predictors in the regression models differ from those with complete data in terms of: reading performance; race/ethnicity of the student, the type of community served by the school (urban, suburban, nonurban), the region of the country, and the type of school (public, private). As a result, the regression analyses conducted for the complete cases are likely to have produced biased results. An example of this

Table 22.3. Selected unweighted regression coefficients predicting narrative scores

	Hot deck imputation		EM algorithm		Available cases		Complete cases	
Predictor variables	*b*	s.e.	*b*	s.e.	*b*	s.e.	*b*	s.e.
Intercept	744.7	22.4	731.0	22.6	726.5	25.0	729.3	27.9
Gender	16.9	2.3	17.2	2.3	17.1	2.5	16.4	2.7
Minority	–36.0	2.7	–36.0	2.7	–36.3	2.9	–36.3	3.3
Father's education < college	15.3	5.0	17.4	5.0	18.0	5.5	16.9	6.0
Father's education college	23.3	4.6	25.4	4.6	25.7	5.1	28.2	5.6
Father absent	18.1	7.5	20.2	7.2	18.7	8.0	31.6	10.3
Family wealth index	9.3	1.2	8.6	1.3	8.5	1.4	7.2	1.5
Extended family	–23.0	2.4	–24.1	2.4	–23.8	2.7	–27.6	2.9

bias is the regression coefficient for "Father absent" in predicting narrative scores shown in Table 22.3. Whereas the other three methods all estimate the coefficient between 18 and 20, the complete cases estimate is 31.6, over 50% higher than any of the other three methods. This increase is greater than a standard error away from the other coefficients. The complete case coefficient for "Extended family" is also greater than one standard error away from the coefficients for all other models. (Reweighting the CC analyses confirmed the findings shown for unweighted analyses.)

The three remaining approaches, the hot deck imputation approach described above, the AC approach, and the EM algorithm [to implement the EM algorithm, software available through ROBMLE (Little, 1988) was used] yielded very similar results in the regression analyses conducted. While we had expected that the hot deck method might have led to sharper attenuation of association than the EM and AC approaches, this is not obvious in the above. Part of this is due to the care taken in the imputation. However, since most of the imputation was done with n-partition hot decks rather than more sophisticated methods designed to preserve covariance structures, we speculate that the favorable results may be due to the low rates of missing data and low frequency of outliers in the variables being imputed.

22.5 VARIANCE ESTIMATION AND RECOMMENDATIONS

We do not have a solution to the problem of how to conduct design-based estimation of variance for multivariate hot deck procedures. In all three of the applications given here, we have been unable to recommend any procedure better than treating the imputed values the same as reported values despite the known attendant underestimation of variances. We could have executed our procedures multiple times and then used Rubin's formula for computing multiple imputation variances, but since the imputation methods are clearly not proper in Rubin's sense, the resulting variance estimates would still be too small. Rather than burden users with multiple imputations when the improvement in the quality of the variance estimates is so uncertain, we have chosen to recommend to users that no adjustments to the variances be made or that consideration be given to dividing naive variance estimates by the item response rates. This is an area that deserves intensive study.

Postimputation variance estimation is a theme for several other chapters in this collection [Lee, Rancourt, and Särndal (Chapter 21), Shao (Chapter 20), and Meng (Chapter 23)], but the approach that we find most promising for multivariate hot decks is a Shao–Sitter bootstrap (Shao and Sitter, 1996). A Shao–Sitter bootstrap would involve repeating the entire imputation process on a series of half samples. Other methods such as the "all-cases" methodology of Montaquila and Jernigan (1997) are intriguing but seem limited to univariate statistics and difficult to extend to the imputation methods considered in this chapter.

For future applications, statisticians will have three broad options. One option will be to publish raw data and let analysts cope with the missingness as best they can. This is probably a poor choice. Users will typically use complete case analysis, resulting in substantial bias and variance penalties. The second option will be to use Bayesian imputation methods. We think these methods may be a good choice on surveys with many continuously distributed variables with complex missing data patterns, but we remain uncomfortable with the specification of priors (proper and improper) and unconvinced that the variance estimates obtained by multiple imputation will always be consistent or conservative for unplanned analyses, or for any analyses on surveys with complex designs. The third option will be to use hot decks or nearest-neighbor (NN) techniques. We believe that hot decks are the best choice (CNP and FICD hot decks in particular) and will remain the best choice for problems involving discontinuous distributions and mixtures of large numbers of categorical variables with a few continuous variables. If there are several continuous variables with important prediction smooth relationships or if there are a large number of low-order associations to preserve, then NN techniques might be the most attractive. Hot decks and NN techniques are generally easier to implement than Bayesian methods, easier to explain to laymen, avoid dependence on prior distributions, and are flexible enough to handle nonstandard distributions.

CHAPTER 23

A Congenial Overview and Investigation of Multiple Imputation Inferences under Uncongeniality

Xiao-Li Meng, *University of Chicago and Harvard University*

23.1 INTRODUCTION

It is common practice, especially with public-use data files, to use imputation to deal with item nonresponse and some other forms of incomplete data in surveys. The most obvious reason for its popularity is that standard complete-data methods and software can be used once the "holes" in a data set are "filled in." Imputation also allows the direct incorporation of the data collector's knowledge, information, and resources into the imputation process. Associated with these advantages, however, is the danger of producing deceptive statistical inferences if the user does not correct for the increased uncertainty due to missing data. Approaches for obtaining valid standard errors from singly imputed data are discussed in Chapters 20 and 21 of this book. In this chapter, we focus on multiple imputation (Rubin, 1987), a method primarily motivated by the desire for reaching valid statistical inferences that account for this increased uncertainty, while not sacrificing the convenience of complete-data analysis tools.

Model-based Bayesian posterior prediction is a general imputation tool that permits a coherent modeling of complex processes, appropriate propagation of uncertainty and variability, and explicit evaluations and diagnostics of assumptions. A common concern with the modeling approach to imputation is the impact of model assumptions on the user's analyses. While concern about incorrect assumptions should be at the core of any scientific study, one simply has to make assumptions in order to meaningfully impute the unobserved values given the observed data. In a broad sense, a statistical model, parametric or nonparametric, is just a set of mathematically compatible assumptions that are perceived to be scientifically and statistically meaningful by the modeler. There is simply no such thing as a completely assumption-free imputation method. For example, hot deck imputation procedures are often presented as objective "model-free" methods, but their validity depends critically on the assumption that the nonrespondents do not differ systematically from respondents after conditioning on the common observed characteristics. In addition, the actual performance of hot deck procedures is difficult or even impossible to evaluate for complex data sets, as these methods often necessarily include components such as ad hoc classification rules that do not permit meaningful probabilistic replications.

Posterior predictive imputation provides a scientific way of addressing these concerns by allowing explicit incorporation and evaluation of assumptions, and it has become much more popular and useful in recent years with the development of multiple imputation and rapid advances in Bayesian computational tools such as Markov chain Monte Carlo methods. With this increased popularity comes increased interest in studying the impact of a Bayesian imputation model on the subsequent analyses conducted by general users, who may have essentially no knowledge of the imputation model other than which values are real and which are imputed. A key aspect of this study is *uncongeniality,* defined in Meng (1994), which is a theoretical concept that formulates the incompatibility between the imputation model and a user's analysis of the imputed data. Because many current analyses are design-based, one needs to link an analysis procedure with a Bayesian model in order to study the impact of the posited Bayesian imputation model. This linkage and the issue of uncongeniality are discussed in Section 23.2, using the case of stratified sampling. Section 23.3 provides a simulation study based on the heteroscedastic regression problem of Robins and Wang (2000). Section 23.4 discusses the connection between biases in confidence coverage and the fraction of missing information under uncongeniality, as well as the intriguing issue of self-inefficiency, which occurs when a user's analysis procedure permits more efficient estimation with incomplete data than with complete data.

23.2 IMPUTATION INFERENCE AND UNCONGENIALITY

23.2.1 Rubin's Repeated-Imputation Inference

Let Z_{com} denote the intended complete data set for the user, Z_{obs} the observed part of Z_{com}, and Z_{mis} the missing part. A multiply imputed data set is produced when an imputer, often the data collector, produces $m(\geq 2)$ imputations of Z_{mis}, say, $\{Z_{\text{mis}}^{(l)}, l = 1, \ldots, m\}$. Ideally, $\{Z_{\text{mis}}^{(l)}, l = 1, \ldots, m\}$ are independent draws from a posterior predictive distribution $p(Z_{\text{mis}}|Z_{\text{obs}}, Z_{\text{add}})$, constructed by the imputer, where Z_{add} represents additional data only available to the imputer (e.g., because of confidentiality constraints). The m sets of the imputations are then combined with Z_{obs} to form m completed datasets, $Z_{\text{com}}^{(l)} \equiv \{Z_{\text{obs}}, Z_{\text{mis}}^{(l)}\}, l = 1, \ldots, m$.

Suppose a user is interested in making an inference about an unknown quantity Q, and his complete data procedure can be summarized as $\mathcal{P}_{\text{com}} = [\hat{Q}(Z_{\text{com}}), U(Z_{\text{com}})]$, where $\hat{Q}(Z_{\text{com}})$ is his complete data point estimator of Q and $U(Z_{\text{com}})$ is the associated variance estimator of $\hat{Q}(Z_{\text{com}}) - Q$. Inference about Q can be based on Rubin's (1987) repeated-imputation method, which consists of the following two steps:

Step 1: Treat each $Z_{\text{com}}^{(l)}$ as a real complete dataset and compute

$$\hat{Q}_l \equiv \hat{Q}(Z_{\text{com}}^{(l)}) \qquad \text{and} \qquad U_l \equiv U(Z_{\text{com}}^{(l)}),\ l = 1, \ldots, m \tag{23.1}$$

Step 2: The repeated-imputation estimate of Q is

$$\overline{Q}_m = \frac{1}{m}\sum_{l=1}^{m} \hat{Q}_l \tag{23.2}$$

with associated variance

$$T_m = \overline{U}_m + (1 + 1/m)B_m \tag{23.3}$$

where $\overline{U}_m$, the average of $\{U_l, l = 1, \ldots, m\}$, estimates the within-imputation variability, and

$$B_m = (m-1)^{-1}\sum_{l=1}^{m}(\hat{Q}_l - \overline{Q}_m)(\hat{Q}_l - \overline{Q}_m)^{\text{T}} \tag{23.4}$$

estimates the between-imputation variability. Here "T" denotes the vector transpose, since Q may be multivariate.

A $100(1-\alpha)\%$ nominal level confidence interval for a univariate Q can be obtained using Rubin's (1987, p. 77) t-approximation:

$$\overline{Q}_m \pm t_\nu(\alpha/2)\sqrt{T_m}, \qquad \text{with } \nu = (m-1)/f_m^2 \tag{23.5}$$

where $t_\nu(\alpha/2)$ is the upper $100(1-\alpha/2)$ percentile of a t distribution with ν degrees of freedom, and $f_m = (1 + 1/m)B_m/T_m$ estimates the fraction of missing information defined in Section 23.4. Confidence regions and hypothesis tests for multivariate Q are given in Li et al. (1991) and Meng and Rubin (1992). All these methods assume that the large-sample normal approximation is adequate with the user's complete-data procedure. When this assumption fails, small-sample procedures (e.g., Barnard and Rubin, 1999) should be used.

23.2.2 Embedding a Design-Based Procedure in a Bayesian Model

The combining rules (23.2)–(23.4) are most easy to justify when the user's analysis procedure is compatible with the imputation model. To quantify this compatibility, which was termed congeniality in Meng (1994), we first need to embed the user's procedure into a Bayesian model. This is not necessarily an easy step, and can even be impossible for some analysis procedures (see Meng, 2000). The following illustrates the embedding, and shows that a variety of design-based analysis procedures are also Bayesian procedures under specific choices of sampling and prior distributions.

Suppose a population of size N is (pre- or post-) stratified into J strata, with size N_j for the jth stratum, $\Sigma_{j=1}^{J} N_j = N$, and $Y_j = \{y_{ij}, i = 1, \ldots, n_j\}$ is a simple random sample taken from the jth stratum, $j = 1, \ldots, J$. Let $\bar{y}_j$ and s_j^2 denote the sample mean and variance in the jth stratum. The standard design-based estimator of population average $Q = \bar{Y}$ is

$$\hat{Q}^{(D)} = \sum_{j=1}^{J} W_j \bar{y}_j \tag{23.6}$$

where $W_j = N_j/N$. Its variance, ignoring finite population corrections, is estimated by

$$U^{(D)} = \sum_{j=1}^{J} W_j^2 \frac{s_j^2}{n_j} \tag{23.7}$$

The inference procedure $\mathcal{P}^{(D)} = [\hat{Q}^{(D)}, U^{(D)}]$ can also be obtained (asymptotically) from the Bayesian perspective. In particular, suppose we model Y_j as independently and identically distributed (i.i.d.) samples from $N(\mu_j, \sigma_j^2)$ and assume the noninformative prior $p(\mu_j, \log \sigma_j^2) \propto 1$, independently for $j = 1, \ldots, J$. Then the posterior distribution of μ_j is a t distribution in the form of

$$\frac{\mu_j - \bar{y}_j}{\sqrt{s_j^2/n_j}} \sim t_{n_j-1} \tag{23.8}$$

(see Gelman et al., 1995, Ch. 2). Note that (23.8) has the same form as the standard sampling-based inference based on t distribution, except the random variable here is μ_j, not $\{\bar{y}_j, s_j^2\}$, which are fixed. Consequently, if $n_j > 3$, the posterior mean and variance of μ_j are respectively

$$E(\mu_j|Y_j) = \bar{y}_j \qquad \text{and} \qquad V(\mu_j|Y_j) = c_{n_j} s_j^2/n_j \tag{23.9}$$

where $c_{nj} = (n_j - 1)/(n_j - 3)$ is the so-called "t-correction factor." The posterior mean of our estimand $Q = \Sigma_{j=1}^{J} W_j \mu_j$ given $Y = \{Y_1, \ldots, Y_J\}$ is then

$$\hat{Q}^{(B)} \equiv E(Q|Y) = \sum_{j=1}^{J} W_j \bar{y}_j \tag{23.10}$$

which is identical to $\hat{Q}^{(D)}$ of (23.6). The posterior variance of Q is

$$U^{(B)} \equiv V(Q|Y) = \sum_{j=1}^{J} c_{n_j} W_j^2 s_j^2/n_j = [1 + \alpha_n/(n_{\min} - 3)]U^{(D)} \tag{23.11}$$

where $0 < \alpha_n \leq 2$ and $n_{\min} = \min\{n_1 \ldots, n_J\}$. Consequently, $\mathcal{P}^{(D)} = [\hat{Q}^{(D)}, U^{(D)}]$ is essentially the same as $\mathcal{P}^{(B)} = [\hat{Q}^{(B)}, U^{(B)}]$ as long as $n_{\min}$ is large enough. The t-correction factor c_{nj} in (23.9) is the consequence of the consistency between the design-based approach and the Bayesian approach at the distributional level, as seen in (23.8).

23.2.3 Complete-Data Congeniality Between a Design-Based Procedure and a Bayesian Model

Instead of assuming $\{\mu_j, j = 1, \ldots, J\}$ are a priori unrelated, as in the previous section, the imputer may want to model the similarity among them via a standard normal hierarchical model that assumes $\mu_1, \ldots, \mu_J \sim_{\text{iid}} N(\mu, \tau^2)$. To avoid inessential complications, we assume σ_j^2 are known (that is, we ignore the variability in s_j^2 for estimating $\sigma_j^2, j = 1, \ldots, J$); this is an adequate approximation when $n_{\min}$ is reasonably large (e.g., Gelman et al., 1995, p. 135). Furthermore, we assume τ^2 is a priori fixed by the modeler as an expression of the prior belief of the degree of similarity among $\{\mu_j, j = 1, \ldots, J\}$. In particular, if the modeler sets $\tau^2 = \infty$, then we are back to the model of Section 23.2.2.

With the constant prior $p(\mu) \propto 1$ on the hyperparameter μ, the posterior mean of our estimand $Q = \Sigma_{j=1}^J W_j \mu_j$ is

$$\hat{Q}^{(B)} = \sum_{j=1}^{J} W_j \hat{\mu}_j = \sum_{j=1}^{J} W_j[\lambda_j \bar{y}_j + (1 - \lambda_j)\hat{\mu}] \tag{23.12}$$

with the corresponding posterior variance

$$U^{(B)} = \sum_{j=1}^{J} W_j^2 p_j + V_\mu \left[\sum_{j=1}^{J} W_j(1 - \lambda_j) \right]^2 \tag{23.13}$$

where $p_j = [(\sigma_j^2/n_j) + \tau^2]^{-1}$, $\hat{\mu} = (\Sigma_{j=1}^J p_j \bar{y}_j)/(\Sigma_{j=1}^J p_j)$, $V_\mu^{-1} = \Sigma_{j=1}^J p_j$, and $\lambda_j = \tau^2/[\tau^2 + (\sigma_j^2/n_j)]$; see Gelman et al., 1995, pp. 138–139. In (23.12), the mean μ_j in stratum j is estimated as a weighted average of the sample mean $\bar{y}_j$ and $\hat{\mu}$, with weights proportional to their precisions, i.e., the reciprocal of the variances.

Meng (2000) compares $\mathcal{P}^{(B)}$ given by (23.12)–(23.13) with $\mathcal{P}^{(D)}$ given by (23.6)–(23.7), and obtains the following results:

Result A: Under the allocation $n_j \propto W_j \sigma_j^2, j = 1, \ldots, J$, the Bayesian procedure $\mathcal{P}^{(B)}$ is the same as $\mathcal{P}^{(D)}$ for any $\tau^2 \geq 0$.

Result B: When $\tau^2 > 0$, $\mathcal{P}^{(B)}$ is asymptotically equivalent to $\mathcal{P}^{(D)}$, in the sense that

$$\lim_{n_{\min}\to\infty} \frac{\hat{Q}^{(B)} - \hat{Q}^{(D)}}{\sqrt{U^{(D)}}} = 0 \quad \text{and} \quad \lim_{n_{\min}\to\infty} \frac{U^{(B)}}{U^{(D)}} = 1 \tag{23.14}$$

Result C: When $\tau^2 = 0$ and n_j is not proportional to $W_j \sigma_j^2$, (23.14) generally does not hold, that is $\mathcal{P}^{(B)}$ differs from $\mathcal{P}^{(D)}$, even asymptotically.

Note that (23.14) defines congeniality as an asymptotic notion, reflecting the large samples typical in survey practice. Nevertheless, it is interesting to observe that under the allocation $n_j \propto W_j \sigma_j^2$, which may remind one of optimal allocation $n_j \propto W_j \sigma_j$ (Cochran, 1977, Chapter 5), complete-data congeniality holds even for small samples. The adjective complete-data is needed here because congeniality as defined in Meng (1994) requires the analysis procedure and the Bayesian imputation model to be asymptotically compatible given both complete data and observed data. Thus, Result A or Result B is only a part of what is needed to guarantee the congeniality as discussed in Section 23.2.4. However, Result C is enough to yield the uncongeniality between the user's analysis procedure and the posited Bayesian (imputation) model.

Result C also shows the danger of using the strong prior $\tau = 0$, which restricts $\mu_1 = \cdots = \mu_J$, i.e. $\bar{Y}_1 = \cdots = \bar{Y}_J$. When this assumption does not hold, the use of $\tau = 0$ not only generally leads to uncongeniality but more importantly to incorrect inferences (e.g., $\hat{Q}^{(B)}$ can be seriously biased when the assumption of common mean is seriously violated). It is interesting to observe that there is one exception to this general conclusion, namely, under the allocation given in Result A, the choice of $\tau = 0$ is acceptable because weighting by the precisions n_j/σ_j^2 is the same as weighting by the strata weights $W_j (j = 1, \ldots, J)$.

23.2.4 Imputation Inference under Uncongeniality

For imputed data sets, the issue of uncongeniality is more complicated because the imputer's assumptions have impact only on the imputed Z_{mis}, but not on Z_{obs}. Consequently, the impact of the uncongeniality on the user's analysis procedure is limited by the amount of missing information (defined in Section 23.4) for estimating Q. To illustrate, suppose the response rate is r_j for jth stratum, and the respondents are a simple random subsample of the intended sample within each stratum. In the terminology of Rubin (1976), the nonresponse mechanism is missing at random (MAR), but not missing completely at random (MCAR), because it depends on the stratifying variable (i.e., r_j can vary with j). The design-based procedure (23.6)–(23.7) becomes

$$\hat{Q}^{(D)}_{\text{obs}} = \sum_{j=1}^{J} W_j \bar{y}_{j,!} \quad \text{and} \quad U^{(D)}_{\text{obs}} = \sum_{j=1}^{J} W_j^2 \frac{s^2_{j,!}}{n_{j,!}} \tag{23.15}$$

where $\{\bar{y}_{j,!}, s^2_{j,!}, n_{j,!}\}$ is the observed-data counterpart of $\{\bar{y}_j, s_j^2, n_j\}$; as in Meng (1994), the subscript "!" added to any quantity means the quantity is being calculated/defined only with respect to the observed data. Under the assumption of known variances, $s^2_{j,!}$ of (23.15) is replaced by $\sigma^2_{j,!} \equiv \sigma_j^2$.

Now suppose a user is provided with the multiply imputed data, imputed via the normal hierarchical model detailed in Section 23.2.3. A user might simply ignore the imputed data and use $\mathcal{P}^{(D)}_{\text{obs}} = [\hat{Q}^{(D)}_{\text{obs}}, U^{(D)}_{\text{obs}}]$ as given in (23.15), or apply $\mathcal{P}^{(D)} = [\hat{Q}^{(D)}, U^{(D)}]$ given in (23.6)–(23.7) to each of the completed data sets and then use combining rule (23.2)–(23.4) to compute $\mathcal{P}_m = [\bar{Q}_m, T_m]$, where m is the number of imputations. Numerically, these two procedures always differ to some degree because $\mathcal{P}_m$ depends on the particular m imputations that were provided. But this source of difference is both expected and well understood, because the m imputations are just simulations from the imputation model, and there are always simulation errors when we cannot use $m = \infty$. The more fundamental comparison is between $\mathcal{P}^{(D)}_{\text{obs}}$ and $\mathcal{P}_\infty = [\overline{Q}_\infty, T_\infty]$, the repeated-imputation inference procedure with $m = \infty$. Meng (1994) defines a difference in these two procedures as inferential uncongeniality. It arises, for example, when the imputer has built data and assessments into the imputation model that are not available to the user, but are helpful for reducing nonresponse bias.

When the user's incomplete-data procedure $\mathcal{P}^{(D)}_{\text{obs}}$ is invalid, inferential uncongeniality is unavoidable unless we want $\mathcal{P}_\infty$ to be also invalid. The interesting question is: when $\mathcal{P}^{(D)}_{\text{obs}}$ is valid, can inferential uncongeniality occur, and if it can, which procedure is better? The answer to the first part of the question is affirmative, as seen below. The second part is a topic studied in a number of papers (e.g., Fay, 1993; Meng, 1994; Rubin, 1996; Robins and Wang, 2000; Meng, 2000), and some main findings are illustrated in the current example and in the example of Section 23.3. For our stratified sample example, Meng (2000) proves the following counterparts to Results A–C of Section 23.2.3:

Result A′: When

$$(n_{j,!}^{-1} - n_j^{-1})W_j\sigma_j^2 = \text{constant} \tag{23.16}$$

$\mathcal{P}_\infty = [\overline{Q}_\infty, T_\infty]$ is the same as $\mathcal{P}^{(D)}_{\text{obs}} = [\hat{Q}^{(D)}_{\text{obs}}, U^{(D)}_{\text{obs}}]$ for any $\tau^2 \geq 0$.

Result B′: When $\tau^2 > 0$, $\mathcal{P}_\infty$ is asymptotically equivalent to $\mathcal{P}^{(D)}_{\text{obs}}$, that is,

$$\lim_{n_{\min,!}\to\infty} \frac{\overline{Q}_\infty - \hat{Q}^{(D)}_{\text{obs}}}{\sqrt{U^{(D)}_{\text{obs}}}} = 0 \quad \text{and} \quad \lim_{n_{\min,!}\to\infty} \frac{T_\infty}{U^{(D)}_{\text{obs}}} = 1 \tag{23.17}$$

Result C′: When $\tau^2 = 0$ and (23.16) does not hold, (23.17) generally does not hold. That is, $\mathcal{P}_\infty$ is different from $\mathcal{P}^{(D)}_{\text{obs}}$, even asymptotically.

These results ensure that as long as $\tau^2 > 0$, the repeated imputation procedure $\mathcal{P}_\infty$ will be congenial, as defined in Meng (1994), with the user's design-based procedure $\mathcal{P}^{(D)}_{\text{obs}}$. Incidentally, when both n_j and $n_{j,!}$ are proportional to $W_j\sigma_j^2$, (23.16) holds, but the converse is not true. Of course, in practice, even when $n_j \propto W_j\sigma_j^2$ holds by design, $n_{j,!} \propto W_j\sigma_j^2$ or condition (23.16) will almost never hold exactly. This example indicates the necessity of formulating uncongeniality as an asymptotic notion, as otherwise we will be distracted by discrepancies that are not of primary importance. When $\mathcal{P}^{(D)}{}_{\text{obs}}$ is valid and $\mathcal{P}_\infty$ is different from it, a question of interest is whether $\mathcal{P}_\infty$ is valid. The validity here refers to whether $T_\infty \geq V(\overline{Q}_\infty)$ (asymptotically when the observed dataset becomes large), which guarantees the confidence validity of $\mathcal{P}_\infty$, e.g., $\overline{Q}_\infty \pm 2\sqrt{T_\infty}$ will have at least 95% coverage. See

Rubin (1996) and Meng (1994) for discussions of confidence validity and why it makes scientific sense, particularly in the context of repeated imputation inference. For our current problem, Result B′ ensures T_∞ is a consistent estimator of $V(\overline{Q}_\infty)$ as long as $\tau^2 > 0$. When $\tau = 0$, Meng (2000) shows that under the condition of either Result A or Result A′, $T_\infty = V(\overline{Q}_\infty)$ exactly. Besides these two special cases, Meng (2000) shows that under conditions that typically apply when response rates do not vary widely with strata, we have

$$V(\overline{Q}_\infty) \leq T_\infty \leq U_{\text{obs}}^{(D)} \tag{23.18}$$

which has the interesting implication that although $\mathcal{P}_{\text{obs}}^{(D)}$ is valid, it is inadmissible as a confidence procedure because it is dominated by $\mathcal{P}_\infty$. That is, the interval from $\mathcal{P}_\infty$ (e.g., $\overline{Q}_\infty \pm 2\sqrt{T_\infty}$) has at least the coverage of the interval from $\mathcal{P}_{\text{obs}}^{(D)}$ with the same nominal level (e.g., $\hat{Q}_{\text{obs}}^{(D)} \pm 2\sqrt{U_{\text{obs}}^{(D)}}$), but its width does not exceed the width of the one from $\mathcal{P}_{\text{obs}}^{(D)}$. See Meng (1994) and Rubin (1996) for general theory and discussions of this phenomenon.

However, it is possible that $T_\infty < V(\overline{Q}_\infty)$ when response rates vary greatly across strata and the strata with high nonresponse rates tend to have smaller variances (see Meng, 2000 for details), implying a great loss of information due to nonresponse. In Section 23.4 we discuss the magnitude of the undercoverage in connection with high fraction of missing information. Before we do so, we provide some simulation results to illustrate the magnitude of the bias in T_∞ for estimating $V(\overline{Q}_\infty)$ in an example of Robins and Wang (2000).

23.3 A SIMULATION STUDY

23.3.1 Complete Data and Missing Data Mechanism

Our simulation study is based on the heteroscedastic regression problem presented in Robins and Wang (2000). The complete data, $Z_{\text{com}} = \{(Y_i, X_i), i = 1, \ldots, n\}$, consist of $n = 150$ pairs from two groups. For the first 100 pairs (i.e., $i = 1, \ldots, 100$), the covariate X_i's are i.i.d. samples from Uniform(0.1, 0.8), and for the other 50 pairs, the X_i's are i.i.d. Uniform(0.8, 2). Given X, the Y_i's for both groups are generated from the normal heteroscedastic regression model

$$Y|X \sim N(X\beta, \sigma^2 X^{\eta^*}) \tag{23.19}$$

where η^* is assumed known, and $\{\beta, \sigma^2\}$ are to be estimated. The missing data are created by randomly deleting π percent of Y's from the second group, so that the overall percentage of missing data is $\pi/3$. This is a case of MAR, but not MCAR, because the missingness depends on X, which is always observed. Let $\mathcal{O}_n$ denote the set of i's with observed Y_i's and $\mathcal{M}_n$ the set of i's with missing Y_i's. Then the observed data are $Z_{\text{obs}} = \{X_i, i = 1, \ldots, n; Y_i, i \in \mathcal{O}_n\}$, and the missing data are $Z_{\text{mis}} = \{Y_i, i \in \mathcal{M}_n\}$.

23.3.2 Imputation Model and Imputed Data Sets

As in Robins and Wang (2000), we multiply impute Z_{mis} by draws from $p(Z_{\text{mis}}|Z_{\text{obs}})$ under the heteroscedastic regression model

$$Y|X \sim N(X\beta, \sigma^2 X^{\eta}), \tag{23.20}$$

with constant prior on $\{\beta, \log \sigma^2\}$, where η is specified by the imputer and may or may not be the same as η^*. That is, we allow the variance structure of the imputation model to be misspecified. Let $\hat{\beta}_{\text{obs}}$ and $\hat{\sigma}^2_{\text{obs}}$ be the standard weighted least squares estimators of β and σ^2 under (23.20); then the joint posterior distribution of (β, σ^2) is given by (see Gelman et al., 1995, pp. 236–237)

$$\sigma^2|Z_{\text{obs}} \sim \hat{\sigma}^2_{\text{obs}} \frac{n_{\text{obs}} - 1}{\chi^2_{n_{\text{obs}}} - 1}, \qquad \beta|\sigma^2, Z_{\text{obs}} \sim N(\hat{\beta}_{\text{obs}}, \sigma^2 V_{\text{obs}}) \tag{23.21}$$

where n_{obs} is the number of observed Y_i's, and $V_{\text{obs}} = [\Sigma_{i \in O_n} X_i^{2-\eta}]^{-1}$. Consequently, to create m (≥ 2) imputations, we repeat the following two steps independently for $l = 1, \ldots, m$ (see Rubin, 1987, p. 167):

Step 1: Draw independently $z \sim N(0, 1)$ and $\chi^2_{n_{\text{obs}}-1}$, and let

$$\sigma^2_* = \hat{\sigma}^2_{\text{obs}} \frac{n_{\text{obs}} - 1}{\chi^2_{n_{\text{obs}}} - 1}, \qquad \beta_* = \hat{\beta}_{\text{obs}} + \sigma_* \sqrt{V_{\text{obs}}} z \tag{23.22}$$

Step 2: Given the draw $\{\beta_*, \sigma^2_*\}$ from Step 1, draw $n_{\text{mis}} = n - n_{\text{obs}}$ independent ε_i's from $N(0, 1)$ and let

$$Y_i^{(l)} = X_i \beta_* + \sigma_* X_i^{\eta/2} \varepsilon_i,\ i \in \mathcal{M}_n \tag{23.23}$$

We thereby create m completed-data sets $\{Z^{(l)}_{\text{com}}, l = 1, \ldots, n\}$, where

$$Z^{(l)}_{\text{com}} = \{Z_{\text{obs}}, Z^{(l)}_{\text{mis}}\} = \{(Y_i, X_i), i \in O_n; (Y_i^{(l)}, X_i), i \in \mathcal{M}_n\}$$

23.3.3 Analysis Procedures

Suppose a user is unaware of the heteroscedastic nature of the data, and estimates β with the unweighted least-squares (LS) estimator. Specifically, with the complete data Z_{com}, the user would estimate β by

$$\hat{\beta}_{\text{com}} \equiv \hat{\beta}(Z_{\text{com}}) = V_{\text{com}} \sum_{i=1}^{n} X_i Y_i, \qquad \text{where } V_{\text{com}} = \left[\sum_{i=1}^{n} X_i^2\right]^{-1} \tag{23.24}$$

Although $\hat{\beta}_{\text{com}}$ is a consistent estimator of β under the heteroscedastic model (23.20), the standard variance estimator,

$$U^{\text{std}} \equiv U^{\text{std}}(Z_{\text{com}}) = V_{\text{com}} \sum_{i=1}^{n} (Y_i - X_i \hat{\beta}_{\text{com}})^2/(n-1) \tag{23.25}$$

is not consistent for the variance of $\hat{\beta}_{\text{com}}$ under (23.20). In particular, U^{std} tends to underestimate or overestimate $V(\hat{\beta}_{\text{com}}|X)$, depending on $\eta^* > 0$ or $\eta^* < 0$. However, the robust "sandwich" estimator

$$U^{\text{rob}} \equiv U^{\text{rob}}(Z_{\text{com}}) = V^2_{\text{com}} \sum_{i=1}^{n} X_i^2 (Y_i - X_i \hat{\beta}_{\text{com}})^2 \tag{23.26}$$

is consistent for $V(\hat{\beta}_{\text{com}}|X)$ even if $\eta^* \neq 0$. It is clear that if the user's complete-data procedure is not valid, then only by luck can combining rules (23.2)–(23.4) still provide a correct inference. However, since in general practice the standard LS estimator U^{std} is much more frequently used than U^{rob} when the user estimates β by (23.24), we include both U^{std} and U^{rob} in our simulation assessments.

23.3.4 Simulation Configurations

As in Robins and Wang (2000), three scenarios were considered:

Scenario 1 (1 level): $\eta = \eta^* = 0$ (analysis model = imputation model = true model)

Scenario 2 (2 levels): $\{\eta = 0, \eta^* = 1\}$ and $\{\eta = 0, \eta^* = -1\}$ (analysis model = imputation model $\neq$ true model)

Scenario 3 (2 levels): $\{\eta = 1, \eta^* = 1\}$ and $\{\eta = -1, \eta^* = -1\}$ (analysis model $\neq$ imputation model = true model)

Each of these scenarios is crossed with a combination of other configurations, as specified in Meng (2000). Due to space limitations, here we only consider the following levels: (i) missing percentage of the second group: $\pi = 0.2, 0.4, 0.6, 0.8$; (ii) number of imputations: $m = 5, 100$ ($m = 100$ is used to approximate $m = \infty$); (iii) user's variance estimator: U^{std} and U^{rob}. Furthermore, in generating simulation replications, we fix the covariate X and the missing-data pattern. The less-conditioned replications yield very similar results, as expected. (See Meng, 2000, for more detailed results.)

23.3.5 Simulation Results

Table 23.1 summarizes some simulation results, with 10,000 replications, under $U = U^{\text{rob}}$ and $m = 100$. It includes (i) $\hat{V}_{\text{MC}}$, the Monte Carlo estimate of $V(\beta_m)$; (ii) $\hat{T}_{\text{MC}}$, the Monte Carlo estimate of $E(T_m)$; and (iii) $\hat{C}_{\text{MC}}$, the Monte Carlo estimate of the actual coverage of the 95% nominal confidence interval from (23.5). The simulation results are consistent with the theoretical large (m, n) approximation (see Meng, 2000, for formulas). It is seen that when $\eta = \eta^* = 0$ (i.e., Scenario 1), the repeated-imputation inference, as expected, provides the exact inference with large m.

Table 23.1. $\hat{V}_{MC} = V_{MC}(\bar{\beta}_m)$, $\hat{T}_{MC} = E_{MC}(T_m)$, $\hat{C}_{MC}$ = **Coverage(%) under** $U = U^{rob}$

Scenario	Sizes		$\pi = 0.2$	$\pi = 0.4$	$\pi = 0.6$	$\pi = 0.8$
$\eta = \eta^* = 0$ Correct imputation model, congenial	Large (m, n) approx.	$\hat{V}_{MC}$	0.009	0.011	0.015	0.22
		$\hat{T}_{MC}$	0.009	0.011	0.015	0.22
		Coverage	95	95	95	95
	$m = 100$, $n = 150$	$\hat{V}_{MC}$	0.0 10	0.011	0.016	0.020
		$\hat{T}_{MC}$	0.010	0.011	0.016	0.021
		Coverage	94.5	94.5	94.8	95.1
$\eta = \eta^* = 1$ Correct imputation model, uncongenial	Large (m, n) approx.	$\hat{V}_{MC}$	0.011	0.013	0.015	0.017
		$\hat{T}_{MC}$	0.012	0.015	0.020	0.025
		Coverage	96.08	96.95	97.64	98.19
	$m = 100$, $n = 150$	$\hat{V}_{MC}$	0.013	0.015	0.019	0.024
		$\hat{T}_{MC}$	0.013	0.016	0.020	0.026
		Coverage	95.7	96.7	97.0	97.8
$\eta = \eta^* = -1$ Correct imputation model, uncongenial	Large (m, n) approx.	$\hat{V}_{MC}$	0.008	0.010	0.013	0.022
		$\hat{T}_{MC}$	0.008	0.009	0.012	0.019
		Coverage	94.74	94.38	93.90	93.25
	$m = 100$, $n = 150$	$\hat{V}_{MC}$	0.009	0.012	0.015	0.030
		$\hat{T}_{MC}$	0.008	0.010	0.012	0.021
		Coverage	94.6	93.8	94.1	93.3
$\eta = 0$, $\eta^* = 1$ Incorrect imputation model, congenial	Large (m, n) approx.	$\hat{V}_{MC}$	0.012	0.014	0.018	0.023
		$\hat{T}_{MC}$	0.010	0.010	0.011	0.013
		Coverage	93.12	90.78	88.16	86.24
	$m = 100$, $n = 150$	$\hat{V}_{MC}$	0.013	0.016	0.019	0.025
		$\hat{T}_{MC}$	0.011	0.011	0.012	0.013
		Coverage	92.0	89.4	87.4	84.1
$\eta = 0$, $\eta^* = -1$ Incorrect imputation model, congenial	Large (m, n) approx.	$\hat{V}_{MC}$	0.008	0.011	0.016	0.029
		$\hat{T}_{MC}$	0.012	0.020	0.033	0.057
		Coverage	98.15	99.12	99.43	99.41
	$m = 100$, $n = 150$	$\hat{V}_{MC}$	0.009	0.013	0.016	0.028
		$\hat{T}_{MC}$	0.012	0.023	0.033	0.058
		Coverage	97.9	99.0	99.4	99.1

For $\eta = \eta^* = 1$, it overcovers, with maximal coverage 98%; and for $\eta = \eta^* = -1$, it slightly undercovers, with minimal coverage 93%. In both cases, the absolute bias in confidence coverage increases with the fraction of missing data, which is not the same but has the same order as the fraction of missing information (see Section 23.4). In contrast, when $\eta \neq \eta^*$, that is, when the imputation model is incorrectly specified, the results are generally much worse, with actual coverage varying from 86% to essentially 100%.

Table 23.2 compares two likely approaches in practice. The first method is to perform the standard least-squares approach with U^{std} to all the observed cases, ig-

Table 23.2. Estimated bias, coverage %, and MC width for least squares with and without imputation ($n = 150$, $m = 5$, $U = U^{std}$)

Scenario		$\pi = 0.2$		$\pi = 0.4$		$\pi = 0.6$		$\pi = 0.8$	
		Without imputation	With imputation	Without imputation	With imputation	Without imputation	With imputation	Without imputation	With imputation
$\eta = \eta^* = 0$	bias	0.001	0.001	–0.001	–0.001	0.000	0.000	–0.001	–0.001
	coverage	94.8	94.3	95.0	94.1	94.8	93.8	95.0	94.0
	width	0.377	0.379	0.401	0.412	0.482	0.532	0.572	0.663
$\eta = \eta^* = 1$	bias	0.001	0.001	0.000	0.000	–0.001	–0.001	0.000	0.000
	coverage	83.3	88.3	83.5	91.8	82.3	93.2	83.8	94.2
	width	0.315	0.355	0.335	0.437	0.378	0.557	0.437	0.707
$\eta = \eta^* = -1$	bias	0.000	0.000	–0.001	–0.001	0.000	0.000	–0.002	–0.002
	coverage	99.7	99.5	99.7	99.1	99.7	98.7	99.5	97.7
	width	0.608	0.557	0.695	0.571	0.807	0.602	1.060	0.764
$\eta = 0,\ \eta^* = 1$	bias	0.002	0.002	0.000	0.000	0.00 1	0.00 1	–0.00 1	–0.001
	coverage	85.4	84.8	84.7	84.5	83.9	84.2	85.5	86.1
	width	0.335	0.337	0.347	0.362	0.371	0.409	0.438	0.518
$\eta = 0,\ \eta^* = -1$	bias	–0.000	–0.000	0.000	0.000	0.000	0.000	–0.00 1	–0.00 1
	coverage	99.7	99.6	99.7	99.5	99.7	99.1	99.5	98.7
	width	0.551	0.554	0.693	0.721	0.827	0.894	0.969	1.146

noring the imputed data. The second method is to apply the same least-squares approach to each of the imputed data sets $Z^{(l)}_{\text{com}}$, and then apply Rubin's repeated-imputation rules; here $m = 5$, which is a realistic value in practice. As expected, in all cases both methods provide unbiased point estimators of β, and when $\eta = \eta^* = 0$, the two approaches produce quite similar interval estimators (when $m \to \infty$, they will give identical results). When $\eta^* \neq 0$ but $\eta = 0$, i.e., when the imputation model is incorrectly specified, the second method generally produces worse results than the first method. This is expected, and illustrates once more the importance of having good imputation models.

The most interesting cases, both practically and theoretically, are when $\eta = \eta^* \neq 0$, that is, when the imputation model is correctly specified but is uncongenial to the user's analysis procedure. It is seen from Table 23.2 that in such cases (i.e., the second and third rows), the imputation approach in general leads to less bias in coverage and when both methods provide at least the nominal coverage, the confidence interval from the imputation approach tends to be shorter due to the efficiency built into the imputed data (see Meng, 1994).

23.4 DISCUSSION

For both of our examples, we noted that, under the correctly specified imputation model, negative bias in T_∞ can occur when the missing data tend to have much smaller variability than the observed data. For example, for the heteroscedastic regression problem, this occurs when $V(Y|X) \propto X^{-1}$. Therefore, the missing data, which are from the second group (i.e., $0.8 < X < 2$), have much less variability than most of observed data, which are from the first group (i.e., $0.1 < X < 0.8$). Consequently, there is a high fraction of missing information (FMI), i.e., a high fraction of increased variance due to the missing data:

$$f_{\hat{Q}} = 1 - \frac{V[\hat{Q}(Z_{\text{com}})]}{V[\hat{Q}(Z_{\text{obs}})]} \tag{23.27}$$

We emphasize that $f_{\hat{Q}}$ depends on the choice of $\hat{Q}$, and it can be very different from the percentage of missing data, as seen in Table 23.3.

Table 23.3 displays the absolute percentage of undercoverage found in Table 23.1 for $\eta = \eta^* = -1$, together with $f_{\hat{Q}}$ for the corresponding LS estimator $\hat{Q} = \hat{\beta}_{\text{LS}}$ [i.e., $\hat{\beta}_{\text{com}}$ of (23.24)]. As a comparison, Table 23.3 also includes the $f_{\hat{Q}}$ for the maximum likelihood estimate $\hat{\beta}_{\text{MLE}}$ under the correct model, namely, the properly weighted least-square estimator under the model (23.20), as well as the efficiency of $\hat{\beta}_{\text{LS}}$ (relative to $\hat{\beta}_{\text{MLE}}$). All results are based on the large (m, n) theoretical calculations given in Meng (2000). We see that (1) the FMIs are much higher than the percentage of missing data; (2) the tiny undercoverage is associated with huge FMI for $\hat{\beta}_{\text{LS}}$; and (3) the FMI for $\hat{Q} = \hat{\beta}_{\text{LS}}$ is higher than that for $\hat{\beta}_{\text{MLE}}$. Note that although theoretically the undercoverage occurs for any $\pi > 0$, it takes an enormous FMI (e.g., over 75%) for the negative bias to have a practically visible effect (e.g.,

Table 23.3. Undercoverage and fraction of missing information (with $\eta = \eta^* = -1$)

	% of Missing data ($100^*\pi/3$)				
	0	6.6	13.3	20	26.6
Undercoverage in absolute %	0	0.26	0.62	1.10	1.75
% of missing information for β_{LS}	0	20.1	39.6	58.4	75.9
% of missing information for β_{MLE}	0	18.3	36.7	55.0	73.4
Efficiency of β_{LS} relative to β_{MLE}	80.9	79.2	77.2	74.8	73.3

93.25% actual coverage for a 95% nominal interval). This is particularly remarkable because the current example was constructed to be extreme, with enormous heterogeneity in $V(Y|X) \propto X^{-1}$ as X varies from 0.1 to 2. Such extreme cases are unlikely in practice, since a scatterplot of the data gives a clear indication of the inappropriateness of the standard LS estimator, even when used with the robust variance estimator. However, the insights we gain from studying such examples may have practical relevance.

For instance, the LS estimator $\hat{\beta}_{\mathrm{LS}}$ under model (23.20) has an interesting property when $\eta^* \neq 0$: because $V(\hat{\beta}_{\mathrm{LS}}|X) = \Sigma X_i^{2+\eta^*}/[\Sigma X_i^2]^2$, it is possible for $\hat{\beta}_{\mathrm{LS}}$ to have *smaller* variance with *less* data! Intuitively, since $\hat{\beta}_{\mathrm{LS}}$ (unlike $\hat{\beta}_{\mathrm{MLE}}$) does not weight the observations properly, its efficiency can be improved by discarding some Y_i's with large variances. If these "bad" Y_i's happen to be missing, a user may be "helped" by the missing data! In this extreme setting the FMI (23.27) is negative, and Rubin's combining rules are obviously inapplicable because the basic variance decomposition rule (23.3):

$$\text{Total Var} = \text{Within-Imputation Var} + \text{Between-Imputation Var}$$

requires f_Q of (23.27) to be nonnegative. Such procedures were classified as self-inefficient in Meng (1994). Current research is investigating to what extent self-inefficiency explains the bias in T_∞, and the impact of other factors such as a high FMI. The ultimate goal of this research is to establish a general paradigm for constructing and comparing practically feasible imputation inference procedures when facing the inevitable separation of the imputer and the user.

ACKNOWLEDGMENTS

The author thanks R. Criau for computational assistance, R. Little for editorial condensing, J. Barnard, J. Eltinge, A. Gelman, R. Little, B. Nandram, D. Rubin, J. Schafer, N. Schenker, and S. Stigler for helpful comments and conversations, and J. Robins and N. Wang for providing a preprint and stimulating exchanges. The research was funded partially by NSF via DMS-9626691. E-mail address: meng@stat.harvard.edu

CHAPTER 24

Multivariate Imputation of Coarsened Survey Data on Household Wealth

Steven G. Heeringa, Roderick J. A. Little, and Trivellore E. Raghunathan, *University of Michigan*

24.1 INTRODUCTION

Survey questions that ask respondents to report currency amounts for financial variables such as household assets, liabilities, income, expenditures, benefits or transfer payments can be subject to missing data rates as high as 20%–40% (Juster and Smith, 1994). One solution to the problem of high missing data rates for financial variables has been to develop questions that capture interval-censored observations (e.g., $5000–$9999) whenever the respondent refuses or is unable to provide an exact response to a question. The use of these "bracketed response" formats for questions in financial surveys has significantly reduced the rates of completely missing data for financial variables, but yields data that are a mixture of actual valued responses, bracketed or interval censored replies, and completely missing data. We call this mixture of data types "coarsened data," after Heitjan and Rubin (1990).

Building on advances in multivariate imputation theory and methods (Schafer, 1997; Kennickell, 1996), we propose a Bayesian approach to multiple imputation of the coarsened data, based on a multivariate mixed normal location model. We apply this approach to coarsened net worth data from the Health and Retirement Study (HRS), and present an empirical comparison of our method with alternative approaches to imputation of the data.

24.2 THE PROBLEM OF COARSENED NET WORTH DATA

In this section, data from Wave 1 of the Health and Retirement Survey (HRS) are used to illustrate the properties of coarsened survey measures of household assets and liabilities, and to motivate our methods.

24.2.1 The Health and Retirement Survey (HRS)

The 1992 Wave 1 HRS was a multistage area probability sample of 7607 U.S. households. The sampling unit was the household and the unit of observation was a household "financial unit"—either a married couple household where one or both spouses were in the eligible age range or a single unmarried adult who was 51 to 61 years old. In households with more than one financial unit, one was chosen at random for interview. In married couple households, the household interview was administered to the individual most knowledgeable about household financial matters.

Each interviewed household is assigned an analysis weight that reflects its selection probability, an adjustment for household unit nonresponse and a poststratification factor. Stratum and primary stage cluster are coded to allow for design-based estimation of standard errors, and missing data values and bracketing information are coded for purposes of imputation.

Bracketed responses were used extensively to collect data on assets and liabilities in the HRS. Figure 24.1 illustrates the format of the HRS Wave 1 bracketed response question sequence for a particular item, the household value of stock and mutual funds assets.

24.2.2 Coarsened Data Patterns for HRS Wave 1 Bracketed Net Worth Variables

Table 24.1 presents rates of bracketed and completely missing data for each of the 23 asset and liability variables needed to compute total net worth for HRS Wave 1 sample households. The upper rows of Table 24.1 summarize the HRS Wave 1 experience for the 10 household asset and liability measures that employed the bracketed response question sequence. The lower rows present missing data outcomes for the 13 additional net worth components that did not use the bracketed response question format of Figure 24.1. Bracketed responses for these variables were obtained using a range card with fixed categories (Heeringa, 1993). The central panel of Table 24.1, labeled "Does item apply?" provides estimates of the percentage of sample households (unweighted) that reported having each asset or liability. For example, 23.1% of the $n = 7607$ respondent households report owning real estate other than their personal residence. The remaining 76.9% of households do not own other real estate and so have an implied value of \$0 for that variable.

For households that report having a particular asset or debt, the right-hand panel of Table 24.1 describes the distribution of response types: actual value, bracketed value, or completely missing. The bracketed values occasionally span two or three

M10. For the next few questions, please exclude any assets held in the form of IRA and Keogh accounts. Do you [or your (husband/wife/partner)] have any shares of stock in publicly-held corporations, mutual funds, or investment trusts?

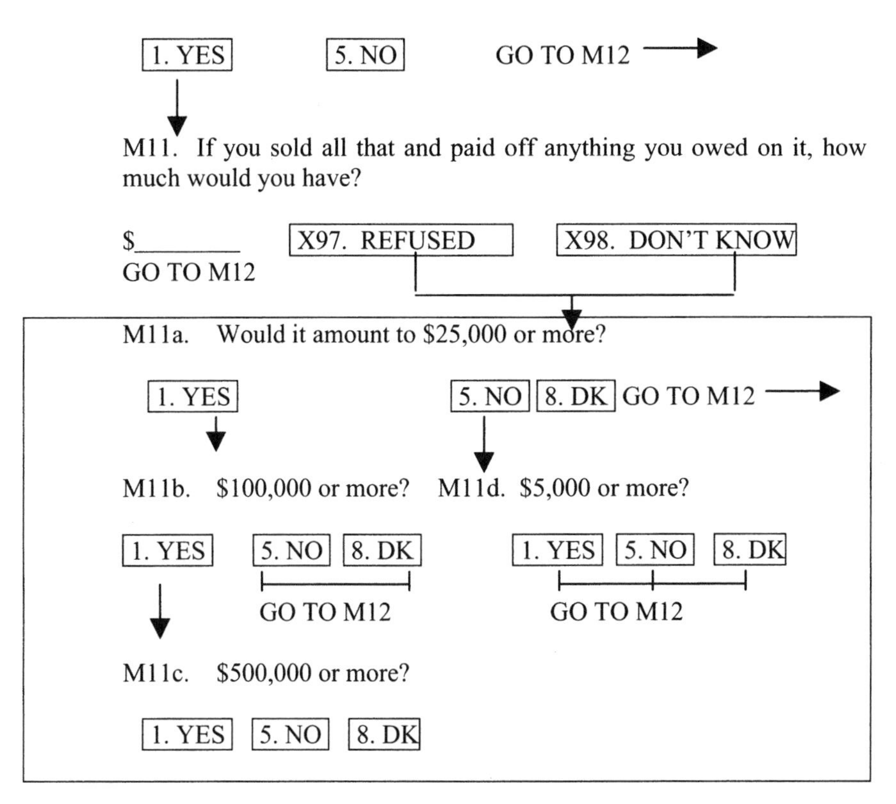

1. YES | 5. NO GO TO M12 →

M11. If you sold all that and paid off anything you owed on it, how much would you have?

$________ GO TO M12 | X97. REFUSED | X98. DON'T KNOW

M11a. Would it amount to $25,000 or more?

1. YES | 5. NO | 8. DK GO TO M12 →

M11b. $100,000 or more? | M11d. $5,000 or more?

1. YES | 5. NO | 8. DK — GO TO M12 | 1. YES | 5. NO | 8. DK — GO TO M12

M11c. $500,000 or more?

1. YES | 5. NO | 8. DK

Figure 24.1. HRS wave 1 bracketed question format.

of the actual bracket ranges for the question item, because of nonresponse or uncertainty. Among financial assets, the percentage of actual value reports ranges from 67.1% for stocks and mutual funds to 96.7% for the value of recreational vehicles owned by the household. Depending on the component, the percentage of bracketed responses ranges from 5.7% for other debt to 26.8% for value of stocks and mutual funds. The rates of missing data with no information on the values range from 1.9% of responses for the vehicle and property value question to 12.6% for value of a third mortgage.

Since the analyses of these data are predominantly multivariate or involve aggregation of several variables, the multivariate pattern of missing data is also relevant. For example, a total of $n = 932$ HRS Wave 1 households required exactly four asset and liability components for the households' net worth computations. Of these, 55.1% reported actual dollar amounts for all four items, 22.2% reported three of

Table 24.1 HRS Wave 1 Net Worth Components: distribution of responses by response type

	Does item apply?				If item applies to household				
HRS Wave 1 household asset or liability	Total	Yes	No	DK	*n*	Total %	Actual value	Bracketed value	Missing Value
Bracketed									
Real estate (not home)	100%	23.1%	76.3%	0.6%	1759	100%	74.7%	21.0%	3.4%
Vehicles, personal property	100%	—	—	—	7607	100%	87.4%	10.6%	1.9%
Business	100%	16.1%	83.4%	0.5%	1226	100%	68.2%	26.0%	5.8%
IRA, Keogh	100%	37.1%	62.2%	0.7%	2825	100%	73.7%	21.5%	4.9%
Stock, Mutual funds	100%	26.2%	73.6%	0.2%	1995	100%	67.1%	26.8%	6.1%
Checking, savings	100%	77.5%	21.5%	1.0%	5895	100%	73.3%	21.5%	5.2%
CDs, savings bonds	100%	24.6%	74.3%	1.1%	1870	100%	70.6%	22.6%	6.8%
Bonds	100%	5.9%	93.2%	1.0%	445	100%	69.7%	19.8%	10.6%
Other valuables	100%	15.0%	83.9%	1.1%	1143	100%	72.1%	21.6%	6.4%
Other debt	100%	38.2%	60.8%	1.0%	2908	100%	90.4%	5.7%	3.9%
Not bracketed*									
Farm value: own all	100%	3.2%	96.8%	0.0%	246	100%	88.2%	7.3%	4.5%
Farm value: own part	100%	0.3%	99.7%	0.0%	23	100%	82.6%	13.0%	4.4%
Mobile home: site	100%	*a*	99.9%	*a*	1	100%	100.0%	0.0%	0.0%
Mobile home: home	100%	2.4%	97.6%	*a*	180	100%	93.9%	0.6%	5.5%
Mobile home: both	100%	2.4%	97.6%	*a*	185	100%	87.6%	5.4%	7.0%
Home value	100%	68.0%	31.9%	0.1%	5176	100%	95.4%	1.2%	3.4%
First mortgage	100%	43.6%	56.0%	0.4%	3314	100%	92.4%	0.8%	6.8%
Second mortgage	100%	5.0%	94.4%	0.6%	376	100%	92.0%	0.8%	7.0%
Third mortgage	100%	2.3%	95.1%	2.6%	174	100%	87.4%	0.6%	12.6%
Home equity loan	100%	7.2%	91.8%	1.0%	550	100%	95.8%	0.9%	3.3%
Other property	100%	13.2%	86.8%	0.2%	998	100%	92.0%	1.4%	6.6%
Other Property debt	100%	5.6%	94.0%	0.4%	431	100%	91.6%	1.9%	6.5%
RV value	100%	8.0%	91.8%	0.2%	607	100%	96.7%	0.7%	2.6%

Note: $n = 7607$ respondent households.

*Cases identified as bracketed resulted from interviewer use of range card.

$a = 1$ case only.

four, 12.2% reported two, 6.9% reported one, and 2.6% did not provide actual values for any of the four required items.

24.2.3 Semicontinuity and Skewness

Our multivariate model for the joint distribution of net worth variables reflects the presence of significant numbers of zero values in the data vector; we use the term "semicontinuous" to describe this feature. As an illustration, Table 24.2 displays the joint distribution of zero and nonzero values for three net worth components. The most common pattern, Pattern 7 (owners of a checking or savings account with no business assets or stock holdings) was observed for 44.2% of sample households. On the other hand, Pattern 2 (owners of business assets and stock who do not have personal checking or savings accounts) was observed for only 0.2% of households.

Another feature of the components of net worth is that the distribution of the untransformed nonzero amounts are highly skewed to the right. Skewness is addressed here by assuming that the logarithms of the nonzero amounts have a multivariate normal distribution. Figures 24.2 and 24.3 illustrate the unweighted empirical distribution of the logarithm of nonzero home values and values of stock and mutual fund portfolios, two assets chosen to represent the extremes of conformity and nonconformity to the log-normal assumption. The histogram of log (home value) appears normal; however, the empirical distribution of log (stock) is left skewed and irregular in the frequencies for adjacent category groupings. The lack of smoothness in these distributions can be attributed to rounding or heaping of actual survey responses at values such as $100,000, $200,000, and $500,000. Our model does not address this heaping of responses, but a useful feature of the imputations from our model is that (unlike a hot deck) they smooth over these artificially heaped values.

The highest coarsened category of an asset or liability is "open-ended," that is, it is bounded below but not above. The distribution of the untransformed continuous

Table 24.2. HRS Wave 1 net worth components patterns of zero/nonzero values for business value, stock and mutual funds, and checking account

	Value of asset				
Pattern(s)	Business	Stock	Checking	n	%
1	>0	>0	>0	481	6.3%
2	>0	>0	0	15	0.2%
3	>0	0	>0	632	8.3%
4	>0	0	0	85	1.1%
5	0	>0	>0	1467	19.3%
6	0	>0	0	65	0.9%
7	0	0	>0	3372	44.2%
8	0	0	0	1490	19.7%
Total	—	—	—	7607	100.0%

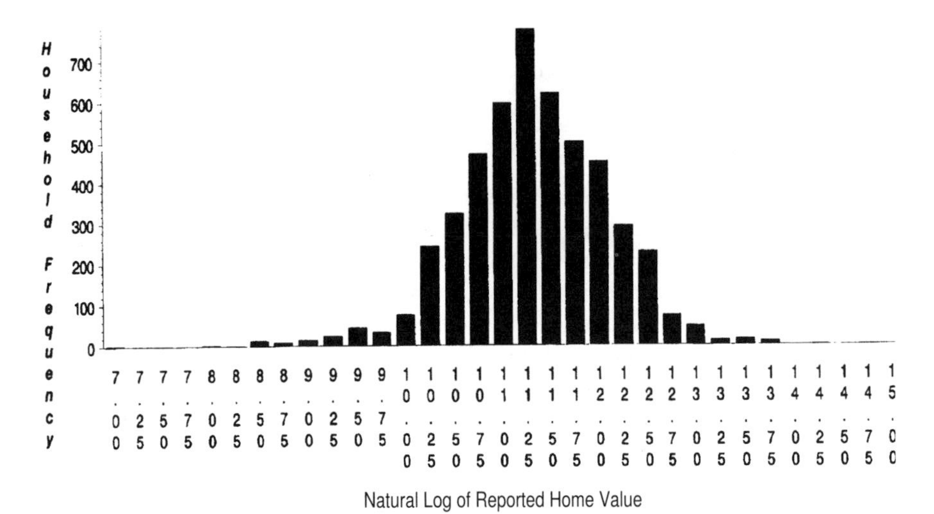

Figure 24.2 HRS Wave 1: distribution of log normal (home value) amounts actual reports >0.

value responses within each of these open-ended brackets is highly skewed, and observations for estimating the distribution here are sparse. Imputations in the open-ended brackets are highly sensitive to model misspecification. Extensions of the log normal model presented here, such as Oh and Scheuren's (1986) methods based on the Pareto distribution, might be developed to fit the right tails of the distributions more closely.

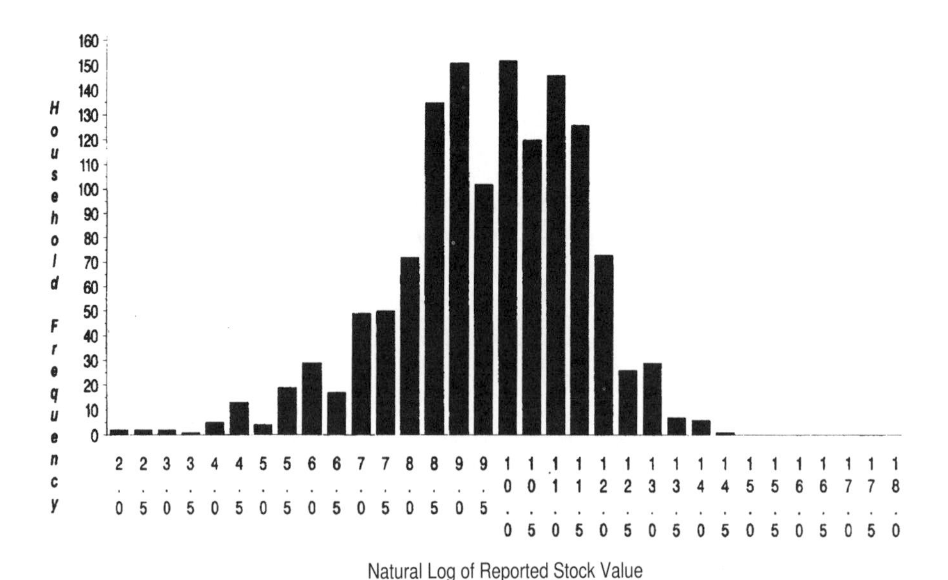

Figure 24.3 HRS Wave 1: distribution of log normal (stock value) amounts actual reports > 0.

24.3 MIXED NORMAL LOCATION MODEL

For simplicity, we describe our multivariate mixed normal location model for bivariate semicontinuous data; our model extends directly to more than two variables. Let $y_i = (y_{i1}, y_{i2})$ denote two asset or liability variables for subject i, and $T_i = (T_{i1}, T_{i2})$ define the presence or absence of these assets or liabilities for unit i, where $T_{ij} = 1$ if $y_{ij} > 0$, $T_{ij} = 0$ if $y_{ij} = 0$.

Define $L_i = (L_{i1}, L_{i2})$ and $U_i = (U_{i1}, U_{i2})$, where L_{ij} is the lower bound for y_{ij} and U_{ij} is the upper bound for y_{ij} based on the observed data. In particular, if y_{ij} is observed, then $L_{ij} = U_{ij} = y_{ij}$; if y_{ij} is completely missing then $L_{ij} = -\infty$, $U_{ij} = \infty$; and if a bounding sequence places y_{ij} between \$5,000 and \$49,999, then $L_{ij} = 5{,}000$, $U_{ij} = 49{,}999$.

Now let $z_i = (z_{i1}, z_{i2})$ be partly observed variables such that

$$(y_{i1}, y_{i2}) = \begin{cases} [(\exp(z_{i1}), & \text{if } T_i = (1, 1) \\ \exp(z_{i2})] & \text{if } T_i = (1, 0) \\ [\exp(z_{i1}), 0] & \text{if } T_i = (0, 1) \\ [0, \exp(z_{i2})] & \text{if } T_i = (0, 0) \end{cases} \tag{24.1}$$

and assume that

$$(z_i|x_i, T_i \in s) \sim N_2(\mathrm{B}_{\{s\}}x_i, \Sigma_{\{s\}}) \tag{24.2}$$

where $\{s\}$ indexes the four semicontinuous patterns defined by the vectors T, x_i is a set of q covariates, and $\mathrm{B}_{\{s\}}$ is a $(2 \times q)$ matrix of regression coefficients. The model (24.2) allows distinct $\mathrm{B}_{\{s\}}$ and covariance matrices $\Sigma_{\{s\}}$ for each semicontinuous pattern, $\{s\}$. More parsimonious forms of the model are obtained by restricting the values of these parameters across the patterns $\{s\}$. One such submodel is the multivariate analysis of variance (MANOVA) model, which assumes a common covariance matrix across all patterns, $\Sigma_{\{s\}} = \Sigma$:

$$(z_i|x_i, T_i \in s) \sim N_2(\mathrm{B}_{\{s\}}x_i, \Sigma) \tag{24.3}$$

We assume (24.3) in our application, while noting that other variants of model (24.2) are possible. Note that the MANOVA model cannot be applied directly to y_i, since the nonzero values are skewed, and the model assumption of a constant Σ is untenable; for example, the variance of a component y_{ij} is zero when $T_{ij} = 0$. The vector z_i is introduced since it can be realistically modeled via (24.3). We treat z_{ij} as unobserved whenever y_{ij} is zero, missing, or known to lie inside a bracket. The (unobserved) components z_{ij} in cells with $T_{ij} = 0$ are not used for imputing y_{ij}, since the latter are set to zero; however, the means μ_{ij} of z_{ij} in cells with $T_{ij} = 0$ are set to zero to avoid superfluous unidentified parameters (Little and Su, 1987).

We assume here that T_i is fully observed. In fact, Table 24.1 shows there are some missing data in the responses to the question of assets or liability ownership, but on average this affects fewer than 1% of households. Our methodology can be extended to handle situations where these indicators also contain missing values, by assuming a multinomial distribution for T_i; when combined with (24.3) this is a general location model (Olkin and Tate, 1961; Little and Schluchter, 1985; Little and Su, 1987; Schafer 1997) for the joint distribution of z_i and T_i.

For our Bayesian analysis, we assume the improper prior:

$$g(\mathrm{B}, \Sigma) \sim |\Sigma|^{-(p+1)/2} \tag{24.4}$$

for (B, Σ), which is a limiting form of the normal-inverse Wishart conjugate prior distribution (Schafer, 1997). The complete-data posterior distribution of (B, Σ) under this model is a normal inverse-Wishart distribution.

Any approach to imputation, estimation and inference requires an implicit or explicit model for the coarsening mechanism that underlies the patterns of bracketing and completely missing data. Throughout the empirical exercises reported in this chapter, we assume the data are coarsened at random (CAR; Heitjan and Rubin, 1991). Heeringa (1999) proposes several models of nonignorable coarsening of net worth components, but these models are not pursued here.

24.4 MULTIPLE IMPUTATION OF THE COARSENED DATA

We created multiple imputations of coarsened assets and liability values using the Gibbs sampler (Gelfand and Smith, 1990), an iterative algorithm that draws imputations from the multivariate predictive distribution of the missing components of Z, Z_{mis}, via a sequence of univariate conditional distributions for the individual components. Let $\phi_j = \{\sigma_{jj\cdot\bar{j}}, \beta_{j\cdot\bar{j}}; j = 1, \ldots, p\}$, where $\sigma_{jj\cdot\bar{j}}$ is the residual variance and $\beta_{j\cdot\bar{j}}$ the vector of regression coefficients for the regression of Z_j on $\{Z_k : k \neq j\}$ and the covariates. The prior (24.4) implies the prior

$$g(\beta_{j\cdot\bar{j}}, \sigma_{jj\cdot\bar{j}}) \propto \sigma_{jj\cdot\bar{j}}^{-1/2}$$

for ϕ_j (Schafer, 1997; Heeringa, 1999). The following sequence of conditional draws defines the Gibbs' sampler for iteration $t + 1$ in terms of these transformed parameters.

$$\begin{aligned}
&\text{For } j = 1, \ldots, p\text{:}\\
&\hat{\sigma}_{jj\cdot\bar{j}}^{(t+1)} \leftarrow p[\sigma_{jj\cdot\bar{j}} | \{(z_{ik}^{(t)}, T_i, x_i): i = 1, \ldots, n\}],\\
&\hat{\beta}_{j\cdot\bar{j}}^{(t+1)} \leftarrow p[\beta_{j\cdot\bar{j}} | \hat{\sigma}_{jj\cdot\bar{j}}^{(t+1)}, \{(\hat{z}_{ik}^{(t)}, T_i, x_i): i = 1, \ldots, n\}],\\
&\text{and for } i = 1, \ldots, n, j = 1, \ldots, p\text{:}\\
&\hat{z}_{ij}^{(t+1)} \leftarrow p[z_{ij} | \hat{\sigma}_{jj\cdot\bar{j}}^{(t+1)}, \beta_{j\cdot\bar{j}}^{(t+1)}, \{(\hat{z}_{ik}^{(t+1)} : k < j, \hat{z}_{ik}^{(t)} : k > j\}, T_i, L_i, U_i, x_i]
\end{aligned} \tag{24.5}$$

Here, $z_{ij}^{(t+1)} = z_{ij}$ if z_{ij} is observed, that is when $T_{ij} = 1$ and the amount y_{ij} is observed; otherwise $z_{ij}^{(t+1)}$ is a random draw from its predictive distribution, given current draws of the parameters. The key feature that makes this Gibbs' approach attractive computationally is that each missing z_{ij} is drawn from a univariate (possibly interval-censored) conditional normal distribution, thus avoiding the need for complex computations involving a multivariate normal distribution with interval-censored components. The draws of the parameters $\{\phi_j\}$ in (24.5) do not involve $\{(L_i, U_i)\}$, since the posterior distribution of (B, Σ) given the filled-in Z matrix $\{z_{ij}^{(t)}\}$ does not involve this interval-censoring information.

The Gibbs' sequence (24.5) converges to a draw from the predictive distribution as $t \rightarrow \infty$. The procedure is repeated with $M = 20$ different random starts to yield 20 multiply imputed data sets. Multiple imputation inference methods described in Rubin (1987) then yield estimates, confidence intervals, and test statistics for model parameters.

In simulation studies reported elsewhere (Heeringa, 1999), this methodology compared favorably with alternative approaches in terms of bias, mean squared error, and confidence interval coverage, for a variety of simulated data sets with different rates and patterns of missing and bracketed data, varying proportions of zeros in the semicontinuous data, a variety of coarsening mechanisms, and normal and nonnormal distributions for the nonzero amounts. In this chapter we focus on the application of the method to the HRS data.

24.5 IMPUTATION METHODS APPLIED TO THE HRS DATA

We now apply the method of Section 24.4 to the problem of estimating the mean and quantiles of total net worth for the HRS survey population, and compare the results with other approaches. HRS Wave 1 household net worth is computed by aggregating the 23 net worth component variables listed in Table 24.1. The 12 most important of the 23 components, namely the 10 bracketed net worth components in Table 24.1 along with values of home and first mortgage, were reimputed; other components were set to their values in the HRS Wave 1 public use data set, which were either observed or imputed by a hot deck. A refinement would have used these items as covariates in the model. Household weights were not used in the imputations of coarsened data. However, the primary factors that explain variation in household weights (age, race, gender, marital status of head) were included as hot deck cell classifiers or as fully observed covariates for the model imputations.

The following missing data methods were applied to the HRS data.

Complete-Case Analysis (No Imputation). A total of 4566 (60.0%) of the 7607 cooperating HRS Wave 1 households provided complete information on holding and amounts for each of the 23 asset and liability components. (Zero values count as nonmissing values here.) Estimates of the distribution of total household net worth were computed from these complete cases.

Mean/Median Substitution. The mean or median of observed values that fell within the known bounds was imputed for bracketed missing values. For example, if a missing observation was reported to fall in the bracket \$5,000–\$49,999, the mean of observed values for this interval (e.g., \$21,000) was imputed. If a value was completely missing (no bracket information), the overall mean or median of the observed values for that variable was imputed.

Multiple Imputation Based on a Univariate Hot Deck. A univariate hot deck method was originally used to produce a single imputation for item missing data in the HRS Wave 1 data set. Each asset and liability variable was imputed independently. All observed and missing cases were assigned to hot deck cells based on the bracket boundaries for the variable and discrete categories defined by covariate information including the age, race, sex, and marital status of the household head. Each missing value was then imputed using the observed value of a randomly selected observed case within the same hot deck cell. If a missing observation had no bracketing information, the random donor was drawn from observed values in the collapsed hot deck cell formed by the discrete cross-classifications of household head's age, race, sex, and marital status. Twenty multiply-imputed data sets were created by repeating this hot deck procedure with different random donors chosen within each adjustment cell. This is by no means the most effective hot deck method that could be envisaged in this setting. Hierarchical hot deck methods (Chapter 22), single case donors for multiple coarsened variables, or predictive mean matching of donors (Chapter 26) have the potential to make more efficient use of multivariate associations in the data.

Multiple Imputation Based on the Mixed Normal Location Model. The Gibbs' sampler algorithm of Section 24.4 was used to impute values for the bracketed and missing observations of the 12 net worth components that were reimputed in this exercise. Age, race, gender, and marital status of the household head were included as fully observed covariates. Twenty multiple imputations were created, although a smaller number like five would probably have sufficed given the fractions of missing information in this application.

To study the effect of constraining imputes in the upper tail of the component distributions, the algorithm was applied with two rules governing maximum imputed values: 1) an essentially *unrestricted* version that limited the imputed amount for cases to be no more than \$99,999,999, and 2) a *restricted* version that limited the imputed amount for cases in the highest bracket to be no greater than the highest observed value in the data set. Also, to assess the influence of the bracketing information on the imputations, a version of the Bayes algorithm was applied to the data that ignored bracketing information in imputing missing values and placed no restriction on maximum values.

Multiple Imputation Based on IVEWARE® Sequential Regression. IVEWARE® is a program for multivariate imputation of mixed categ-orical and continuous variables, developed by the Survey Methodology Program of the University of Michigan. The program is based on a sequential regression algorithm that approximates the Bayes method of Section 24.4. The algorithm takes into account semicontinuous variables and bracketing information, and is multivariate in the sense that it conditions on the information for each case (complete or incomplete). See Raghunathan et al. (1997) for details of the algorithm and software. As with the Bayes method, "unrestricted" and "restricted" versions of the algorithm were applied, depending on whether imputations of components in the highest open interval were unrestricted, or restricted not to exceed the highest observed value in the data set.

Results. Table 24.3 compares population-weighted point estimates and standard errors for means and quantiles of the distribution of total net worth. Estimates for the hot deck, Bayes, and sequential regression method are averaged over the 20 multiply imputed data sets. Standard errors for all the estimates were estimated using the jackknife repeated replications (JRR) method (Wolter, 1985), and reflect the influences of weighting, stratification, and clustering of the complex multistage HRS sample design. Standard errors for the mean and median imputation methods do not account for imputation uncertainty. Standard errors for the univariate hot deck, Bayes, and sequential regression methods were computed using the multiple imputation formulae in Rubin (1987), with the within-imputation variance being the design-based JRR variance estimate. These variances incorporate estimates of imputation uncertainty, as well as the effects of the complex sample design.

Complete-case analysis appears to markedly underestimate the distribution of household net worth for HRS households, yielding lower estimates of the overall mean, standard deviation, and quantiles. Compared to stochastic imputation alternatives, mean and median substitution also appear to underestimate the mean and percentiles of the full net worth distribution. As expected, the standard deviation of the imputed household net worth distribution produced by these deterministic imputation methods is attenuated compared to standard deviation in net worth amounts imputed by the stochastic hot deck, Bayes, and sequential regression alternatives (Kalton, 1981).

Since the univariate hot deck method can never impute an asset value greater than the largest observed value, it makes sense to compare this method with the "restricted" version of the Bayes algorithm. As seen in panels (4) and (6) of Table 24.3, the hot deck method produces lower estimated values for the mean and upper quantiles of the distribution. This finding may relate to how bracketed and completely missing data are treated under the two imputation alternatives. In cases of completely missing data—from 2% to 12% of all cases depending on the net worth component—the hot deck method performs a random draw from cells that provide no bracketing information values for the single variable. Unlike the Bayes method, the hot deck imputations do not utilize information (including bracketing) for other variables in the multivariate vector of net worth components. The estimated net worth distribution based on the Bayes method and the IVEWARE® sequential regression imputations are similar, with the Bayes estimates of the overall mean, median, and extreme quantiles being slightly higher.

As expected, relaxing the restrictions on the maximum imputed values from the

Table 24.3. HRS Wave 1 net worth imputations: estimated distribution for total household net worth (in thousands of dollars) under imputation alternatives

	1. Complete data		2. Mean imputation		3. Median imputation		4. Hot deck method		
N	4566		7607		7607		7607		
	Estimate	JRR SE	Estimate	JRR SE	Estimate	JRR SE	MI Estimate	MI SE	Imp Eff
Mean	186.8	9.1	213.5	7.7	195.9	7.3	232.5	9.4	1.20
SD	417.3	27.5	443.7	29.4	413.0	29.3	491.1	42.8	1.75
Q25	15.3	2.8	28.4	2.0	27.8	2.8	29.5	2.4	1.02
Q50	78.0	3.3	97.3	4.4	90.2	2.3	100.8	5.0	1.01
Q75	195.5	10.0	218.0	7.0	203.3	6.7	240.4	8.6	1.04
Q90	408.5	15.3	471.6	23.6	424.0	29.7	515.0	30.9	1.02
Q95	663.0	28.0	779.6	33.2	697.3	27.9	839.8	56.6	1.14
Q99	1995.0	107.1	2142.1	62.0	1837.0	133.1	2317.8	216.0	1.40
Max	6202.0	322.0	9096.5	469.4	8632.3	402.3	9644.8	3070.0	1.91

	5. Gibbs sampler—unrestricted			6. Gibbs sampler—restricted			7. Gibbs sampler—no brackets		
N	7607			7607			7607		
	Estimate	MI SE Error	Imp Eff	Estimate	MI SE Error	Imp Eff	Estimate	MI SE Error	Imp Eff
Mean	256.4	17.9	1.66	247.9	10.6	1.14	231.5	11.7	1.26
SD	827.8	494.5	2.84	598.6	77.2	1.52	625.6	219.6	1.79
Q25	28.9	2.2	1.01	28.9	2.2	1.01	27.8	2.5	1.02
Q50	99.7	4.4	1.01	99.7	4.4	1.01	96.2	5.1	1.02
Q75	240.1	9.8	1.01	240.1	9.8	1.01	230.0	8.8	1.02

Q90	537.1	25.3	1.07	537.1	25.2	1.07	493.7	19.1	1.05
Q95	902.5	54.6	1.22	902.5	54.7	1.23	800.8	38.9	1.16
Q99	2,646.5	272.0	1.29	2,642.3	264.1	1.30	2,380.3	243.3	1.38
Max	32,918.8	38,601.0	2.95	15,664.0	9,458.7	1.57	21,359.3	22,000.0	1.96

	8. Sequential regression—unrestricted			9. Sequential regression—restricted		
N		7607			7607	
	Estimate	MI SE	Imp Eff	Estimate	MI SE Error	Imp Eff
Mean	249.5	13.3	1.41	240.7	9.5	1.15
SD	695.7	227.3	2.56	526.4	41.3	1.36
Q25	28.4	2.1	1.01	28.4	2.1	1.01
Q50	99.4	4.6	1.01	99.3	4.6	1.02
Q75	241.1	10.0	1.03	241.1	9.8	1.03
Q90	538.8	23.8	1.04	537.6	24.4	1.04
95	897.3	59.2	1.07	895.9	59.7	1.09
Q99	2576.5	245.5	1.36	2505.9	229.5	1.38
Max	24,458.1	15,700.0	1.80	10,907.4	2951.0	1.48

Bayes algorithm increases the estimated mean, standard deviation, and maximum value of total net worth (panel 5 of Table 24.3). Relaxing the restrictions also produces far greater instability in the estimates of these statistics. The estimated quantiles Q25, Q50, Q75, Q90, Q95, and even Q99 are very similar for the restricted and unrestricted imputations.

The importance of the bracketing information to the multivariate imputation of net worth components is clearly seen in panel (7) of Table 24.3. The estimates of the mean, median, and other quantiles for the Bayes algorithm that does not use the bracketed information are considerably smaller than the estimates that use this information. Since the bracketing information is clearly useful for prediction, these are probably underestimates.

The columns in Table 24.3 labeled "imputation effect" are estimates of the factor by which variances of parameter estimates are increased by the uncertainty in the imputation of bracketed and missing values. The imputation effect is calculated as the ratio of the total multiple imputation variance to the within-imputation component of variance (the average estimated variance present in single imputation replicates). Imputation effects are modest for low percentiles (for example 1.01 to 1.02 for the 25th percentile), but are substantial for high percentiles of the distribution, means, and standard deviations, reflecting the influence of large imputation uncertainty in the upper tails.

24.6 DISCUSSION AND SUMMARY

We have described a method for multiple imputation of coarsened, semicontinuous multivariate household net worth components based on a mixed normal location model. Simulation experiments reported elsewhere (Heeringa, 1999) have demonstrated the utility of the method. The results of the empirical comparison of Section 24.5 show that complete-case analysis and simple deterministic imputation by mean or median substitution result in a serious underestimation of the mean and percentiles of the distribution of household net worth. The hot deck method produces some shrinkage of the distribution toward the overall mean, a result of the weak predictive power of the univariate hot deck method when no bracketing information is available. In contrast, the Bayes and sequential regression methods draw predictive strength from the other net worth components to impute completely missing observations. Although multivariate hot decks could be developed with this feature, we believe that models provide more principled and flexible ways of creating multiple imputations that condition on a large amount of predictive information. Other models, such as multivariate generalizations of Tobin's (1958) model, were also considered, but the univariate Tobin model did not fit our empirical asset and liability distributions well. The IVEWARE® sequential regression imputation program yielded similar results to those from the Gibbs' sampler, suggesting that it is a useful approximation of the fully Bayes method. This suite of SAS-compatible programs for multivariate imputation is currently available to researchers at the web site www.isr.umich.edu/src/smp/ive.

Future empirical work should examine the applicability of multiple imputation based on the mixed normal location model to a wider range of financial survey variables that are semicontinuous in nature, such as transfer payments. Future research might also focus on refinements of the distributional assumptions in the upper tails, extensions to multivariate repeated-measures data of this type, and extensions to nonignorable coarsening models.

CHAPTER 25

Modeling Nonignorable Attrition and Measurement Error in Panel Surveys: An Application to Travel Demand Modeling

David Brownstone and Thomas F. Golob, *University of California, Irvine*
Camilla Kazimi, *San Diego State University*

25.1 INTRODUCTION

Panel surveys frequently suffer from high and likely nonignorable attrition, and transportation surveys suffer from poor travel time estimates. This paper examines new methods for adjusting forecasts and model estimates to account for these problems. The methods we describe are illustrated using a new panel survey of 1500 commuters in San Diego, California. These data are collected to evaluate a federally funded "Congestion Pricing" experiment investigating the impacts of allowing solo drivers to pay to use freeway carpool lanes. The panel survey, begun in Fall 1997, collects data on travel behavior and attitudes at 6 month intervals through telephone interviews. The panel sample is refreshed with new respondents at each wave to counteract attrition between waves. Both the original and refreshment samples are stratified on commuters' mode choice (solo drive in free lanes, pay to solo drive in the carpool lanes, or carpool for free in carpool lanes) to insure sufficient sample size for estimating our models.

We illustrate this methodology using a standard conditional logit model of commuters' mode choice (solo drive in free lanes, pay to solo drive in the carpool lanes, or carpool for free in carpool lanes). The basic model is documented in Kazimi et al. (2000), and it is summarized in Sections 25.2 and 25.6 of this chapter. Our model is calibrated from the third wave of the panel study which was collected in Fall, 1998.

We use data from the second wave to estimate an attrition model and then use this model to predict attrition probabilities as described in Section 25.5. We expect nonignorable attrition because commuters who use the carpool lanes are more interested in the survey questions. It turns out that attrition does not significantly bias key parameter estimates, even though there is some indication that it is nonignorable.

We also have potentially nonignorable measurement error in the time saved by using the carpool lane. Objective measurements of time savings are available from two types of data on speeds. First, floating car observations were obtained by driving cars down the corridor at frequent intervals and recording the actual travel times. During wave 3 of the panel survey, these floating car measurements were carried out for 5 days, but the panel survey data collection involved reported travel behavior over two months. Second, point speeds derived from magnetic loop detectors placed along the corridor for general traffic counting purposes were available during the entire data collection period, but these data are subject to significant errors as described in Section 25.4.

We built a model to predict the floating car data from the loop detector data. This model fits well (R-squared of 0.9), and we use it to predict the time savings for each survey respondent as a function of the date and time of entry into the corridor. We use multiple imputations to account for the component of error in our estimates and predictions from this imputation model. Correcting for measurement error leads to significant differences in key model estimates, as reported in Section 25.6.

We view measurement error as formally equivalent to nonresponse for the true key time savings variable. Since we have external data, we can model the measurement error process and use multiple imputation (which was originally devised to handle nonresponse) to correct for measurement error. In our nonlinear model, measurement error in any independent variable causes all of the parameter estimates to be inconsistent. Even if measurement error does not bias some mean estimates in simpler models, it almost always biases standard inferences. In our application, the measurement error occurs in engineering data, but measurement errors are endemic in sample surveys (see Biemer et al., 1991; Groves, 1989; Lessler and Kalsbeek, 1992). Fuller (1987) and Carroll et al. (1995) review methods for modeling measurement error and correcting its effect on inference.

This chapter examines the impact of a number of common survey problems (nonrandom sampling, panel attrition, and measurement error) on estimates from a nonlinear model. Although only one of these problems (measurement error) led to significant biases, there is no way to know which source(s) of error are important empirically without carefully analyzing the impacts of all possible problems.

25.2 THE SAN DIEGO CONGESTION PRICING PROJECT

The pricing demonstration project (referred to as FasTrak) allows solo drivers to pay to use an eight-mile stretch of reversible high occupancy vehicle (HOV) lanes along Interstate Route 15 (I-15). The combination of free HOV use and priced solo driver use is generally referred to as high-occupancy toll (HOT) lanes. The HOT

Lanes are operated in the southbound (inbound) direction for four hours in the morning and in the northbound (outbound) direction for four hours in the afternoon and evenings. The per-trip fee for solo drivers is posted on changeable message signs upstream from the entrance to the lanes, and may be adjusted every 6 minutes to maintain free-flowing traffic in the HOT lanes. Solo drivers who subscribe to the FasTrak program are issued windshield-mounted transponders for automatic vehicle identification. Each time they use the lanes, their accounts are automatically debited the per-trip fee. This is a dynamic form of voluntary congestion pricing, in which solo drivers can choose to pay to reduce their travel time, and the payment is related to the level of congestion.

25.2.1 The Panel Survey

The panel survey consists of three samples of approximately equal size: 1) FasTrak program subscribers and former subscribers, 2) other I-15 users, and 3) a control group of users of another freeway corridor (I-8) in the San Diego area. The analysis in this chapter excludes the I-8 control group. The first wave of the panel was conducted prior to per-trip pricing. The second wave of the panel was conducted in Spring 1998, during the first few months of dynamic pricing. For the purposes of this analysis, we focus primarily on program subscribers and other I-15 users in the third wave of panel data, collected during the Fall of 1998 (October through November). During this time period, dynamic per-trip congestion pricing was well established.

FasTrak subscribers were picked at random from a list maintained by the billing agency, and the remaining respondents were recruited using random digit dialing (RDD) of residential areas along the respective corridors. In the initial wave of the panel, a partial quota sampling procedure was used to increase the number of carpoolers in nonsubscriber parts of the sample. Panel attrition is about 33% per wave, and the sample is refreshed at each wave with a new random sample of FasTrak subscribers as well as I-15 and I-8 commuters recruited using RDD sampling. The partial quota sampling procedure implies that the resulting sample is choice-based and weights are needed to represent the population of regular I-15 corridor users. We estimated weights from traffic counts carried out during the survey period.

Survey respondents were queried for detailed information about their most recent inbound trip along I-15 if that trip was made during the hours of operation of the HOT facility and covered the portion of I-15 corresponding to the facility. By design, trip lengths must be at least eight miles long (the length of the facility). There were 699 I-15 respondents with full information on morning peak-period inbound trips, divided into three modes: 1) 304 solo drivers in the main lanes, 2) 279 solo drivers using FasTrak transponders to travel in the HOT facility, and 3) 116 carpoolers who also travel the HOT facility for free.

25.2.2 Dynamic Per-Trip Tolls

Solo drivers face tolls that are a function of arrival time at the HOT facility. The level of congestion in the HOT facility determines the toll (i.e., tolls increase to avoid exceeding preset capacity constraints). While program subscribers are provided with a profile of maximum tolls that vary by time-of-day, actual tolls may be less than the maximum tolls, depending upon usage of the facility.

In October and November 1998 (excluding Thursday and Friday of Thanksgiving weekend), the actual maximum toll by time of day is flat at $0.50 before 6:30 a.m., rising in an approximately linear fashion to $4.00 over the 6:30 to 7:30 period. It stays at $4.00 in the 7:30 to 8:30 period, then falls back down to $1.00 by about 8:45 and $0.75 by about 9:30. The average actual toll paid by the survey respondents who chose FasTrak varies by time of day in a similar manner from $0.50 to a maximum of approximately $3.50 in the 7:45 to 8:00 period. Average tolls are similar across the days of the week. (Kazimi et al., 2000).

Based on the estimated arrival time at the HOT lanes, each survey respondent is assigned a toll price for that specific arrival time and date of travel. For respondents who choose to drive alone in the HOT lanes, this represents actual price paid. For solo drivers in the regular lanes and those who carpool, this represents the price they would have paid had they chosen to use FasTrak.

Arrival time at the HOT lanes is determined using a combination of information from the panel survey and speed estimates for the upstream portion of I-15. The panel survey queried respondents for on-ramp used in the morning commute and arrival time at that on-ramp. Travel time from the on-ramp to the beginning of the HOT lanes is estimated using time-of-day point speeds calculated from California Department of Transportation (CALTRANS) loop detectors embedded in the roadway. These loop detector data are computed every 6 minutes. Point speeds at loop detector locations are converted into speeds on intervening roadway segments using an algorithm that assumes that the point speed at the beginning of the segment applies to the first half of the segment and the point speed at the end applies to the second half of the segment (van Grol, 1997). Since loop detectors are placed near on-ramps, the freeway is effectively broken into segments from on-ramp to on-ramp.

25.2.3 Time Savings From HOT Lane Use

For mode choice modeling, we determine possible time saving from travel in the HOT lanes for all respondents regardless of mode choice. Time saving is defined as the difference in travel time in the HOT lanes and travel time in the parallel main lanes. Both are a function of when commuters arrive at the facility, speeds in the HOT lanes, and speeds in the main lanes. Speed in the HOT facility is assumed to be 70 miles per hour based on several days of floating car experiments. Speeds on the main lanes are estimated every 6 minutes during the entire survey period using the loop detector data. These speeds were also estimated by driving along the roadway every 15 minutes for 1 week in the middle of the survey period (referred to as floating car measurements).

The median time saving, based solely on loop detector speed measurements by time of arrival at the HOT facility, peaks at about 7 minutes in the same time period (7:30–8:00 a.m.) in which average tolls peak at $4.00. Considerable variation occurs within each half-hour time period, as indicated by the divergence among the median, 90th percentile, and 10th percentile time savings. Details are provided in Kazimi et al. (2000).

Those entering I-15 at one particular on-ramp (the Ted Williams Parkway on-ramp at the north end of the HOT lanes) may also benefit from a special dedicated entrance to the HOT facility that avoids a congested main-lane onramp with a ramp-meter traffic signal. We estimated this additional time savings for each time interval from floating car observation of queuing times, and added it to the estimated time savings from use of the HOT lanes for those respondents entering I-15 at this location (approximately 36% of the sample). These additional time savings ranged up to 5 minutes (Kazimi et al., 2000).

25.3 MODE CHOICE AND VALUE OF TIME

The key ingredient in evaluating projects designed to reduce travel time is commuters' willingness to pay for these reductions. If commuters value highly the time saved, then it may be worthwhile to make investments in new transportation infrastructure. This section reviews the model structure and estimation methods that transportation economists use to estimate value of time (VOT) from reducing delays.

25.3.1 Conditional Logit Mode Choice Models

Suppose that respondent n faces a choice of three modes for travel to work indexed by j. In this chapter, the modes are drive alone, pay to drive alone in the HOT lanes (FasTrak), or carpool in the HOT lanes. In most previous studies, the modes are automobile, bus, or subway. The conditional logit model assumes that the probability that respondent n takes mode j conditional on observed variables x_j is given by

$$P_{jn} = \frac{\exp(\theta x_{jn})}{\sum_{i=1}^{3} \exp(\theta x_{in})} \tag{25.1}$$

The value of time saved (VOT) equals the increase in cost required to keep P_{jn} constant after a small decrease in travel time. If time and cost only enter as linear terms in x, then the VOT equals $\theta_{\text{time}}/\theta_{\text{cost}}$.

Small (1992) and Wardman (1998) provide comprehensive reviews of VOT studies, and Gonzalez (1997) provides a review of the theory of consumer choice and its connection to value of time and mode choice modeling. Based on his review, Small (1992) suggests that 50% of gross wage rate is a reasonable value of time estimate. On the higher end of previous studies, Cambridge Systematics (1977) estimates that VOT for commuters in Los Angeles is 72% of gross hourly wage. These previous studies are based upon mode choice models that consider differences between transit and automobile travel, and to the extent that differences between transit and private automobiles are not captured, the results will be biased. In more recent work, Calfee and Winston (1998) attempt to avoid this problem by using stated preference data that only considers the tradeoff between travel by automobile in slower, free lanes and travel by automobile in faster, priced lanes. Their results indicate that commuters have a lower VOT than previously estimated (roughly \$3.50 to \$5.00 per hour or 15 to 25% of hourly wage). Calfee and Winston rely upon stated preference data because they lack revealed preference data for the choices involved with congestion pricing. Our results are not subject to the same potential biases associated with stated preference data as we use revealed preference data.

Given a random sample of N commuters, the model in equation (25.1) is typically estimated by maximizing the log-likelihood function

$$L = \sum_{n=1}^{N} \sum_{i=1}^{3} D_{in} \log(P_{in}) \tag{25.2}$$

where $D_{in} = 1$ if respondent n chooses mode i and zero otherwise. See Train (1986) for more information about this model and its application to transportation problems.

25.3.2 Choice-Based Sampling

It is very common for one mode to have a very low market share, which makes collecting a random sample with a reasonable sample size for each mode very expensive. For example, in the I-15 corridor, the FasTrak users account for only 3.5% of the inbound peak period trips. To reduce data collection costs, most transportation surveys stratify on mode choice, which results in a nonignorable sampling scheme.

Maximizing a random-sample likelihood function as in equation (25.2) with a choice-based sample will generally yield inconsistent parameter estimates. McFadden (see proof in Manski and Lerman, 1977) shows that for the conditional logit model with a full set of mode-specific constants, only the parameter estimators associated with these mode-specific constants are inconsistent. Scott and Wild (1986) provide similar results and give links to case-control sampling schemes. These results imply that we can use the unweighted maximum likelihood estimator for our conditional logit model. However, it is useful to consider alternative estimators that are consistent for more general choice models such as the nested logit model (see Train, 1986).

An estimator that yields consistent estimates under choice-based sampling was developed by Manski and Lerman (1977). Their weighted exogenous sample maximum likelihood estimator (WESMLE) is the maximand of the weighted likelihood function

$$\sum_{n} \omega_n L_n(\theta, x_n) \tag{25.3}$$

where L_n is the log likelihood function for the nth observation and the sampling weight, and ω_n is the inverse of the probability that the nth individual would be chosen from the population. This estimator is known as the "pseudomaximum likelihood estimator" in the survey sampling literature (Skinner, 1989). If the sampling scheme were completely random, then all of the sampling weights would be equal and the WESMLE would equal the usual maximum likelihood estimator. The WESMLE is inefficient, but Imbens (1992) gives an efficient method of moments estimator for choice-based samples.

25.4 MEASUREMENT MODEL

The loop detector data described in Section 25.2.3 can give inaccurate estimates of the actual time savings commuters get from taking the HOT lanes. Depending on the traffic flows between the loop detectors (which are miles apart on the I-15 corridor), actual speeds can be either over- or underpredicted. Since these measurement errors will generally be larger when the road is congested; the measurement errors in time savings are likely to be larger for FasTrak and carpool lane users. Since time saved using the HOT lanes is a key independent variable in the choice models in Section 25.6, this measurement error will bias key parameter estimates.

We use the 5 days during the survey period for which we have both floating car and loop detector data available to fit a model that we use to predict floating car travel time for the other 7 weeks of the survey period. These predicted floating car data are then used to fit mode choice models in Section 25.6. This approach assumes that the floating car data are correct, and we will use multiple imputations to correct for the measurement error caused by imperfect predictions.

The floating car data are collected at 15 minute intervals, whereas the loop detector data are at 6 minute intervals. To make these data compatible, we interpolated the floating car data into 6 minute intervals. The floating car estimates over the morning commutes from October 26 through October 30, 1998 are generally more than twice as large as the loop detector time savings. The median floating car time savings is 8.5 minutes, while the median loop detector time savings is 2.2 minutes. Obviously, the loop detector estimates are badly biased for this corridor.

Table 25.1 shows the best-fitting linear regression model for predicting floating car HOT lane time savings. To avoid unreasonable predictions, we first transform both time savings measures to keep them bounded between 0 and 35 minutes, which is the maximum observed loop detector time savings. The exact transformation for both time savings variables is given by the following transformed logit:

$$\log\left(\frac{\frac{t}{35}}{\left(1-\frac{t}{35}\right)}\right) \tag{25.4}$$

We tried a number of different specifications including higher-order terms in loop detector time savings and toll variables, but none of them significantly improved the fit of the model. Since the purpose of this model is accurate prediction, we are looking for the most parsimonious model with the best fit. Although including the main effect is traditional when including interactions in a regression model, including the logit of loop detector time savings results in a coefficient of 0.06 with a standard error of 0.22. Since no other coefficients were changed, we deleted the main effect to avoid inflating the variance of the model's predictions. We also experimented with lagged values, but the cubic polynomial in time effectively removes the autocorrelation in the time savings measures (residual first-order autocorrelation is 0.08).

Although the variables involving the tolls are not individually significant, they are jointly significantly different from zero at the 1% level. If they are excluded from the model, then the R^2 drops slightly to 0.89. However, excluding the loop detector data reduces the R^2 to 0.82 and increases the MSE of the residuals to 0.46.

There are two general approaches for estimating a behavioral model with measurement error in the explanatory variables: joint maximum likelihood of the behavioral and measurement models, or Rubin's multiple imputation approach. Joint maximum likelihood would be very difficult for the model in Section 25.6 since the actual explanatory variables are complicated non-differentiable transformations of the variable explained by the measurement model in Table 25.1. We will therefore implement the multiple imputation approach as given in Rubin (1987, 1996). Brownstone (1998) gives more detail using the same notation as this section. Rubin developed his methodology for missing data, and in our application floating car time savings are missing for approximately 80% of our respondents.

Suppose we want to estimate an unknown parameter vector θ. If no data are missing, then we would use the estimator $\tilde{\theta}$ and its associated covariance estimator $\tilde{\Omega}$. If we have a model for predicting the missing values conditional on all observed data, then we can use this model to make independent draws for the missing data. If m independent sets of draws are selected and m corresponding parameter and covariance matrix estimators, $\tilde{\theta}_j$ and $\tilde{\Omega}_j$, are computed for each of these imputed data sets, then Rubin's multiple imputation estimators are given by:

$$\hat{\theta} = \sum_{j=1}^{m} \tilde{\theta}_j / m \tag{25.5}$$

$$\hat{\Sigma} = U + (1 + m^{-1})B \tag{25.6}$$

where

$$B = \sum_{j=1}^{m} \frac{(\tilde{\theta}_j - \hat{\theta})(\tilde{\theta}_j - \hat{\theta})'}{m - 1} \tag{25.7}$$

$$U = \sum_{j=1}^{m} \frac{\tilde{\Omega}_j}{m} \tag{25.8}$$

Note that B is an estimate of the covariance among the m parameter estimates for each independent simulated draw for the missing data, and U is an estimate of the covariance of the estimated parameters given a particular draw. B can also be inter-

Table 25.1. Imputation model for floating car HOT lane time savings

Dependent variable: logit of floating car time savings			$R^2 = 0.90$ Root MSE = 0.36
Independent variables:	Coefficient	SE	t
Logit of loop detector time savings × minutes past 5:00 a.m.	0.0029	0.00031	9.3
Minutes past 5:00 a.m.	0.222	0.0149	14.8
(Minutes past 5:00 a.m.)^2	–0.00138	0.000121	–11.4
(Minutes past 5:00 a.m.)^3	2.73E-06	2.91E-07	9.38
Toll	–0.229	0.188	–1.22
Toll × minutes past 5:00 a.m.	0.00222	0.00126	1.77
Constant	–11.4	0.52	–22.1

preted as a measure of the covariance caused by the nonresponse (or measurement error) process.

Rubin (1987) shows that for a fixed number of draws, $m \geq 2$, $\hat{\theta}$ is a consistent estimator for θ and $\hat{\Sigma}$ is a consistent estimator of the covariance of $\hat{\theta}$. Of course, B will be better estimated if the number of draws is large, and the factor $(1 + m^{-1})$ in equation (25.6) compensates for the effects of small m. Rubin (1987) shows that as m gets large, then the Wald test statistic for the null hypothesis that $\theta = \theta^0$,

$$(\theta = \theta^0)' \hat{\Sigma}^{-1} (\theta = \theta^0) \tag{25.9}$$

is asymptotically distributed as an F random variable with K (the number of elements in θ) and ν degrees of freedom. The value of ν is:

$$\nu = (m - 1)(1 + r_m^{-1})^2$$

and

$$r_m = (1 + m^{-1}) \, \text{Trace}(BU^{-1})/K \tag{25.10}$$

This suggests increasing m until ν is large enough (e.g., 100) so that the standard asymptotic chi-squared distribution of Wald test statistics applies. We used this stopping rule and found that the models in Section 25.6.2 required $m = 20$ multiple imputations. Although this is more than the four to five multiple imputations used in most applications, recall that the proportion of missing floating car data is 80% in our application. Meng and Rubin (1992) show how to perform likelihood ratio tests with multiply imputed data. Their procedures are useful in high-dimensional problems where it may be impractical to compute and store the complete covariance matrices required for the Wald test statistic (equation 25.9).

To draw one set of imputed values for the missing floating car data, first draw one set of slope and residual variance parameters from the asymptotic distribution of the linear regression estimators from Table 25.1. The slope parameters are drawn from the joint normal distribution centered at the parameter estimates with covariance given by the usual least squares formula, $s^2(X'X)^{-1}$. The residual variance, σ_*^2, is drawn by dividing the residual sum of squares by a draw from an independent χ_d^2 distribution, where d is the residual degrees of freedom. An imputed residual vector is then drawn from independent normal distributions with mean zero and variance equal to σ_*^2. The imputed values are then computed by adding this imputed residual to the predicted value from the regression using the imputed slope parameters. Additional sets of imputed values are drawn the same way beginning with independent draws of the slope and residual variance parameters. Observations where floating car data are observed are fixed at these observed values across all imputations. This imputation method, which Schenker and Welsh (1988) call the "full normal imputation" procedure, is equivalent to drawing from the Bayesian predictive posterior distribution when the dependent variable and the regressors follow a joint normal distribution with standard uninformative priors.

For each imputed value, we add the mean time savings for those respondents entering the I-15 at Ted Williams Parkway. The medians and 90th percentiles across each month are computed for each 6 minute time interval. These medians and the difference between the 90th percentiles and the medians are then used to estimate the parameters of the choice model in Section 25.6.2. The multiple imputation procedure described here has been implemented in STATA, and it could be programmed in most modern statistical packages.

25.5 ATTRITION MODEL

The 39% attrition rate between waves 2 and 3 of our panel is not unusual for transportation panel surveys (Raimond and Hensher, 1997). The high attrition might be due to the required detailed questions about the commute trip, which respondents may find intrusive or difficult to answer. Although new respondents (the refreshment sample) are recruited for each wave to maintain sample size, it is crucial to account for attrition when analyzing these data. Once the data are collected, there is nothing to be done about the loss of efficiency due to the decreased sample size, but there are flexible modeling techniques to identify and correct for nonignorable attrition.

The simplest approach is to compare the panel sample with the refreshment sample. There do not appear to be striking differences in the distribution of key variables across these samples, but the panel sample exhibits slightly higher income and longer commute distance. Since the samples are approximately equal in size, it is also possible to fit the choice model in Section 25.6.1 separately for each sample. The hypothesis that attrition is ignorable is then equivalent to the hypothesis that the coefficients of the choice model are equal across the samples. A standard likelihood ratio test shows that this hypothesis cannot be rejected at any reasonable significance level for these data.

If there is no reasonable size refreshment sample, or if the data are used for dynamic analysis, then the attrition process can be modeled using the initial wave of the panel. The results from fitting a binomial logit attrition model show that the only significant predictors of attrition are refusal to disclose income, distance, and proportion of FasTrak use during the previous week. Commute distance enters as a quadratic term that has a maximum negative effect on attrition at 42 miles. This implies that for the relevant range of the data, longer-distance commuters are less likely to attrite. Proportion of FasTrak use is an endogenous variable in our choice models, so its significance in the attrition model implies that the attrition process is nonignorable. The higher attrition of FasTrak users might be related to the substantial number of additional survey questions administered to this group.

Unless there are significant interactions between the dependent variable and other independent variables, the attrition process described above is just another form of choice-based sampling. Therefore, unweighted maximum likelihood estimates of the conditional logit model will be consistent except for the alternative-specific constants. In our application, there are no significant interactions in the attrition model, so we will base our estimates in Section 25.6 on unweighted estimates. If there are significant interactions, then Brownstone (1998) and Brownstone and Chu (1997) show that the WESMLE estimator can be used with multiply imputed weights from the attrition model to get consistent inference.

25.6 CHOICE MODEL RESULTS

Sections 25.6.1 and 25.6.2 compare mode choice model estimates using uncorrected loop detector data and correcting for measurement error. We use a model derived from the specification in Kazimi et al. (2000). The main difference in the specifications is that here we include a variable identifying sample respondents who do not pay their own tolls. Any teenager knows that if someone else is paying (here, typically the employer), then they will be less sensitive to the price.

In addition to the parameter estimates, we also report value of time (VOT) estimates for the models in Sections 25.6.1 and 25.6.2. Since toll enters the specification both linearly and interacted with variability (the difference between the 90th percentile and the median of time saved by taking the HOT lane over the month), the VOT in dollars per hour is

$$\frac{60 \times \theta_{\text{timesavings}}}{\theta_{\text{toll}} + \theta_{\text{toll}\times\text{variability}} \times \text{variability}} \tag{25.11}$$

Since VOT varies across respondents, we give the distribution across respondents weighted by the choice-base sampling weights to match the population of morning commuters. We also give this VOT evaluated at the weighted mean of variability. This latter quantity is useful for comparison with other studies that typically do not report the variable in equation (25.11). Our definition of variability is based on the notion that commuters are much more concerned about unexpected delays than about unexpected speedy trips.

25.6.1 Loop Detector Time Savings

The left panel of Table 25.2 gives parameter estimates for the mode choice model using loop detector time savings. High-income, home-owning, middle-aged females with a graduate degree are the most likely group to pay for FasTrak. Large

Table 25.2. Conditional logit mode choice model estimates

	Loop detector data			Corrected data		
	Coefficient	SE	t	Coefficient	SE	t
FasTrak choice						
Constant	–5.978	1.994	–3.00	–7.179	3.342	–2.15
Income ≥ \$100K + refused to answer*	0.855	0.183	4.68	0.830	0.271	3.06
Income < \$40K*	–0.621	0.505	–1.23	–0.591	0.536	–1.10
Female*	0.730	0.183	3.98	0.704	0.251	2.81
Age between 35 & 45*	0.423	0.179	2.36	0.445	0.210	2.12
Has graduate degree*	0.741	0.195	3.80	0.747	0.266	2.81
Household owns home*	0.754	0.293	2.57	0.812	0.355	2.29
Distance (miles)	0.019	0.010	1.86	0.015	0.011	1.39
Toll paid by someone else*	1.747	0.454	3.85	1.816	0.633	2.87
Toll (\$/trip)	–0.787	0.220	–3.58	–0.600	0.387	–1.55
Median total time savings for commuters	0.182	0.047	3.87	0.074	0.037	2.04
Median total time savings for noncommuters	0.417	0.216	1.93	0.297	0.200	1.49
Toll × variability	0.135	0.035	3.83	0.090	0.053	1.69
Commute trip*	3.395	1.939	1.75	4.495	3.004	1.50
Carpool choice						
Constant	–2.265	1.006	–2.25	–2.139	1.145	–1.87
Workers per vehicle	1.005	0.366	2.74	0.982	0.435	2.26
Distance (miles)	0.102	0.056	1.82	0.099	0.060	1.64
Distance squared	–0.001	0.001	–1.27	–0.001	0.001	–1.23
Single worker household*	–0.973	0.350	–2.78	–1.005	0.426	–2.36
Two worker household*	–0.522	0.289	–1.81	–0.548	0.318	–1.72
Commute trip*	–1.762	0.414	–4.25	–1.747	0.588	–2.97
Median total time savings	0.144	0.045	3.19	0.056	0.033	1.71
Carpool ramp bypass*	0.556	0.278	2.00	0.634	0.315	2.01
Variability of solo drive time	0.098	0.076	1.29	0.039	0.076	0.51
	Pseudo R^2 = 0.21			Pseudo R^2 = 0.20		
	Log likelihood = –606.56			Log likelihood = –611.27		

Note: The number of observation is 699.

*These are dummy variables defined to equal one if the condition is true and zero otherwise.

households with more workers than cars are most likely to carpool. Both carpoolers and FasTrak users have similar positive coefficients for time savings, but the reduction in variability from HOT lane use is not significant. However, if variability is removed from the model then the toll coefficient drops and becomes insignificant. Relative to solo driving, commute trip drivers are more likely to choose FasTrak and noncommute trip drivers are more likely to carpool.

The middle column of Table 25.3 gives various VOT estimates (computed from equation 25.11) from the model using loop detector data. Note that the distribution is skewed and there is substantial variance across the population. The median values are much higher than Calfee and Winston's (1998) estimates, and they are on the high end of the estimates reviewed in Small (1992). These medians are similar to equation (25.11) evaluated at the weighted sample mean variability (labeled "VOT at mean variability"). This is the number typically presented in studies where VOT varies according to observed variables. Since this is just a scalar, it is straightforward to estimate the standard error of this estimate (caused by parameter estimation error) using the delta method. Although this estimate is significantly different from zero, the standard error is large enough to include almost all previous estimates. Calfee and Winston do not report standard errors for their VOT estimate of $5.00, but the $26/hour estimate in Table 25.3 is more than two standard errors away from their point estimate.

25.6.2 Predicted Floating Car Time Savings

The right panel of Table 25.2 gives the results of estimating the choice model using the predicted floating car data and multiple imputation algorithm described in Section 25.4. The coefficient estimates are roughly similar to the uncorrected loop detector estimates, but the key coefficients of toll and time savings for commuters are reduced in magnitude and significance. Overall the standard errors are considerably larger than the uncorrected loop detector estimates. This is due to the component of error caused by the error in the prediction model used to generate the predictions.

Since the floating car time savings are generally larger than the corresponding loop detector measures, we would expect that the value of time estimates would drop relative to the uncorrected loop detector estimates. The third column of Table 25.3

Table 25.3. Value of time saved estimates

Value of time (VOT) ($/hour)	Loop detector	Corrected
90th Percentile	73.63	72.12
50th Percentile	23.37	18.71
10th Percentile	14.43	-20.72
Mean	32.64	25.63
Standard deviation	94.29	74.75
VOT at mean variability	25.96	18.63
Standard deviation of VOT at mean variability	7.70	13.88

confirms this and shows that the VOT estimates have dropped \$5–\$7. While this change is quite significant from a policy perspective, it is not statistically significant, given the large standard errors of these measures.

If the error in the prediction model is ignored and only one set of imputed floating car time savings is used, then the standard errors are downward biased by over 50% for this model. Even though the prediction model fits very well, the prediction error is still an important component of the total estimation error.

25.7 CONCLUSION

This paper reviews techniques for handling attrition, choice-based sampling, and measurement error in panel surveys. Although we concentrate on commuter surveys and value of time measurement, the techniques are general and can be applied in other settings. It turns out that only measurement error is a serious problem in our application, although there is no way to know this without first carefully modeling the attrition and sampling process.

Section 25.4 shows that measurement error in travel time is a serious problem for mode-choice models. The relatively cheap measures of loop detectors and respondents' perceptions of time savings are both badly biased. When we collect additional data on all respondents' perceptions, then we can add these perceptions to our imputation models. In any case, the multiple imputations approach used here to integrate the measurement error and choice models is a good general tool for these sorts of problems. Ignoring the component of error in the choice model parameters caused by the prediction model leads to serious underestimates of the precision of the choice model parameters.

The substantive conclusions from the models in Section 25.6 are largely negative. We cannot estimate value of travel time reduction accurately enough to resolve current controversies. In particular, our confidence bands cover most existing estimates, and the differences among these estimates, are important for planning new transportation infrastructure investments. Additional work is required to combine perceived time savings loop detector savings, and floating car savings using data from more recent waves of the I-15 panel. Stated preference questions have been added to the survey so that we can jointly model responses to hypothetical and real situations. We hope the enhanced models will shed more light on the problem of evaluating time savings.

ACKNOWLEDGMENTS

John Eltinge, Rod Little, Don Rubin, and Doug Wissoker provided many useful comments on an earlier draft. We would like to acknowledge financial support from the U.S. Department of Transportation and the California Department of Transportation through the University of California Transportation Center. Arindam

Ghosh provided excellent research assistance, and Dirk van Amelsfort calculated the loop detector time savings. Additional thanks go to Jackie Golob of Jacqueline Golob Associates, Kim Kawada of San Diego Association of Governments (SANDAG), and Kathy Happersett of the Social Science Research Laboratory of San Diego State University. None of these people or agencies is responsible for any errors or omissions.

CHAPTER 26

Using Matched Substitutes to Adjust for Nonignorable Nonresponse through Multiple Imputations

Donald B. Rubin, *Harvard University*
Elaine Zanutto, *University of Pennsylvania*

26.1 INTRODUCTION

Unit nonresponse occurs in sample surveys when some sampled units do not provide any survey information (e.g., Cochran, 1983; Kalton and Kasprzyk, 1986). One approach to unit nonresponse is to replace each nonresponding unit with a new matching unit, called a survey substitute. These substitutes are contacted after originally selected units fail to respond. However, since substitutes are, by definition, respondents, they may differ systematically from the nonrespondents they replace and, as a result, may bias survey estimates if used literally as substitutes. To capitalize on the similarity between matched substitutes and nonrespondents, without requiring that substitutes be perfect replacements for nonrespondents, here we propose a new method using data from matched substitutes to help generate multiple imputations for the missing data. This methodology adjusts for systematic differences between nonrespondents and substitutes, and moreover can be used, in some cases, to adjust for nonignorable nonresponse, which occurs when the missing data mechanism depends on unobserved responses or covariates (Little and Rubin, 1987, Chapter 11).

Methods for adjusting for nonignorable nonresponse often involve strong modeling assumptions or sensitivity analysis (Rubin, 1977; Greenlees et al., 1982; Nordheim, 1984; Fay, 1986; Rubin, 1987, Chapter 6; Baker and Laird, 1988; Robins, 1997; Rotnitzky and Robins, 1997). Our methodology focuses on the sample design and collects more data in reaction to nonresponse, as in Glynn et al. (1993). However, in contrast to this related work, which assumes follow-up respondents are a random sample of initial nonrespondents, we assume only additional data from substitute respondents. Our methodology can be useful when the population has a clustered structure that is not necessarily observable and clusters have characteristics that are related to both nonresponse and outcomes. In this case, our methodology can produce less biased estimates than other nonresponse adjustment methods. Korn and Graubard (1998) examine the impact of hidden clustering on variance estimation.

After a review of the existing research on survey substitutes, we describe our proposed method and evaluate it by simulations. We conclude with suggestions of applications to proxy responses and administrative records.

26.2 RESEARCH ON SURVEY SUBSTITUTES

Survey substitutes, also called field substitutes (Chapman, 1983; Vehovar, 1999), are additional responding survey units who are either randomly selected according to a probability sampling design, often the original survey design ("random substitutes"), or chosen to match the nonrespondents on one or more characteristics ("matched substitutes"). Traditionally, substitutes' data replace data of unit nonrespondents, and the resulting dataset is analyzed as if there had been complete response.

A related approach, typically not considered as substitution, is "supplementary sampling" (Kish, 1965, p. 278), which inflates the sample size in anticipation of nonresponse so that the desired number of respondents is ultimately obtained. Respondent data are subsequently reweighted or the nonrespondent missing data are imputed to adjust for nonresponse (e.g., Kalton and Kasprzyk, 1986). Because of its similarity to random substitution, we characterize supplementary sampling as a substitution method for purposes of comparison with other methods.

A variety of surveys have used substitutes' data to replace nonrespondents' data, including the National Longitudinal Study (Chapman, 1983; Williams and Folsom, 1977), the Michigan Survey of Substance Use (Sirken, 1975), surveys conducted by Westat (Waksberg, 1985), and a variety of academic and official surveys in Europe (Vehovar, 1999). The main advantage of substitution is that it can target specific areas of the population found to have low response rates to help ensure appropriate representation of each part of the population. This aspect can be important for complex designs, especially when a small number of units is ultimately selected from each stratum or cluster (Kish, 1965, p. 558; Chapman, 1983; Lessler and Kalsbeek, 1992, p. 177; Vehovar, 1999).

The most commonly cited criticism of substitution is that it does not necessarily eliminate nonresponse bias (Birnbaum and Sirken, 1950; Kish, 1965; Nathan, 1980; Chapman, 1983; Lessler and Kalsbeek, 1992; Fowler, 1993, p. 143; Vehovar, 1999). However, this criticism can also be made of essentially all other nonresponse adjustments, certainly if they are based only on respondents. Also, as noted in Lessler and Kalsbeek (1992 p. 177), if field substitutes can be matched to nonrespondents on known correlates of the response variable, the biasing effects of nonresponse will be diminished for the same reasons that weighting class adjustments or hot deck imputation using respondents' data reduce bias.

Other disadvantages to substitution have also been cited. For example, an interviewer's effort to obtain responses from originally sampled units may decrease if it is known that a substitute can be used instead (Hansen, 1975; Chapman, 1983; Chapman and Roman, 1985; Lessler and Kalsbeek, 1992). However, Sudman (1966) reports that this problem with interviewers does not necessarily occur, and Chapman and Roman (1985) note that this issue does not arise in samples that use centralized computer-assisted telephone-interviewing system since the interview procedures for substitutes are the same as those for the original sample units. Another criticism of substitutes is that their use may prolong the field work for a survey if substitutes are contacted only after a substantial effort has been made to obtain the cooperation of the originally selected unit (Hansen, 1975; Vehovar, 1999; Waksberg, 1985). Such delays may not be the case, however, in situations such as medical studies in which eligible units are collected over time (e.g., as they present themselves). In these cases, the search for substitutes for early refusals can coincide with the search for additional primary survey units.

The topic of survey substitutes rarely appears in the technical statistical literature. Textbooks about survey sampling mention substitutes only very briefly (Kish, 1965; Lessler and Kalsbeek, 1992; Fowler, 1993, p. 51 and p. 143) or not at all (Cochran, 1977; Groves, 1989; Särndal et al., 1992). Kish and Hess (1959) apparently propose one of the earliest substitution methods. Kish (1965, p. 549), however, later advises against field substitution, stating that the substitutes resemble the respondents rather than the nonrespondents.

In a summary article on the use of substitutes, Chapman (1983) found just four empirical studies of substitution procedures (Durbin and Stuart, 1954; Cohen, 1955; Sirken, 1975; Williams and Folsom, 1977). These studies attempt to examine the bias associated with one or more substitution procedures, but only Durbin and Stuart (1954) include a comparison with the estimated biases for one or more alternative imputation procedures. In that study, there was little difference between key estimates based on the gold-standard unlimited callback procedure and those based on the best substitution procedure (choosing substitutes if no response was obtained after the third call), although the authors noticed that the substitutes tended to be somewhat younger and less affluent than the respondents who cooperated after three calls.

The studies by Cohen (1955) and Sirken (1975) suggest that there can be significant differences between substitutes and nonrespondents. However, these differences may be due in part to the differences between early and late responders. Cohen selected a substitute if the primary unit did not respond after the first call. Sirken made only one call to each substitute before approaching another potential substitute.

Williams and Folsom (1977) also found significant differences between survey estimates using substitution with weighting and estimates using follow-up data from the nonrespondents. Unfortunately, the impact of the use of substitutes was not separated from the impact of the nonresponse weighting adjustments.

More recently, Biemer, Chapman, and Alexander (1985) studied the use of substitutes in a random-digit-dial (RDD) telephone survey. A substitute was chosen for each nonrespondent from within the same primary sampling unit as the nonrespondent. They found that the variance estimate for a substitution-based estimator was smaller than the variance estimate for a weight-adjustment-based estimator from an alternative equal-cost sample that did not use substitutes. However, a follow-up sample of nonrespondents found that substitute households contained, on average, a higher proportion of female and older occupants than the corresponding nonrespondents. This may have occurred as a result of time constraints in the substitution collection phase that made substitutes more like early than late responders in the original sample. Chapman and Roman (1985) report similar results in their study of the use of substitutes in an RDD survey. Vehovar (1999) reports that substitutes introduced a "considerable bias" in the Slovenian Labour Force Survey and the Slovenian General Social Survey. He also reports similar problems with the use of substitutes in the Italian Family Expenditure Survey.

In summary, empirical research suggests that substitutes are clearly not perfect replacements for nonrespondents. Moreover, to the extent that substitutes resemble early responders, substitution methods may introduce more bias than other nonresponse adjustment methods, such as weighting adjustments (that reweight the original sample, which contains both early and late responders).

Nathan (1980) and Vehovar (1999) examine the use of substitutes from a more analytic perspective. Nathan (1980) compares a sample size inflation method to a simple random field substitution method for the estimation of a population proportion. He shows that both methods require approximately the same sample size to attain the same accuracy and result in the same expected costs. However, the substitution method provides tighter budgetary control and does not rely on having an accurate estimate of the nonresponse rate at the sample design stage.

Vehovar (1999) presents an analytic and empirical comparison of the variance of estimators of the population mean using a two-stage cluster sample with random field substitution and, alternatively, sample size inflation combined with a sample weighting adjustment. The substitution method replaces each nonrespondent with a randomly selected substitute from the same cluster. The sample-size-inflation method increases the initial sample size in each cluster to obtain, in expectation, the desired sample size and then adjusts the sample weights of the respondents within a cluster to be inversely proportional to the estimated response rate in that cluster. Assuming that the data is missing completely at random (Little and Rubin, 1987, p. 14) within clusters so that there is no response bias, Vehovar showed that the improved precision obtained by using random field substitutes is relatively small except for situations with a large nonresponse rate and small average sample size within clusters. Based on empirical evaluations and the assumption that time constraints yielded early responders as substitutes, he concluded that the potential bias introduced by substitutes (relative to using sample size inflation in combination with a sample weighting adjustment) dominates the negligible increase in precision.

Apparently, therefore, existing analytic research is limited to the comparison of random substitution to similar sample size inflation methods and suggests that using random substitutes can lead to moderate gains in efficiency, which, however, are typically overshadowed by potential biases.

Titterington and Sedransk (1986) evaluate an imputation method that replaces each nonrespondent's missing response, y, with the response from a donor respondent that matches the nonrespondent on a continuous background covariate, x, available for all units. A key conclusion from their simulations is that, in terms of absolute expected bias and mean squared error of the estimated population mean, replacing missing responses with donor responses generally works well as long as the donors are well matched to the nonrespondents. If only poor matches are available (as may be the case with survey substitutes, since substitutes are often early responders), then a model, such as using linear regression, that imputes y based on a nonrespondent's x-value may improve the imputations, especially if this model is fit using those respondents that were chosen to be imputation donors.

Although this research suggests that substitutes' data should not be used to replace nonrespondents' data, it does not imply that substitutes are completely useless. Intuitively, if data from substitutes were available for nonrespondents, it is not clear that the only reasonable approach would be to ignore this extra information and use one of the traditional nonresponse adjustment methods such as reweighting or imputation (e.g., Kalton and Kasprzyk, 1986; Rubin, 1987). In particular, there may be important similarities between nonrespondents and substitutes that can be exploited through statistical modeling to help impute for nonresponse. The goal of our research is to develop such imputation methods.

26.3 MULTIPLE IMPUTATION USING SURVEY SUBSTITUTES

We propose a method, which we call "matching, modeling, and multiple imputation" (MMM), that uses matched substitutes to help impute for missing data due to nonresponse. In this approach, substitutes are matched to nonrespondents to be similar with respect to background covariates that are both available prior to the survey and convenient for matching ("matching covariates"). As a result of this matching, nonrespondents and their substitutes may share similar values of other covariates that are only implicitly observed and are therefore not available for data analysis ("field covariates"). For example, if neighbors are substituted for nonrespondent households, they may have similar socioeconomic characteristics, experience similar levels of crime, or have similar access to public transportation, even though these characteristics may not be recorded. Rather than the usual approach of using the substitutes directly to replace nonrespondent data, we use them together with respondent information and background covariates available for both the nonrespondents and substitutes ("modeling covariates") to build a model to multiply impute nonrespondent data. To help fit this model, substitutes are also chosen for some respondents that resemble the nonrespondents. Once the nonrespondents' data are multiply imputed, all data from the substitutes (for both respondents and nonrespondents) are discarded. The use of multiple imputation (Rubin, 1987) facilitates point and variance estimation, even for complicated estimands and complex surveys (e.g., Schafer et al., 1993, 1996).

A simple example of our methodology is given here and evaluated through simulation in Section 26.4. More complex models may often be required in practice, but the example conveys the basic idea. Suppose that the following relationship holds in the population,

$$y_i = \beta_0 + \beta_1 x_{\text{model},i} + \beta_2 x_{\text{field},i} + \varepsilon_i \tag{26.1}$$

where x_{model} is a modeling covariate, x_{field} is a field covariate, and ε is a random error term, $\varepsilon_i \sim N(0, \sigma^2)$. We define a modeling covariate to be a variable that can be included in statistical models to adjust for observed differences between nonrespondents and their substitutes but that may not be available or used for matching. For example, when age and address covariates are available for all units in the population prior to sampling, it may be difficult to match on both age and address. Therefore, we may choose to match only on address (e.g., choosing a neighbor to be a substitute), but we still can adjust for systematic age differences between nonrespondents and substitutes through statistical modeling. In other examples, modeling covariates may not be available for all units prior to the survey, but are obtained during the survey for respondents, nonrespondents, and substitutes. For example, the initial questionnaire in a survey such as the National Health and Nutritional Examination Survey (U.S. Department of Health and Human Services, 1994) provides modeling covariates for participants who fail to complete the subsequent physical examination. Similarly, the first phase of a two-phase sampling design can also provide modeling covariates.

We define a field covariate to be a covariate that can be used in matching but cannot be included in statistical models. For example, if a substitute is chosen by matching on address, then the nonrespondent and substitute may share other characteristics related to their neighborhood (e.g., they are exposed to the same general neighborhood noise level). These characteristics may be related to both survey response and the probability of response. Traditional nonresponse adjustments, such as weighting and imputation, may fail to take many of these field covariates into account because such information is not recorded; for example, it may be impossible to identify high-noise and low-noise areas for use in nonresponse weighting adjustments. Furthermore, attempts to include detailed factors, such as address, in statistical models through the use of indicator variables to represent classes of units, such as neighborhoods, are equivalent to coarsened versions of the substitution matching procedure; in this case, substitutes are selected from the pool of survey respondents by matching on class index rather than detailed address.

Even if model (26.1) is known to hold in the population, it cannot be used to impute for nonresponse because x_{field} is not directly observed. However, since substitutes are chosen from the collection of unsampled survey units, it follows that the relationship in (26.1) also holds for the substitutes, and, as a result, imputations can be based on the following model:

$$y_i = y_i^s + \beta_1(x_{\text{model},i} - x^s_{\text{model},i}) + \varepsilon_i', \tag{26.2}$$

where y_i^s, $x^s_{\text{model},i}$, and $x^s_{\text{field},i}$ denote the y, x_{model}, and x_{field} values, respectively, of the substitute for unit i. Here we have assumed that $x^s_{\text{field},i} = x_{\text{field},i}$ as a result of the matching process. In practice, this will not always be true, and future research should more systematically examine the relationship of the resulting bias to cluster size and response propensities within the clusters. In the simple case where x_{field} is an indicator variable denoting cluster membership (as in the simulations in Section 26.4), the chance of a mismatch in x_{field} between pairs of nonrespondents and their substitutes should decrease as the cluster size increases and as the probability of the response for members of the cluster increases. In some surveys, it may be possible to find primary survey respondents who match the nonrespondents in terms of x_{field}, making it unnecessary to obtain substitutes for these nonrespondents.

To fit model (26.2) we choose substitutes for a few respondents who are similar to the nonrespondents in terms of their distribution on x_{model}. (Simulations in this paper choose substitutes for $n^* = 0.3n_m$ respondents, where n_m is the number of nonrespondents. The choice of n^* affects both the cost of the survey and the precision of the estimates. Future research will explore this cost–variance trade-off and will develop guidelines for the choice of n^*.) Assuming the same relationship holds between the differences in y for pairs of (a) respondents and substitutes and (b) pairs of nonrespondents and substitutes, we can then multiply impute the missing data based on a model, such as

$$y_i \sim N(y_i^s + b_0 + b_1(x_{\text{model},i} - x^s_{\text{model},i}), \sigma^2). \tag{26.3}$$

An intercept parameter, b_0, is included in the above model to help reduce bias due to possible model misspecification. Ideally, to minimize the impact of our modeling assumptions, substitutes should be chosen for respondents who are similar to the nonrespondents in terms of their distribution on both x_{model} and x_{field}, since this limits the covariate space over which we rely on the model. Because using well-matched substitutes should make it unnecessary to include x_{field} in model (26.3), it is most important to choose respondents who are similar to the nonrespondents in terms of the distribution of x_{model}.

26.4 A SIMULATION STUDY

26.4.1 Three Artificial Populations

To illustrate the performance of our new methodology relative to other commonly used nonresponse adjustment methods, we summarize a few simulation results. We focus on estimates of the population mean, $\overline{Y}$, using estimates based on the sample mean under simple random sampling from three simulation populations, each with 10,000 units generated according to the outcome (y) and response models in Table 26.1. Here I_{field} denotes a dichotomous field covariate; x_{model} denotes a modeling

Table 26.1. Populations used for simulations

Population	Outcome model	Response model
1	$y = x_{\text{model}} + 5I_{\text{field}} + \varepsilon$	$p = \dfrac{0.95}{1 + \exp(-1.25 + x_{\text{model}} + I_{\text{field}})} + 0.05$
2	$y = x_{\text{model}} + \varepsilon$	$p = \dfrac{0.95}{1 + \exp(-0.95 + x_{\text{model}})} + 0.05$
3	$y = x_{\text{model}} + 5I_{\text{field}} + z + \varepsilon$	$p = \dfrac{0.95}{1 + \exp(-1.35 + x_{\text{model}} + I_{\text{field}} + z)} + 0.05$

covariate; z denotes a covariate that is not available for either modeling or matching; and p denotes the probability of response. In these simulations, x_{model}, z, and are independent $N(0, 1)$ random variables. Populations 1 and 3 contain hidden clusters, represented by the indicator variable, I_{field}, which has a value of one for 100 clusters of 25 consecutive units in the population, and a value of zero for the remaining units (i.e., the population list contains 75 units with $I_{\text{field}} = 0$, followed by 25 units with $I_{\text{field}} = 1$, followed by 75 units with $I_{\text{field}} = 0$, and so on.) This field covariate can be used by matching on index number (i.e., the simulation equivalent of "address"). Population units who are close in index number are likely to share the same value of I_{field}. In these simulations, a substitute is the closest unsampled unit (in terms of absolute difference in index numbers) to the nonrespondent who turns out to be a respondent.

The relative performance of the nonresponse adjustment methods studied here depends on the relative contributions of x_{model} and I_{field} to y and to p. Population 1 was chosen to represent a situation that is well suited to the use of our MMM methodology. Population 2 represents a difficult situation for MMM. Finally, none of the estimation methods is ideally suited for Population 3. The response models were chosen to give an average response rate of 70% in each population. Similar response rates are reported in NSF's 1993 National Survey of Recent College Graduates (75%), NORC's 1996 General Social Survey (76%), and the U.S. Census Bureau's 2000 decennial census mail returns (65%).

26.4.2 Six Nonresponse Adjustment Methods

We compare six nonresponse adjustment methods. Each method uses data from a total of $n = 300$ respondents on average, including initial respondents, substitutes (if any), and follow-up respondents (if any). Because the last three methods are designed to deal with nonignorable nonresponse, they are expected to produce less biased estimates than the first three methods. These first three methods are included to illustrate several strengths and weaknesses of survey substitutes. For comparison, a "pseudoestimate" of the population mean based on the sample mean assuming there is complete response is calculated to measure the length of a 95% confidence interval under ideal conditions. A detailed description of each of the six methods follows.

1. Sample Size Inflation (INF). To obtain n respondents, on average, when the average response rate is $\overline{p}$, sample $n' = n/\overline{p}$ units. (We use the known value of $\overline{p}$ in these simulations). Then $\hat{\overline{Y}}_{\text{INF}}$ is the mean of the respondents, and its variance is estimated by

$$Y(\hat{\overline{V}}_{\text{INF}}) = \frac{s_r^2}{n_r'}\left(1 - \frac{n_r'}{N}\right) \tag{26.4}$$

where s_r^2 denotes the sample variance of y for the n_r' respondents and $1 - n_r'/N$ is the finite population correction factor.

2. Substitute Single Imputation (SSI). This is the traditional use of substitutes. Select an initial sample of size n. Then, for each nonrespondent, choose one substitute from the pool of unsampled units by selecting the unit who is closest to the nonrespondent in terms of index number. If this unit does not respond, contact the next closest unsampled unit, repeating this process until a substitute who responds is found. Then, treating the resulting data set as if there had been complete response, calculate the usual estimate of the population mean and corresponding variance estimate [as in (26.4)].

3. Multiple Imputation Using Respondents (MI-0). Sample $n/\overline{p}$ units to yield n respondents on average. Use the data from the respondents to fit a normal linear regression model:

$$y_i \sim N(\beta_0 + \beta_1 x_{\text{model},i}, \sigma^2) \tag{26.5}$$

with a noninformative prior distribution $p(\beta_0, \beta_1, \sigma^2) \propto \sigma^{-2}$ on the parameters, to create m sets of multiple imputations for the missing data using the algorithm given by Rubin (1987, p. 167) ($m = 10$ in these simulations). Each set of imputations forms a completed data set. By calculating the mean of the completed data set and the corresponding variance estimate, each completed data set yields an estimate of the population mean and the standard error of this estimate. Combining these estimates using the multiple imputation combining rules (Rubin 1987, p. 76) gives one overall estimate of the population mean and the standard error of this estimate.

4. Multiple Imputation Using Complete Follow-Up on a Random Sample of Nonrespondents (MI-1). Select an initial sample size of $n/[\overline{p} + (1 - \overline{p})(0.3)]$. Follow up on a random sample of 30% of the nonrespondents. We assume, unrealistically, a 100% response rate to the follow-up survey, but relax this assumption in the MI-.7 method (described below). Following the proposal by Glynn et al. (1993), use only the follow-up data to fit a normal linear regression model (26.5) to create

$m = 10$ multiple imputations for the nonrespondents who were not included in the follow-up sample.

5. Multiple Imputation Using Incomplete Follow-Up on a Random Sample of Nonrespondents (MI-.7). Select an initial sample size of $n/[\bar{p} + (1 - \bar{p})(0.3)(0.7)]$. Follow up on a random sample of 30% of the nonrespondents. Unlike the MI-1 method, we assume, realistically, that only some of the nonrespondents selected for follow-up actually respond to the follow-up survey: specifically, we assume that the 70% of the follow-up sample with the largest probabilities of response respond during follow-up. Using data from the successful follow-ups, fit model (26.5) to create $m = 10$ imputations for each of the remaining nonrespondents. The names MI-1, MI-.7, and MI-0 denote that these are multiple imputation methods that use follow-up data from 100%, 70%, and 0%, respectively, of the units in a random sample of initial nonrespondents.

Because more intensive effort is usually used to obtain responses during follow-up than during the initial survey, it may appear conservative to assume that the follow-up response rate is no greater than the initial nonresponse rate (which averages 70% in this simulation). However, the follow-up sample consists of initial nonrespondents who may be systematically more difficult to reach than the initial respondents. Therefore, we believe that 70% is a reasonable estimate for the follow-up response rate.

6. Matching, Modeling, and Multiply Imputing (MMM). Select an initial sample of size $n/[1 + 0.3(1 - \bar{p})]$. As in the SSI method, choose one substitute for each nonrespondent by matching on index number. Also choose n^* substitutes for a subset of the respondents (in these simulations, we use $n^* = 0.3n_m$ where n_m is the number of nonrespondents). Then create $m = 10$ sets of multiple imputations for the missing data using the MMM method described in Section 26.3.

26.4.3 Measures of Performance

An estimate of the population mean is calculated from each of 5000 samples for each of the six nonresponse adjustment methods. The performances of the six methods are evaluated by calculating three measures of error:

1. Percent reduction in the average bias of $\hat{\bar{Y}}$, relative to the bias of the estimate resulting from using respondents from an inflated sample (the INF method).
2. Actual coverage of the nominal 95% confidence interval for $\bar{Y}$.
3. Percent increase in the average length of a 95% confidence interval for $\bar{Y}$ relative to the length of the confidence interval obtained from the analogous estimate (and corresponding standard error) when there is complete response. A t-distribution with degrees of freedom given by the multiple imputation combining rules (Rubin, 1987, p. 77; Barnard and Rubin, 1999) is used for the multiple imputation methods. A normal approximation is used for all other methods.

Table 26.2 shows the relative performance of the six nonresponse adjustment methods in Populations 1, 2, and 3. The average percent reduction in average bias relative to the INF method and the average actual coverage of a nominal 95% confidence interval for the population mean are presented for each method in each population. The methods are ordered in the table approximately according to their average actual confidence coverage: methods with good coverage appear towards the bottom of the table. The MI-1 results are presented for comparison only because unrealistic assumptions (i.e., 100% follow-up response) must hold in order to achieve the favorable results shown in this table. Comparisons of the average length of a 95% confidence interval for the population mean are restricted to methods that provide essentially unbiased estimates of the population mean and are presented in the text summarizing the simulation results.

As a measure of the precision of this simulation, the 5000 samples yield estimated coefficients of variation for the percent reduction in the average bias of $\hat{\bar{Y}}$ relative to the bias of $\hat{\bar{Y}}_{\text{INF}}$ of less than 1.5%.

26.4.4 Population 1 Results

In Population 1, the INF estimate has a large bias, and, therefore, disastrous confidence coverage. Because respondents tend to have small values of x_{model} and I_{field} and, as a result, small values of y, the INF estimate underestimates the population mean. The SSI estimate is biased because substitutes may match nonrespondents on I_{field}, but substitutes are, by definition, respondents and so tend to have small values of x_{model} and, therefore, small values of y. The MI-0 estimate is also biased and has poor confidence coverage because it uses only the information in x_{model} and not I_{field}.

Table 26.2. Simulation results for estimates of $\bar{Y}$ using six estimation methods in three artificial populations

	Population 1		Population 2		Population 3	
	Percent reduction in average bias	Actual coverage of 95% C.I.	Percent reduction in average bias	Actual coverage of 95% C.I.	Percent reduction in average bias	Actual coverage of 95% C.I.
INF	0*	7	0*	13	0*	1
SSI	46	55	1	14	27	17
MI-0	44	52	100**	95	27	14
MI-.7	41	70	100**	95	60	75
MMM	95	95	100**	96	59	81
MI-1***	100**	95	100**	96	100**	95

*Zero by definition.

**100 by construction, for the simple estimand $\bar{Y}$.

***MI-1 results shown for comparison only, since this method assumes 100% nonresponse follow-up, which is usually not possible in practice.

Both the MMM and MI-1 estimates are essentially unbiased in this population and yield confidence intervals with actual coverages of approximately 95%. However, the average MMM confidence interval is 39% wider than the average interval that would be obtained if there were complete response (which is, of course, impossible), whereas the average MI-1 confidence interval is only 25% wider. The width of the MMM confidence interval can be decreased by increasing n^*, but with the accompanying financial costs. Nevertheless, the advantage of the MMM method is clear when compared to the MI-.7 method, which assumes that it is not possible to contact all nonrespondents in the follow-up sample. The MI-.7 method leads to only a 41% average reduction in bias in the estimate of the population mean relative to using only sample size inflation (INF), which is much less than the 95% average reduction using the MMM method. This difference occurs in this population because the nonresponding units in the follow-up sample are high leverage points.

Furthermore, graphs of the resulting imputed datasets show that, in general, the MMM method more accurately reflects the distribution of the two groups in the population (defined by I_{field}) compared to the MI-1 and MI-.7 methods. Although this difference may not be important for simple estimands such as the population mean, it may be important when using the data for more complex statistical modeling.

26.4.5 Population 2 Results

In Population 2, the INF estimate is again badly biased because the average outcome of the respondents underestimates the population mean. The SSI estimate also greatly underestimates the population mean because in this population the substitutes are essentially just additional randomly chosen respondents. The remaining estimates are all essentially unbiased with actual confidence interval coverages of approximately 95%. (The MMM and MI-1 intervals have coverages that are slightly larger than 95% in these results due to simulation variability.) However, the average length of the MMM confidence interval is 89% wider than the average length of the interval that would be obtained under complete response, whereas in these simulations, the average MI-.7 interval is only 36% wider, the average MI-1 interval is only 8% wider, and the average MI-0 interval is 4% shorter. The MI-0 method is essentially equivalent to complete response in this case because the multiple imputation model agrees exactly with the model that generated the population, but has the advantage of using x_{model} data from an average of 129 additional units compared to the complete-response method, resulting in a slightly shorter interval than the complete-response interval. Again, increasing n^* would reduce the width of the MMM confidence interval, but with an associated increase in costs. Therefore, when substitutes are chosen by matching on a variable that is unrelated to both the response and the probability of response, the MMM method leads to a loss in efficiency, but no increase in bias.

26.4.6 Population 3 Results

In Population 3, there are unobservable systematic differences between respondents and nonrespondents due to the unobserved covariate, z. As expected, the MI-1 estimate of the population mean is essentially unbiased in this population. The MMM procedure, on the other hand, removes only 59% of the bias relative to using only sample size inflation in these simulations. Thus, in this situation, the MMM method is more sensitive to model misspecification than the MI-1 method, which assumes 100% follow-up response. However, the reduction in bias with the more realistic MI-.7 method is only 60% (similar to the MMM method), showing that the success of using follow-ups depends on complete follow-up of the random sample of nonrespondents.

26.4.7 Simulation Summary

These simulations suggest that the MMM method can be used to adjust for nonignorable nonresponse caused by hidden clustering, when suitable matching and modeling covariates are available. Compared to following up on nonrespondents, which is the closest competitor to MMM in these simulations, the MMM method has the advantage of being applicable even in cases where it is not possible to follow up a random sample of nonrespondents, and also has the flexibility to incorporate information from a variety of types of substitutes. Practical limitations of MMM include the availability of matching variables that are correlated with the field covariates, and modeling covariates that can account for systematic differences between substitutes and nonrespondents. Furthermore, the cost and logistics of obtaining responses from the required number of substitutes need to be considered when evaluating the potential benefits of this method in practice.

26.5 ADMINISTRATIVE RECORDS AND PROXY RESPONSES AS SUBSTITUTES

Our MMM methodology can also be applied to incorporate information from administrative records or proxy respondents into survey estimates. Like substitutes, proxy respondents are contacted when the primary sample unit fails to respond, but, unlike substitutes, they provide information about the primary sample unit rather than about themselves. Similarly, like substitutes, administrative records can be consulted, if a suitable administrative records database is accessible, when the original unit fails to respond to the survey. More important, both of these sources are like substitutes in that the information they provide may be related to the primary sample unit, but may also contain systematic errors (e.g., Boyle and Brann, 1992; Magaziner et al., 1996, 1997; Zanutto and Zaslavsky, Chapter 27, this volume). Following our general methodology, when using these sources of information to impute for nonresponse, proxy respondents or administrative records should be selected for some of the respondents in addition to those selected for the nonrespondents. Then a multiple imputation model can be fit to impute the missing data, avoiding the bias introduced by the traditional method of using administrative records or proxy responses to replace nonrespondents' missing information.

Zanutto and Zaslavsky (1996, 1997, Chapter 27, this volume) describe a specific example of this type of statistical modeling to impute the characteristics of nonrespondents to the U.S. Decennial Census using information in administrative records. This example is simpler than the simulations of the previous section in that there are no modeling covariates, but the statistical model used is more complex.

Our methodology is particularly well suited for incorporating proxy responses into survey estimates in cases where proxy responses for some survey respondents occur naturally. For example, in telephone surveys, a proxy response may be obtained for a primary sampling unit who is unavailable at the time the call is made. However, a response from the primary unit may be obtained from later callbacks, resulting in a recorded response from both the original unit and the proxy. Previous research has focussed on evaluating whether proxy responses can substitute for primary responses to questions on specific topics such as diet, smoking, occupational exposure, and overall health (e.g., Boyle and Brann, 1992; Magaziner et al., 1996, 1997). Our methodology extends the potential usefulness of proxy responses to cases where they are not perfect replacements for the responses of the originally selected units.

CHAPTER 27

Using Administrative Records to Impute for Nonresponse

Elaine Zanutto, *University of Pennsylvania*
Alan Zaslavsky, *Harvard University*

27.1 INTRODUCTION

Administrative records are a relatively inexpensive source of detailed information, especially as technology increases our ability to manipulate large data sets. In particular, they can be used to impute for both unit and item nonresponse. However, since administrative records can differ in coverage, content, and reference period from the survey, simply substituting administrative records for the missing responses could introduce bias into the survey estimates. Instead, we propose using administrative records as covariates. We model the relationship between survey and administrative data for respondents and use the estimated relationship to impute data for the nonrespondents based on their administrative records. Where the administrative data can be regarded as an imperfect measure of the values elicited on the survey, this process corrects for systematic differences between the two data systems. We begin with a general discussion of administrative records and their statistical utility. We then give a theoretical framework for using them to impute for survey nonresponse and conclude with an example.

The broad range of functions assumed by government and other large organizations and their growing information-processing capability generate increasingly many forms of administrative records. Brackstone (1987) summarized six broad purposes for administrative records:

1. Records maintained to regulate the flow of goods and people across borders, including import, export, immigration, and emigration records
2. Records resulting from legal requirements to register events such as birth, deaths, marriages, divorces, business incorporations, and licensing

3. Records needed to administer benefits or obligations such as unemployment insurance, pensions, health insurance, and taxes
4. Records needed to administer public institutions such as schools, courts, prisons, and health institutions
5. Records arising from government regulation of industry such as banking and telecommunications
6. Records arising from the provision of utilities such as electricity, telephone, water, and mail delivery

Another important class of record system is the population register, based on requirements in many countries for legal residents to register with local authorities.

27.2 STATISTICAL USES OF ADMINISTRATIVE RECORDS

Administrative records have a broad range of potential statistical uses. Their characteristics as information sources are determined by their primary use for administrative purposes such as enforcement, regulation, accounting, and service delivery. For these purposes, identification of individual units and a specific set of characteristics associated with them is crucial. Hence, administrative records can be useful wherever nonsampled coverage is important, or where the item content of the record matches an information need.

Administrative records can be used to create, supplement, or update frames for surveys. See, for example, Brackstone (1987) for applications to business surveys, or U.S. Bureau of the Census (1978) for applications to the Current Population Survey.

In Denmark, Finland, Norway, and Sweden, population registers can be used as the primary source for conducting a population census (Myrskylä, 1991; Redfern, 1989). In the United States, where population registration is not practiced even locally, researchers have proposed linkage of government databases with high population coverage, such as Internal Revenue Service (tax) and Social Security Administration (retirement insurance) files, to form the basis for an administrative record census (Alvey and Scheuren, 1982; Marquis et al., 1996; Sailer et al., 1993). However, questions remain about the accuracy and coverage of the combined database, the technical feasibility of a record linkage operation of this magnitude, and maintaining confidentiality.

Administrative records have been used to evaluate population estimates, using a sample of records. Zaslavsky and Wolfgang (1993) found that administrative records, used as a third source in triple-system estimation to evaluate census coverage, identified a substantial number of people missed both by census and the Post-Enumeration Survey.

Administrative records play an important role in intercensal estimates. Tax data in adjacent years can be used to estimate population migration (Verma and Parent,

1985; Kozielec, 1995). Birth and death records are also used to estimate age-specific components of local population change (Long, 1990, 1996; Batutis, 1993).

Many key measures of the population and economy, such as estimates of births and deaths and components of national accounts, are largely based on aggregates of administrative records. More complex uses of such aggregates involve modeling, in combination with other data. Intercensal small-area estimates of the numbers of school-age children in poverty in the United States by county are predicted from three county-level aggregates based on administrative records (the number of food stamp recipients and the numbers of child exemptions reported on tax returns by all families and by families in poverty) (Citro et al., 1997; Siegel, 1995).

Administrative aggregates can also be used in models to improve the quality of survey-based databases. Dorinski (1995) and Huggins and Fay (1988) describe adjustment of data from the Survey of Income and Program Participation (SIPP) to control totals from the Internal Revenue Service and the Social Security Administration, which reduced the variance of estimates for most income and program participation variables. The MATH microsimulation model (Schechter et al., 1997; Thurston and Zaslavsky, 1996; Lewis, 1999) adjusts for underreporting of income assistance recipiency in surveys (SIPP or CPS) by randomly imputing enough survey-reported nonrecipients to be recipients so aggregates agree with administrative totals.

Administrative records provide a wealth of information that can be used as primary sources for research. Such studies are common in epidemiology, public health and health services research, and in the social sciences. Records from different administrative records databases can be linked together or to national surveys to obtain more data on individuals than either source alone can provide (e.g., Waien, 1997; Chamberlayne et al., 1998). Lillard and Farmer (1997) describe linkages among the National Death Index, Medicare administrative data, the Longitudinal Study of Aging, and the Panel Study of Income Dynamics to study interactions of health status, insurance coverage, and utilization of health care services among the elderly. File linkage is straightforward where there is a common unique identifier, such as Social Security number. Otherwise, probabilistic record linkage methodology can identify records from different databases that are likely to correspond to the same individual (Newcombe et al., 1959; Fellegi and Sunter, 1969; DeGroot, 1986; Newcombe, 1988; Winkler, 1995).

Record linkage can help to evaluate the accuracy of survey responses, when the information in the record is more reliable than that in the survey. Moore and Marquis (1989) describe a comparison of administratively recorded payments from nine federal programs with those reported in SIPP. Fowles et al. (1997) and Kashner et al. (1999) describe the use of medical records to validate self-reported data from interviews of patients.

Administrative records of sufficient quality and coverage can be used to reduce response burden. For examples involving Canadian tax data, see Hidiroglou et al. (1995) and Royce et al. (1997). Arguably, it is unethical to collect survey data that duplicates information that is already available (Gordon Brackstone, personal communication, August, 1999).

27.3 ADVANTAGES AND LIMITATIONS OF ADMINISTRATIVE RECORDS

Administrative records have several distinct advantages as data sources. Because the costs of collecting administrative records are borne by the administrative agency, they can be less expensive than primary data. Geographically coded records can provide small-area detail. Administrative records avoid some measurement problems typical of surveys, such as interviewer effects and recall biases. Their use can reduce respondent burden in the primary survey.

A major limitation of administrative records is that their purpose is primarily administrative, rather than statistical. This determines the design of the record system and the trade-offs of accuracy and completeness against cost (to the data collector) and respondent burden (to the provider of data). These considerations affect coverage, content, accuracy, timeliness, and accessibility for statistical uses.

Completeness of coverage may be extremely important for administrative records because they are used as a basis for action concerning individuals. If an individual unit (person, household, building, firm) is omitted from the record, then the appropriate action (e.g., collecting taxes, providing services, verifying a person's date and place of birth) cannot be performed. This may give both the provider and collector of data an incentive to ensure completeness. On the other hand, individuals may have an incentive to avoid inclusion, e.g., to evade taxes or to conceal an unsanctioned activity (such as illegal construction, or immigration without proper documents). Furthermore, the scope of coverage of the system is limited to those for whom the administrative function is relevant. For example, many households have insufficient income to file a tax return, and school systems have no record of students who are schooled privately or at home. Furthermore, the unit for an administrative record may differ from that which is of interest for statistical purposes. Business units that file corporate income tax might not be the operational units of interest in an economic analysis.

Content for administrative records is typically limited to what is required for accomplishing the intended purposes, hindering some unplanned statistical uses. In particular, race is an essential demographic item for the census but is absent from most administrative records. Furthermore, unless the statistical use of the records is recognized explicitly, the content of the records may change to meet the needs of the administrative programs, hindering consistent use of records over time. For example, Verma and Parent (1985) describe how proposed changes in the Canadian Family Allowance Program would necessitate an entirely new methodology for Canadian migration estimates.

Accuracy, like content, is driven by the primary uses of the system. Some items may have many errors or substantial missing data. Definitions of items may differ, explicitly or in practice, from those used in surveys. In particular, a household survey may use addresses corresponding to place of residence, whereas administrative records may carry filing addresses or post office boxes. Administrative records often have substantial temporal lags relative to surveys. Income tax returns in the United States for a given year's income are filed in the following year and may not be processed until late in that year. Other systems may contain data of widely varying vintage if the data are updated only when a transaction takes place. Survey items that depend upon subtle distinctions in wording are unlikely to agree with any administrative record. Even for apparently simple facts, administrative records databases may disagree substantially with survey responses. For example, Rogers et al. (1997) investigated how age and race are misreported on death certificates by matching a health survey to the National Death Index. Other studies have found varying rates of agreement between medical records and hospital administrative records (Faciszewski et al., 1997; Rawson and D'Arcy, 1998).

Finally, administrative records often are governed by confidentiality restrictions that accompany the requirement to provide the information. Use of these records for statistical purposes may be restricted to analyses performed within the record-keeping agency, or files moved outside the agency may be have to be deidentified in ways that limit their utility for linking to survey data. Also, respondents' consent may be required before their records can be used for research (Jacobsen et al., 1999; Leigh, 1998) or before they can be linked to other data (Cox and Boruch, 1998).

There are two basic approaches to using administrative records to impute for nonresponse. The first substitutes administrative records for the missing responses. This approach is critically dependent on the accuracy of the records. The second uses administrative records as a covariate in a statistical model, either for weighting or to impute data for nonrespondents. We consider each of these approaches in turn.

An important difference between imputation for nonresponse and the use of administrative records as a primary data source described in the previous section is that in the former case, the survey-described responses are the desired target of inference. Consequently, if the two sources systematically differ, a response similar to that which would have been obtained from the survey is what is desired. Most of the theory in Section 27.3.2 applies equally well, however, if the administrative record is the primary source and a survey is conducted to complete the missing data in the records.

27.3.1 Substituting Administrative Records for Missing Responses

When an administrative records database is available that has the necessary content and good coverage of the nonrespondents, a natural strategy is to substitute values from the records to impute for nonresponse. For example, England et al. (1994) report that in the Medicare Current Beneficiary Survey, missing prescription drug prices were imputed using administrative data on average wholesale prices and estimated mark-up rate whenever the respondent reported the drug name, strength, and volume ("exogenous imputation"). This approach assumes that the administrative records contain the "truth." Systematic differences in content between the administrative records and the survey responses due to inaccuracy, time lags, or definitional differences in the administrative records can introduce bias into survey estimates if this imputation strategy is used. Orr et al. (1996) found that in the National Job Training Partnership Act Study commissioned by the U.S. Department of Labor in 1986, unemployment insurance wage records systematically reported lower earnings for survey participants than participants reported for themselves. Therefore, any imputations that used the unemployment insurance wage data to substitute for the missing survey data about earnings would introduce bias into the data. Similarly, substitution of data from records for census nonrespondents would introduce large biases (Section 27.4.3).

When several administrative record sources are available for the same items, we can try to determine which of the available records is most likely to be accurate and substitute from the selected record for each nonrespondent. Furthermore, data that are corroborated by several sources may be more likely to be accurate. In evaluations for U.S. decennial census planning, Vacca et al. (1996, p. 42) found that information found on two or more administrative files was more than twice as likely to be accurate (i.e., match information reported on the census questionnaire) as information from a single source.

A more sophisticated substitution strategy attempts to determine when the administrative record is likely to be accurate and to substitute only for a subset of the cases (Larsen, 1998). For example, Wurdeman and Pistiner (1997) report that due to the undercoverage of children in the administrative records sources evaluated for possible use in the U.S. Decennial Census, administrative records are more likely to report size accurately for 1- or 2-person households than for larger households. Further research would be required to apply this strategy here and in other applications.

27.3.2 Using Administrative Records through Statistical Modeling

A fundamentally different approach is to use administrative records as a covariate in statistical models. Supposing that administrative data are available for the entire population, there are at least three distinct approaches to using them. One approach treats the missing data process as part of the sample selection (quasirandomization), modeling the probability of response conditional on administrative data, through either a weighting class adjustment or some more complicated model for nonresponse weights (Little and Rubin, 1987, Chapter 4). Administrative records can then be used to define classes or to provide model covariates. The inference relies on the assumption that nonresponse is independent of characteristics conditionally on the covariates of the nonresponse model (e.g., within weighting classes). Because a single weight must be applied to each unit, this method generally applies only to unit nonresponse.

A second approach, the primary focus of this chapter, is to impute missing values for nonrespondents in the sample and then perform a design- or model-based inference with the completed data. It also depends on the validity of the model used for imputation, but not critically so if the amount of missing data is modest.

The last approach resembles the second, except that missing values are imputed for the entire population; the completed-data inference is then a simple tabulation. With this entirely model-based approach, the inference is maximally dependent on the validity of the models. Nevertheless, it may be the preferred methodology under certain circumstances because it makes use of all of the administrative records and not just those for the sample. At the aggregate level, this inference is loosely similar to regression (covariance) adjustment, in which the administrative records play the role of the covariate with known population distribution. Hence, it is most useful when the survey sample is relatively small but the relationship between survey and administrative variables is strong and fairly consistent across small areas. For example, if the administrative data source has a substantial but stable bias relative to the survey, a small survey may provide enough information to make possible correction of this bias across the entire population covered by the administrative data system. This approach also offers a solution to the small-area estimation problem, i.e., making estimates for many small domains, each of which may contain few or no sample units. The availability of administrative data in each domain makes possible what might be regarded as a microdata version of the small-area estimation methods described in Section 27.2.

We apply this last approach to imputing for census nonresponse in Section 27.4. Statistics Canada similarly uses "mass imputation" in surveys or censuses where some items are missing by design for a large portion of the population in an effort to reduce the overall response burden (Whitridge et al., 1990; Royce et al., 1997; Rancourt and Hidiroglou, 1998). Administrative records provide basic information about all units (or a first phase sample) and more detailed information is obtained for a sample of the population (or a subsample of the primary sample) through a survey. A model fit to the subsample is used to impute the missing responses of the remaining units based on their administrative records. Chapter 25 in this volume uses a similar approach to multiply impute traffic speed, calibrating relatively extensive but inaccurate data on traffic speed from magnetic loop detectors by a model based on a single week of more accurately measured data. Orr et al. (1996) impute missing earnings based on unemployment insurance records and a model fit to a subsample of the data for which both sources are available.

27.3.3 Model-Based Imputation Using Administrative Records

We now lay out the theoretical framework for model-based imputation. The essence of the approach is to regard the missing responses (whether due to item or unit nonresponse) as missing data, adopting the general framework of Little and Rubin (1987), and perform single or multiple imputation (Rubin, 1987) for missing values.

Suppose that variables Y are measured from the survey and Z represents administrative record data relevant to Y; Z and Y could measure the same or related variables. Note that Z may include variables that were available in the sampling frame and could have been used in the sample design. For our purposes, the distinction between "frame variables" and other administrative records is irrelevant.

We are interested in estimating $h(Y_U)$, a function of the Y-values for the entire population U, by calculating $\hat{h}(Y_S, D)$, a function of the data for sample S and design information D, such as weights. (In the special case of a census rather than sample, or when using the third of the strategies described in Section 27.4.2, $S = U$.) Let R_S be the collection of indicators for response by each sampled case; if Y is multivariate and there is item nonresponse, R_S is a matrix. The survey data consists of R_S, $Y_{S,\text{obs}}$, the response indicators and the completed items from sampled respondents, where we partition $Y_S = (Y_{S,\text{obs}}, Y_{S,\text{mis}})$ to represent the observed and missing data.

Broadly speaking, our objective is to complete the data set by imputing values $Y^*_{S,\text{mis}}$ for the missing data, so we can analyze the completed data $Y^*_S = (Y_{S,\text{obs}}, Y^*_{S,\text{mis}})$ using standard analytical methods, i.e., as if there had been no missing data. In order to impute, we need a model that relates the observed values, including the values of Z for the units with missing Y, to the missing values. A wide range of model formulations are possible, and the choice of model must be based on the nature of the available data and on the strength and form of the relationships between Y and Z. A typical approach is to regress Y on Z among the responding units, and then to predict Y for the missing units from the regression model, using the known values of Z (and the observed part of Y in case of item nonresponse). Hot deck imputation may be regarded as a "nonparametric" approach to regression, in which the predictive distribution for Y (possibly a complex multivariate outcome) is defined by drawing from responding cases with identical or similar values of Z.

A simplified approach is to impute the missing values once and then analyze them as if the imputed values had been observed. Although in many cases the resulting point estimates will be acceptable, standard errors from these procedures tend to underestimate the uncertainty of the results because they ignore the fact that the true values of the imputed data are unknown. Multiple imputation involves imputing the missing values several times, each time drawing stochastically from the predictive distributions under the model. Variances are obtained by a simple rule (Rubin, 1987, Section 3.1) that combines within-imputation variance (obtained from the standard analysis of the completed data) with between-imputation variance (calculated using the estimates from the several imputed data sets, representing the uncertainty due to missing data). The amount of additional variance due to imputation depends in a complex way both on the amount and disposition of the missing data and on the strength of the relationships with the predictors Z. Consequently, multiple imputation or some other special analysis is required to determine how much the missing data affect the precision of the inference.

More formally, from a Bayesian point of view our objective is to calculate $E[\hat{h}(Y_S, D)|Y_{S,\text{obs}}, R_S, Z]$ and an appropriate standard error. Modeling parametrically, we write this as

$$\int\int \hat{h}(Y_S, D)P(Y_{S,\text{mis}}|Y_{S,\text{obs}}, R_S, \theta, Z)P(\theta|Y_{S,\text{obs}}, R_S, Z)dY_{S,\text{mis}}d\theta \qquad (27.1)$$

where θ is the parameter that relates the variables of Y and Z. The factor $\hat{h}(Y_S, D)$ represents the complete-data inference, $P(Y_{S,\text{mis}}|Y_{S,\text{obs}}, R_S, \theta, Z)$ represents the predictive density of the missing data given observed data, parameters, missingness pattern, and administrative records, and $P(\theta|Y_{S,\text{obs}}, R_S, Z) \propto P(Y_{S,\text{obs}}, R_S, Z|\theta)\pi(\theta)$ represents the posterior density of the parameters given the survey responses and administrative data, where $\pi(\theta)$ is the prior density. The double integral is typically approximated by repeatedly drawing θ and then $Y_{S,\text{mis}}$.

A common assumption is that the data are missing at random (MAR) (Rubin, 1976; Little and Rubin, 1987, sections 1.5–1.6), meaning that nonresponse is unrelated to the values of the data that are not observed, conditional on observed values. (This assumption can be varied for sensitivity analysis.) Under this assumption,

(27.1) simplifies by dropping R_S from the second and third factors. Furthermore, under a sufficiently richly parameterized imputation model or with strongly predictive administrative data, the dependency of $Y_{i,\text{mis}}$ on $(Y_{S,\text{obs}}, Z)$ may be only through θ and $(Y_{i,\text{obs}}, Z_i)$, i.e., the missing data depends only on the observed survey data (vacuous in the case of unit nonresponse) and supplementary data for the same case, and on the parameters. Then the second factor decomposes into separate factors for each case, so each can be imputed separately after the parameters are drawn.

The propriety of the imputation depends on the correctness of the models used. A complex design is typically a sign that the population structure is also complex. In principle, this suggests that to yield "proper" imputations that fully represent the contribution to variance of missing data, a model should include covariates corresponding to the stratification and random effects for clusters. [See, for example, discussion by Little (1993b) of Belin et al. (1993).] Although requirements for the relationship between the missing-data and complete-data inferences have been hotly debated (Meng, 1994; Fay, 1994, 1996; Rubin, 1996; Meng, 2000), the precise form of the imputation model may have relatively little impact on either point or interval estimates if the proportion of missing information is small. Furthermore, it makes no difference in principle whether the underlying parametric model is expressed as a conditional (regression) model $P(Y|Z, \theta)$ or a joint model $P(Y, Z|\theta)$.

Given our model selection, we can simplify inference by approximations to the integral (27.1). First, we can replace draws from the distribution of θ with a single estimate (posterior mode or maximum likelihood estimate). Second, we can perform a single imputation, either stochastically or substituting a posterior mean. If $\hat{h}$ is linear in $Y_{S,\text{mis}}$ (for example, a weighted mean), the latter corresponds to mean imputation for $\hat{h}(Y_S, D)$; this is often at least approximately true even for nonlinear estimators. While these approaches do not yield proper imputations, they may be good enough with relatively small amounts of missing information. The full modeling framework (27.1) provides a principled basis for imputation, and then the simplifications described here and other ad-hoc implementation details can be justified as approximations to the full model-based strategy.

27.4 EXAMPLE: ESTIMATING NONRESPONDENT CHARACTERISTICS IN THE U.S. DECENNIAL CENSUS

Sampling for nonresponse followup (NRFU) was an innovation considered for the U.S. Decennial Census in 2000. In place of the traditional method of sending field enumerators to follow up on all households that did not respond to the census mailout questionnaire, this proposal required contact with only a random sample of the nonresponding households. Although sampling could reduce costs substantially, it would also create an unprecedented amount of missing data. Consequently, the characteristics of the nonrespondent households omitted from the follow-up sample would need to be estimated. Zanutto and Zaslavsky (1996, 1997; henceforth "ZZ") proposed a loglinear model to estimate the relationship between administrative and NRFU records. Estimates from this model can be used to impute the characteristics of households at nonsample nonresponding addresses based on their administrative records. Although sampling for NRFU was not used in the census in 2000, the ZZ methodology examplifies use of administrative records to impute for nonresponse.

27.4.1 General Estimation and Imputation Procedure

The core of the proposed methodology is a hierarchical loglinear model that estimates the number of nonsample nonrespondent households that are of each "type" in each block, using data from administrative records and from NRFU respondents. (A block is a geographical unit roughly corresponding to a city block or a compact rural area, averaging 15 households.) ZZ define 18 household types by cross-classifying race of the household (Black, non-Black Hispanic, Other), number of adults (0–1, 2, or 3 or more adults), and whether or not children are present in the household. The estimated counts of the number of nonsample nonrespondent households of each type produced by this model are rounded, using an unbiased random rounding procedure, to obtain integer counts. Households are then imputed into the nonsample nonrespondent units in numbers corresponding to these rounded counts by household types, producing a completed roster that is suitable for preparing tabulations or microdata samples. These imputations complete the nonrespondent household characteristics that are not explicitly included in the loglinear model.

To focus on the potential benefits of using administrative records to estimate and impute the characteristics of the census nonrespondents that are not in the NRFU sample, the remainder of this discussion is limited to the loglinear model and its alternatives. We omit discussion of the treatment of vacant households, the methodology for rounding the estimated counts, and the specification of the imputation procedure.

ZZ's loglinear model uses low-dimensional covariates at the block level (the level at which we have the least information) and more detailed covariates at more aggregated levels of geography (for which we have more information). Heuristically, the distribution of household types observed in the administrative records is shifted, using information about the biases of the administrative records as observed in the NRFU sample in the surrounding area. Most importantly, this model uses administrative records to predict the characteristics of nonsample nonrespondents while constraining tract level estimates to equal their unbiased estimates from the NRFU sample. (A tract is a unit of census geography averaging about 140 contiguous blocks.) Using maximum likelihood estimation through a modified iterative proportional fitting (IPF) algorithm described in Zanutto and Zaslavsky (1995a,b), model predictions for the effects included in the loglinear model are constrained to agree with observed rates estimated from the NRFU sample, so that the corresponding estimates are approximately unbiased [see also Purcell and Kish (1980)].

To estimate the number of nonsample nonrespondent households of each type in each block, ZZ use the following model, expressed in the standard generalized linear models notation of Wilkinson and Rogers (1973):

$$\log E\, n(i, j, d) \sim i + d + i*d + i*x_2 + t*d*x_1 \quad (27.2)$$

The left side is the logarithm of the expected number of households in block i of household type j and data source d, where d indicates either a census record for a nonrespondent (obtained through NRFU) or an administrative record for a nonrespondent. The right side represents a linear predictor determined by the block index i, data source indicator d, tract index $t = t(i)$, and $x_1 = x_1(j)$ = household type, $x_2 = x_2(j)$ = race for household type j. They modeled all households in a District Office (DO, a collection of about 125,000 households).

In the notation of Section 27.3.3, ZZ jointly model Y, the vector of the number of households in each block of each type according to NRFU, and Z, the corresponding vector according to the administrative records database (incomplete because some addresses are not covered in that database). Finally, θ consists of the parameters of the loglinear model, which is highly parameterized with effects down to the block level. The objective of the census is small-area estimation of counts for every block in the country, and therefore we impute all cases, not just a sample. We estimate parameters by maximum likelihood and then perform single mean imputation; the final rounding step is stochastic, but not proper multiple imputation. This procedure corresponds to the Census Bureau's need for a single file with minimal variation to use for tabulations. For research purposes, proper multiple imputations can be obtained by using Bayesian IPF (Gelman et al., 1995, pp. 400–401) and drawing from a multinomial predictive distribution.

This model can also be used to estimate the number of nonsample nonrespondent households of each type in each block using respondents as predictors (ignoring administrative records). To do this, d is redefined as response status (respondent or nonrespondent). In this case, estimates of the number of nonsample nonrespondent households depend on the characteristics of respondents in the same block and nonrespondents in the NRFU sample in the same tract. This application of the model is useful when administrative records are not available (Zanutto and Zaslavsky, 1995a,b; Zanutto, 1998).

27.4.2 Modeling Strategies

Our imputation strategy is unavoidably complicated by the incomplete coverage of the combined administrative record databases, requiring use of a two-part model. We first divide nonrespondent households into those that can (Group A) and cannot (Group B) be linked to administrative records. To estimate household types in Group A, we fit loglinear model (27.2), in which the two sources are administrative records and census returns (d = census, administrative records), the latter being available only from the NRFU follow-up sample. From the fitted model, we predict the types of nonrespondent households that have administrative records but are not in the NRFU sample. We fit loglinear model (27.2) to Group B using census records from respondent households, and census records from the NRFU sample for households that do not have administrative records (d = respondent, nonrespondent). Combining the estimates for Groups A and B gives estimates for all nonsample nonrespondents. This strategy uses administrative records, whenever they are available, as predictors of the characteristics of the nonrespondents, and census respondents otherwise. We call this the "two model method."

This proposal uses administrative records through modeling. We compare this strategy, through simulation, to two others: one that uses administrative records without modeling, and one that uses modeling without administrative records. In the "substitution method," we substitute household types from administrative records, as in Section 27.3.1, for all households in Group A, with model-based estimation for Group B. The "one model method" ignores administrative records and fits a single loglinear model to both groups, using census respondents to predict the number of nonsample nonrespondent households of each type in each block.

To compare the three estimation methods, we calculate the root mean squared error (RMSE) of the estimates of demographic aggregates (such as number of households by race, number of adults, and number of children) at the block, tract, and DO levels of geography. Lacking reliable analytical estimates of these quantities, we performed the evaluations through simulation.

Using data (described below), for which we know the characteristics of both respondents and nonrespondents, we simulated NRFU sampling by drawing a 1 in 3 simple random sample of nonrespondent households in each tract. (ZZ also applied these methods using census blocks as the primary sampling unit.) We fit the models and estimated the number of nonsample nonrespondent households of each type in each block. We compared aggregates at the block, tract and DO levels to the truth. We repeated these steps 30 times to obtain sufficiently accurate estimates of RMSE, as defined in ZZ, where desirable properties of this measure are described. These measures may be interpreted as average errors for percentages in a category over geographical units (and over simulations, if applicable).

We used data from two 1995 Census Test sites (Oakland, California; Paterson, New Jersey) and the associated administrative records databases (Neugebauer et al., 1996; Wurdeman and Pistiner, 1997). These databases combine records from the Social Security Administration, Internal Revenue Service, Department of Housing and Urban Development, Medicare, food stamp programs, drivers license files, school enrollment files, voter registration files, and parolee/probationer files. Our data included all blocks containing nonrespondent households in the NRFU sample, which was selected at a rate of roughly 30%. The simulation populations consist of a total of 58,387 households in Oakland and 11,096 in Paterson.

A comparison of the distributions of the basic household characteristics in the administrative records and the census NRFU sample, for households where both sources of information available, illustrates some of the common problems with administrative records. Only 50.9% of the nonrespondents in Oakland and 21.5% in Paterson had usable administrative records. In both data sets, the administrative records severely understate the proportion of households with children, a consequence of reliance on sources that contain few or no children. In Oakland, the proportion of households with 3 or more adults was overstated in the administrative records, because many of the records were outdated, so both current and previous occupants were listed at the same address. In Paterson, the proportion of households with 0–1 adults was overstated, due to undercoverage of adults when records contained information for only a single household member. On the other hand, the administrative records agreed fairly well with the census on the distribution of households by race.

Similar patterns were found for agreement between census and administrative data for individual households. The agreement rates in Oakland and Paterson were 29.9% and 24.7% on household type, 84.0% and 74.8% on race, 42.7% and 48.9% on adult count category, and 77.0% and 56.0% on child count category.

27.4.3 Simulation Results

We summarize here the key results from ZZ. Substitution produces block-level estimates with substantially larger RMSE than the other methods for the children and adult categories, which are critical because they determine total population. These effects of bias are even more dramatic at the tract and DO levels, where sampling error is a smaller component of error.

The model-based methods have smaller RMSE than substitution for almost all household characteristics at all levels of geography. We compare these methods only at the block level, because the loglinear model constrains tract- and DO-level estimates to equal the same unbiased estimates from the NRFU sample. Use of administrative records reduces RMSE for the race categories ($p < 0.0001$), due to a smaller bias component, but has little effect on RMSE for the children and adult categories.

The reduction in RMSE of block level race estimates through using administrative records may appear small (e.g., RMSE of 3.3%, 2.9%, and 2.9% for the Black, Hispanic, and Other race categories for the two-model method in Oakland compared to 3.8%, 3.1%, and 3.3%, respectively, for the one-model method, where the simulation standard errors for these values are all less than 0.015 percentage points.). However, potential gains from records are restricted to nonsampled nonrespondents with matching records, only 9.8% of all households in Oakland. If more nonrespondents had administrative records, the difference between the two methods would be larger.

27.4.4 Conclusions

This example illustrates that using administrative records through modeling can improve, albeit modestly in this application, the accuracy of estimates at very detailed levels of geography. Direct substitution of administrative records for missing data can engender large biases. When an administrative records database with fairly complete and consistent records is developed, we can overcome concerns about bias relative to the gold standard of the survey estimates, because our methodology uses information from administrative records to impose estimates at detailed levels of geography while constraining these estimates to agree with unbiased survey estimates at higher levels. This methodology is widely applicable because administrative records databases usually will disagree systematically with census or survey information.

CHAPTER 28

Imputation for Wave Nonresponse: Existing Methods and a Time Series Approach

Danny Pfeffermann and Gad Nathan, *Hebrew University, Israel*

28.1 INTRODUCTION AND LITERATURE REVIEW

Longitudinal studies have recently become a mainstay of sample survey practice (Binder, 1998). Following Duncan and Kalton (1987), we include in this term any type of survey for which at least some of the units are measured more than once. These include traditional panel surveys, with fixed or rotating panels, retrospective longitudinal studies, and cohort follow-ups. The imputation methods discussed in the following are, in general, equally applicable to data obtained from administrative sources and to nonsurvey data (as in the medical and biological sciences) as to sample surveys proper.

We focus primarily on missing data resulting from wave nonresponse; data are available for some points in time and missing for others. Different patterns of wave nonresponse to be considered are attrition (no observations from some time point onwards), missing for a single time or for a continuous period, and intermittent dropout. For all these patterns, the existence of observations for some points in time for the same unit suggests the consideration of plausible relationships over time between individual measurements for more efficient imputation. This is in contrast to the treatment of missing data in cross-sectional settings and requires more elaborate modeling efforts.

When studying the properties of imputation methods, the relationships between the missing data mechanism and the missing and observed data need to be examined; see, e.g., Rubin (1976), Little (1982), Little and Rubin (1987), Little (1995) and Chapter 1 of this monograph. An important distinction is between the mechanisms of missing completely at random (MCAR), missing at random (MAR), and not missing at random (NMAR) or informative missingness. Imputation methods have been proposed for dealing with both ignorable and nonignorable mechanisms. The applications are generally to cross-sectional data, though Little and Rubin (1987, pp. 161–168) also consider missing data in univariate and bivariate time series. Little (1995) considers complete dropout in a longitudinal study.

The specific treatment of wave nonresponse in panel surveys has been addressed in a randomization framework in a series of papers by Kalton et al. (1985), Kalton (1986), Kalton and Miller (1986), and Lepkowski (1989). The methods proposed use imputation and weighting based on regression models. The models incorporate known auxiliary data, including response to other waves, and take into account cross-sectional and longitudinal interrelationships.

The analysis of longitudinal data has received widespread attention in the medical sciences; see, e.g., Laird and Ware (1982) and Jennrich and Schluchter (1986). Recently, the analysis of longitudinal data has been largely influenced by the introduction of generalized estimating equations (GEE) by Zeger and Liang (1986); see, e.g., Laird (1988), Diggle and Kenward (1994), Diggle et al. (1994), Murphy and Li (1995), Paik (1997), and Schafer and Schenker (2000). In addition, Rotnizky et al. (1998) consider the use of semiparametric regressions for the treatment of informative nonresponse. This estimation procedure can be viewed as an extension of the GEE method that allows for informative nonresponse.

In this chapter, we propose to use time series models with hierarchical structures to take into account time series relationships between lower level observations and higher-level group effects (e.g., household effects). The models combine standard multilevel (mixed linear) models operating at given time points with time series state–space models for the random group effects and the individual measurements. The use of unit level time series models for the analysis of unequally spaced longitudinal data has been considered by Jones and Boadi-Boateng (1991), Jones and Vecchia (1993), and Jones (1993). Chi and Reinsel (1989) consider first-order autoregressive models for within-individual errors and develop a score test for autocorrelation, on which they base an explicit maximum likelihood estimation procedure. However, none of these studies considers explicitly the effects of missing data and they do not account for hierarchical population structures. Goldstein et al. (1994) consider the analysis of repeated measurements using a two-level hierarchical model, with individuals as second levels and the repeated measurements as the first levels. The model permits the first-level measurements to be correlated over time.

The following section reviews possible mechanisms for wave nonresponse and the methods considered for imputation of the missing data. Section 28.3 sets up the proposed multilevel time series modeling approach. Section 28.4 reports the results of a simulation study and an analysis of empirical data. The final section mentions possible ramifications and extensions.

28.2 NONRESPONSE MECHANISMS AND IMPUTATION METHODS

We consider longitudinal data as obtained from retrospective longitudinal studies, cohort studies, or from panel surveys with fixed or rotating panels. Our interest is in wave nonresponse, wherein data are missing from one or more waves of a multiwave survey either according to design or due to common reasons for nonresponse such as "not contacted," "refusal," etc. For modeling purposes, wave nonresponse is regarded as resulting from a random mechanism, which may be related to the outcome variables. If it can be shown that no such relationship exists, the nonresponse mechanism is MCAR. Even if such a relationship does exist, it may still be due to a mechanism which is MAR, wherein the probability of the observed response pattern, given the missing and observed data, does not depend on the missing values. If the nonresponse probabilities depend also on the values of the missing data, then they define an informative missing or NMAR data mechanism. Precise definitions of these terms can be found in Chapter 1.

In what follows we consider several imputation methods for wave nonresponse, with emphasis on methods that utilize the time series structure of longitudinal data and the hierarchical structure of the population. An important outcome of the present study is that by considering the interrelationships among observations related to the same natural group, the imputation bias induced by informative nonresponse can be largely reduced, even when ignoring the response process. The methods we consider assume the existence of (fully observed) auxiliary variables, which bear some relationship to the response variable. Below is a brief description of the methods considered (the first three serve as benchmarks). Exact specifications under the proposed model are given in Section 28.3.

1. *Mean imputation.* Homogeneous imputation groups are created on the basis of values of the auxiliary variables and/or outcomes from other waves. Missing values are imputed as the mean of the reported values in the corresponding imputation group.
2. *Nearest neighbor.* A distance measure is defined in the auxiliary variable space and the missing value is imputed as the reported value nearest to the missing point. In longitudinal studies, the outcome obtained in another wave for the same individual may be defined as the "nearest neighbor" after appropriate modifications to account for fixed time effects.
3. *Simple regression imputation.* A regression relationship is estimated between the response variable and the auxiliary variables. The regression coefficients are estimated by ordinary least squares (OLS). The regression predicted value serves as the imputed value. A variation of this method is to add a random residual to the imputed value so as to better reflect the variation in the data.
4. *Augmented regression imputation.* The regression prediction is extended by adding a correction term that accounts for the existing correlations between the observed and the missing data. The unknown regression coefficients are estimated by generalized least squares (GLS). The imputation of missing data is based on all the observations for all the time periods. Two variants are considered: (a) only the observed data for the individual with the missing data are used for the imputation; (b) all the observed data for all individuals in the same hierarchical group are used for the imputation. As in (3), random residuals can be added to the imputed values.

5. *State–space model imputation.* A state–space model combining multilevel models operating at given points in time is postulated. Predictions obtained under the model are used as the imputed values. The unknown parameters of the combined model are estimated by MLE.

28.3 TIME-SERIES MODELS FOR LONGITUDINAL DATA

28.3.1 Model Specification

Following Feder et al. (2000), we seek a model that encompasses the hierarchical nature of many human populations and the time series relationships between repeated measurements and between the random effects of higher-level groups. The proposed model combines separate two-level mixed linear models (Goldstein, 1986, 1995), operating at given points in time, using a state–space model that represents the time series relationships of the random group effects and the individual measurements. Basic notation and assumptions follow. We refer for convenience to the higher-level groups as "households" and to the lower-level units as "individuals." Let y_{hjt} define the value of the response variable at time $t = 1, \ldots, T$, for individual $j = 1, \ldots, n_h$, belonging to household $h = 1, \ldots, N$. The measurements y_{hjt} are assumed to follow the hierarchical two-level linear model:

$$y_{hjt} = \mathbf{x}'_{hjt}\mathbf{b}_t + \mathbf{z}'_{ht}\mathbf{v}_t + \mathbf{z}'_{ht}\mathbf{u}_{ht} + e_{hjt} \tag{3.1.1}$$

where $\mathbf{x}_{hjt}$ is a p-dimensional vector of individual level explanatory variables values, $\mathbf{z}_{ht}$ is a q-dimensional vector of household level explanatory variables, $\mathbf{b}_t$ and $\mathbf{v}_t$ are fixed vector coefficients of appropriate orders, $\mathbf{u}_{ht}$ is a $(q \times 1)$ vector of household-level random effects, and e_{hjt} is an individual-level random error. The random household effects represent specific household characteristics not represented by the fixed effects.

The individual- and household-level random errors are assumed to follow independent first-order autoregressive models:

$$\mathbf{u}_{ht} = \mathbf{A}\mathbf{u}_{ht-1} + \mathbf{d}_{ht}; \qquad \mathbf{d}_{ht} \sim \mathrm{N}(\mathbf{0}_q, \mathbf{D}); \tag{3.1.2}$$

$$e_{hjt} = \rho e_{hjt-1} + \varepsilon_{hjt}; \qquad \varepsilon_{hjt} \sim \mathrm{N}(0, \sigma^2_\varepsilon). \tag{3.1.3}$$

In the following, we assume for convenience that $\mathbf{A}$ and $\mathbf{D}$ are diagonal, implying independence of the random household effects. We also assume $|\mathbf{A}_{ii}| < 1$ and $|\rho| < 1$ to ensure stationarity. More elaborate models could be considered, depending on the number of observations per unit. It follows from (3.1.2) and (3.1.3) that for any given time t

$$\mathbf{u}_{ht} \sim \mathrm{N}(\mathbf{0}_q, \mathbf{D}^*); \qquad \mathbf{D}^* = (\mathbf{I}_q - \mathbf{A}^2)^{-1}\mathbf{D} \tag{3.1.4}$$

$$e_{hjt} \sim \mathrm{N}(0, \sigma^2_e); \qquad \sigma^2_e = (1 - \rho^2)^{-1}\sigma^2_\varepsilon \tag{3.1.5}$$

Thus, the models operating at the various time points are standard multilevel models with fixed variances for the random first- and second-level effects.

28.3.2 Model-Based Imputation Methods

In this section, we illustrate the application of the imputation methods described in Section 28.2 under the model defined in Section 28.3.1. We assume for convenience that the longitudinal measurements are taken over a fixed period $t = 1, \ldots, T$, with some of the measurements possibly missing, and that all the model parameters are known. The unknown model parameters are replaced in practice by sample estimates. Parameter estimation is considered in Section 28.3.3.

1. *Mean imputation* (described in Section 28.2).
2. *Nearest neighbor.* Here we restrict to the case where the nearest neighbor is defined by the nearest observation in time obtained for the same individual. If t is the time point with missing value and t^* is the nearest time with an observation, the imputed value is defined as

$$y_{hjt}^{(nn)} = \mathbf{x}'_{hjt}\mathbf{b}_t + \mathbf{z}'_{ht}\mathbf{v}_t + (y_{hjt^*} - \mathbf{x}'_{hjt^*}\mathbf{b}_{t^*} - \mathbf{z}'_{ht^*}\mathbf{v}_{t^*}) \tag{3.2.1}$$

3. *Simple regression imputation.* The imputed values are obtained as the simple regression predictions:

$$y_{hjt}^{(P)} = \mathbf{x}'_{hjt}\mathbf{b}_t + \mathbf{z}'_{ht}\mathbf{v}_t \tag{3.2.2}$$

4. *Augmented regression.* Let $\tilde{\mathbf{Y}}_{hj} = (y_{hj1}, \ldots, y_{hjT})'$ represent the generic vector of complete values (observed and missing) for individual j in household h, with variance–covariance (V–C) matrix, $\mathbf{S}_h$ (defined by the parameters contained in $\mathbf{A}$, $\mathbf{D}$, ρ, σ_e^2). Let $\mathbf{Q}_{hj}$ define the response indicator matrix of size $t_{hj} \times T$ corresponding to unit hj, (t_{hj} is the number of times that unit hj is observed), such that the observed values are $\mathbf{Y}_{hj} = \mathbf{Q}_{hj}\tilde{\mathbf{Y}}_{hj}$. Similarly, denote by $\overline{\mathbf{Q}}_{hj}$ the indicator matrix for the missing values, of size $\bar{t}_{hj} \times T$, ($\bar{t}_{hj} = T - t_{hj}$) such that the missing values are $\mathbf{Y}_{hj}^{(m)} = \overline{\mathbf{Q}}_{hj}\tilde{\mathbf{Y}}_{hj}$. The imputed values, based only on data for the same individual (method 4a) are the augmented (BLU) regression predictions (Pfeffermann, 1988):

$$\hat{\mathbf{Y}}_{hj}^{(m)} = \overline{\mathbf{Q}}_{hj}\tilde{\mathbf{Y}}_{hj}^{(p)} + \overline{\mathbf{Q}}_{hj}\mathbf{S}_h\mathbf{Q}'_{hj}(\mathbf{Q}_{hj}\mathbf{S}_h\mathbf{Q}'_{hj})^{-1}(\mathbf{Y}_{hj} - \mathbf{Q}_{hj}\tilde{\mathbf{Y}}_{hj}^{(p)}) \tag{3.2.3}$$

where $\tilde{\mathbf{Y}}_{hj}^{(p)} = (y_{hj1}^{(p)}, \ldots, y_{hjT}^{(p)})'$ is the complete vector of regression predictions defined by (3.2.2), $\overline{\mathbf{Q}}_{hj}\mathbf{S}_h\mathbf{Q}'_{hj} = \mathrm{Cov}(\mathbf{Y}_{hj}^{(m)}, \mathbf{Y}_{hj})$ and $\mathbf{Q}_{hj}\mathbf{S}_h\mathbf{Q}'_{hj} = \mathrm{V}(\mathbf{Y}_{hj})$. Similarly, we denote by $\tilde{\mathbf{Y}}_h = (\tilde{\mathbf{Y}}'_{h1}, \ldots, \tilde{\mathbf{Y}}'_{hn_h})'$, the generic vector of complete values for all the individuals belonging to household h, with V–C matrix, $\tilde{\mathbf{S}}_h$. Let $\mathbf{Q}_h$ and $\overline{\mathbf{Q}}_h$ define the household indicator matrices for observed and missing values of orders $\Sigma_j t_{hj} \times n_h T$ and $\Sigma_j \bar{t}_{hj} \times n_h T$, respectively, so that $\mathbf{Y}_h = \mathbf{Q}_h\tilde{\mathbf{Y}}_h$ and $\mathbf{Y}_h^{(m)} = \overline{\mathbf{Q}}_h\tilde{\mathbf{Y}}_h$ are the corresponding vectors of observed and missing values. The imputed values, based on all the observed data for all the individuals in the household (method 4b) are the augmented regression predictions:

$$\hat{\mathbf{Y}}_h^{(m)} = \overline{\mathbf{Q}}_h \widetilde{\mathbf{Y}}_h^{(P)} + (\overline{\mathbf{Q}}_h \widetilde{\mathbf{S}}_h \mathbf{Q}_h')(\mathbf{Q}_h \widetilde{\mathbf{S}}_h \mathbf{Q}_h')^{-1}(\mathbf{Y}_h - \mathbf{Q}_h \widetilde{\mathbf{Y}}_h^{(P)}) \tag{3.2.4}$$

The imputations defined by (3.2.3) and (3.2.4) are the conditional expectations of the missing values given the observed values. They are the best linear unbiased predictors, even without normality of the residual terms.

5. *State–space model imputation.* The model (3.1.1)–(3.1.3) is written in state–space form with observation equation

$$[\mathbf{Q}_{ht}\mathbf{Y}_{ht}] = [\mathbf{Q}_{ht}\widetilde{\mathbf{X}}_{ht}]\widetilde{\boldsymbol{\beta}}_t + [\mathbf{Q}_{ht}\widetilde{\mathbf{Z}}_{ht}]\boldsymbol{\alpha}_{ht} \tag{3.2.5}$$

and transition equation

$$\boldsymbol{\alpha}_{ht} = \mathbf{T}_h \boldsymbol{\alpha}_{h,t-1} + \boldsymbol{\nu}_{ht} \tag{3.2.6}$$

where $[\mathbf{Q}_{ht}\mathbf{Y}_{ht}]$ denotes the observed values for household h at time t ($\mathbf{Y}_{ht}$ defines the generic vector at time t of complete values for all the individuals in household h of order n_h and $\mathbf{Q}_{ht}$ is the corresponding response indicator matrix), $[\mathbf{Q}_{ht}\widetilde{\mathbf{X}}_{ht}]$ and $[\mathbf{Q}_{ht}\widetilde{\mathbf{Z}}_{ht}]$, with

$$\widetilde{\mathbf{X}}_{ht} = \begin{pmatrix} \mathbf{x}'_{h1t} & \mathbf{z}'_{ht} \\ \vdots & \vdots \\ \mathbf{x}'_{hn_ht} & \mathbf{z}'_{ht} \end{pmatrix}$$

and

$$\widetilde{\mathbf{Z}}_{ht} = \begin{pmatrix} \mathbf{z}'_{ht} & \\ \vdots & \mathbf{I}_{nh} \\ \mathbf{z}'_{ht} & \end{pmatrix}$$

are the design matrices of the explanatory variables, $\widetilde{\beta}_t = (\mathbf{b}'_t, \mathbf{v}'_t)'$ is the $(p+q) \times 1$ vector of fixed parameters, $\boldsymbol{\alpha}_{ht} = (\mathbf{u}'_{ht}, \mathbf{e}'_{ht})'$ is the $(q + n_h) \times 1$ state vector with $\mathbf{e}_{h_t} = (e_{h1t}, \ldots, e_{hn_ht})'$, $\mathbf{T}_h = \mathbf{A} \oplus \rho\mathbf{I}_{n_h}$ is the transition matrix (a block-diagonal matrix with $\mathbf{A}$ and $\rho\mathbf{I}_{n_h}$ as the two blocks), and $\boldsymbol{\nu}_{ht} = (\boldsymbol{d}'_{ht}, \boldsymbol{\varepsilon}'_{ht})'$ is a vector of random errors with V-C matrix: $V(\boldsymbol{\nu}_{ht}) = \mathbf{R}_h = \mathbf{D} \oplus \sigma^2_\varepsilon \mathbf{I}_{nh}$.

Under the model (with known parameters), the random components $\mathbf{u}_{ht}$ and $\mathbf{e}_{ht}$ can be predicted either by application of the Kalman filter, if only current and past observations are available, or by an appropriate smoothing filter if data for subsequent time periods are known; see Harvey (1989) and De Jong (1989) for details. Starting values for the filters at time $t = 1$ are defined by (3.1.4) and (3.1.5). The imputed values under the model are obtained as

$$\hat{y}_{hjt}^{(M)} = y_{hjt}^{(P)} + \mathbf{z}'_{ht}\hat{\mathbf{u}}_{ht} + \hat{e}_{hjt} \tag{3.2.7}$$

where $y_{hjt}^{(P)}$ is defined by (3.2.2) and $\hat{\mathbf{u}}_{ht}$ and $\hat{e}_{hjt}$ are the predicted values obtained from the Kalman filter or the smoothing algorithm.

It should be noted that the difference between the augmented regression imputations defined by (3.2.4) and the corresponding state–space imputations defined by (3.2.7) is only in the estimation of the unknown parameters (see below). Both procedures use the same data, the same model, and the same imputation criterion.

28.3.3 Estimation of Model Parameters

The use of the various imputation methods requires the estimation of the model parameters. For the simple regression imputation (method 3), the coefficients are estimated by OLS. For the augmented regression and state–space modeling approaches, we consider MLE under the model (3.1.1)–(3.1.3). Direct maximization of the likelihood under augmented regression (methods 4 and 5) is complicated due to the complex structure of the V-C matrix $\tilde{\mathbf{S}}_h = \mathrm{V}[\tilde{\mathbf{Y}}_h]$. This matrix takes the form $\tilde{\mathbf{S}}_h = \mathbf{I}_{n_h} \otimes \sigma_e^2 \tilde{\mathbf{R}} + \mathbf{J}_{n_h} \otimes (\mathbf{Z}_h \tilde{\mathbf{A}} \mathbf{Z}_h')$, where $\otimes$ denotes the Kronecker product; $\mathbf{J}_{n_h} = \mathbf{1}_{n_h} \mathbf{1}_{n_h}'$, with $\mathbf{1}_{n_h}$ defining the unit vector of order n_h; $\mathbf{Z}_h = [\mathbf{z}_{h1}, \ldots, \mathbf{z}_{hT}]$; $\tilde{\mathbf{R}}$ is $T \times T$ with $\tilde{\mathbf{R}}_{t,t'} = \rho^{|t-t'|}$; and $\tilde{\mathbf{A}}$ is a $qT \times qT$ block matrix, whose (t, t')-th block is $\mathbf{A}^{|t-t'|}\mathbf{D}^*$, $t, t' = 1, \ldots, T$. The V-C matrix for the observed data in household h is $\mathbf{Q}_h \tilde{\mathbf{S}}_h \mathbf{Q}_h'$.

For the simulation study in the next section, we used the method of iterative generalized least squares (IGLS), as proposed by Anderson (1973). The iterations alternate between estimation of the fixed parameters, $\tilde{\boldsymbol{\beta}} = (\tilde{\boldsymbol{\beta}}_1', \ldots, \tilde{\boldsymbol{\beta}}_T')'$, where $\tilde{\boldsymbol{\beta}}_t' = (\mathbf{b}_t', \mathbf{v}_t')'$, and estimation of the elements of $\tilde{\mathbf{S}}_h$.

To simplify the computations and also obtain estimators that are more robust to possible model misspecifications, we use a more flexible definition of the matrix $\tilde{\mathbf{A}}$, whereby the (t, t')-th block of $\tilde{\mathbf{A}}$ is $\mathbf{D}^*_{|t,-t'|}$, where $\mathbf{D}^*_0, \ldots, \mathbf{D}^*_{T-1}$ are arbitrary diagonal matrices of order q. Note that the number of unknown parameters in $\tilde{\mathbf{A}}$ is increased this way from $2q$ to Tq.

For the state–space model imputations (method 5), we maximize the likelihood by use of the method of scoring. Let $\mathbf{Y}_{ht}^{(r)} = \mathbf{Q}_{ht} \mathbf{Y}_{ht}$ denote the vector of r_{ht} observed values for household h at time t. Let $\tilde{\mathbf{X}}_{ht}^{(r)} = \mathbf{Q}_{ht} \tilde{\mathbf{X}}_{ht}$ and $\tilde{\mathbf{Z}}_{ht}^{(r)} = \mathbf{Q}_{ht} \tilde{\mathbf{Z}}_{ht}$ be the corresponding $r_{ht} \times (p + q)$ and $r_{ht} \times (q + n_h)$ design matrices. The state vector estimate at time t, $\hat{\boldsymbol{\alpha}}_{ht}$, and its V-C matrix, $\mathbf{P}_{ht}$, are computed by the Kalman filter (Harvey, 1989), with initial values set as $\hat{\boldsymbol{\alpha}}_{h0} = 0$ and $\mathbf{P}_{h0} = \boldsymbol{D}^* \oplus \sigma_e^2 \mathbf{I}_{nh}$, where $\mathbf{D}^*$ and σ_e^2 are defined by (3.1.4) and (3.1.5). The likelihood is computed as follows: at time t, the predicted state vector, given all data till time $t - 1$, is $\hat{\boldsymbol{\alpha}}_{ht|t-1} = \mathbf{T}_h \hat{\boldsymbol{\alpha}}_{ht-1}$, with prediction error V-C matrix $\mathbf{P}_{ht|t-1} = \mathbf{T}_h \mathbf{P}_{ht-1} \mathbf{T}_h' + \mathbf{R}_h$, where $\mathbf{R}_h = \mathbf{D} \oplus \sigma_\varepsilon^2 \mathbf{I}_{n_h}$. The predicted value of $\mathbf{Y}_{ht}^{(r)}$ at time $(t - 1)$ is $\hat{\mathbf{Y}}_{ht|t-1}^{(r)} = \tilde{\mathbf{X}}_{ht}^{(r)} \boldsymbol{\beta}_t + \tilde{\mathbf{Z}}_{ht}^{(r)} \boldsymbol{\alpha}_{ht|t-1}$, with prediction error V-C matrix $\mathbf{F}_{ht} = \tilde{\mathbf{Z}}_{ht}^{(r)} \mathbf{P}_{ht|t-1} \tilde{\mathbf{Z}}_{ht}^{(r)\prime} = \mathrm{V}(\tilde{\mathbf{e}}_{ht})$, where $\tilde{\mathbf{e}}_{ht} = \mathbf{Y}_{ht}^{(r)} - \hat{\mathbf{Y}}_{ht|t-1}^{(r)}$ is the prediction error. The contribution to the log-likelihood from household h is:

$$L_h(\theta) = -\left\{ \sum_t \ln(|\mathbf{F}_{ht}|) + \sum_t \tilde{\mathbf{e}}_{ht}' \mathbf{F}_{ht}^{-1} \tilde{\mathbf{e}}_{ht} \right\} \qquad (3.3.1)$$

28.3.4 Estimation of Biases and Mean Square Errors

Though imputation methods are aimed primarily at completing the sample data and improving point estimation, the variance of estimators that use the observed and imputed values is needed for inference and interval estimation. The model considered in this chapter permits model-based estimation of imputation bias and mean square error (MSE). The appropriate expressions are presented in the Appendix, assuming known parameter values and that the model holds for both the missing and the observed data. The contribution to the variance from parameter estimation can be ignored in practice since it is ordinarily of lower order than the imputation variance.

28.4 SIMULATIONS AND EMPIRICAL EXAMPLE

To evaluate and compare the performance of the various imputation methods, we generated populations of size $N = 1000$ (500 for the state–space imputation method) from the model (3.1.1)–(3.1.3), for four time points. Household sizes, n_h, were randomly assigned the values 2 or 3. The parameters and regressors are: $\mathbf{b}_t = (1, 2)'$, $\mathbf{v}_t = (3, 4)'$, $\mathbf{A} = \text{diag}[0.7, 0.7]$, $\mathbf{D} = \text{diag}[0.5, 0.5]$, $\rho = 0.4$, $\sigma_\varepsilon^2 = 4$, and $z_{ht1} \equiv 5$. The values x_{hjtl} ($l = 1, 2$) and z_{ht2} were selected independently from the uniform distribution U(1, 10) for each set of indices.

We simulated three different response mechanisms with an expected nonresponse rate of $P_0 = 0.2$ for each time point. The mechanisms are defined by the distribution of the response indicator:

1. MCAR: $R_{hjt} \overset{\text{ind}}{\sim} \text{Bernoulli}(1 - P_0)$
2. MAR:
   For $t = 1$: $R_{hj1} \overset{\text{ind}}{\sim} \text{Bernoulli}(1 - P_0)$
   For $t = 2$: $R_{hj2} = \text{I}(R_{hj1}e_{hj1} \geq \sigma_e Z_q)$, where I( ) is the indicator variable, e_{hj1} is the residual defined by (3.1.1), $q = P_0/(1 - P_0)$, and Z_q is the $100q$ percentile of the standard normal distribution. Units missing at time 1 are observed at time 2, since $Z_q < 0$ for $P_0 < 0.5$.

$$\text{For } t > 2\text{: } R_{hjt} = \begin{cases} \text{I}(R_{hj2}e_{hj2} \geq \sigma_e Z_{P_0}) & \text{if } R_{hj1} = 0 \\ \text{I}(R_{hj1}e_{hj1} \geq \sigma_e Z_{P_0}) & \text{if } R_{hj1} = 1 \end{cases}$$

3. NMAR: $R_{hjt} = \text{I}(e_{hjt} \geq \sigma_e Z_{P_0})$

Under the second and third mechanisms, nonresponse occurs for large negative values of the outcome variable. In the case of MAR, nonresponse is determined by a previously observed outcome, whereas in the case of NMAR, nonresponse depends on the missing outcome.

The five imputation methods (and their variants) specified in Section 28.3 were applied to each simulated sample. For method 1 (mean imputation), eight imputation groups were formed by the cubes created by dividing the range [1, 10] of each of the auxiliary variables, x_{hjt1}, x_{hjt2}, and z_{ht2} into two equal parts (separately for each t). The group mean of the reported values was used for the imputation.

The imputed values, $y_{hjt}^{(I)}$, were compared to the true missing values, y_{hjt}, to obtain empirical estimates of bias and root mean square error (RMSE), as averages over all the missing values in the four time points. The bias and RMSE are defined as:

$$BIAS = \Sigma_{hjt}(1 - R_{hjt})(y_{hjt}^{(I)} - y_{hjt})/\Sigma_{hjt}(1 - R_{hjt})$$

$$RMSE = [\Sigma_{hjt}(1 - R_{hjt})(y_{hjt}^{(I)} - y_{hjt})^2/\Sigma_{hjt}(1 - R_{hjt})]^{1/2}$$

One hundred samples were generated for each method. Table 28.1 shows the average relative bias and RMSE over the 100 samples. The values in parentheses are the corresponding empirical standard errors.

The main outcomes emerging from Table 28.1 are as follows. All the methods yield effectively unbiased imputations under MCAR and the last three methods are also unbiased under MAR. The bias of the first three methods under MAR is explained by the fact that the MAR response is determined by the observed outcome on a previous occasion and this outcome is not used for the first three imputation methods (except in some cases for the nearest neighbor method). All the methods yield biased imputations under NMAR.

Next consider the relative RMSE that shows much more pronounced differences between some of the methods. As expected, the methods 4b and 5b, which use all the observed household data for all the time points, perform best. However, method 5a, which likewise uses the data observed for other household members, also per-

Table 28.1. Simulation results: relative bias and RMSE under different imputation methods (%)

		Relative bias			Relative RMSE		
No.	Imputation method	MCAR	MAR	NMAR	MCAR	MAR	NMAR
1	Mean imputation	0.12	0.92	6.14	20.13	20.11	20.63
		(0.05)	(0.05)	(0.05)	(0.72)	(0.80)	(0.77)
2	Nearest neighbor	0.07	-3.44	6.15	14.79	16.53	15.96
		(0.08)	(0.09)	(0.09)	(0.33)	(0.38)	(0.35)
3	Simple regression	0.06	0.94	6.10	16.22	16.17	16.81
		(0.02)	(0.03)	(0.02)	(0.12)	(0.22)	(0.19)
4a	Augmented regression,	0.06	-1.13	6.17	12.14	13.58	13.84
	individual	(0.02)	(0.03)	(0.02)	(0.03)	(0.03)	(0.21)
4b	Augmented regression,	0.03	-0.05	7.10	6.21	6.73	8.66
	household	(0.03)	(0.04)	(0.03)	(0.02)	(0.03)	(0.02)
5a	State–space, unsmoothed	0.03	-0.01	7.31	6.79	7.05	9.09
		(0.03)	(0.03)	(0.02)	(0.04)	(0.05)	(0.03)
5b	State–space, smoothed	0.01	-0.03	7.13	6.20	6.75	8.67
		(0.03)	(0.04)	(0.02)	(0.07)	(0.08)	(0.09)

forms well, despite the fact that it only employs past and current data. On the other hand, method 4(a), which uses only the data observed for the same individual, performs much worse, although it still dominates the first three methods. These results illustrate the potential benefits from accounting for the relationships between individual measurements within households and the relationships between measurements over time.

Table 28.2 compares the average imputation MSE estimators presented in the Appendix (with true parameter values replaced by the sample estimates), with the corresponding empirical MSE (same as in Table 28.1). The values shown indicate a very close agreement between the estimated and empirical RMSE under the MCAR and MAR response mechanisms. This is true for all the imputation methods. The RMSE estimators underestimate the true RMSE for NMAR, since they fail to account for the imputation bias inherent under this mechanism.

Finally, we applied the imputation methods to data extracted from the Israeli Labour Force Survey (ILFS) for Jerusalem for the years 1990–1994. This survey uses a rotation pattern of two quarters in the sample, two quarters out of the sample, and two quarters in again. The data contain complete records for 567 individuals in 475 households, with each individual observed for four quarters, according to the rotation pattern described above. The dependent variable is the number of hours worked during the week preceding the interview [$\bar{y} = 39.8$; $\text{sd}(y) = 14.8$, calculated over all individuals and all time periods]. The household level auxiliary variables are $z_1 = 1$ and z_2 = number of employed persons in the household [$\bar{z}_2 = 1.48$; $\text{sd}(z_2) = 0.56$]; the individual level auxiliary variables are x_1 = years of education [$\bar{x}_1 = 13.4$; $\text{sd}(x_1) = 4.8$] and x_2 = gender (41% females). The estimation methods described in Section 28.3.3 are easily modified to handle the missing values implied by the two quarters interview gap. We generated missing values using the same response mechanisms as used for the simulation study (a single set of missing data under each of the three response mechanisms).

The model parameters were estimated by use of only the state–space modeling

Table 28.2. Empirical and estimated relative RMSE (%)

		MCAR		MAR		NMAR	
No.	Imputation method	Estimate	Empirical	Estimate	Empirical	Estimate	Empirical
1	Mean imputation	20.13	20.13	20.13	20.11	20.11	20.63
2	Nearest neighbor	14.77	14.76	16.24	16.59	15.19	15.99
3	Simple regression	16.20	16.22	16.19	16.17	16.18	16.81
4a	Augmented regression, individual	12.18	12.14	13.56	13.58	12.85	13.84
4b	Augmented regression, household	6.29	6.21	6.75	6.73	6.09	8.66
5a	State–space, unsmoothed	6.79	6.79	7.10	7.05	6.05	9.09
5b	State–space, smoothed	6.28	6.20	6.87	6.75	5.89	8.67

Table 28.3. ILFS data set: relative bias and RMSE (%)

		Relative bias			Relative RMSE		
No.	Imputation method	MCAR	MAR	NMAR	MCAR	MAR	NMAR
1	Mean imputation	–1.83	21.95	74.18	32.57	45.50	81.61
2	Nearest neighbor	1.39	-14.41	57.65	33.94	43.18	73.33
3	Simple regression	–2.23	22.51	74.58	32.79	46.86	82.13
4a	Augmented regression, individual	–1.57	14.73	49.84	31.14	41.57	60.41
5a	State–space, unsmoothed	1.20	15.38	56.60	35.39	43.99	65.08
5b	State–space, smoothed	1.38	15.03	56.77	35.80	43.68	65.20

approach. We encountered convergence problems and negative variance estimators with the IGLS approach. For this reason, we did not apply method 4b (yields the same imputations as method 5b when using the same parameter values), whereas for method 4a we used the parameter estimates obtained for method 5. The average relative bias and RMSE over all the missing values are shown in Table 28.3.

The results here are much more erratic than the results obtained for the simulated data. The large RMSEs can be explained by three main reasons: a) the fit of the model is no longer "perfect"; b) in almost all the households, there is only one member, so that the clear advantages noted in Table 28.1 from borrowing information within households cannot be utilized here; and c) the average sample size available for parameter estimation is almost 80% smaller than the sample size used for the simulation study (567×0.8 compared to 2500×0.8). We mention in this regard that when estimating the model parameters for method 5b from all the data and not just the responses, the RMSEs are already reduced by about 40% under MCAR, 25% under MAR, and 5% under NMAR.

28.5 RAMIFICATIONS AND EXTENSIONS

The model defined by (3.1.1)–(3.1.3) refers to the population measurements, and the question arising is whether it holds equally for the sample data. As is often the case, the first- and/or second-level units are selected with unequal probabilities, and when the selection probabilities are related to the values of the outcome variable, even after conditioning on the model explanatory variables, the model holding for the sample data could be distorted by the sampling process. For the two-level model operating at given time points (equations 3.1.1, 3.1.4, and 3.1.5), Pfeffermann et al. (1998) propose a weighting procedure that accounts for the sampling effects and guarantees consistent estimators for all the model parameters. Feder et al. (2000) extend this procedure to the model underlying the present study by weighting the time series likelihoods defined by (3.3.1). Finally, we mention possible extensions to discrete or multivariate outcomes, and the problem of the robustness of the proposed imputation methods to model misspecification.

ACKNOWLEDGMENT

The authors thank Mr. Israel Einot of the Department of Statistics, Hebrew University of Jerusalem, who wrote the computer programs used for the numerical work reported here.

APPENDIX: MODEL-BASED ESTIMATION OF BIAS AND MSE

Under the model, $V(y_{hjt}) = V(\mathbf{z}'_{ht}\mathbf{u}_{ht} + e_{hjt}) = C_{ht} + \sigma_e^2 = V_{yht}$, where $C_{ht} = \mathbf{z}'_{ht}\,\mathbf{D}^*\mathbf{z}_{ht}$ $= \mathrm{Cov}(y_{hjt}, y_{hj't})$, $(j \neq j')$. Let the missing outcome be $y_{h_0 j_0 t}$.

1. *Mean imputation.* Define by g_t the group to which unit (h_0, j_0) belongs at time t. Let $m_{ght}(\leq n_h)$ be the number of individuals j in household h belonging to g_t with mean value $\bar{y}_{ght}$. The group mean for imputation is $\bar{y}_{gt} = \Sigma_{h=1}^{N} m_{ght}\bar{y}_{ght}/m_{gt}$, where m_{gt} $= \Sigma_{h=1}^{N} m_{ght}$ and $V(\bar{y}_{gt}) = [\Sigma_{h=1}^{N} m_{ght}^2 C_{ht}/m_{gt}^2] + [\sigma_e^2/m_{gt}]$. Let $\bar{\mathbf{x}}_{ght}^* = \Sigma_{j=1}^{nh}\tilde{x}_{hjt}/m_{ght}$ and $\bar{\mathbf{x}}_{gt}^*$ $= \Sigma_{h=1}^{N} m_{ght}\bar{\mathbf{x}}_{ght}^*/m_{gt}$ where $\tilde{x}_{hjt} = (x'_{hjt}, z'_{ht})'$. The imputation bias is:

$$B_1(h_0, j_0, t) = (\tilde{\mathbf{x}}_{h_0 j_0 t} - \bar{\mathbf{x}}_{gt}^*)'\tilde{\beta}_t \tag{A.1}$$

$$\mathrm{MSE}_1(h_0, j_0, t) = V_{yh_0 t} + V(\bar{y}_{gt}) + B_1^2(h_0, j_0, t) \tag{A.2}$$

The variance computation assumes given imputation groups and $y_{h_0 j_0 t}$ independent of $\bar{y}_{gt}$. For the remaining imputation methods there is no imputation bias under the model .

2. *Nearest neighbor.* (imputed value defined by nearest observed value $y_{h_0 j_0 t^*}$)

$$V_2(h_0, j_0, t) = V_{yh_0 t^*} + V_{yh_0 t} - 2\mathbf{z}'_{h_0 t}\mathbf{D}^*\mathbf{A}^{|t-t^*|}\mathbf{z}_{h_0 t^*} - 2\sigma_e^2\rho^{|t-t^*|} \tag{A3}$$

3. *Simple regression*

$$V_3(h_0, j_0, t) = E(y_{h_0 j_0 t} - \tilde{\mathbf{x}}'_{h_0 j_0 t}\tilde{\boldsymbol{\beta}}_t)^2 = V_{yh_0 t} \tag{A.4}$$

4. *Augmented regression*

4a. Based on observed data for the same individual. Let $\boldsymbol{\Lambda}_{h_0 j_0} = \mathbf{Q}_{h_0 j_0}\mathbf{S}_{h_0}\mathbf{Q}'_{h_0 j_0}$ where $\mathbf{S}_{h_0} = V[\tilde{\mathbf{Y}}_{h_0 j_0}]$. Denote by $\bar{\mathbf{q}}'_{h_0 j_0 t}$ the row of the missing indicator matrix $\bar{\mathbf{Q}}_{h_0 j_0}$ with 1 in column t and define $\boldsymbol{\lambda}'_{mt} = \bar{\mathbf{q}}'_{h_0 j_0 t}\,\tilde{\mathbf{S}}_{h_0}\mathbf{Q}'_{h_0 j_0}$.

$$V_4^a(h_0, j_0, t) = V_{yh_0 t} - \boldsymbol{\lambda}'_{mt}\boldsymbol{\Lambda}_{h_0 j_0}^{-1}\boldsymbol{\lambda}_{mt} \tag{A.5}$$

4b. Based on all the household data. Let $\boldsymbol{\Lambda}_{h_0} = \mathbf{Q}_{h_0}\tilde{\mathbf{S}}_{h_0}\mathbf{Q}'_{h_0}$ where $\tilde{\mathbf{S}}_{h_0} = V[\tilde{\mathbf{Y}}_{h_0}]$. Denote by $\bar{\mathbf{q}}^{*\prime}_{h_0 j_0 t}$ the row of $\bar{\mathbf{Q}}_{h_0}$ (indicator matrix for the missing household data) with 1 in the column corresponding to the missing value $y_{h_0 j_0 t}$ and zeroes elsewhere, and define $\boldsymbol{\lambda}^{*\prime}_{mt} = \bar{\mathbf{q}}^{*\prime}_{h_0 j_0 t}\,\tilde{\mathbf{S}}_{h_0}\mathbf{Q}'_{h_0}$.

$$V_4^b(h_0, j_0, t) = V_{yh_0 t} - \boldsymbol{\lambda}_{mt}^{*\prime} \boldsymbol{\Lambda}_{h_0}^{-1} \boldsymbol{\lambda}_{mt}^{*} \tag{A.6}$$

5. *State-space model imputation*

5a. Based only on current and past data (unsmoothed). By (3.2.6): $\hat{y}_{h_0 j_0 t}^{(M)} - y_{h_0 j_0 t} = (\mathbf{z}_{h_0 t}^{\prime}, 1)(\hat{\boldsymbol{\alpha}}_{h_0 t} - \boldsymbol{\alpha}_{h_0 t}) = \tilde{\mathbf{z}}_{h_0 t}^{\prime}(\hat{\boldsymbol{\alpha}}_{h_0 t} - \boldsymbol{\alpha}_{h_0 t})$ where $\boldsymbol{\alpha}_{h_0 t} = (u_{h_0 t}^{\prime}, e_{h_0 j_0 t})$. Hence,

$$V_5^a(h_0, j_0, t) = \tilde{\mathbf{z}}_{h_0 t}^{\prime} \mathbf{P}_{h_0 t} \tilde{\mathbf{z}}_{h_0 t}^{\prime}, \tag{A.7}$$

where $\mathbf{P}_{h_0 t} = E[(\boldsymbol{\alpha}_{h_0 t} - \hat{\boldsymbol{\alpha}}_{h_0 t})(\boldsymbol{\alpha}_{h_0 t} - \hat{\boldsymbol{\alpha}}_{h_0 t})^{\prime}]$ is obtained from the Kalman filter.

5b. Based on all the data for all the time periods (smoothed). Same as in (A.7), with appropriate modficiations. The smoothed predictor $\hat{\boldsymbol{\alpha}}_{h_0 t}^{sm}$ and the corresponding V-C matrix $\mathbf{P}_{h_0 t}^{sm}$ are obtained from the smoothing algorithm.

CHAPTER 29

Diagnostics for the Practical Effects of Nonresponse Adjustment Methods

John L. Eltinge, *U.S. Bureau of Labor Statistics and Texas A&M University*

29.1 INTRODUCTION

Recall from Chapter 1 that evaluation of the effects of nonresponse, and of potential nonresponse adjustment methods, requires balanced consideration of several issues, including the following.

1. Biases in point estimators
2. Inflation of the variances of point estimators
3. Biases in customary variance estimators intended for estimation of (2)

Several chapters of this book have addressed various aspects of these issues and have presented a suite of nonresponse adjustment methods intended to account for them.

Application of these methods to a given survey generally will involve two sets of diagnostic tools that are focused, respectively, on the nonresponse model and on the practical effects of a given nonresponse adjustment method for a specific survey and population. First, any adjustment method is based explicitly or implicitly on models for the survey items and the response probabilities. Consequently, it is important to validate these models thoroughly, within the limits of available data. In addition, such models are broadly acknowledged to be approximations to the true relationships between survey items and response probabilities. Thus, it can be important to evaluate the extent to which the properties of nonresponse adjustment methods are sensitive to deviations from assumed conditions. Section 29.2 provides limited literature references for this first set of diagnostics, but a detailed discussion is beyond the scope of the present work.

Second, implementation of a given nonresponse adjustment method for a specific survey will frequently involve nontrivial burdens due to required changes in data management methods, software, or institutional culture. Consequently, a survey organization will be much more likely to implement a proposed new method if it has evidence that the new method is substantially better than current methods for use in inference for some important estimands of the specific survey in question. In this sense, selection of a nonresponse adjustment method can be viewed as one of several features of the design of a full survey procedure, where we define a survey procedure to include the sample design, questionnaire, fieldwork, and construction of point estimators and inferential statistics adjusted for nonresponse. For this purpose, one may use both tools for direct evaluation of estimation bias and tools traditionally associated with sample design, e.g., design effects and power curves. This chapter explores this idea with special emphasis on comparison of confidence intervals and power curves within the framework of customary large-sample inference.

Using the notation developed in Chapter 1, consider again inference for a finite population mean $\mu = N^{-1}\Sigma_{i=1}^{N} Y_i$ from a sample S selected through a complex design D, and consider an associated full-sample point estimator $\hat{\mu}_F$. Nonresponse described by the quasirandomization model (M.1) of Chapter 1 leads to a set R of respondents obtained from S. Then one may consider estimation of μ through the unadjusted estimator $\hat{\mu}_{UA}$, an imputation-based estimator $\hat{\mu}_I$ considered further below, or the reweighted estimator

$$\hat{\mu}_p = \left(\sum_{i \in R} w_{pi} \right)^{-1} \sum_{i \in R} w_{pi} Y_i \tag{29.1}$$

where $w_{pi} = w_i/p_i$ is a weight that has been adjusted by the inverse of the selection probability for the unit.

In practice, one generally does not know the response probabilities p_i, and one must estimate them, either explicitly or implicitly, from the observed response indicators r_i and available element-level auxiliary variables X_i, say. For example, in some cases one uses a logistic regression model

$$\ln(p_i/1 - p_i) = X_i\beta \tag{29.2}$$

computes an estimator $\hat{\beta}$ of β through customary pseudolikelihood methods, and then computes estimators $\hat{p}_i$ of p_i through substitution of $\hat{\beta}$ for β into expression (29.2). If model (29.2) is correctly specified, then the resulting version of estimator (29.1) is approximately unbiased for μ.

In other cases, one forms groups of the sampled units into adjustment cells. Within a given cell c, say, each unit i has its response probability estimator $\hat{p}_i$ set equal to $(\Sigma_{i \in S \cap c} w_i)^{-1} \Sigma_{i \in R \cap c} w_i$, the weighted sample proportion of respondents in cell c. In general, the cells c may be formed by grouping together units with similar values of the logistic regression based estimates of p_i, or through ad hoc combinations of available classificatory variables X_i, like age, race, or gender. Under mild regularity conditions, if the finite-population correlation of Y_i and p_i within each cell c is approximately equal to zero, then the resulting cell-based version of the estimator (29.1) is approximately unbiased for μ. For some general discussion of weighting adjustment cells see, e.g., Little (1986), Czajka et al. (1992), Eltinge and Yansaneh (1997) and references cited therein; and the related work by Rosenbaum and Rubin (1983, 1984) on response propensity.

A second approach to nonresponse adjustment is based on imputation. Imputation can be implemented in many forms. For a detailed discussion see, e.g., Little (1988), Fay (1996), Rao (1996), Rubin (1996) and references cited therein. For this discussion, we restrict attention to a simple form of hot deck single imputation. Specifically, assume that the population is partitioned into C cells, $c = 1, 2, \ldots, C$, say. Next, consider a given nonresponding sample unit i contained in cell c. From among the responding units in cell i, randomly select one unit to be the "donor" for the missing unit i, with selection probabilities proportional to the initial sample selection probabilities π_j. Given a selected donor j, say, define the imputed value $Y_i^* = Y_j$. In addition, for all responding sample units i, define $Y_i^* = Y_i$. Then an imputed-data estimator of μ is

$$\hat{\mu}_I = \left(\sum_{i \in S} w_i \right)^{-1} \sum_{i \in S} w_i Y_i^* \tag{29.3}$$

As with the weighting adjustment cell-based estimator, the imputation-cell estimator (29.3) is approximately unbiased for μ if the finite-population correlation of Y_i and p_i in each cell c is approximately equal to zero.

In evaluation of the practical effects of nonresponse adjustments based on weighting or imputation, each of issues (1) through (3) above can play an important role. First, as indicated above, unbiasedness of the adjusted point estimators $\hat{\mu}_p$ and $\hat{\mu}_I$ depends on specific conditions on the response probabilities p_i and their relationship with the survey items Y_i. Second, the adjusted estimators $\hat{\mu}_p$ and $\hat{\mu}_I$ generally will have higher variances than the idealized full-sample point estimator $\hat{\mu}_F$. Third, direct application of standard full-sample variance estimation methods to $\hat{\mu}_p$ or $\hat{\mu}_I$ generally will lead to negatively biased variance estimators. Section 29.2 reviews these issues and outlines some diagnostics that the literature has developed to quantify the magnitude of each problem in a given application.

Although each of issues (1) through (3) can be important in itself, a given adjustment method generally will not perform uniformly well according to all three criteria. Consequently, it is important to have diagnostic methods that provide a balanced display of the competing effects of (1) through (3) in a given application. Section 29.3 suggests that one way to achieve this balanced approach is to consider the effect of nonresponse adjustments on standard large-sample inference methods. Specifically, Section 29.3.1 shows how estimators of point estimator and variance estimator bias and variance inflation lead to relatively simple methods for assessment of confidence interval coverage rates and widths. Section 29.3.2 develops related methods for estimation of the power curves of hypothesis tests. Section 29.4 illustrates some of the ideas of Section 29.3 with an application to the U.S. Third National Health and Nutrition Examination Survey. Section 29.5 discusses some possible extensions of the proposed diagnostics.

29.2 LIMITATIONS OF ADJUSTMENT METHODS

29.2.1 Point Estimator Bias

Several related diagnostics can help one assess the point estimator bias issues raised in Section 29.1. Misspecification of the models relating the probability of response (29.2) (for weighting) or the observed outcome (for imputation) to observed covariates is likely to result in bias in the resulting point estimators. Hence, diagnostics to check the specification of these models are important. For example, one may consider standard goodness-of-fit diagnostics for the logistic regression model (29.2) or for other binary outcome models applied to the quasirandomization approach (M.1). See, e.g., Hosmer and Lemeshow (1989) for an accessible review of some practical methods. Of special interest is the Hosmer and Lemeshow (1980) method of grouping units according to their $\hat{p}_i$ estimates and then comparing the means of $\hat{p}_i$ and of the response indicators within each group. Similarly, in developing imputation cells based on predicted-Y grouping of sample units, it is worthwhile to apply standard regression diagnostics to test the adequacy of the specific models used to predict Y values. In addition, automatic interaction detection (e.g., Scott and Knott, 1976; Steel and Boal, 1988) and related regression tree methods (Alexander and Grimshaw, 1996, and references cited therein) can be useful in ad hoc formation of adjustment cells.

Second, one can consider variants on methods developed to test for the informativeness of a sample design; see, e.g., DuMouchel and Duncan (1983), Fuller (1984), and Pfeffermann (1993, 1996). To simplify notation, we will restrict attention to estimation of a population total $Y = \Sigma_{i=1}^{N} Y_i$; similar ideas apply to means or other smooth nonlinear functions of a vector of population totals. We consider two estimators, $\hat{Y}_A = \Sigma\, w_{Ai} Y_i$ and $\hat{Y}_M = \Sigma\, w_{Mi} Y_i$ based on responses Y_i, $i \in R$ and two sets of weights $\{w_{Ai}\}$ and $\{w_{Mi}\}$ obtained from two different sets of nonresponse adjustment factors. We will use the working assumption that the adjusted weights w_{Ai} are correct in the sense that $\hat{Y}_A$ is approximately unbiased for Y, and that the weights w_{Mi} may be misspecified. For example, the application in Section 29.4 will consider weights w_{Ai} based on a logistic regression model that includes several main effects and interactions, and will consider alternative weights w_{Mi} based on simpler logistic models involving only main effects or (in an extreme case) only an intercept term. For such cases, even if logistic regression interaction terms are statistically significant, the simpler weighting method for w_{Mi} may be of serious interest (for reasons of estimator stability or a general preference for parsimony), provided $\hat{Y}_M$ is approximately unbiased for Y.

To test this unbiasedness condition, note that

$$\hat{Y}_M - \hat{Y}_A = \sum_{i \in R} (w_{Mi} - w_{Ai}) Y_i = \sum_{i \in R} w_{Ai} \tilde{Y}_i \tag{29.4}$$

where $\tilde{Y}_i = (w_{Ai}^{-1} w_{Mi} - 1) Y_i$. Under the working assumption that $\hat{Y}_A$ is approximately unbiased for Y, one may test for bias in $\hat{Y}_M$ by testing the null hypothesis that the expectation of expression (29.4) equals zero. A standard error $se(\Sigma_{i \in R} w_{Ai} \tilde{Y}_i)$ can be computed through customary design-based methods (e.g., Wolter, 1985). This in turn leads to the t type test statistic $\Sigma_{i \in R} w_{Ai} \tilde{Y}_i / se(\Sigma_{i \in R} w_{Ai} \tilde{Y}_i)$. Large absolute values of this t statistic would indicate that $\hat{Y}_M$ is biased. See also the work by Little et al. (1997) on assessment of the impact of different weights on a real survey.

Third, recall from Section 29.1.2 the cell-formation method based on grouping units according to their $\hat{p}_i$ values. In keeping with results reviewed in Section 29.1.2, the resulting point estimator may have some residual nonresponse bias if the cells employed are excessively coarse. Eltinge and Yansaneh (1997) developed some simple methods for identification and refinement of such "problem cells." These methods used variants on the $\tilde{Y}$ based tests reviewed above, and were applied to income nonresponse in the U.S. Consumer Expenditure Survey.

29.2.2 Variance Inflation

Due to the reduction in the available number of respondents, a point estimator based on the respondent group R generally will have a higher variance than the idealized full-sample estimator $\hat{\mu}_F$.

To quantify this variance inflation and place it in a broader context, one may apply the ideas of classical design effects as developed in Kish (1965, 1995). Recall from Chapter 1 and Section 29.1 that our work has considered a sample of n units selected through a complex design, and leading (under full response) to the full-sample point estimator $\hat{\mu}_F$. Now consider a hypothetical alternative sample of n units selected through simple random sampling with full response; let $\hat{\mu}_{SRS}$ be the resulting estimator of μ, and let V_{SRS} be the variance of $\hat{\mu}_{SRS}$, evaluated with respect to the hypothetical simple random sample design. Then the design effect for $\hat{\mu}_F$ under the original complex design and full response is

$$\Delta_F = V(\hat{\mu}_F)/V_{SRS} \tag{29.5}$$

where $V(\hat{\mu}_F)$ is the variance of $\hat{\mu}_F$ evaluated with respect to the original complex design. Similarly, define the design effects for the weighting adjusted estimator,

$$\Delta_p = V(\hat{\mu}_p)/V_{SRS} = [V(\hat{\mu}_p)/V(\hat{\mu}_F)][V(\hat{\mu}_F)/V_{SRS}] \tag{29.6}$$

and for the imputation estimator,

$$\Delta_I = V(\hat{\mu}_I)/V_{SRS} = [V(\hat{\mu}_I)/V(\hat{\mu}_F)][V(\hat{\mu}_F)/V_{SRS}] \tag{29.7}$$

For these definitions, recall that $V(\hat{\mu}_p)$ was evaluated with respect to both the sample design and the quasirandomization model (M.1), and that $V(\hat{\mu}_I)$ was evaluated with respect to those two components, and the random selection of imputed values within cells.

Note that the design effects defined in (29.5)–(29.7) are measures of the efficiency of specific point estimators under a given set of design and response conditions. Within this framework, comparison of the factors $V(\hat{\mu}_p)/V(\hat{\mu}_F)$ and $V(\hat{\mu}_I)/V(\hat{\mu}_F)$ with the full-sample design effect Δ_F gives an indication of the impact of nonresponse, and nonresponse adjustments, on survey efficiency relative to the impact of the full-sample design features.

In addition, one could extend design effect ideas to cover biased estimators $\hat{\mu}_M$, say, through examination of the mean squared error ratio

$$MSE(\hat{\mu}_M)/V_{SRS} = [\{E(\hat{\mu}_M - \mu)\}^2 + V(\hat{\mu}_M)]/V(\hat{\mu}_F)][V(\hat{\mu}_F)/V_{SRS}] \qquad (29.8)$$

This ratio is of practical interest for cases in which bias–variance trade-offs are of serious concern, e.g., when $\hat{\mu}_M$ has a bias that is detectable through tests based on (29.4), but is moderate relative to the standard error of $\hat{\mu}_M$.

29.2.3 Variance Estimator Bias

The sample survey literature also addresses issues of variance estimator bias arising in nonresponse work. For instance, consider a dataset with nonresponses singly imputed using the cell-based hot deck method described in Section 29.1.2. Then direct application of full sample variance estimation methods (e.g., Wolter, 1985) to this imputed dataset generally will produce negatively biased variance estimators. See, e.g., Rubin (1987) and Rao and Shao (1992). In addition, Fay (1993, 1996) has noted that in some cases, multiple imputation (Rubin, 1987) methods can produce variance estimators with a positive bias. If the point estimator itself is not seriously biased, this yields confidence intervals likely to have conservative coverage. Overestimation of variance may be regarded as aceptable if it leads to conservative tests or confidence intervals, provided the conservatism does not entail a major loss of efficiency. On the other hand, underestimation of the variance is generally unacceptable, since it leads to confidence intervals and tests that do not appropriately reflect the true variability of the point estimator.

Problems with variance estimation bias also arise in other areas of survey work, e.g., stratum collapse. To address this problem, the sample survey literature has developed "misspecification effect" measures, defined as follows. First, let $\hat{\mu}$ be an approximately unbiased point estimator of a parameter μ and let $\hat{V}(\hat{\mu})$ be an approximately unbiased estimator of $V(\hat{\mu})$. In addition, let $\hat{V}_{\text{mis}}$ be a biased estimator of $V(\hat{\mu})$. In some applications, $\hat{V}_{\text{mis}}$ is considered a serious candidate for use with $\hat{\mu}$, due to computational convenience, greater stability than $\hat{V}(\hat{\mu})$, or inaccessibility of data required to compute $\hat{V}(\hat{\mu})$ in routine survey production.

Following Skinner (1989, pp. 24ff.) define the mispecification effect $meff = \{E(\hat{V}_{\text{mis}})\}^{-1}V(\hat{\mu})$. A simple estimator of $meff$ is $\widehat{meff} = (\hat{V}_{\text{mis}})^{-1}\hat{V}(\hat{\mu})$. Previous literature has used misspecification effect estimates in two complementary ways. First, $\widehat{meff}$ may be viewed as a general diagnostic to assess the extent to which use of $\hat{V}_{\text{mis}}$ may cause serious problems in analysis and inference work. For example, Skinner (1989, p. 29, Table 2.1) examined confidence intervals for μ based on $\hat{\mu}$ and $\hat{V}_{\text{mis}}$, and displayed a direct linkage between coverage rates and the misspecification effect for $\hat{V}_{\text{mis}}$. Holt et al. (1980) and Scott and Holt (1982, Section 4, Table 1) examined related problems with the sizes of hypothesis tests. In general, values of $\widehat{meff}$ less than unity indicate that $\hat{V}_{\text{mis}}$ is a conservative variance estimator. Conversely, values of $\widehat{meff}$ greater than unity indicate that $\hat{V}_{\text{mis}}$ tends to underestimate the true variance of $\hat{\mu}$. Standard inferential reasoning indicates that in the latter case the analyst has a strong incentive to avoid the use of $\hat{V}_{\text{mis}}$. In place of $\hat{V}_{\text{mis}}$, the analyst would use an alternative variance estimator, e.g., $\hat{V}(\hat{\mu})$. Second, several authors have used estimated misspecification effects to adjust test statistics or confidence sets for μ based on $\hat{V}_{\text{mis}}$. See, e.g., Holt et al. (1980), Rao and Scott (1981), Rao and Thomas (1989), Graubard and Korn (1993) and references cited therein. These adjustment methods could also be of interest in nonresponse problems, especially those involving inference for multivariate parameters μ, but details of this are beyond the scope of the present work.

29.3 USE OF CONFIDENCE INTERVAL PERFORMANCE AND POWER CURVES FOR COMBINED ASSESSMENT OF LIMITATIONS

As indicated in Section 29.2.3, moment-based properties of point estimators and variance estimators have important implications for the operating characteristics of standard large-sample inference methods. This is of practical interest because conceptual or methodological distinctions among competing survey analyses are most likely to receive attention when they lead to nontrivial differences in the scientific or policy conclusions drawn from these competing analyses. Subsections 29.3.1 and 29.3.2 explore these ideas in additional detail for confidence intervals and hypothesis tests, respectively.

29.3.1 Confidence Interval Coverage Rates and Widths

In general, the frequentist operating characteristics of interval estimators of μ are acknowledged to be of broad interest, even if the original development of the interval estimator arose in a nonfrequentist setting. See, e.g., Rubin (1996), citing Neyman (1934). In such evaluations, principal attention focuses on whether the intervals have a true repeated-sampling coverage rate greater than or equal to their nominal rate, $1 - \alpha$, say. Conditional on having satisfactory coverage rates, a secondary criterion is confidence interval width, with narrower intervals preferred.

The Skinner (1989) work reviewed in Section 29.2 noted that variance estimator bias can have a serious effect on confidence interval coverage rates. To extend these results to account for both point estimation and variance estimation bias, consider a nominal $(1 - \alpha)100\%$ confidence interval $(\hat{\mu}_L, \hat{\mu}_U)$ defined by

$$\hat{\mu}_M \pm z_{1-\alpha/2}(V_{\text{mis}})^{1/2} \tag{29.9}$$

where $z_{1-\alpha/2}$ is the $1-\alpha/2$ quantile of the standard normal distribution. Let $1-\gamma$ be the true coverage rate of $(\hat{\mu}_L, \hat{\mu}_U)$ as an interval estimator of the true μ. In keeping with Scott and Holt (1982, p. 850) and Skinner (1986, 1989), we assume that effective sample sizes are large enough to ensure that $\hat{V}_{\text{mis}}$ is approximately equal to its expectation. Thus, this subsection will not address questions involving the stability of $\hat{V}_{\text{mis}}$.

Routine standardization arguments then show that

$$1-\gamma = P(\hat{\mu}_L < \mu < \hat{\mu}_U) \cong P(B_L < Z < B_U) = \Phi(B_U) - \Phi(B_L) \tag{29.10}$$

where $(B_L, B_U) = \{V(\hat{\mu})\}^{-1/2}[-\{E(\hat{\mu}_M) - \mu\} \pm z_{1-\alpha/2}\{E(\hat{V}_M)\}^{1/2}]$, Z is a standard normal random variable, $\Phi(\cdot)$ is a standard normal distribution function and the approximation (29.10) is based on a routine normal approximation. Direct substitution of $\hat{V}_{\text{mis}}$ for its expectation, $\hat{V}(\hat{\mu})$ for $V(\hat{\mu})$, and $\hat{\mu}_M - \hat{\mu}$ for $E(\hat{\mu}_M) - \mu$ leads to

$$(\hat{B}_L, \hat{B}_U) = [\hat{V}(\hat{\mu})]^{-1/2}[-(\hat{\mu}_M - \hat{\mu}) \pm z_{1-\alpha/2}(\hat{V}_M)^{1/2}]$$

and thus to an estimator $1-\hat{\gamma} = \Phi(\hat{B}_U) - \Phi(\hat{B}_L)$ of the true coverage rate.

In addition, note that if $E(\hat{\mu}_M) = \mu$, then B_L and B_U are simple functions of *meff*, and specific values of *meff* translate directly into a true coverage rate $1-\gamma$; see, e.g., Skinner (1989, Table 2.1). However, if $E(\hat{\mu}_M) \neq \mu$, then the combined effects of the biases of $\hat{\mu}_M$ and $\hat{V}_{\text{mis}}$ contribute to $1-\gamma$ in a somewhat more complex form. Also, see Bethlehem and Kersten (1985) for a previous example of a nonresponse sensitivity analysis that evaluates the effects of point estimation bias on confidence interval coverage rates.

Finally, recall that conditional on the true coverage rate $1-\gamma$ being greater than or equal to the nominal rate $1-\alpha$, one generally will prefer a confidence interval with smaller mean widths. These confidence interval width comparisons are of interest in two cases. First, consider the case in which two competing nonresponse adjustment methods each produce approximately unbiased point estimators and variance estimators. Then under mild regularity conditions, true coverage rates will be approximately equal to their nominal values, and comparison of these two methods will focus on comparison of their confidence interval widths. In particular, note that these widths will be proportional to the square roots of the design effect ratios reviewed in Section 29.2. Second, consider the case in which $\hat{\mu}_M$ is approximately unbiased and $\hat{V}_{\text{mis}}$ has a moderate positive bias. Then the resulting confidence intervals are conservative, and practical interest focuses on the extent, if any, to which the positive bias in $\hat{V}_{\text{mis}}$ may inflate confidence interval widths, relative to a competing method based on unbiased point estimators and variance estimators. In the special case in which the point estimator $\hat{\mu}_M$ is more efficient than its competitor, it is possible for the "misspecified" analysis to produce confidence intervals that are narrower, on average, than the intervals produced by the competing "unbiased" analysis. See, e.g., Rubin (1996) for a specific example involving multiple imputation.

29.3.2 Power Curves

Power curves are widely acknowledged to be useful tools in the design of survey procedures. For instance, they frequently are used in the selection of sample sizes and related sample design work. However, as noted in Section 29.1, they also can be useful in exploration of all aspects of the design of a full survey procedure, including nonresponse adjustments. In particular, power curves allow one to evaluate inferential performance in a form that is linked directly with specific differences in a parameter μ that are large enough to be of substantive interest. Consider the hypotheses H_0: $\mu = \mu_0$ and H_1: $\mu \neq \mu_0$ for a prespecified null value μ_0. Given a point estimator $\hat{\mu}_M$ and a variance estimator $\hat{V}_{\text{mis}}$, one may compute the customary test statistic $t_0 = (\hat{\mu} - \mu_0)/(\hat{V}_{\text{mis}})^{1/2}$ and compare $|t_0|$ with $t_{d,1-\alpha/2}$, the $1 - \alpha/2$ quantile of a t distribution on d degrees of freedom. For a given true value μ, the probability of rejection is then equal to

$$P(|t_0| > t_{d,1-\alpha/2}|\mu) \tag{29.11}$$

Calculations similar to those for expression (29.10) allow estimation of (29.11) for specified μ and μ_0. A plot of these estimates for an interval of values of μ_0 around a specified true value μ then yields an estimated power curve. A specific example is presented in Section 29.4 below.

Power curves are useful because they provide a broad range of data users (including those who are not specialists in survey inference) with a concrete visualization of the linkage between statistical operating characteristics and practical differences in parameter values. In particular, readily observed features of power curves have direct links with the point estimation bias, variance inflation, and variance estimation bias issues reviewed in Chapter 1 and Section 29.1 above. For example, bias in a point estimator leads to a corresponding horizontal shift in the associated power curve. If this shift is large enough to degrade the power of the test to detect deviations from H_0 that are of serious substantive interest, then one has a strong case for considering alternative point estimators that are unbiased. Similarly, a positive bias in a variance estimator will produce a downward vertical shift in the power curve, reflecting the conservative property of the associated test. On the other hand, a negative bias in a variance estimator will produce an upward vertical shift in the power curve, reflecting the fact that the associated test has a type I error rate in excess of its nominal level α. In addition, inflation in the true variance of a point estimator will attenuate the slope of the power curve at a given distance from the true value μ. This loss of efficiency is of practical concern for cases in which there is a strong scientific or policy interest in the specific alternatives against which one has lost a substantial amount of power.

It should be noted that there are important conceptual and foundational limitations of the use of a hypothesis testing framework for inference; see, e.g., Berger and Sellke (1987) and references cited therein. However, despite these limitations, power curves are useful in the present context for two reasons. First, despite the abovementioned limitations, hypothesis testing is used (both explicitly and implicitly) in many survey applications. Consequently, it is worthwhile to evaluate the operating characteristics of these methods applied to nonresponse-adjusted data.

Second, in many cases confidence intervals are constructed through test inversion methods, or are closely approximated by related test inversion confidence intervals; see, e.g., Binder and Patak (1993). For a given univariate test, the vertical and horizontal locations of its power curve are closely related to the coverage rate of the associated test-inversion confidence intervals. In addition, more attenuated slopes of power curves (reduced power against a specified alternative hypothesis test) correspond to higher mean or median widths of test inversion confidence intervals. Consequently, power curves for hypothesis tests provide a useful graphical summary of the principal operating characteristics of the associated test inversion confidence intervals.

29.4 APPLICATION TO NHANES III

To illustrate some of the ideas developed in this chapter, we consider an application to the U.S. Third National Health and Nutrition Examination Survey (NHANES III) that originally motivated some of this work. For some general background on NHANES III, see National Center for Health Statistics (1996). This survey was carried out for the U.S. National Center for Health Statistics and was intended to assess the health and nutritional status of the noninstitutionalized civilian population of the United States. Analyses generally treat the design as involving 49 strata, with two primary sample units selected from each stratum with possibly unequal probabilities and with replacement. Additional levels of sampling lead to selection of individual persons who are asked to participate in an interview and a medical examination.

The current analysis will focus on measurements of femur neck bone mineral density (BMD, measured in units of grams per centimeter squared) collected in the medical examination. The process of collecting these BMD measurements is potentially prone to error, so the National Center for Health Statistics sought remeasurements on some sample persons, in order to obtain an indication of the magnitude of the variance of the measurement error in the BMD data. In principle, it would have been preferable to use a standard two-phase sample design to select sample persons for remeasurement. However, this was not feasible due to administrative and fieldwork constraints. In particular, anecdotal evidence indicated that nonparticipation in the remeasurement process was essentially equivalent to nonresponse, with a serious potential for concern regarding differences in response rates across different demographic groups. Out of 16,416 persons aged 20 and above who participated in the general NHANES III medical examination, only 1108, or 6.75%, provided both initial and replicate measurements of bone density.

Eltinge et al. (1997) reported results from fitting a logistic regression model (29.2) for response probabilities (the probability of providing both initial and replicate BMD measurements, conditional on participation in the general NHANES III medical examination). Following exclusion of effects that were not significant at the $\alpha = 0.05$ level, the selected model included main effects for region, race–ethnic classification, gender, and decade-level age group membership; and also included interactions between age group and gender. However, the main effects for the 50–59 and 60–69 age groups were dominant, while the remaining age main effects and interactions were not as pronounced. This led to consideration of four possible weighting adjustments.

a. Weighting adjustments based on the full logistic regression model, including interactions.
b. Weighting adjustment based on a refit logistic regression model that included only the main effects from (1).
c. Weighting adjustment based on a refit logistic regression model that included only an intercept term and an indicator variable for membership in the 50–69 age group.
d. No explicit weighting adjustment, equivalent to treating all response probabilities as being equal.

The mean μ of interest for this analysis is the mean of the square of the difference between the first and second bone mineral density measurements. All four methods (a)–(d) were implemented in a form that did not involve anticipated variance estimation bias, so the principal issues of interest were potential point estimation bias and inflation of point estimator variance.

Moment-based diagnostics described in Section 29.2 did not give any strong indications of point estimation bias for the three simplified weighting methods (b) through (d). Consequently, we turn to evaluation of efficiency, as displayed in power curves. Figure 29.1 displays four overlaid power curves associated with methods (a) through (d), respectively. In each case, the true value for μ was set equal to the mean estimate computed from method (a). The horizontal axis of Figure 29.1 displays different potential null hypothesis values μ_0, and the corresponding estimates of power are displayed on the vertical axis. In comparison of the four estimated power curves, note especially that there are moderate horizontal shifts among curves (a) through (d), reflecting moderate differences in the point estimator values. In addition, note that option (d) has a somewhat steeper power curve, reflecting a somewhat smaller variance of the associated point estimator.

29.5 DISCUSSION

In closing, we note four additional points. First, the discussion of inferential performance has taken at face value the assumption that pivotal quantities followed the specified t or normal distributions. While this assumption is used widely and may be reasonable in many applications, it can be problematic in some cases. For example, in problems involving skewed underlying distributions, rare subpopulations or extreme weights, the numerator and denominator of a t statistic may fail to satisfy

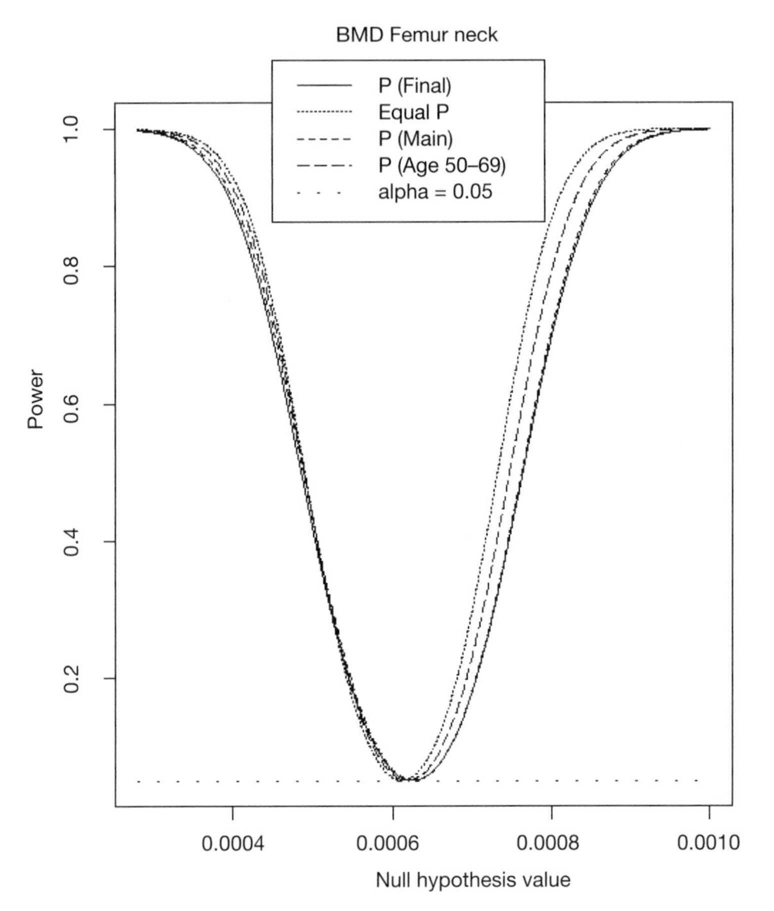

Figure 29.1 Estimated power curves for femur neck bone mineral density.

customary approximate-independence assumptions. This in turn can have a nontrivial effect on the properties of the associated confidence intervals and test statistics. See, e.g., Casady et al. (1998) and references cited therein.

Second, Section 29.1 noted that a survey organization is more likely to adopt a specific nonresponse adjustment method if there is strong empirical evidence that its currently employed nonresponse adjustment methods lead to nontrivial problems in inference for specific parameters that are of central interest. However, if application of diagnostics to a finite set of variables does not identify serious inferential problems with current nonresponse adjustment methods, there can still be inferential problems for other variables that were not studied. In addition, any specific set of diagnostics can be imperfect in the sense that they are intended to identify specific types of deviations from assumed conditions of an underlying explicit or implicit model, and may have relatively little power to detect other types of deviations that may also cause problems in inference for the parameter of principal interest, μ.

(See, e.g., Hansen et al. (1983) for related comments on the potential limitations of model diagnostics, developed outside of the nonresponse context.) Consequently, if a limited diagnostic study does not identify major inferential problems with current nonresponse adjustment methods, that does not by any means imply a uniform endorsement of the current methods.

Third, this chapter has focused on large-sample inference for a given parameter. In some cases, data users may have a considerably broader interest in the overall distribution of a given survey item. For such cases, it is useful to extend the moment-based methods of Section 29.2 to estimated distribution functions of Y, or of related auxiliary variables X observed for both respondent and nonrespondent units. For example, in work with the 1995 Survey of Consumer Finances, Kennickell (1997) used quantile–quantile plots to explore differences between response groups (classified according to the number of contacts); and Kennickell (1999) used average shifted histograms to compare respondent and refusal units. See also Rosenbaum (1995) for somewhat related ideas in work with observational data. In addition, one may construct quantile–quantile or offset-function plots to compare quantile estimators obtained from competing nonresponse-adjusted methods, e.g., the no-adjustment, main-effects, and full-logistic-model options discussed in Section 29.4.

Finally, this chapter focused on the inferential effect of nonresponse adjustment methods. Some of the general ideas presented here also could be applied to evaluation of changes in questionnaire design or fieldwork methods. For example, a change in a nonresponse follow-up method may be intended to increase a unit response rate, and thus to reduce the potential for nonresponse bias. However, changes in nonresponse follow-up frequently involve nontrivial administrative burdens. Consequently, these changes are more likely to be implemented if the survey organization has evidence that the resulting bias reduction, at least for some important estimands, is not small relative to current estimator standard errors, and relative to differences in the true parameter values that would be of serious substantive interest.

ACKNOWLEDGMENTS

The author thanks R. M. Groves, R. J. A. Little, T. E. Raghunathan, and D. B. Rubin for helpful comments on an earlier version of this chapter; C. Johnson, A. Looker, V. Parsons, and W. Sribney for many useful discussions of the issues covered in this paper; and S. Heo and S. R. Lee for performing the computations reported in Section 29.4. This work was supported in part by the U.S. National Center for Health Statistics. The opinions expressed in this paper are those of the author and do not necessarily represent the policies of the U.S. Bureau of Labor Statistics or the U.S. National Center for Health Statistics.

References

Abbott, Andrew (1988), *The System of Professions: An Essay on the Division of Expert Labor.* Chicago: University of Chicago Press.

Abreu, D. A., Martin, E., and Winters, F. (1999), "Money and Motive: Results of an Incentive Experiment in the Survey of Income and Program Participation," paper presented at the International Conference on Survey Nonresponse, Portland, Oregon.

Adams, J. S. (1965), "Inequity in Social Exchange," in L. Berkowitz, *Advances in Experimental Social Psychology,* New York: Academic. Vol. 2 pp. 267–299.

Aigner, D. J. Goldberger, A. S., and Kalton, G. (1975), "On the Power of Dummy Variance Regressions," *International Economic Review*, **16,** 2, pp. 503–510.

Akkerboom, H. and DeHue, F. (1997), "The Dutch Model of Data Collection Development for Official Survey,"*International Journal of Public Opinion Research,* **9,** 2, pp. 126–145.

Alexander, W. P. and Grimshaw, S. D. (1996), "Treed Regression," *Journal of Computational and Graphical Statistics,* **5,** pp. 156–175.

Allen, M., Ambrose, D., and Atkinson, P. (1997), "Measuring Refusal Rates," *Canadian Journal of Marketing Research,* **16,** pp. 31–42.

Alvey, W. and Scheuren, F. (1982), "Background for an Administrative Record Census," *ASA Proceedings of the Social Statistics Section,* Alexandria, VA: American Statistical Association. pp. 137–146.

Alwin, D. F. and Krosnick, J. A. (1991), "The Reliability of Survey Attitude Measurement: The Influence of Question and Respondent Attributes,"*Sociological Methods and Research,* **20,** pp. 139–181.

Amemiya, T. (1984), "Tobit Models: A Survey," *Journal of Econometrics,* **24,** pp. 3–61.

American Association for Public Opinion Research, AAPOR, (1997), *Best Practices for Survey and Public Opinion Research.*

American Association for Public Opinion Research, AAPOR (2001), Standard Definitions: *Final Disposition of Case Codes and Outcome Rates for RDD Telephone Surveys and In-person Household Surveys.*

American Statistical Association, (1974), "Report of the ASA Conference on Surveys of Human Populations," *American Statistician,* **28,** pp. 30–34.

Ammar, N. (1992), "Coverage Differences in a Mixed Neighborhood in Hartford Connecticut. Ethnographic Evaluation of the 1990 Census," Report #21. Prepared under Joint Statistical Agreement 89–35 with the Bureau of the Census. Washington, D.C.: Bureau of the Census. (http://www.census.gov/srd/papers/pdf/ev92–21.pdf).

Anderson, B. A., Silver, B. D., and Abramson, P. R. (1988), "The Effects of Race of the Interviewer on Race-Related Attitudes of Black Respondents in SRC/CPS National Election Studies," *Public Opinion Quarterly,* **52,** pp. 289–324.

Anderson, C. and Nordberg, L. (1994), "A Method for Variance Estimation of Non-linear Function of Totals in Surveys – Theory and a Software Implementation," *Journal of Official Statistics,* **10,** pp. 395–405.

Anderson, T. W. (1973), "Asymptotically Efficient Estimation of Covariance Matrices with Linear Structure," *The Annals of Statistics,* **1,** pp. 135–141.

Andrews, F. M. (1984), "Construct Validity and Error Components of Survey Measures: A Structural Modeling Approach,"*Public Opinion Quarterly,* **48,** pp. 409–442.

Aneshensel, C. S., Becerra, R. M., Fielder, E. P., and Schuler, R. H. (1989), "Participation of Mexican American Female Adolescents in a Longitudinal Panel Survey," *Public Opinion Quarterly,* **53,** pp. 548–562.

Aneshensel, C. S., Estrada, A. L., Hansell, M. J., and Clark, V. A. (1987), "Social Psychological Aspects of Reporting Behavior: Lifetime Depressive Episode Reports," *Journal of Health and Social Behavior,* **28,** pp. 232–246.

Aneshensel, C. S., Frerichs, R. R., Clark, V. A., and Yokopenic, P. A. (1982), "Telephone Versus in Persaon Surveys of Community Health Status," *American Journal of Public Health,* **72,** pp. 1017–1021.

Aquilino, W. S. (1992), "Telephone Versus Face-to-Face Interviewing for Household Drug Use Surveys," *International Journal of the Addictions,* **27,** pp. 71–91.

Aquilino, W. S. (1994), "Interview Modes Efects in Surveys of Drug and Alcohol Use: A Field Experiment," Public Opinion Quarterly, **58,** pp. 210–240.

Arbuckle, J. L. (1996), "Full Information Estimation in the Presence of Incomplete Data," *Advanced Structual Equation Modeling,* Mahwah, NJ: Erlbaum.

ARF (1999), *Guidelines for Conducting Marketing and Opinion Research.* ARF's Online Research Day—Towards Validation, New York: Advertising Research Foundation, pp. 53–54.

Argyle, M (1969), *Social Interactions,* London: Methuen.

Armstrong, J. S. (1975), "Monetary Incentives in Mail Surveys," *Public Opinion Quarterly,* **39,** pp. 111–116.

Armstrong, J. S. and Overton, T. (1977), "Estimating Nonresponse Bias in Mail Surveys. "*Journal of Marketing Research,* **14,** pp. 396–402.

Aschenbrenner, J. (1990), "A Community-Based Study of the Census Undercount in a Racially Mixed Area. Ethnographic Evaluation of the 1990 Census," Report #1. Prepared under Joint Statistical Agreement 89–44 with the Bureau of the Census. Washington, D.C.: Bureau of the Census (http://www.census.gov/srd/papers/pdf/ev91-01.pdf).

Asch, D. A., Cristakis, N. A., and Ubel, P. A. (1998), "Conducting Physician Mail Surveys on a Limited Budget." *Medical Care,* **36,** 1, pp. 95–99.

Assael, H. and Keon, J. (1982), "Nonsampling vs. Sampling Errors in Survey Research," *Journal of Marketing.* ***46,*** pp. 114–123.

Atrostic, B. K. and Burt, G. (1999), "What We have Learned and a Framework for the Future," in *Seminar on Interagency Coordination and Cooperation, Statistical Policy Working Paper 28,* Washington, D.C.: FCSM.

Ayidiya, S. A. and McClendon, M. J. (1990), "Response Effects in Mail Surveys," *Public Opinion Quarterly,* **54,** pp. 229–247.

Bailar, B. A. and Lanphier, C. M. (1978), *Development of Survey Methods to Assess Survey Practices,* Washington, D.C.: American Statistical Association.

Baker, R. P. (1998), "The CASIC Future," in M. P. Couper, R. P. Baker, J. Bethlehem, C. Z. F. Clark, J. Martin, W. L. Nicholls II, and J. M. O'Reilly (eds.), *Computer Assisted Survey Information Collection,* New York: John New York: Wiley, pp. 583–605.

Baker, S. G. and Laird, N. M. (1988), "Regression Analysis for Categorical Variables with Outcome Subject to Nonignorable Monresponse," *Journal of the American Statistical Association,* **83,** pp. 62–69.

Balakrishnan, P. V., Chawla, S. K., Smith, M. F., and Micholski, B. P. (1992), "Mail Survey Response Rates Using a Lottery Prize Giveaway Incentive." *Journal of Direct Marketing,* **6,** pp. 54–59.

Balden, W. A. (1999), "Project Landmark. A Comprehensive Study to Determine if the Internet Presents a Valid Data Collection Alternative for Mainstream Consumer Goods and Services," ARF's Online Research Day—Towards Validation. New York: Advertising Research Foundation, pp. 14–27.

Banks, R. (1998), "The Internet and Market Research: Where can we go Today?" Paper presented at the Internet, Marketing & Research 5, a seminar organized by Computing Marketing & Research Consultancy Ltd. (CMR), London, UK. http://www.cmrgroup.com/bimr501.htm.

Barnard, J. and Rubin, D. B. (1999), "Small Sample Degrees of Freedom with Multiple Imputation," *Biometrika,* **86,** pp. 948–955.

Barone, M., Ujifune, G., and Mathews D. (1999), *Almanac of American Politics 2000,* Washington, D.C.: National Journal.

Batagelj, Z., Lozar, K., and Vehovar, V. (1998), "Respondent's Satisfaction in WWW Surveys," paper presented at the International Conference on Methodology and Statistics, Preddvor, Slovenia. http://www.ris.org/preddvor/1998/.

Batagelj, Z. and Vehovar, V. (1999), "Web Surveys: Revolutionising the Survey Industry or (Only) Enriching its Spectrum?" Proceedings of the ESOMAR Worldwide Internet Conference Net Effects, Amsterdam: ESOMAR, pp. 159–176.

Batutis, M. J. (1993), "Evaluation of the 1990 Population Estimates and the Future of the Census Bureau Subnational Estimates Program," *Proceedings of the Section on Social Statistics,* American Statistical Association, pp. 123–130.

Bauer, R. K. and Meissner, F. (1963), "Structures of Mail Questionnaires: Test of Alternatives," *Public Opinion Quarterly,* **27,** pp. 308–311.

Baumgartner, R. and Rathbun, P. (1997), "Prepaid Monetary Incentives and Mail Survey Response Rates," paper presented at the Annual Conference of the American Association of Public Opinion Research, Norfolk, Virginia.

Baumgartner, R., Rathbun, P., Boyle, K., Welsh, M., and Laughland, D. (1998), "The Effect of Prepaid Monetary Incentives on Mail Survey Response Rates and Response Quality," paper presented at the Annual Conference of the American Association of Public Opinion Research, St. Louis, Missouri.

Bay, D. E. (1999), "Establishment Nonresponse Section," in Seminar on Interagency

Coordination and Cooperation, Statistical Policy Working Paper 28. Washington, D.C.: FCSM.

Beatty, P. and Herrmann, D. (1995), "A Framework for Evaluating Don't Know Responses in Surveys," *Proceedings of the Survey Research Methods Section, American Statistical Association.*

Beatty, P., Herrmann, D., Puskar, C., and Kerwin, J. (1998), "Don't Know Responses in Surveys: Is What I know What You Want to Know, and Do I Want You to Know it?" *Memory,* **6,** pp. 407–426.

Beckenbach, A. (1995), "Computer Assisted Questioning: The New Survey Methods in the Perception of the Respondents," *BMS,* **48,** pp. 82–100.

Belin, T., Diffendal, D., Mack, S., Rubin, D. B., Schafer, J. L., and Zaslavsky, A. (1993), "Hierarchical Logistic Regression Models for Imputation of Unresolved Enumeration Status in Undercount Estimation," *Journal of the American Statistical Association,* **88,** pp. 1149–1159.

Bell, P. A. (1992), *Racial/Ethnic Homogeneity of Neighborhoods and Variation in Census Coverage of African Americans.* Washington, D.C.: Bureau of the Census (http://www.census.gov/srd/papers/pdf/ev93–39.pdf).

Berger, J. M., Zelditch, M. Jr., Anderson, B., and Cohen, B. (1972), "Structural Aspects of Distributive Justice: A Status Value Formulation," in J. Berger, M. Zelditch, Jr., and B. Anderson (eds.), *Sociological Theories in Progress,* **2,** Boston: Houghton Mifflin. pp. 119–146.

Berger, J. O. and Sellke, T. (1987), "Testing a Point Null Hypothesis: The Irreconcilability of P Values and Evidence," *Journal of the American Statistical Association,* **82,** pp. 112–122.

Berger, P. K. and Sullivan, J. E. (1970), "Instructional Set, Interview Context, and the Incidence of "Don't Know" Responses," *Journal of Applied Psychology,* **54,** pp. 414–416.

Berlin, Martha, Mohadjer, Leyla, Waksberg, Joseph, Kolstad, Andrew, Kirsch, Irwin, Rock, D., and Yamamoto, Kentaro, (1992), "An Experiment in Monetary Incentives," *Proceedings of the Survey Research Methods Section of the American Statistical Association,* pp. 393–398.

Berry, J. and Kanouse, D. (1987), "Physician Response to a Mailed Survey," *Public Opinion Quarterly,* **54,** pp. 102–114.

Bethlehem, J. G. (1988), "Reduction of Nonresponse Bias Through Regression Estimation," *Journal of Official Statistics,* **4,** pp. 251–260.

Bethlehem, J. G. (1996), *Bascula for Weighting Sample Survey Data, Reference Manual,* Statistics Netherlands, Statistical Informatics Department, Vorburg/Heerlen, The Netherlands.

Bethlehem, J. G and Keller, W. J. (1987), "Linear Weighting of Sample Survey Data," *Journal of Official Statistics,* **3,** pp. 141–153.

Bethlehem, J. G. and Kersten, H. M. P. (1985), "On the Treatment of Nonresponse in Sample Surveys," *Journal of Official Statistics,* **1,** pp. 287–300.

Biemer, P., Chapman, D. W., and Alexander, C. (1985), "Some Research Issues in Random-Digit-Dialing Sampling and Estimation," *Proceedings of the U.S. Bureau of the Census Annual Research Conference,* U.S. Bureau of the Census, Washington, D.C., pp. 71–86.

Biemer, P. P., Groves, R. M., Lyberg, L., Mathiowetz, N. A., and Sudman, S. (1991), *"Measurement Errors in Surveys,"* New York:Wiley.

Binder, D. A. (1983), "On the Variances of Asymptotically Normal Estimators from Complex Surveys," *International Statistical Review,* **51,** pp. 279–292.

Binder, D. A. (1998), "Longitudinal Surveys: Why Are These Surveys Different from All Other Surveys?" *Survey Methodology,* **24,** pp. 101–108.

Binder, D. A. and Patak, Z. (1994), "Use of Estimating Functions for Estimation from Complex Surveys," *Journal of the American Statistical Association,* **89,** pp. 1035–1043.

Binder, D. A. and Sun, W. (1996), "Frequency Valid Multiple Imputation for Surveys with a Complex Design," *Proceedings of the Section on Survey Research Methods of the American Statistical Association,* pp. 281–286.

Biner, P. M. and Kidd, H. J. (1994), "The Interactive Effects of Monetary Incentive Justification and Questionnaire Length on Mail Survey Response Rates," *Psychology and Marketing,* **11,** pp. 483–492.

Birnbaum, Z. W. and Sirken, M. G. (1950), "Bias Due To Non-Availability In Sampling Surveys," *Journal of the American Statistical Association,* **45,** pp. 98–111.

Bischoping, K. and Schuman, H. (1992), "Pens and Polls in Nicaragua: An Analysis of the 1990 Preelection Surveys," *American Journal of Political Science, **36,*** pp. 331–350.

Bishop, G. F., Oldendick, R. W., and Tuchfarber, A. J. (1980), "Experiments In Filtering Political Opinions,"*Political Behavior,* **2,** pp. 339–369.

Bishop, G. F., Oldendick, R. W., Tuchfarber, A. J., and Bennett, S. E. (1979)," Effects Of Opinion Filtering and Opinion Floating: Evidence From A Secondary Analysis,"*Political Methodology,* **6,** pp. 293–309.

Bishop, G. F., Tuchfarber, A. J., and Oldendick, R. W. (1986), "Opinions on Fictitious Issues: the Pressure to Answer Survey Questions," *Public Opinion Quarterly,* **50,** pp. 240–250.

Bishop, G. F., Oldendick, R. W, Tuchfarber, A. J., and Bennett, S. E. (1980), "Pseudo-Opinions on Public Affairs," *Public Opinion Quarterly,* **44,** pp. 198–209.

Black, G. S. (1998), "Internet Surveys—A Replacement Technology," paper presented at the AAPOR '98 Conference, St. Louis, Missouri.

Bogardus, E. S. (1925), "Measuring Social Distance," *Journal of Applied Sociology,* **9,** pp. 299–308.

Bogart, L. (1972), *Silent Politics: Polls and The Awareness Of Public Opinion,* New York: Wiley-Interscience.

Bogen, K., Lee, M., and DeMaio, T. (1996), "Report of Cognitive Testing of Decennial Long Form Developed by 212 Associates," Center for Survey Methods Report, U.S. Bureau of the Census. Bollen, K. A. (1989), *Structural Equations with Latent Variables,* New York: Wiley.

Bond, D., Cable, G., and Andrews, S. (1993), "Improving Response Rates by Touchtone Data Entry on the Manufacturers' Shipments, Inventories, and Orders Survey," *Proceedings of the International Conference on Establishment Surveys, American Statistical Association,* pp. 484–489.

Bond, R. and Smith, P. B. (1996), "Culture And Conformity: A Meta-Analysis Of Studies Using Asch's (1952b, 1956) Line Judgment Task". *Psychological Bulletin,* **119,** pp. 111–137.

Bowen, G. L. (1994), "Estimating the Reduction in Nonresponse Bias from Using a Mail Survey as a Backup for Nonrespondents to a Telephone Interview Survey," *Research on Social Work Practice,* **4,** pp. 115–128.

Bowman, P. J. (1991), "Race, Class and Ethics in Research: Belmont Principles to Functional Relevance," *Black Psychology*, R. L. Jones (ed.), Berkeley, pp. 747–766.

Boyle, C. A. and Brann, E. A. (1992), "Proxy Respondents and the Validity of Occupational and Other Exposure Data," *American Journal of Epidemiology,* **136,** pp. 712–721.

Brackstone, G. J. (1987), "Issues in the Use of Administrative Records for Statistical Purposes," *Survey Methodology,* **13,** pp. 29–43.

Bradburn, N. M. (1992), "Presidential Address: A Response to the Non-Response Problem," *Public Opinion Quarterly,* **56,** pp. 391–398.

Bradburn, N. M., Rips, L. J., and Schevell, S. K. (1987), "Answering Autobiographical Questions: The Impact of Memory and Inference on Surveys," *Science,* **239,** pp. 157–161.

Bradburn, N. M. and Sudman, S. (1979), "Improving Interview Method and Questionnaire Design," San Francisco, Jossey-Bass.

Bradburn, N. M. and Sudman, S. (1991), "The Current Status of Questionnaire Research," in P. Briemer, R. M. Groves, L. Lyberg, N. A. Mathiowetz, and S. Sudman (eds.), *Measurement Errors in Surveys,* New York: Wiley.

Brazziel, W. F. (1973), "White Research in Black Communities: When Solutions Become a Part of the Problem," *Journal of Social Issues,* **29,** pp. 41–44.

Brehm, John (1994a), "Stubbing our Toes for a Foot in the Door? Prior Contact, Incentives, and Survey Response," *International Journal of Public Opinion Research,* **6,** 1, pp. 45–63.

Brehm, J., (1994b), *The Phantom Respondents: Opinion Surveys and Political Representation,* Ann Arbor, University of Michigan Press.

Brennan, M. and Hoek, J. (1992), "Behavior Of Respondents, Nonrespondents And Refusers Across Mail Surveys," *Public Opinion Quarterly,* **56,** pp. 530–535.

Brennan, M., Hoek, J., and Astridge, C. (1991), "The Effects of Monetary Incentives on the Response Rate and Cost Effectiveness of a Mail Survey." *Journal of the Market Research Society,* **33,** pp. 229–241.

Brick, J. M. and Kalton, G. (1996), "Handling Missing Data in Survey Research," *Statistical Methods in Medical Research,* **5,** pp. 215–238.

Brick, J. M. and Morganstein, (1996), "WesVarPC: Software for Computing Variance Estimates from Complex Designs," *Proceedings of the 1996 Annual Research Conference,* US Bureau of the Census, Washington, D.C., pp. 861–866.

Broman, C. L., Hoffman, W. S., and Hamilton, V. L. (1994), "Impact of Mental Health Services Use on Subsequent Mental Health of Autoworkers,"*Journal of Health and Social Behavior,* **35,** pp. 80–94.

Bromley, C., Bryson, C., Jarvis, L., Park, A., Stratford, N., and Thomson, K. (2000), *British Social Attitudes and Young Peoples Social Attitudes surveys 1998—Technical Report.* London: National Centre for Social Research.

Brooks, J. E. (1990), "The Opinion-Policy Nexus in Germany," *Public Opinion Quarterly,* **54,** pp. 508–529.

Brown, R., and McNeil, D. (1977), "The 'Tip of the Tonque' Phenomenon," *Journal of Verbal Learning and Verbal Behavior,* **5,** pp. 325–337.

Brownstone, D. (1998), "Multiple Imputation Methodology For Missing Data, Non-Random Response, and Panel Attrition," in T. Gärling, T. Laitila and K. Westin (eds.), *Theoretical Foundations of Travel Choice Modeling,* Amsterdam: Elsevier, pp. 421–450.

Brownstone, D. and X. Chu (1997), "Multiply-Imputed Sampling Weights for Consistent Inference with Panel Attrition," in T. F. Golob, R. Kitamura and L. Long (eds.), *Panels for Transportation Planning,* Boston: Kluwer Academic Publishers, pp. 261–273.

Bruzzone, D. (1999), "The Top 10 Insights About the Validity of Conducting Research Online That Came Out of the Advertising Research Foundation's 'The Future of Research: Online'," January 25, 1999, Los Angeles, CA. http://www.arfsite.org/Webpages/onlineresearch99/LA_99_top10. htm.

Busch, E. M. (1990), "Multiple and Replicate Item Imputation in a Complex Sample Survey,"*Proceedings of the Sixth Annual Research Conference,* U.S. Bureau of the Census, pp. 655–665.

Caldwell, J. G. et al. (1973), "Aortic Regurgitation in the Tuskegee Study of Untreated Syphilis," *Journal of Chronic Disease*, **26,** 187–194.

Calfee, J. and Winston, C. (1998), "The Value Of Automobile Travel Time: Implications For Congestion Policy," *Journal of Public Economics,* **69,** 83–102.

Cambridge Systematics, Inc. (1977), The Development of a Disaggregate Behavioral Work Mode Choice Model, prepared for California Department of Transportation and Southern California Association of Governments, Cambridge, MA: Author.

Cameron, T., Shaw, W., and Ragland, S. (1999), "Nonresponse Bias in Mail Survey Data: Salience Vs Endogenous Survey Complexity," in J. A. Herriges and C. L. Kling (eds.), *Valuing the Environment using Recreation Demand Models,* Edward Elgar Publisher.

Campanelli, P., Sturgis, P. and Purdon, S. (1997),*Can You Hear Me Knocking: An Investigation into the Impact of Interviewers on Survey Response Rates,* London: National Centre for Social Research.

Campbell, A., Converse, P. E., Miller, W. E., and Stokes, D. E. (1960), *The American Voter,* New York: Wiley.

Cannell, C. F. and Henson, R. (1974), "Incentives, Motives, and Response Bias," *Proceedings of the Section on Survey Methods Research, American Statistical Association,* pp. 425–430.

Cannell, C. F., Miller, P. V., and Oksenberg, L., (1981), "Research on Interviewing Techniques," in S. Leinhardt (ed.), *Sociological Methodology,* San Francisco: Jossey-Bass.

Cannell, C. F., Oksenberg, L., and Converse, J. M. (1979), *Experiments in Interviewing Techniques.* Ann Arbor, MI: Institute for Social Research, The University of Michigan.

Carlin, J. B., Wolfe, R., Coffey, C., and Patton, G. C. (1999), *"Analysis of Binary Outcomes in Longitudinal Studies Using Weighted Estimating Equations and Discrete-Time Survival Methods"* (Tutorial in Biostatistics), *Statistics in Medicine,* **18,** in press.

Carroll, R. J., Ruppert, D., and Stefanski, L. A. (1995), *"Measurement Error in Nonlinear Models,"* London: Chapman and Hall.

Casady, R. J., Dorfman, A., and Wang, S. (1998), "Confidence Intervals for Sub-Domain Means and Totals," *Survey Methodology,* **24,** pp. 57–67.

Casper, R. A. (1992), "Follow-up of Nonrespondents in 1990," in C. F. Turner, J. T. Lessler and J. C. Gfroere (eds.), *Survey Measurement of Drug Use: Methodological Studies,* Rockville, MD: National Institute on Drug Abuse, pp. 155–173.

Cassel, C. M., Särndal, C. E., & Wretman, J. . H. (1977), "Foundations of Inference in Survey Sampling," New York: Wiley.

Chamberlayne, R., B., Green, M. L., Barer, C., Hertzman, W., Lawrence, J., and Sheps, S. B.

(1998), "Creating a Population-Based Linked Health Database: a New Resource for Health Services," *Research. Canadian Journal of Public Health,* **89,** pp. 270–273.

Chapman, D. W. and Roman, A. (1985), "An Investigation of Substitution for an RDD Survey," *Proceedings of the Section on Survey Research Methods, American Statistical Association,* pp. 269–274.

Chapman, D. W. (1983), "The Impact of Substitution on Survey Estimates,". In W. G. Madow, I. Olkin, and D. B. Rubin (eds.) *Incomplete Data In Sample Surveys, Vol. II, Theory and Bibliographies,* New York: Academic Press.

Chen, H. C. K. (1996), "Direction, Magnitude and Implications of Non-Response Bias in Mail Surveys," *Journal of the Market Research Society,* **38,** pp. 267–276.

Chen, K. and Kandel, D. B. (1995), "The Natural History of Drug Use from Adolescence to the Mid-Thirties in a General Population Sample," *American Journal of Public Health,* **85,** pp. 41–57.

Cheng, S. (1998), "Who are the Reluctant or Rarely-at-home Respondents?" *Survey Methods Centre Newsletter,* **18,** pp. 8–11.

Chi, E. M. and Reinsel, G. C. (1989), "Models for Longitudinal Data with Random Effects and AR(1) Errors," *Journal of the American Statistical Association,* **84,** pp. 452–459.

Chisholm, J. (1998), *Using the Internet to Measure Customer Satisfaction and Loyalty,* the Worldwide Internet Seminar 1998 in Paris, France, Amsterdam: ESOMAR. http://www.customersat.com/uni/whiteframe.html.

Choldin, H. M. (1994), *Looking for the Last Percent: The Controversy over Census Undercounts,* New Brunswick, NJ: Rutgers University Press.

Christianson, A. and Tortora, R. D. (1995), "Issues in Surveying Businesses: An International Survey," in B. G. Cox, D. A. Binder, B. N. Chinnappa, A. Christianson, M. J. Colledge, and P. S. Kott (eds.), *Business Survey Methods,* New York: Wiley.

Chromy, James R., and Horvitz, Daniel G. (1978), "The Use of Monetary Incentives in National Assessment Household Surveys," *Journal of the American Statistical Association,* **73,** pp. 473–478.

Church, A. H. (1993), "Estimating the Effect of Incentives on Mail Survey Response Rates: A Meta-Analysis," *Public Opinion Quarterly,* **57,** pp. 62–79.

Cialdini, R. B. (1987), "Compliance and Principles of Compliance Professionals: Psychologists of Necessity," in M. P. Zanna, J. M. Olson and C. P. Herman (eds.), *Social Influence: The Ontario Symposium,* Hillsdale, New Jersey: Erlbaum, Vol. 5, pp. 165–184.

Cialdini, R. B. (1988), *Influence: Science and Practice.* Glenview, IL: Scott, Foresman.

Citro, C. F., Cohen, M. L., Kalton, G., and West, K. W. (1997), *Small-Area Estimates of School-Age Children in Poverty, Interim Report I. Evaluation of 1993 County Estimates for Title 1 Allocations,* Washington, D.C.: National Academy Press.

Clausen, J. A. and Ford, R. N. (1947), "Controlling Bias in Mail Questionnaires," *Journal of the American Statistical Association,* **42,** pp. 497–511.

Clayton, R. L. and Werking, G. S. (1998), "Business Surveys of the Future: The World Wide Web as a Data Collection Methodology," In M. P. Couper, R. P. Baker, J. Bethlehem, C. Z. F. Clark, J. Martin, W. L. Nicholls II, and J. M. O'Reilly (eds.), *Computer Assisted Survey Information Collection,* New York: Wiley, pp. 543–563.

Coale, A. J. and Rives, N. W. (1973), "A Statistical Reconstruction of the Black Population of the United States, 1880–1970: Estimates of True Numbers by Age and Sex, Birth Rates, and Total Fertility," *Population Index,* **39,** pp. 3–36.

Cochran, W. G. (1977), *Sampling Techniques* (3rd ed.), NewYork: Wiley.

Cochran, W. G. (1983), "Historical Perspective. In Incomplete Data in Sample Surveys," in W. G. Madow, I. Olkin, and D. B. Rubin (eds.), *Theory and Bibliographies,* New York: Academic Press.

Cohen, R. (1955), "An Investigation of Modified Probability Sampling Procedures in Interview Surveys," Master's thesis, The American University, Washington, D.C.

Cole, S., Kusch, G., Berry, J., and Hoy, C. E. (1993), "Studies of Nonresponse in Industrial Surveys," paper presented at the International Conference on Establishment Surveys, Buffalo, New York.

Colombo, R. A. (1992), "Using Call-Backs to Adjust for Nonresponse Bias," in A. Westlake et al. (eds.), *Survey and Statistical Computing,* Amsterdam: Elsevier.

Colsher, P. L. and Wallace, R. B. (1989), "Data Quality and Age: Health and Psychobehavioral Correlates of Item Nonresponse and Inconsistent Responses," *Journal of Gerontology,* **44,** pp. 45–52.

Comley, P. (1996), "The Use of Internet as a Data Collection Method," paper presented at the ESOMAR Conference, Edinburg, UK. http://www.sga.co.uk/esomar.html.

Comley, P. (1997), *The Use of the Internet for Opinion Polls. Learning from the Future: Creative Solutions for Marketing,* the 50th ESOMAR Marketing Research Congress, Amsterdam: ESOMAR. http://www.virtualsurveys.com/papers/Poll.htm.

Comley, P. (1998), "On-Line Research: Some Methods, Some Problems, Some Case Studies in New Methods in Survey Research," A. Westlake et al. (eds.), *Proceedings of the ASC 1998 International Conference, ASC.* http://www.virtualsurveys.com/papers/ASC.htm.

Converse, J. M. (1976), "Predicting No Opinion In The Polls,"*Public Opinion Quarterly,* **40,** pp. 515–530.

Converse, J. M. and Presser, S. (1986), *Survey Questions: Handcrafting the Standardized Questionnaire,* Beverly Hills, CA: Sage.

Converse, J. M. and Schuman, H. (1984), "The Manner of Inquiry: An Analysis of Survey Question Form Across Organizations and Over Time," In C. F. Turner and E. Martin (eds.), *Surveying Subjective Phenomena* (Vol. 2), New York: Russell Sage.

Converse, P. E. (1964), "The Nature of Belief Systems in the Mass Public," in D. E. Apter (ed.), *Ideology and Discontent,* New York: Free Press, pp. 206–261.

Converse, P. E. and Markus, G. B. (1979), "Plus Ca Change . . . The New CPS Election Study Panel," *The American Political Science Review,* **73,** pp. 32–49.

Coomber, R. (1997), "Using the Internet for Survey Research," Sociological Research Online, **2,** p. 2. http://www.socresonline.org.uk/socresonline/2/2/coomber.html.

Coombs, C. H. and Coombs, L. C. (1975), "'Don't know': Item Ambiguity or Respondent Uncertainty?" *Public Opinion Quarterly,* **40,** pp. 497–514.

Cottler, L. B., Zipp, J. F., Robins, L. N. and Spitznagel, E. L. (1987), "Difficult-to-Recruit Respondents and Their Effect on Prevalence Estimates in an Epidemiological Survey," *American Journal of Epidemiology*, **125,** pp. 3–36.

Couper, M. P. (1997), "Survey Introductions and Data Quality," *Public Opinion Quarterly,* **61,** pp. 317–338.

Couper, M. P. and Groves, R. M. (1992), "The Role of the Interviewer in Survey Participation," *Survey Methodology,* **18,** pp. 263–278.

Couper, M. P. Groves, R. M., and Raghunathan, T. E. (1996), *"Nonresponse in the Second*

Wave of a Longitudinal Survey," Paper Presented at the International Workshop on Household Survey Nonresponse, Rome, October.

Couper, M. P. and Nicholls, W. L. II (1998), "The History and Development of Computer Assisted Survey Information Collection Methods," in M. P. Couper, R. P. Baker, J. Bethlehem, C. Z. F. Clark, J. Martin, W. L. Nicholls II, and J. M. O'Reilly (eds.), *Computer Assisted Survey Information Collection,* New York: Wiley.

Cox, E. P. (1976), "A Cost/Benefit View of Prepaid Monetary Incentives in Mail Questionnaires." *Public Opinion Quarterly,* **40,** pp. 101–104.

Cox, L. H. and Boruch, R. F. (1998), "Record Linkage, Privacy, and Statistical Policies," *Journal of Official Statistics,* **4,** pp. 3–16.

Craig, C. S. and McCann J. M. (1978), "Item Nonresponse in Mail Surveys: Extent and Correlates,"*Journal of Marketing Research,* **15,** pp. 285–289.

Crespi, I. (1998), "Ethical Considerations When Establishing Survey Standards," *International Journal of Public Opinion Research,* **10,** pp. 75–82.

Cronbach, L. J. (1950), "Further Evidence on Response Sets and Test Design," *Educational and Psychological Measurement,* **10,** pp. 3–31.

Crosby, F., Bromley, S., and Saxe, L. (1980), "Recent Unobtrusive Studies of Black and White Discrimination and Prejudice: A Literature Review," *Psychological Bulletin,* **87,** pp. 546–563.

Culpepper, I. J., Smith, W. R., and Krosnick, J. A. (1992), "The Impact of Question Order on Satisficing in Surveys," paper presented at the Midwestern Psychological Association Annual Meeting, Chicago, Illinois.

Czajka, J. L., Hirabayashi, S. M., Little, R. J. A., and Rubin, D. B. (1992), "Projecting From Advance Data Using Propensity Modeling: An Application to Income and Tax Statistics,"*Journal of Business and Economic Statistics,* **10,** pp. 117–131.

Dalenius, T. (1983), "Some Reflections on the Problem of Missing Data," In: W.G. Madow and I. Olkin (eds.), *Incomplete Data in Sample Surveys*, 3, New York, Academic Press, pp. 411–413.

Daniel, W. W. (1975), "Nonresponse in Sociological Surveys: A Review of Some Methods for Handling the Problem,"*Sociological Methods & Research,* **3,** pp. 291–307.

Darden, J., Jones, L. and Price, J. (1992), "Ethnographic Evaluation of the Behavioral Causes of Undercount in a Black Ghetto of Flint, Michigan. Ethnographic Evaluation of the 1990 Census, Report #24," prepared under Joint Statistical Agreement with the Bureau of the Census. Washington, D.C.: Bureau of the Census. (http://www.census.gov/srd/papers/pdf/ev92–24.pdf).

Darity, W. A. and Turner, C. B. (1972), "Family Planning, Race Consciousness and the Fear of Genocide," *American Journal of Public Health,* **62,** pp. 1454–1459.

David, M. H, Little, R. J. A, Samuhel, M. E, and Triest, R. K. (1986), "Alternative Methods for CPS Income Imputation," *Journal of the American Statistical Association,* **82,** pp. 29–41.

Davidson, A. C. and Hinkly, D. V. (1997), *Bootstrap Methods and Their Application,* New York: Cambridge University Press.

Davidson, A. R., Kalmuss, D., Cushman, L. F., Romero, D., Heartwell, S., and Rulin, M. (1997), "Injectable Contraceptive Discontinuation and Subsequent Unintended Pregnancy among Low-Income Women," *American Journal of Public Health,* **87,** pp. 1532–1534.

Davis, W. (1999), *Evaluation of the Mail Return Questionnaires. Census 2000 Dress Rehearsal Evaluation Memorandum A2,* U.S. Bureau of the Census, Washington, D.C.

Day, G. S. (1975) "The Threats to Market Research," *Journal of Marketing Research,* **12,** pp. 462–467.

de Heer, W. (1999), "International Response Trends: Results of an International Survey," *Journal of Official Statistics,* **15,** pp. 129–142.

de Heer, W. and Jargels, A. (1992), "Response Trends in Europe," Paper Presented at the 152nd Conference of the American Statistical Association, Boston.

De Jong, P. (1989), "Smoothing and Interpolation with the State–Space Model," *Journal of the American Statistical Association,* **84,** 408, pp. 1085–1088.

De la Puente, M. (1993), *Using Ethnography to Explain Why People Are Missed or Erroneously Included by the Census: Evidence from Small Area Ethnographic Studies,* Washington, D.C.: Bureau of the Census.

de la Puente, M. (1993), *A Multivariate Analysis of the Census Omission of Hispanics and Non-Hispanic Whites, Blacks, Asians and American Indians: Evidence from Small Area Ethnographic Studies,* Washington, D.C.: Bureau of the Census. (http://www.census.gov/srd/papers/pdf/ev93-38.pdf).

DeGroot, M. H. (1986), "Record Linkage and Matching Systems," *Encyclopedia of Statistical Sciences,* **7,** pp. 649–654.

Del Valle, M., Morgenstern, H., Rogstad, T., Albright, C., and Vickrey, B. (1997), "A Randomized Trial of the Impact of Certified Mail on Response Rate to a Physician Survey and a Cost Effectiveness Analysis,"*Evaluations and the Health Professions,* **20,** pp. 389–406.

de Leeuw, E. D. and Hox, J. J. (1996), "The Effect of an Interviewer on the Decision to Cooperate in a Survey of the Elderly," In S. Laaksonen (ed.), *International Perspectives on Nonresponse Proceedings of the Sixth International Workshop on Household Survey Nonresponse,* Helsinki: Statistics Finland.

de Leeuw, E. D. and Nicholls, W. L. II (1996), "Technological Innovations in Data Collection: Acceptance, Data Quality and Costs," *Sociological Research Online,* **1,** p. 4. (http://www.socresonline.org.uk/socresonline/1/4/leeuw.html.

de Leeuw, E. D., Hox, J. J., Snijkers, G., and de Heer, W. (1997), "Interviewer Opinions, Attitudes and Strategies Regarding Survey Participation and Their Effect on Response,"*Nonresponse in Survey Research,* ZUMA Special, 4, Mannheim.

de Leeuw, E. D. (1999), "Preface: Special Issue on Survey Nonresponse," *Journal of Official Statistics,* **15,** 2, pp. 127–128.

DeMaio, T. J. (1980), "Refusals: Who, Where and Why," *Public Opinion Quarterly,* **44,** pp. 223–233.

DeMaio, T. J. (1984), "Social Desirability and Survey Measurement: A Review," in C. Turner, and E. Martin (eds.), *Surveying Subjective Phenomena,* **2,** New York: Russell Sage, pp. 257–282.

Deming, W. E. (1965), "Principles of Professional Statistical Practice," *Annals of Mathematical Statistics,* **26,** pp. 1883–1993.

Deming, W. E. and Stephan, F. F. (1940), "On a Least Squares Adjustment of a Sampled Frequency Table When the Expected Marginal Tables Are Known," *The Annals of Mathematical Statistics,* **11,** pp. 427–444.

Dempster, A. P., and Raghunathan, T. E. (1987), Using A Covariate for Small Area Estimation: A Common Sense Bayesian Approach. In *Small Area Statistics: An International Symposium,* ed.

Dempster, A. P., Laird, N. M., and Rubin, D. B. (1977), Maximum Likelihood From Incomplete Data Via the EM Algorithm. *Journal of the Royal Statistical Society,* B, **39,** 1–38.

Derlega, V. J. and Stelien, E. G. (1977), "Norms Regulating Self-Disclosure among Polish University Students," *Journal of Cross-Cultural Psychology,* **8,** pp. 369–376.

Deville, J. C. and Särndal, C. E. (1992), "Calibration Estimators in Survey Sampling," *Journal of the American Statistical Association,* **87,** pp. 376–382.

Deville, J. C., Särndal, C. E., and Sautory, O. (1993), "Generalized Raking Procedures in Survey Sampling," *Journal of the American Statistical Association,* **88,** pp. 1013–1020.

Dickinson, J. R. and Kirzner, E. (1985), "Questionnaire Item Omission as A Function of Within-Group Question Position," *Journal of Business Research,* **13,** pp. 71–75.

Diggle, P. and Kenward, M. G. (1994), "Informative Dropout in Longitudinal Data Analysis," *Applied Statistics,* **43,** pp. 49–93.

Diggle, P. J., Liang, K-Y., and Zeger, S. L. (1994), *Analysis of Longitudinal Data,* Oxford: Clarendon Press.

Dijkstra, W. (1999a), "A New Method for Studying Verbal Interactions in Survey-Interviews," *Journal of Official Statistics,* **15,** pp. 67–85.

Dijkstra, W. (1999b), *Sequence: The Next Step,* Amsterdam: Profile.

Dillman, D. A., Christenson, J. A., Carpenter, E. H., and Brooks, R. (1974), "Increasing Mail Questionnaire Response: A Four-State Comparison," *American Sociological Review,* **39,** pp. 744–756.

Dillman, D. A. (1978), *Mail and Telephone Surveys: The Total Design Method,* New York: Wiley.

Dillman, D. A. (1991), "The Design and Administration of Mail Surveys," *Annual Review of Sociology,* **17,** pp. 225–249.

Dillman, D. A., Clark, J. R., and Treat, J. B. (1994), "Influence of 13 Design Factors on Completion Rates to Decennial Census Questionnaires," paper presented at the 1994 Annual Research Conference of the U.S. Bureau of the Census, Arlington, Virginia.

Dillman, D. A. (1996), "Token Financial Incentives and Reduction of Nonresponse Error in Mail Surveys," *Proceedings of the Government Statistics Section, American Statistical Association.*

Dillman, D., Jenkins, C., Martin, E., and DeMaio, T. (1996), "Cognitive and Motivational Properties of Three Proposed Decennial Census Forms," report prepared for the Bureau of the Census, Washington. D. C.

Dillman, D. A. (1998), "Mail and Other Self Administered Surveys in the 21st Century; The Beginning of a New Era," http://survey. sesrc. wsu. edu/dillman/papers. htm.

Dillman, D., Carley-Baxter, L., and Jackson, A. (1999a), "Skip Pattern Compliance in Three Test Forms: A Theoretical and Empirical Evaluation." *SESRC Technical Report # 99-01,* Social and Economic Science Research Center, Washington State University, Pullman.

Dillman, D., Redline, C., and Carley-Baxter, L. (1999b), "Influence of Type of Question on Skip Pattern Compliance in Self-Administered Questionnaires," paper prepared for presentation at the American Statistical Society, Baltimore, MD.

Dillman, D. A., Tortora, R. D., and Bowker, D. (1999c), "Principles for Constructing Web Survey," http://survey. sesrc. wsu. edu/dillman/papers/websurveyppr. pdf.

Dillman, D. A. (2nd ed.) (2000), *Mail and Internet Surveys. The Tailored Design Method.* New York: Wiley.

Dodd, S. C. and Svalastoga, K. (1952), "On Estimating Latent from Manifest Undecidedness: The "Don't Know" Percent as a Warning of Instability Among the Knowers," *Journal of Educational and Psychological Measurement,* **12,** pp. 467–471.

Donald, M. N. (1960), "Implications of Nonresponse for the Interpretation of Mail Questionnaire Data," **18,** pp. 40–52.

Donsbach, Wolfgang, (1997), "Survey Research at the End of the Twentieth Century: Theses and Antitheses," *International Journal of Public Opinion Research,* **9,** pp. 17–28.

Dorinski, S. M. (1995), "Continuing Research on Use of Administrative Data in SIPP Longitudinal Estimation," *Proceedings of the Survey Research Methods Section, American Statistical Association,* pp. 233–238.

Downs, A. (1957), *An Economic Theory of Democracy,* New York: Wiley.

Drew, J. H. and Fuller, W. A. (1980), "Modelling Nonresponse in Surveys with Callbacks," *Proceedings of Survey Research Methods section, American Statistical Association,* pp. 639–642.

DuMouchel, W. H. and Duncan, G. J. (1983), "Using Sample Survey Weights in Multiple Regression Analyses of Stratified Samples," *Journal of the American Statistical Association,* **78,** pp. 535–543.

Duncan, G. J. and Kalton, G. (1987), "Issues of Design and Analysis of Surveys Across Time," *International Statistical Review,* **55,** pp. 97–117.

Duncan, O. D. and Stenbeck, M. (1988), "No Opinion or not Sure?" *Public Opinion Quarterly,* **52,** pp. 513–525.

Dunkelberg, W. and Day, G. S. (1973), "Nonresponse Bias and Callbacks in Sample Surveys," Journal of Marketing Research, **10,** pp. 160–172.

Dunlap, J. W., De Mello, A., and Cureton, E. E. (1929), "The Effects of Different Directions and Scoring Methods on the Reliability of a True-False Test," *School and Society,* **30,** pp. 378–382.

Durand, R. M. and Lambert, Z. V. (1988), "Don't Know Responses in Surveys: Analyses and Interpretational Consequences," *Journal of Business Research,* **16,** pp. 169–188.

Durant, T. and Jack, L. (1993), "Undercount of Black Inner City Residents of New Orleans, Louisiana," *Ethnographic Evaluation of the 1990 Census, Report # 27,* Bureau of the Census.

Durbin, J. and Stuart, A. (1951), "Differences in Response Rates of Experienced and Inexperienced Interviewers," *Journal of the Royal Statistical Society,* Series A, **114,** pp. 163–205.

Durbin, J. and Stuart, A. (1954), "Callbacks and Clustering in Sample Surveys: An Experimental Study," *Journal of the Royal Statistical Society,* Series A (General), **117,** pp. 387–418.

Eaton, W. W., Anthony, J. C., Tepper, S., and Dryman, A. (1992), "Psychopathology and Attrition in the Epidemiologic Catchment Area Surveys," *American Journal of Epidemiology,* **135,** pp. 1051–1059.

Edwards, W. S. and Cantor, D. (1991), "Toward a Response Model in Establishment Surveys," in P. B. Biemer, R. M. Groves, L. E. Lyberg, N. A. Mathiowetz, and S. Sudman (eds.), Measurement Error in Surveys, New York: Wiley.

Efron, B. (1979), Bootstrap Methods: "Another Look at the Jackknife."*The Annals of Statistics,* **7,** 1–26.

Efron, B. (1994), "Missing Data, Imputation and the Bootstrap." *Journal of the American Statistical Association,* **89,** 2, pp. 463–479.

Ehrlich, H. J. (1964), "Instrument Error and the Study of Prejudice," *Social Forces,* **43,** pp. 197–206.

Eichman, C. (1999), "Research Methods on the Web," *Proceedings of the ESOMAR Worldwide Internet Conference Net Effects,* Amsterdam: ESOMAR, pp. 69–76.

Eisenhower, D., Mathiowetz, N. A., and Morganstein, D., (1991), "Recall Error: Sources and Bias Reduction Techniques," in P. Biemer, R. M. Groves, L. Lyberg, N. A. Mathiowetz, and S. Sudman (eds.), *Measurement Errors in Surveys,* New York: Wiley.

Eisenberg, P. and Wesman, A. G. (1941), "Consitency in Response and Logical Interpretation of Psychoneurotic Inventory Items,"*Journal of Education Psychology,* **32,** pp. 321–338.

Elder, A. (1999), "Power Buyers or Cautious Shoppers? Segmenting the E-Commerce Market," *Proceedings of the ESOMAR Worldwide Internet Conference Net Effects,* Amsterdam: ESOMAR, pp. 15–34.

Elliott, M. R. and Little, R. J. A, (2000), "Model-Based Alternatives to Trimming Survey Weights." *Journal of Official Statistics,* **16,** 3, pp. 191–209.

Elliott, M. R, Little, R. J. A, and Lewitsky, S. (2000)," Subsampling Callbacks to Improve Survey Effeciency," *Journal of the American Statistical Association,* **95,** pp. 730–738.

Ellis, R. A., Endo, C. M., and Armer, J. M. (1970), "The Use of Potential Nonrespondents for Studying Nonresponse Bias," *Pacific Sociological Review,* **13,** pp. 103–109.

Eltinge, J. L. (1992), "Conditions for Approximation of the Bias and Mean Squared Error of a Sample Mean Under Nonresponse," *Statistics and Probability Letters,* **15,** pp. 267–276.

Eltinge, J. L., Heo, S., and Lee, S. R. (1997), "Use of Propensity Methods in the Analysis of Subsample Re-Measurements for NHANES III,"*Proceedings of the Annual Meeting of the Statistical Society of Canada.*

Eltinge, J. L. and Yansaneh, I. S. (1997), "Diagnostics for Formation of Nonresponse Adjustment Cells, With an Application to Income Nonresponse in the U. S. Consumer Expenditure Survey," *Survey Methodology,* **23,** pp. 33–40.

Enander, J. and Sajti, A. (1999), "Online Survey of Online Customers, Value-Added Market Research through Data Collection on the Internet," *Proceedings of the ESOMAR Worldwide Internet Conference Net Effects,* Amsterdam: ESOMAR, pp. 35–52.

England, A., Hubbell, K., Judkins, D., and Ryaboy, S. (1994), "Imputation of Medical Cost and Payment Data," *Proceedings of the Section on Survey Research Methods of the American Statistical Association,* pp. 406–411.

ESOMAR (1997), "Market Research and the Internet," ESOMAR Position Paper. http://www.esomar.nl/guidelines/position(paper.html/.

ESOMAR (1998), "Conducting Marketing and Opinion Research Using the Internet," ESOMAR Guidelines. http://www.esomar.nl/guidelines/internet_guidelines.htm.

Estevao, V., Hidiroglou, M. A. and Sarndal, C-E. (1995), "Methodological Principles for a Generalized Estimation System at Statistics Canada," *Journal of Official Statistics*, **11,** pp. 181–204.

Everett-Church, R. (1999a), "Spam Law," *OnTheInternet, Journal of the Internet Society,* May/June, p. 21.

Everett-Church, R. (1999b), "Why Spam is a Problem," *OnTheInternet, Journal of the Internet Society,* May/June, pp. 16–21.

Ezzati-Rice,T., Johnson, W., Khare, M., Little, R., Rubin, D., and Schafer, J. (1995), "A Stimulation Study to Evaluate the Performance of Model-Based Multiple Imputations in NCHS Health Examination Surveys,"*Proceedings of the 1995 Annual Research Conference,* U.S. Bureau of the Census, pp. 257–266.

Faciszewski, T., Broste, S. K., and Fardon, D. (1997), "Quality of Data Regarding Diagnoses of Spinal Disorders in Administrative Databases," *The Journal of Bone and Joint Surgery,* **79,** pp. 1481–1488.

Fahimi, M., Judkins, D., Khare, M., and Ezzati-Rice, T. M. (1993), "Serial Imputation of NHANES III with Mixed Regression and Hot -deck Techniques," *Proceedings of the Section on Survey Research Methods of the American Statistical Association,* pp. 292–296.

Farmer, T. (1998), "Using the Internet for Primary Research Data Collection," *Market Research Library.* http://www.researchinfo.com/library/infotek/index.shtml.

Faulkenberry, G. D. and Mason, R. (1978), "Characteristics of Nonopinion and No Opinion Response Groups," *Public Opinion Quarterly,* **42,** pp. 533–543.

Fay, R. E. (1986), "Causal Models for Patterns of Nonresponse." *Journal of the American Statistical Association,* **81,** 354–365.

Fay, R. E. (1991), "A Design-based Perspective on Missing Data Variance," *Proceedings of the 1991 Annual Research Conference*, U.S. Bureau of the Census, pp. 381–440.

Fay, R. E. (1992), "When Are Inference from Multiple Imputation Valid?" *Proceedings of the Section on Survey Research Methods*, American Statistical Association, pp. 227–232.

Fay, R. E. (1993), "When are Inference From Multiple Imputation Valid?" *Proceedings of the Survey Research Methods Section, American Statistical Association,* pp. 227–232.

Fay, R. E. (1994), "Comment on 'Multiple-Imputation Inferences with Uncongenial Sources of Input'," *Statistical Science,* **9,** pp. 558–560.

Fay, R. E. (1996), "Alternative Paradigms for the Analysis of Imputed Survey Data," *Journal of the American Statistical Association,* **91,** pp. 490–498.

Fay, R. E. and Herriot, R. A. (1979), "Estimates of Income for Small Places: An Application of James-Stein Procedures to Census Data," *Journal of the American Statistical Association,* **74,** pp. 269–277.

Fay, R. E., Bates, N. and Moore, J. (1993), "Lower Mail Response in the 1990 Census: A Preliminary Interpretation," *Proceedings from the 1991 Annual Research Conference, Washington, D.C.: Bureau of the Census*, pp. 3–32.

Featherston, F. and Moy, L. (1990), "Item Nonresponse in Mail Surveys," paper presented at the International Conference of Measurement Errors in Surveys, Tucson, Arizona.

Fecso, R. and Tortora, R. D. (1981), *Farmers' Attitudes Toward Crop and Livestock Surveys: A Collection of Papers Related to the Analysis of the Survey of Dakota Farmers and Ranchers,* SRS Staff Report No. AGES811007, Research Division, Statistical Reporting Service, U. S. Department of Agriculture, Washington D. C.

Feder, M., Nathan, G. and Pfeffermann, D. (2000), "Time Series Multilevel Modelling of Longitudinal Data from Complex Surveys," *Technical Report.*

Feick, L. F. (1989), "Latent Class Analysis of Survey Questions That Include Don't Know Responses," *Public Opinion Quarterly,* **53,** pp. 525–547.

Fein, D. J. (1990), "Racial and Ethnic Differences in U.S. Census Omission Rates," *Demography*, **27,** pp. 285–302.

Feldman, J. and Lynch, J., (1988), "Self-Generated Validity and Other Effects of Measurement on Belief, Attitude, Intention, and Behavior," *Journal of Applied Psychology,* **73,** pp. 421–435.

Fellegi, I. P. and Sunter, A. B. (1969), "A Theory for Record Linkage," *Journal of the American Statistical Association,* **64,** pp. 1183–1210.

Ferber, R. (1948), "The Problem of Bias in Mail Returns: A Solution," *Public Opinion Quarterly,* **12,** pp. 669–672.

Ferber, R. (1966), "Item Nonresponse in a Consumer Survey," *Public Opinion Quarterly,* **30,** pp. 399–415.

Filion, F. L. (1976a), "Exploring and Correcting for Nonresponse Bias Using Follow-Ups of Nonrespondents," *Pacific Sociological Review,* **19,** pp. 401–408.

Filion, F. L. (1976b), "Estimating Bias Due To Nonresponse in Mail Surveys," *Public Opinion Quarterly,* **39,** pp. 482–492.

Findlater, A. and Kottler, R. E. (1998), *Web Interviewing. Validating the Application of Web Interviewing Using a Comparative Study on the Telephone,* The Worldwide Internet Seminar 1998 in Paris, France, Amsterdam: ESOMAR. http://www.quantime.co.uk/web_bureau/knowledge/reedesomar98.htm.

Finkelhor, D., Asdigian, N., and Dziuba-Leatherman, J. (1995), "Victimization Prevention Programs for Children: A Follow-Up," *American Journal of Public Health,* **85,** pp. 1684–1689.

Fitzgerald, R. and Fuller, L. (1982), "I Hear You Knocking But You Can't Come In: The Effect of Reluctant Respondents and Refusers on Sample Survey Estimates," *Sociological Methods and Research,* **11,** pp. 3–32.

Fitzgerald, Gottschalk, and Moffitt (1998), "An Analysis of Sample Attrition in Panel Data: The Michigan Panel Study of Income Dynamics," *The Journal of Human Resources,* **33**(2), pp. 251–299.

Flay, B. R., McFall, S., Burton, D., Cook, T. D., and Warnecke, R. B. (1993), "Health Behavior Changes Through Television: The Roles of De Factor and Motivated Selection Processes," *Journal of Health and Social Behavior,* **34,** pp. 322–335.

Flemming, G. and Sonner, M. (1999), "Can Internet Polling Work? Strategies for Conducting Public Opinion Surveys Online," paper presented at the 1999 AAPOR Conference, St. Petersburg, Florida.

Fonda, C. P. (1951), "The Nature and Meaning of the Rorschach White Space Response," *Journal of Abnormal and Social Psychology,* **46,** pp. 367–377.

Ford, B. M. (1983), "An Overview of Hot-deck Procedures," in W. G. Madow, L. Okin, and D. B. Rubin (eds.), *Incomplete Data in Sample Surveys,* New York: Academic Press.

Ford, N. M. (1968), "Questionnaire Appearance and Response Rates in Mail Surveys,"*Journal of Advertising Research,* **8,** pp. 43–45.

Foster, J. J. (1979), "The Use of Visual Cues in Text," *Processing of Visible Language,* **1,** pp. 189–203.

Foster, K. (1997), "The Effect of Call Patterns on Non-Response Bias in Household Surveys," *Survey Methodology Bulletin,* **41,** pp. 37–47.

Foster, K. (1998), "Evaluating Non-Response on Household Surveys," GSS Methodology Series no. **8,** London: Government Statistical Service.

Fowler, F. J. (1995), *Improving Survey Questions,* Thousand Oaks, CA: Sage.

Fowler, F. J. (2nd ed.) (1993), *Survey Research Methods,* Newbury Park, CA: Sage.

Fowles, J. B., Fowler. E., Craft, C., and McCoy, C. E. (1997), "Comparing Claims Data and Self-Reported Data with the Medical Record for Pap Smear Rates," *Evaluation and the Health Professions,* **20,** pp. 324–342.

Fox, R. J., Crask, M. R., and Kim, J. (1988), "Mail Survey Response Rate: A Meta-Analysis of Selected Techniques for Inducing Response," *Public Opinion Quarterly,* **52,** pp. 467–491.

Francis, J. B. and Busch, J. A. (1975), "What We Now Know about 'I Don't Knows'?," *Public Opinion Quarterly,* **39,** pp. 207–218.

Franzen, R. and Lazarsfeld, P. F. (1945), "Mail Questionnaires as a Research Problem," *The Journal of Psychology,* **20,** pp. 293–320.

Freidson, E. (1984), "The Changing Nature of Professional Control," *Annual Review of Sociology,* **10,** pp. 1–20.

Freidson, E. (1994), *"Professionalism Reborn: Theory, Prophecy, and Policy,"* Chicago: University of Chicago Press.

Frey, J. H. (2nd ed.) (1989), *Survey Research by Telephone,* Newbury Park, CA: Sage.

Fry, G. M. (1984), *Night Riders in Black Folk History,* Knoxville: University of Tennessee Press.

Fuller, C. H. (1974), "Weighting to Adjust for Survey Nonresponse," *Public Opinion Quarterly,* **38,** pp. 239–246.

Fuller, W. A. (1984), "Least Squares and Related Analyses for Complex Survey Designs," *Survey Methodology,* **10,** pp. 97–118.

Fuller, W. A. (1987), *"Measurement Error Models,"* New York: Wiley.

Gagnon, F., Lee, H., Provost, M., Rancourt, E. and Särndal, C-E. (1997), "Estimation of Variance in Presence of Imputation," *Proceedings of Statistics Canada Symposium 97: New Directions in Surveys and Census,* Statistics Canada, pp. 273–277.

Gagnon, F., Lee, H., Rancourt, E. and Särndal, C-E. (1996), "Estimating the Variance of the Generalized Regression Estimator in the Presence of Imputation for the Generalized Estimation System," *Proceedings of the Survey Methods Section, Statistical Society of Canada*, pp. 151–156.

Gamble, V. N. (1997), "Under the Shadow of Tuskegee: African Americans and Health Care," *American Journal of Public Health*, **87,** pp. 1773–1778.

Gannon, M., Northern, J., and Carrol, S. Jr. (1971), "Characteristics of Non-Respondents among Workers," *Journal of Applied Psychology,* **55,** pp. 586–588.

Gates, R. and Helton A. (1998), "The Newset Mousetrap: What Does it Catch?? Internet Versus Telephone Data Collection: A Case Study," *The Book of Papers from the Worldwide Internet Seminar in Paris, France,* January 1998, ESOMAR.

Gelfand, A. E., Hills, S. E., Racine-Poon, A., and Smith, A. F. M. (1990), "Illustration of Bayesian Inference in Normal Data Models Using Gibbs Sampling," *Journal of the American Statistical Association,* **85,** pp. 972–985.

Gelfand, A. E. and Smith, A. F. M. (1990), "Sampling-based Approaches to Calculating Marginal Densities," *Journal of the American Statistical Association*, 85, no. 410, pp. 398–409.

Gelman A. and King, G., (1993), "Why are American Presidential Election Campaign Polls so Variable When Votes are so Predictable?" *British Journal of Political Science,* **23,** pp. 409–451.

Gelman, A., Carlin, J. B., Stern, H. S, and Rubin, D. B. (1995), *Bayesian Data Analysis,* London: Chapman & Hall.

Gelman, A. and Little, T. C. (1997), "Poststratification into many Categories Using Hiearchical Logistics Regression", *Survey Methodology,* **23,** pp. 127–135.

Gelman, A. and Little, T. C. (1998), "Improving Upon Probability Weighting for Household Size," *Public Opinion Quarterly,* **62,** 398–404.

Gelman A., King G., and Liu, C. (1998), "Not Asked and Not Answered: Multiple Imputation for Multiple Surveys (with discussion), *Journal of the American Statistical Association,* **93,** pp. 846–874.

Gfroerer, J., Gustin, J., and Turner, C. F. (1992), "Introduction," in Turner, C. F., Lessler, J. T., and Gfroerer, J. C. (eds.), *Survey Measurement of Drug Use: Methodological Studies,* Rockville, MD: U. S. Department of Health and Human Services.

Ghosh, M., Natarajan, K., and Maiti, T. (1998), "Bayesian Small Area Estimation with Binary Data," Presentation at the 1998 Joint Statistical Meetings in Dallas, Texas.

Gilljam, M. and Granberg, D. (1993), "Should We Take Don't Know for an Answer?" *Public Opinion Quarterly,* **57,** pp. 348–357.

Glass, A. L. and Holyoak, K. J. (1986), *Cognition,* New York: Random House.

Gluksberg, S. and McCloskey, M. (1981), "Decisions About Ignorance: Knowing that You Don't Know," *Journal of Experimental Psychology, Human Learning and Memory,* **7,** pp. 311–325.

Glynn, R. J., Laird, N. M., and Rubin. D. B. (1993), "Multiple Imputation In Mixture Models For Nonignorable Nonresponse With Follow-Ups," *Journal of the American Statistical Association,* **88,** pp. 984–993.

Glynn, R. J. and Rubin, D. B., (1986), "Selection Modeling Versus Mixture Modeling with Nonignorable Nonresponse, in *Drawing Inferences from Self-Selected Samples,"* H. Wainer, ed., New York: Springer-Verlag, pp. 119–146.

Goldstein, H. (1986), "Multilevel Mixed Linear Model Analysis Using Iterative Generalized Least Squares," *Biometrika,* **73,** pp. 43–56.

Goldstein, H. (2nd ed.) (1995), *Multilevel Statistical Models,* London: Arnold/ New York: Halsted.

Goldstein, H., Healy, M. J. R., and Rasbash, J. (1994), "Multilevel Time Series Models with Applications to Repeated Measures Data," *Statistics in Medicine,* **13,** pp. 1643–1655.

Gonier, D. E. (1999), *The Emperor Gets New Clothes. ARF's Online Research Day—Towards Validation.* New York: Advertising Research Foundation, pp. 8–13.

Gonzalez, M. E., Ogus, J. L., Shapiro, G., and Tepping, B. J., "Standards for Discussion and Presentation of Errors in Survey and Census Data," *Journal of the American Statistical Association,* **70,** pp. 5–23.

Gonzalez, R. M. (1997), "The Value Of Time: A Theoretical Review,"*Transport Reviews,* **17,** pp. 245–266.

Gouldner, A. W. (1960), "The Norm of Reciprocity: A Preliminary Statement," *American Journal of Sociology,* **25,** pp. 161–178.

Gower, A. R. and Dibbs, R. (1989), "Cognitive Research: Designing a Respondent Friendly Questionnaire for the 1991 Census," *Proceeding of the Bureau of the Census Fifth Annual Research Conference,* pp. 257–266.

Goyder, J. C. (1985a), "Face-to-Face Interviews and Mail Questionnaires: The Net Difference in Response Rate," *Public Opinion Quarterly,* **49,** pp. 234–252.

Goyder, J. (1985b), "Nonresponse in Surveys: A Canada-United States Comparison," *Canadian Journal of Sociology,* **10,** pp. 231–251.

Goyder, J. C. (1987), *The Silent Minority: Nonrespondents on Sample Surveys,* Boulder: Westview Press.

Goyder, J. C. and Warriner, K. (1999), "Measuring Socioeconomic Bias in Surveys: Toward Generalzation and Validation," paper presented at International Conference on Household Nonresponse, Portland, Oregon, October.

Gray, F. D. (1998), *The Tuskegee Syphilis Study.* Montgomery, AL: Black Belt Press.

Greenlees, J. S., Reece, W. S., and Zieschang, K. D. (1982), "Imputation of Missing Values When the Probability of Response Depends on the Variable Being Imputed," *Journal of the American Statistical Association,* **77,** pp. 251–261.

Graubard, B. I. and Korn, E. L. (1993), "Hypothesis Testing with Complex Survey Data: The Use of Classical Quadratic Test Statistics with Particular Reference to Regression Problems," *Journal of the American Statistical Association,* **88,** pp. 629–641.

Grotzinger, K. M., Stuart, B. C., and Ahern, F. (1994), "Assessment and Control of Nonresponse Bias in Survey of Medicine Use by the Elderly," *Medical Care,* **32,** pp. 989–1003.

Groves, R. M. and Kahn, R. (1979), *Surveys by Telephone: A National Comparison with Personal Interviews,* New York: Academic Press.

Groves, R. M. and Fultz, N. H. (1985), "Gender Effects among Telephone Interviewers in a Survey of Economic Attitudes," *Sociological Methods and Research,* **14,** pp. 31–52.

Groves, R. M. (1989), *Survey Errors and Survey Costs,* New York: Wiley.

Groves, R. M., Cialdini, R. B., and Couper, M. P. (1992), "Understanding the Decision to Participate in a Survey," *Public Opinion Quarterly,* **56**(4), pp. 475–495.

Groves, R. M. and Couper, M. P. (1992), "Respondent–Interviewer Interactions in Survey Introductions," paper presented at the 3th International Workshop on Household Survey Nonresponse, CBS, Voorburg, Holland.

Groves, R. M., Cantor, D., Couper, M. P., Levin, K., McGonagle, K., and Singer, E. (1997), "Research Investigations in Gaining Participation from Sample Firms in the Current Employment Statistics Program,"*Proceedings of Survey Research Methods Section, American Statistical Association,* pp. 289–294.

Groves, R. M. and Couper, M. P. (1998), *Nonresponse in Household Interview Surveys,* New York: Wiley.

Groves, R. M. and Hansen, S. E. (1996), *Survey Design Features to Maximize Respondent Retention in Longitudinal Surveys,* Survey Research Center, University of Michigan. Report to National Center for Health Statistics.

Groves, R. M., Singer, E., Corning, A. D., and Bowers, A. (1999a), "A Laboratory Approach to Measuring the Effects on Survey Participation of Interview Length, Incentives, Differential Incentives, and Refusal Conversion." *Journal of Official Statistics,* **15,** pp. 251–268.

Groves, R. M., Singer, E., and Corning, A. D. (1999b), "Leverage-Saliency Theory of Survey Participation: Description and an Illustration," *Public Opinion Quarterly,* **64,** pp. 299–308.

Guadagnoli, E. and Cunningham, S. (1989), "The Effects of Nonresponse and Late Response on a Survey of Physician Attitudes," *Evaluation and the Health Profession,* **12,** pp. 318–328.

Guadagnoli, E. and Cleary, P. D. (1992), "Age-related Item Nonresponse in Surveys of Recently Discharged Patients," *Journal of Gerontology,* **3,** pp. 207–212.

Gudykunst, W. B. (1983), "Uncertainty Reduction and Predictability of Behavior in Low and High Contact Cultures," *Communications Quarterly*, **31,** pp. 49–55.

Gudykunst, W. B., Gao, G., Schmidt, K. L., Nishida, T., Bond, M. H., Leung, K., Wang, G., and Barraclough, R. (1992), "The Influence of Individualism-Collectivism on Communication in Ingroup and Outgroup Relationships," *Journal of Cross-Cultural Psychology,* **23,** pp. 196–213

Gudykunst, W. B. (1997), "Cultural Variability in Communication: An Introduction,"*Communication Research,* **24,** pp. 327–348.

Gudykunst, W. B. and Kim, Y. Y. (3rd eds.) (1997), *Communicating with Strangers: An Approach to Intercultural Communication,* New York: McGraw-Hill.

Gudykunst, W. B. (1998), Individualistic and Collectivistic Perspectives on Communication: An Introduction," *International Journal of Intercultural Relations,* **22,** pp. 107–134.

Guenzel, P. J., Berckmans, T. R., and Cannell, C. F. (1983), *General Interviewing Techniques,* Ann Arbor, MI: Survey Research Center, The University of Michigan.

Gunn, W. and Rhodes, I. (1981), "Physician Response Rates to a Telephone Survey: Effects of Monetary Incentive Level," *Public opinion Quarterly,* **45,** pp. 109–115.

Gupta, V. K. and Nigam, A. K. (1987), "Mixed Orthogonal Arrays for Variance Estimation with Unequal Numbers of Primary Selections Per Stratum," *Biometrika,* **74,** pp. 735–742.

Gurney, M. and Jewett, R. S. (1975), "Constructing Orthogonal Replications for Standard Errors," *Journal of the American Statistical Association,* **70,** pp. 819–821.

GVU (1994–1999), GVU's User Surveys. http://www.gvu. gatech.edu/user_surveys.

GVU (1998), GVU's 10th WWW User Survey. http://www.cc. gatech.edu/gvu/user_surveys/survey–1998–10/.

Gwiasda, V., Taluc, N., and Popkin, S. J. (1997), "Data Collection in Dangerous Neighborhoods: Lessons from a Survey of Public Housing in Chicago," *Evaluation Review,* **21,** pp. 77–93.

Hall, E. T. (1966), *The Hidden Dimension,* New York: Doubleday.

Hamid, A. (1991), "Ethnographic Follow-Up of a Predominantly African American Population in a Sample Area in Central Harlem, New York City: Behavioral Causes of the Undercount of the 1990 Census," Ethnographic Evaluation of the 1990 Census, Report #11, prepared under Joint Statistical Agreement with the Bureau of the Census. Washington, D.C.: Bureau of the Census (http://www.census. gov/srd/papers/pdf/ev91–11.pdf).

Hansen, M. H. (1975), "Comment: 'Accuracy in market surveys. I: Nonresponse levels and effects'," *Proceedings of Business and Economic Statistics Section, American Statistical Association,* pp. 90–92.

Hansen, M. H. and Hurvitz, W. N. (1946), "The Problem of Nonresponse in Sample Surveys," *Journal of the American Statistical Association,* **41,** pp. 517–529.

Hansen, M. H., Madow, W. G., and Tepping, B. J. (1983), "An Evaluation of Model-Dependent and Probability-Sampling Inferences in Sample Surveys." *Journal of the American Statistical Association,* **78,** pp. 776–807 (with discussion).

Hansen, R. A. (1980), "A Self-Perception Interpretation of the Effect of Monetary and Nonmonetary Incentives on Mail Survey Respondent Behavior," *Journal of Marketing Research,* **17,** pp. 77–83.

Hansen, R. H. (1978), "The Current Population Survey: Design and Methodology,"*Technical Paper No. 40,* U.S. Bureau of the Census.

Harris-Kojetin, B. A. and Tucker, C. (1998), "Longitudinal Nonresponse in the Current Population Survey (CPS)," *ZUMA Nachtrichten Spezial,* **4,** pp. 263–272.

Harris-Kojetin, B. and Tucker, C. (1999), Exploring the Relations of Economic and Political Conditions with Refusal Rates to a Government Survey, *Journal of Official Statistics,* **15,** 2, pp. 167–184.

Hartley, J. (1981), "Eighty Ways of Improving Instruction Text," *IEEE Transactions on Professional Communication,* **24,** pp. 17–27.

Harvey, A. C. (1989), *Forecasting, Structural Time Series Models, and the Kalman Filter.* Cambridge: Cambridge University Press.

Harville, D. A. (1977), Maximum Likelihood Approaches to Variance Component Estimation and to Related Problems (with discussion), *Journal of the American Statistical Association,* **72,** 320–340.

Hasher, L. and Griffin, M. (1978), "Reconstructive and Reproductive Process in Memory," *Journal of Experimental Psychology: Human Learning and Memory,* **4,** pp. 318–330.

Hawkins, D. F. (1975), "Estimation of Nonresponse Bias," *Sociological Methods and Research,* **3,** pp. 461–487.

Hawkins, D. J. and Coney, K. A. (1981), "Uninformed Response Error in Survey Research," *Journal of Marketing Research,* **18,** pp. 370–374.

Heberlein, T. A. and Baumgartner, R. M. (1978), "Factors Affecting Response Rates to Mailed Questionnaires: A Quantitative Analysis of the Published Literature." *American Sociological Review,* **43,** pp. 447–462.

Heckman, J. (1976), "The Common Structure of Statistical Models of Truncation, Sample Selection and Limited Dependent Variables, and a Single Estimator for Such Models," *Annals of Economic and Social Measurement,* **5,** pp 475–492.

Hecht, M. L., Anderson, P. A., and Ribeau, S. A. (1989), "The Cultural Dimensions of Nonverbal Behavior," in M. K. Asante and W. B. Gudykunst (eds.), *Handbook of International and Intercultural Communication,* Newbury Park, CA: Sage, pp. 163–185.

Hedeker, D. (1993), *MIXOR: a Fortran Program for Mixed-Effects Ordinal Probit and Logistic Regression, Prevention Research Center,* University of Illinois at Chicago, Chicago, Illinois, 60637.

Hedge, A. and Yousif, Y. H. (1992), "Effects of Urban Size, Urgency, and Cost on Helpfulness: A Cross-Cultural Comparison between the United Kingdom and the Sudan," *Journal of Cross-Cultural Psychology,* **23,** pp. 107–115.

Heeringa, S. G. (1993), "Imputation of Item Missing Data in the Health and Retirement Survey," *Proceedings of the Survey Methods Section, American Statistical Association,* pp. 107–116.

Heeringa, S. G. (1999), "Multivariate Imputation and Estimation for Coarsened Survey Data on Income and Wealth," Unpublished Ph. D. thesis, Ann Arbor (MI): Department of Biostatistics, University of Michigan.

Heitjan, D. F. and Rubin, D. B. (1990), "Inference from coarse data via multiple imputation with application to age heaping," *Journal of the American Statistical Association,* **85,** 410, pp. 304–314.

Heitjan, D. F. and Rubin, D. B. (1991), "Ignorability and Course Data," *Annals of Statistics,* **19,** pp. 2244–2253.

Heitjan, D. F. (1989), "Inference from Grouped Continuous Data: A Review," *Statistical Science,* **4,** pp. 164–183 (with discussion).

Heitjan, D. F. (1994), "Ingnorability in General Complete-Data Models," *Biometrika,* **81,** pp. 701–708.

Herrmann, D. J. (1986), "Remebering past Experiences: Theoretical perspectives Past and Present," in T. Schlecter and M. Toglia (eds.), *New Direction in Cognitive Science,* Norwood, New Jersey: Ablex.

Herrmann, D. J. (1994), "The Validity of Restrospective Reports as a Function of the Directness of the Retrieval Process," in Schwartz, N. and Sudman, S. (eds.), *Autobiographical Memory and the Validity of Retrospective Reports,* New York: Springer Verlag.

Herrmann, D. J. (1995), "Reporting Current, Past, and Changed Health Status: What We Know about Distortion," *Medical Care,* **33,** pp. 89–94.

Herzog, A. R., Rogers, W. L., and Kulka, R. A. (1983), "Interviewing Older Adults: A Comparison of Telephone and Face-to-Face Modalities," *Public Opinion Quarterly,* **47,** pp. 405–418.

Hidiroglou, M. A., Latouche, M., Armstrong, B., and Gossen, M. (1995), "Improving Survey Information Using Administrative Records: The Case of the Canadian Employment Survey," *Proceedings of the Bureau of the Census Annual Research Conference,* Suitland, MD, pp. 171–197.

Hidiroglou, M. A., Drew, J. D., and Gray, G. B. (1993), "A Framework for Measuring and Reducing Nonresponse in Surveys," *Survey Methodology,* **19,** pp. 81–94.

Hilgard, E. R. and Payne, S. L. (1944), "Those Not at Home: Riddle for Pollsters," *Public Opinion Quarterly,* **8,** pp. 254–261.

Hill, D. H. and Willis, R. J. (1998), *"Reducing Panel Attrition: A Search for Effective Policy Instruments."* Paper Prepared for the Conference on Data Quality in Longitudinal Surveys, Ann Arbor, MI, October.

Hippler, H. J. and Schwarz, N. (1989), "'No-Opinion' Filters: A Cognitive Perspective,"*International Journal of Public Opinion Research,* **1,** pp. 77–87.

Hippler, H. J., Schwartz, N., and Sudman, S. (eds.) (1987), *Social Information Processing and Survey Methodology,* New York: Springer Verlag.

Hochstim, J. R. (1962), "Comparisons of Three Information-Gathering Strategies in a Population Study of Sociomedical Variables," *Proceedings of the Social Statistics Section, American Statistical Association,* Washington, D.C.: American Statistical Association, pp. 154–159.

Hofstede, G. (1980a), *Culture's Consequences: International Differences in Work-Related Values,* Beverly Hills, CA: Sage.

Hofstede, G. (1980b), *Masculinity and Femininity: The Taboo Dimension of National Cultures,* Thousand Oaks, CA: Sage Publications.

Hofstede, G. (1991), *Cultures and Organizations: Software of the Mind.* London: McGraw-Hill.

Hofstede, G. and Bond, M. H. (1984), "Hofstede's Culture Ddimensions: An Independent Validation Using Rokeach's Value Survey," *Journal of Cross-Cultural Psychology,* **15,** pp. 417–433.

Hogan, H. (1993), "The 1990 Post-enumeration Survey: Operations and Results," *Journal of the American Statistical Association*, **88,** pp. 1047–1060.

Hollander, S. (1992), "Survey Standards," in P. B. Sheatsley and W. J. Mitofsky (eds.), *A Meeting Place: The History of the American Association for Public Opinion Research,* n.p.: American Association for Public Opinion Research.

Hollis, N. S. (1999), "Can a Picture Save 1,000 Words? Integrating Phone and Online Methodologies," *ARF's Online Research Day—Towards Validation,* New York: Advertising Research Foundation, pp. 41–49.

Holt, D. and Smith, T. M. F. (1979), "Post Stratification," *Journal of the Royal Statistical Society,* **142,** pp. 33–46.

Holt, D., Scott, A. J., and Ewings, P. D. (1980), "Chi-Squared Tests with Survey Data,"*Journal of the Royal Statistical Society, Series A,* **143,** pp. 303–320.

Homans, G. C. (1961, 1974), *Social Behavior: Its Elementary Forms.* New York: Harcourt, Brace, Jovanovitch.

Hopkins, K. D. and Gullickson, A. R. (1992), "Response Rates in Survey Research: A Meta-Analysis of Monetary Gratuities,"*Journal of Experimental Education,* **61,** pp. 52–56.

Hopkins, K. D., Hopkins, B. R., and Schon, I. (1988), "Mail Surveys of Professional Populations: The Effects of Monetary Gratuities on Return Rates," *Journal of Experimental Education,* **56,** pp. 173–175.

Horvitz, D. G. and Thompson, D. J. (1952), "A Generalization of Sampling Without Replacement from a Finite Universe," *Journal of the American Statistical Association,* **47,** pp. 663–685.

Hosmer, D. W. and Lemeshow, S. (1980), "A Goodness-of-Fit Test for the Multiple Logistic Regression Model," *Communications in Statistics,* **A10,** pp. 1043–1069.

Hosmer, D. W. and Lemeshow, S. (1989), *Applied Logistic Regression,* New York: Wiley.

Houston, M. J. and Jefferson, R. W. (1975), The Negative Effects of Personalization on Response Patterns in Mail Surveys," *Journal of Marketing Research,* **11,** pp. 114–117.

Houston, M. J. and Nevin, J. R. (1977), The Effects of Source and Appeal on Mail Survey Response Patterns, *Journal of Marketing Research,* **14,** pp. 374–378.

Houtepen, H. (2000), Demografisch-Economisch Landenprofiel [In Dutch: Demographic-economic profile of 25 countries], in S. Gorseling, W. Graafmans, P. Langley, F. J. Louwen, and M. Temminghoff (eds.), *Gfk Jaargids 2000 [Gfk Yearbook 2000],* Utrecht: Geografiek, pp. 24–27.

Howell, F. M. and Frese, W. (1983), "Size of Place, Residential Preferences and the Life Cycle: How People Come to Like Where They Live," *American Sociological Review,* **48,** pp. 569–580.

Hox, J. J., de Leeuw, E., and Kreft, G. G., (1991), The Effect of Interviewer and Respondent Characteristics on the Quality of Survey Data: a Multilevel Model, *Measurement Errors in Surveys,* New York: Wiley.

Hox, J. J. and de Leeuw, E. (1999), "The Influence of Interviewers' Attitude and Behavior on Household Survey Nonresponse: An International Comparison," paper presented at the International Conference on Survey Nonresponse, Portland, Oregon.

Hox, J. J. and de Leeuw, E. D. (1994), "A Comparison of Nonresponse in Mail, Telephone, and Face-to-Face Surveys. Applying Multilevel Modeling to Meta-Analysis," *Quality and Quantity,* **28,** pp. 329–344.

Hox, J. J., de Leeuw, E. D., and Vorst, H. (1996), "A Reasoned Action Explanation for Survey Nonresponse," in S. Laaksonen (ed.), *International perspectives on nonresponse,* Helsinki: Statistics Finland.

Huang, E. T. and Fuller, W. A. (1978), "Nonnegative Regression Estimation for Sample Survey Data," *Proceedings of the Social Statistics Section, American Statistical Association,* pp. 300–305.

Hubbard, R. and Little, E. L. (1988), "Promised Contributions to Charity and Mail Survey Responses: Replication with Extension," *Public Opinion Quarterly,* **52,** pp. 223–230.

Hudgins, J. L., Holmes, B. J., and Locke, M. E. (1991), "The Impact of Family Structure Variations Among Black Families on the Underenumeration of Black Males, Part Two: Focus Group Research," Ethnographic Evaluation of the 1990 Census, Report #14, prepared under Joint Statistical Agreement with the Bureau of the Census. Washington, D.C.: Bureau of the Census (http://www.census.gov/srd/papers/pdf/ex90–14.pdf).

Huggins, V. J. and Fay, R. E. (1988), "Use of Administrative Data in SIPP Longitudinal Estimation," *Proceedings of the Section of Survey Research Methods, American Statistical Association,* pp. 354–359.

Humphrey, D. C. (1973), "Dissection and Discrimination: The Social Origins of Cadavers in America, 1760–1915," *Bulletin of the New York Academy of Medicine,* **49,** pp. 819–827.

Hurh, W. M. and Kim, K. C. (1982), "Methodological Problems in the Study of Korean Immigrants: Conceptual, Interactional, Sampling, and Interviewer Training Difficulties," in W. T. Liu (ed.)*Methodological Problems in Minority Research,* Chicago: Pacific/Asian American Mental Health Research Center, pp. 61–80

Hyman, H. H. (1954), *Interviewing in Social Research,* Chicago: University of Chicago Press.

Iannacchione, V. G. (1998), "Location and Response Propensity Modeling for the 1995 National Survey of Family Growth," *Proceedings of the American Statistical Association, Survey Research Methods Section,* pp. 523–528.

Imbens, G. (1992), "An Efficient Method of Moments Estimator for Discrete Choice Models with Choice-Based Sampling,"*Econometrica,* **60,** pp. 1187–1214.

Interagency Group on Establishment Nonresponse (IGEN) (1998), *Establishment Nonresponse: Revisiting the Issues and Looking to the Future. Statistical Policy Working Paper 28: Seminar on Interagency Coordination and Cooperation,* Washington, D.C.: Federal Committee on Statistical Methodology, pp. 181–227.

Jackson, C., Henriksen, L., Dickinson, D., and Levine, D. W. (1997), "The Early Use of Alcohol and Tobacco: Its Relation to Children's Competence and Parents' Behavior," *American Journal of Public Health,* **87,** pp. 359–364.

Jacobsen, S. J., Xia, Z., Campion, M. E., Darby, C. H., Plevak, M. F., Seltman, K. D., and Melton, L. J. (1999), "Potential Effect of Authorization Bias on Medical Record Research," *Mayo Clinic Proceedings,* **74,** pp. 330–338.

James, J. M. and Bolstein, R. (1990), "The Effect of Monetary Incentives and Follow-up Mailings on the Response Rate and Response Quality in Mail Surveys," *Public Opinion Quarterly,* **54,** pp. 346–361.

James, J. M. and Bolstein, R. (1992), "A Large Monetary Incentives and Their Effect on Mail Survey Response Rates," *Public Opinion Quarterly,* **56,** pp. 442–53.

James, T. (1997), "Results of the Wave 1 Incentive Experiment in the 1996 Survey of Income and Program Participation." *Proceedings of the Survey Research Section, American Statistical Association.*

Jansen, J. H. (1985), "Effect of Questionnaire Layout and Size and Issue-Involvement on Response Rates in Mail Surveys," *Perceptual and Motor Skills,* **61,** pp. 139–142.

Japec, L, Lundqvist, P., and Wretman, J. (1998), "Interviewer Strategies: How do Interviewers Schedule their Call Attempts," *Paper Presented at the 9th International Workshop on Household Survey Nonresponse,"* Bled, Slovenia.

Jeavons, A. and Bayer, L. (1997), "The Harris Poll Online," paper presented at The Internet, Marketing and Research 4, a seminar organized by Computing Marketing and Research Consultancy Ltd. (CMR), London, UK. http://www.cmrgroup.com/bimr489.htm.

Jenkins, C. and Ciochetto, S. (1993), *Results of Cognitive Research on the Multiplicity Question from the 1991 Schools and Staffing Student Records Questionnaire,* Center for Survey Methods Report: U.S. Bureau of the Census.

Jenkins, C. R. and Dillman, D. A. (1995), "The Language of Self-Administered Questionnaires as Seen Through the Eyes of Respondents," *Statistical Policy Working Paper 23: New Directions in Statistical Methodology: U. S. Office of Management and Budget,* **3,** pp. 470–516.

Jenkins, C. R. and Dillman, D. A. (1997), "Towards a Theory of Self-Administered Questionnaire Design," in L. Lyberg, P. Biemer, M. Collins, L. Decker, E. DeLeeuw, C. Dippo, N. Schwarz, and D. Trewin (eds.), *Survey Measurement and Process Quality,* New York: Wiley, pp. 165–196.

Jennrich, R. I. and Schluchter, M. D. (1986), "Unbalanced Repeated-Measures Models with Structured Covariance Matrices," *Biometrics,* **42,** pp. 805–820.

Hox, J. J., de Leeuw, E. D., and Kreft, G. G. (1991), "The Effect of Interviewer and Respondent Characteristics on the Quality of Survey Data: a Multilevel Model," in: P. P. Biemer, R. M. Groves, L. E. Lyberg, N. A. Mathiowetz, and S. Sudman (eds.)*Measurement Errors in Surveys,* New York: Wiley.

Jobe, J. B. and Herrmann, D. J. (1996), "No Opinion Filters: A Cognitive Perspective,"*International Journal of Public Opinion Research,* **1,** pp. 77–87.

Johanson, G. A., Gips, C. J., and Rich, C. E. (1993), "If You Can't Say Something Nice: A Variation on the Social Desirability Response Set," *Evaluation Review,* **17,** pp. 116–122.

Johnson, A. E., Botman, S. L., and Basiostis, P. (1994), "Nonresponse in Federal Demographic Surveys, 1981–1991," paper presented to the American Statistical Association, Toronto.

Johnson, T. P. (1988), *The Social Environment and Health,* unpublished Ph. D. dissertation, Lexington: University of Kentucky.

Johnson, T. P., O'Rourke, D., Chavez, N., Sudman, S., Warnecke, R., Lacey, L., and Horm, J. (1997), "Social Cognition and Responses to Survey Questions among Culturally Diverse Populations," in L. Lyberg, P. Biemer, M. Collins (eds.), *Survey Measurement and Process Quality,* New York: Wiley.

Jones, C., Sheatsley, P. B., and Stinchcombe, A. L. (1979),*Dakota Farmers and Ranchers Evaluate Crop and Livestock Surveys,* Chicago: National Opinion Research Center.

Jones, E. L. (1963), "The Courtesy Bias in South-East Asian Surveys," *International Social Science Journal,* **25,** pp. 70–75.

Jones, J. H. (1993), *Bad Blood: The Tuskegee Syphilis Experiment,* New York: Free Press.

Jones, R. H. and Vecchia, A. V. (1993), "Fitting Continuous ARMA Models to Unequally Spaced Spatial Data," *Journal of the American Statistical Association,* **88,** pp. 947–954.

Jones, R. H. (1993), *Longitudinal Data with Serial Correlation A State-space Approach,* New York: Chapman.

Jones, R. H. and Boadi-Boateng (1991), "Unequally Spaced Longitudinal Data with AR(1) Serial Correlation," *Biometrics,* **47,** pp. 161–175.

Jones, W. H. and Lang, J. R. (1978), "Sample Composition Bias and Response Bias in a Mail Survey: A Comparison of Inducement Methods," *Journal of Marketing Research,* **17,** pp. 69–76.

Jordan, L. A., Marcus, A. C., and Reeder, L. G. (1980), "Response Styles in Telephone and Household Interviewing: A Field Experiment," *Public Opinion Quarterly,* **44,** pp. 210–222.

Josephson, E. (1970), "Resistance to Community Surveys," *Social Problems,* **18,** pp. 116–129.

Judkins, D. R. (1996), "Discussion (of Articles by Rao, Fay, and Rubin on Variance Estimation in the Aftermath of Imputation)," *Journal of the American Statistical Association,* **91,** pp. 507–510.

Judkins, D. R. (1997), "Imputing for Swiss Cheese Patterns of Missing Data," *Proceedings of Statistics Canada Symposium 97, New Directions in Surveys and Censuses,* pp. 143–148.

Judkins, D. R. (1998), "Discussion (of article by Gelman, King and Liu on "not Asked and Not Answered: Multiple Imputation for Multiple Surveys")," *Journal of the American Statistical Association,* **93,** pp. 861–864.

Judkins, D. R., Hubbell, K. A., and England, A. M. (1993), "The Imputation of Compositional Data," *Proceedings of the Section on Survey Research Methods of the American Statistical Association,* pp. 458–462.

Juster, F. T. and Suzman, R. (1995), "An Overview of the Health and Retirement Survey," *Journal of Human Resources,* **30** (Supplement 1995), pp. 9-56.

Juster, F. T. and Smith, J. P. (1994), "Improving the Quality of Economic Data: Lessons from the HRS," HRS Working Paper Series #94-027, presented at NBER Summer Institute on Health and Aging. Cambridge (Massachusetts), July.

Kahn, D. F. and Hadley, J. M. (1949), "Factors Related to Life Insurance Selling," *Journal of Applied Psychology,* **33,** pp. 132–140.

Kahneman, D. (1973), *Attention and Effort,* Englewood Cliffs, New Jersey: Prentice Hall.

Kahneman, D. and Tversky, A. (1973), "On the Psychology of Prediction," *Psychological Review,* **80,** pp. 327–351.

Kalton, G. (1981), *Compensating for Missing Survey Data,* Ann Arbor (MI): Survey Research Center, The University of Michigan.

Kalton, G. and Kasprzyk, D. (1982), "Imputing for Missing Survey Nonresponse," *Proceedings of the Survey Research Methods Section, American Statistical Association,* pp. 22–31.

Kalton, G. (1986), "Handling Wave Nonresponse in Panel Surveys," *Journal of Official Statistics,* **2,** pp. 303- 314.

Kalton, G. and Kasprzyk, D. (1986), "The Treatment of Missing Survey Data," *Survey Methodology,* **12,** pp. 1–16.

Kalton, G. and Miller, M. E. (1986), "Effects of Adjustments for Wave Nonresponse on Panel Survey Estimates," *Proceedings of the Survey Research Methods Section, American Statistical Association,* pp. 194–199.

Kalton, G., Lepkowski, J., and Lin, T. (1985), "Compensating for Wave Nonresponse in the 1979 ISDP Research Panel," *Proceedings of the Survey Research Methods Section, American Statistical Association,* pp. 372ú377.

Kalton, G., Roberts, J., and Holt, D. (1980), "The Effects of Offering a Middle Response Option with Opinion Questions,"*The Statistician,* **29,** pp. 65–79.

Kalton, G., Lepkowski, J., Montanari, G. E., and Maligalig, D. (1990), "Characteristics of Second Wave Nonrespondents in a Panel Survey,"*Proceedings of the American Statistical Association, Survey Research Methods Section,* pp. 462–467.

Kalton, G. and Maligalig, D. S. (1991), "A Comparison of Methods of Weighting Adjustment for Nonresponse," *Proceedings of the 1991 Annual Research Conference,* U.S. Bureau of the Census, pp. 409–428.

Kane, E. W. and Macaulay, L. J. (1993), "Interviewer Gender and Gender Attitudes," *Public Opinion Quarterly,* **57,** pp. 1–28.

Kanuk, L. and Berenson. C. (1975), "Mail Surveys and Response Rates: A Literature Review," *Journal of Marketing Research,* **12,** pp. 440–453.

Kasprzyk, D. and Kalton, G., (1998), "Measuring and Reporting the Quality of Survey Data," *Proceedings of Statistics Canada Syposium 97, New Directions in Surveys and Censuses.* Ottawa: Statistics Canada.

Kashner, T. M., Suppes, T., Rush, A. J., and Altshuler, K. Z. (1999), "Measuring Use of Outpatient Care Among Mentally Ill Individuals: A Comparison of Self Reports and Provider Reports," *Evaluation and Program Planning,* **22,** pp. 31–39.

Kasse, M., (1999),*Quality Criteria for Survey Research,* Berlin: Akademie Verlag.

Katosh, J. P. and Traugott, M. W. (1981), "The Consequences of Validated and Self-reported Voting Measures," *Public Opinion Quarterly,* **45,** pp. 519–535.

Kazimi, C., Brownstone, D., Ghosh, A., Golob, T. F., and Van Amelsfort, D. (2000), "Willingness-to-Pay to Reduce Commute Time and Its Variance: Evidence from the San Diego I–15 Congestion Pricing Project," *Presented at Annual Meeting of Transportation Research Board, National Research Council,* January 9–13, Washington, D.C.

Kehoe, C. M. and Pitkow, J. E. (1996), "Surveying the Territory. GVU's Five WWW User Surveys," *The World Wide Web Journal,* **1,** p. 3. http://www.cc.gatech.edu/gvu/user_surveys/papers/w3j.html.

Kelly, S. Jr. and Mirer, T. W. (1974), "The Simple Act of Voting," *The American Political Science Review,* **68,** pp. 572–591.

Kennickell, A. B. (1997), "Using range techniques with CAPI in the 1995 Survey of Consumer Finances," *Proceedings of the Section on Survey Research Methods, American Statistical Association.*

Kerachsky, S. H. and Mallar, C. D. (1981), "The Effects of Monetary Payments on Survey Responses: Experimental Evidence from a Logitudinal Study of Economically Disadvantaged Youths," *Proceedings of the Survey Research Methods of the American Statistical Association,* pp. 258–263.

Kersten, H. M. P. and Bethlehem, J. G. (1984), "Exploring and Reducing the Nonresponse Bias by Asking the Basic Question," *The Statistical Journal of the United Nations Commission for Europe,* **2,** pp. 369–380.

Kim, K. C. Lee and Y. Whang. (1995), "The Effect of Respondent Involvement in Sweepstakes on Response Rates in Mail Surveys," *Proceedings of the Section on Survey Research Methods, American Statistical Association,* pp. 216–220.

King, C. and Kornbau, M. (1994), *Inventory of Economic Area Statistical Practices: Phase 3: Survey Data Quality,* Economic Statistical Methods Report Series ESMD–9042, U.S. Bureau of the Census, Washington D. C.

King, J. (1998), "Nonresponse Bias in a Household Expenditure Survey: A Study Using Geodemographic Codes," paper presented at the 9th International Workshop on Household Survey Nonresponse, Bled, Slovenia, September.

Kennickell, A. B. (1997), "Analysis of Nonresponse Effects in the 1995 Survey of Consumer Finances," *Proceedings of the Section on Survey Research Methods, American Statistical Association,* pp. 377–382.

Kennickell, A. B. (1999), "Analysis of Nonresponse Effects in the 1995 Survey of Consumer Finances," *Journal of Official Statistics,* **15,** pp. 283–303.

Kish, L. (1965), *Survey Sampling,* New York: Wiley.

Kish, L. (1998), *"Quota Sampling: Old Plus New Thought,"* working paper, University of Michigan.

Kish, L. (1995), "Methods for Design Effects," *Journal of Official Statistics,* **11,** pp. 55–77.

Kish, L. and Frankel, M. R. (1970)," Balanced Repeated Replication for Standard Errors," *Journal of the American Statistical Assocation,* **65,** pp. 1071–1094.

Kish, L. and Hess, I. (1959), "A 'Replacement' Procedure for Reducing the Bias of Nonresponse," *The American Statistician,* **13,** pp. 17–19.

Klare, G. R. (1950), "Understandability and Indefinite Answers to Public Opinion Questions,"*International Journal of Opinion and Attitude Research,* **4,** pp. 91–96.

Klopfer, F. J. and Madden, T. M. (1980), "The Middlemost Choice on Attitude Items: Ambivalence, Neutrality, or Uncertainty,"*Personality and Social Psychology Bulletin,* **6,** pp. 97–101.

Kochman, T. (1981), *Black and White: Styles in Conflict,* Chicago: University of Chicago Press.

Kojetin, B. A. and Tucker, C. (1999), "Exploring the Relation of Economical and Political Conditions with Refusal Rates to a Government Survey," *Journal of Official Statistics,* **2,** pp. 167–184.

Kojetin, B. A., Borgida, E., and Snyder, M. (1993), "Survey Topic Involvement and Nonresponse Bias," *Proceedings of the Survey Research Methods Section, American Statistical Association,* **II,** pp. 838–843.

Koriat, A. and Lieblich, I. (1974), "What Does a Person in a 'TOT' State Know That a Person in a 'Don't Know' State Doesn't Know," *Memory and Cognition,* **2,** pp. 647–655.

Kormendi, E. (1988), "The Quality of Income Information in Telephone and Face-to-Face Surveys," in R. M. Groves, P. Biemer, L. Lyberg, J. Massey, W. Nicholls, and J. Waksberg (eds.), *Telephone Survey Methodology,* New York: Wiley.

Korn, E. I. and Graubard, B. I. (1998), "Variance Estimation for Superpopulation Parameters,"*Statistica Sinica,* **8,** pp. 1131–1151.

Kott, P. S, (1994)," A Note of Handling Nonresponse in Surveys," *Journal of the American Statistical Association,* **89,** pp. 693–696.

Kottler, R. E. (1997a), "Search No Further. Yahoo!'s Audience Analysis Project—A Case

Study," *Quirk's Marketing Research Review,* June/July. http://www.quantime.co.uk/web_bureau/knowledge/yahooqq.htm.

Kottler, R. E. (1997b), "Web Surveys—The Professional Way," paper presented at the Advertising Research Foundation Conference, New York, USA. http://www.quantime.co.uk/web_bureau/knowledge/websurv.htm

Kottler, R. E. (1998), "Sceptics Beware! Web Interviewing Has Arrived and is Established. Embrace it or be Left Behind," paper presented at Annual Conference of the Market Research Society, Birmingham, UK. http://www.quantime.co.uk/web_bureau/knowledge/mrs98.htm.

Kozielec, J. (1995), "The Tax Return: A Unique Data Source for Tracking Migration," *Turning Administrative Systems into Information Systems,* U.S. Government Printing Office.

Kreiger, A. M. and Pfeffermann, D. (1992), "Maximum Likelihood Estimation from Complex Sample Surveys," *Survey Methodology*, 18, pp. 225.

Krenzke, T., Mohadjer, L. and Montaquila, J. (1998), "Generalizing the Imputation Error Variance in the Alcohol and Drug Services Study," *Proceedings of the Biometrics Section, American Statistical Association,* pp. 118–123.

Krosnick, J. A. and Berent, M. K. (1990), "The Impact of Verbal Labeling of Response Alternatives and Branching on Attitude Measurement Reliability in Surveys," paper presented at the American Association for Public Opinion Research Annual Meeting, Lancaster, Pennsylvania.

Krosnick, J. A. and Milburn, M. A. (1990), "Psychological Determinants of Political Opinionation,"*Social Cognition,* **8,** pp. 49–72.

Krosnick, J. A. (1991), "Response Strategies for Coping with the Cognitive Demands of Attitude Measure in Surveys,"*Applied Cognitive Psychology,* **5,** pp. 213–236.

Krosnick, J. A. (1999), "Survey Research," *Annual Review of Psychology,* **50**.

Krosnick, J. A., Carson, R. T., Hanemann, W. M., Kopp, R. J., Mitchell, R. C., Presser, S., Rudd, P. A., Smith, V. K., Berent, M. K., Conway, M., and Martin K. (1999)," The Impact of No-opinion Response Options on Data Quality: Prevention on Non-attitude Reporting or an Invitation to Satisfice," Unpublished Manauscript, Ohio State University, Columbus, Ohio.

Kulka, R. A. (1994), "The Use of Incentives to Survey 'Hard-to-Reach' Respondents: A Brief Review of Empirical Research and Current Practice," paper prepared for Seminar on New Directions in Statistical Methodology, Bethesda, MD.

Kviz, F. J., (1977), "Toward a Standard Definition of Response Rate," *Public Opinion Quarterly,* **41,** pp. 265–267.

Kydoniefs, L. and Stanley, J. (1999), "Establishment Non-Response: Revisiting the Issues and Looking into the Future," in Seminar on Interagency Coordination and Cooperation, Statistical Policy Working Paper 28, Washington, D.C.: FCSM.

Laird, N. M. (1988), "Missing Data in Longitudinal Studies," *Statistics in Medicine,* **7,** pp. 305–315.

Laird, N. M. and Ware, J. H. (1982)," Random-Effects Models for Longitudinal Data, *Biometrics,* **38,** pp. 963–974.

Lankford, S. V., Buxton, B. P., Hetzler, R., and Little, J. R. (1995), "Response Bias and Wave Analysis of Mailed Questionnaires," *Journal of Travel Research,* **33,** pp. 8–30.

Larsen, M. D. (1998), "Predicting the Residency Status for Administrative Records That Do

Not Match Census Records," *Technical Report Administrative Records Research Memorandum #20,* United States Department of Commerce, Bureau of the Census.

Larson, R. F. and Catton, W. R. (1959), "Can the Mail-Back Bias Contribute to a Study's Validity?" *American Sociological Review,* **24,** pp. 243–245.

Lau, R. R. (1995), "An Analysis of the Accuracy of 'Trial Heat' Polls during the 1992 Presidential Election," *Public Opinion Quarterly,* **58,** pp. 2–20.

Laurie, H., Smith, R., and Scott, L. (1999), "Strategies for Reducing Nonresponse in a Longitudinal Panel Survey," *Journal of Official Statistics,* **15**(2), pp. 269–282.

Lavrakas, P. J., Settersten, R. A., and Maier, R. A. (1991), "RDD Panel Attrition in Two Local Area Surveys," *Survey Methodology,* **17,** pp. 143–152.

Lazzeroni, L. C, Schenker, N., and Taylor, J. M. G, (1990), "Robustness of Multiple Imputation Techniques to Model Specification," *Proceedings of the Section on Survey Research Methods, American Statistical Association,* pp. 260–265.

Lee, H., Rancourt, E., and Särndal, C.-E. (1994), "Experiment with Variance Estimation from Survey Data with Imputed Values," *Journal of Official Statistics,* **10,** pp. 231–243.

Lee, H., Rancourt, E., and Särndal, C.-E. (1995), "Jackknife Variance Estimation for Data with Imputed Values," *Proceedings of the Survey Methods Section, Statistical Society of Canada,* pp. 111–115.

Lee, H., Rancourt, E., and Särndal, C.-E. (2000), "Variance Estimation from Survey Data Under Single Imputation," Working Paper, Statistics Canada, ISMD-2000-006E.

Lehtonen, R. (1996), "Interviewer Attitudes and Unit Nonresponse in Two Different Interviewing Schemes," in S. Laaksonen (ed.), *International Perspectives on Nonresponse; Proceedings of the Sixth International Workshop on Household Survey Nonresponse,* Helsinki: Statistics Finland

Leigh, W. A. (1998), "Participant Protection with the Use of Records: Ethical Issues and Recommendations," *Ethics and Behavior,* **8,** pp. 305–319.

Lengacher, J. E., Sullivan, C. M., Couper, M. P., and Groves, R. M. (1995), "Once Reluctant, Always Reluctant? Effects of Differential Incentives on Later Survey Participation in a Longitudinal Study," paper presented at the Annual Conference of the American Association for Public Opinion Research, Fort Lauderdale, Florida.

Lent, J., and Miller., S., Cantwell, P., and Duff, M. (1999), "Effects of Composite Weights on Some Estimates from the Current Population Survey," *Journal of Official Statistics,* **15,** 3, pp. 431–448.

Lepkowski, J. (1989), Treatment of wave nonresponse in panel surveys," in D. Kasprzyk, G. Duncan, G. Kalton and M. P. Singh (eds.), *Panel Surveys.* New York: Wiley, pp. 348–374.

Lepkowski, J. M. and Couper, M. P. (1999), "Nonresponse in Longitudinal Household Surveys," paper presented at the International Conference on Survey Nonresponse, Portland, Oregon.

Lessler, J. T. and Kalsbeck, W. D. (1992), *Nonsampling Error in Surveys,* New York: Wiley.

Levine, S. and Gordon, G. (1958), "Maximizing Returns on Mail Questionnaires," *Public Opinion Quarterly,* **22,** pp. 568–575.

Lewis, K. (1999), *Creation of the 1998 CPS-Based MATH Model and Database,* Washington, D.C.: Mathematica Policy Research, Inc.

Li, K. H., Raghunathan, T. E., and Rubin, D. B. (1991), "Largesample Significance Levels from Multiply Imputed Data Using Moment-Based Statistics and an *f* Reference Distribution," *Journal of the American Statistical Association,* **86,** pp. 1065–1073.

Lievesley, D. (1988), "Unit Nonresponse in Interview Surveys," Unpublished Working Paper, London: Social and Community Planning Research.

Lillard, L. A. and Farmer, M. A. (1997), "Linking Medicare and National Survey Data," *Annals of Internal Medicine,* **127,** pp. 691–695.

Lillard, L. A. and Panis, C. W. A. (1998), "Panel Attrition from the Panel Study of Income Dynamics: Household Income, Marital Status, and Mortality. "*The Journal of Human Resources,* **33**(2), p. 437.

Lilley, S-J, Brook, L., Park, A., and Thomson, K. (1997), *British Social Attitudes and Northern Ireland Social Attitudes 1995 Surveys—Technical Report,* London: National Centre for Social Research.

Lin, I. F. and Schaeffer, N. C. (1995), "Using Survey Participants to Estimate the Impact of Nonparticipation," *Public Opinion Quarterly,* **59,** pp. 236–258.

Linsky, A. S. (1975), "Stimulating Responses to Mailed Questionnaires: A Review," *Public Opinion Quarterly,* **39,** pp. 82–101.

Little, R. J. A. (1982), "Models for Nonresponse in Sample Surveys," *Journal of the American Statistical Association,* **77,** pp. 237–250.

Little, R. J. A. (1986), "Survey Nonresponse Adjustments for Estimates of Means," *International Statistical Review,* **54,** pp. 139–157.

Little, R. J. A. (1988a), *ROBMLE User Notes,* Unpublished Manuscript.

Little, R. J. A. (1988b), "Robust Estimation of the Mean and Covariate Matrix from Data with Missing Values," Applied Statistics, **37,** pp. 23–38.

Little, R. J. A. (1991), "Inference with Survey Weights," *Journal of Official Statistics,* **7,** pp. 405–424.

Little, R. J. A. (1992), "Missing Data Adjustment in Large Surveys," *Journal of Business and Economic Statistics, American Statistical Association,* **6,** pp. 287–301.

Little, R. J. A. (1993a), Pattern-Mixture Models for Multivariate Incomplete Data," *Journal of the American Statistical Association,* **88,** pp. 125–134.

Little, R. J. A. (1993b), "Comment on 'Hierarchical Logistic Regression Models for Imputation of Unresolved Enumeration Status in Undercount Estimation'," *Journal of the American Statistical Association,* **88,** pp. 1159–1161.

Little, R. J. A. (1995)," Modeling the Dropout Mechanism in Repeated Measures Studies," *Journal of the American Statistical Association,* **90,** pp. 1112–1121.

Little, R. J. A. and Rubin, D. B. (1987), *Statistical Analysis with Missing Data,* New York: Wiley.

Little, R. J. A., (1997), "Biostatical Analysis with Missing Data," *Encyclopedia of Biostatics,* P. Armitage and T. Colton (eds.), London: Wiley.

Little, R. J. A. and Schluchter, M. D. (1985), "Maximum Likelihood Estimation for Mixed Continuous and Categorical Data with Missing Values," *Biometrika,* **72,** 3, pp. 492–512.

Little, R. J. A. and Su, H. L. (1987), "Missing Data Adjustments for Partially Scaled Variables," *Proceedings of the Section on Survey Research Methods," American Statistical Association,* pp. 644–649.

Little, R. J. A. and Wu, M. M. (1991), "Models for Contingency Tables With Known Margins When Target and Sampled Populations Differ," *Journal of the American Statistical Association,* **86,** pp. 87–95.

Little, R. J. A. and Vartivarian, S. A., (2001), "Don't Weigh the Rates in Nonresponse Weights," Unpublished Manuscript.

Little, R. J. A., Lewitzky, S., Heeringa, S., Lepkowski, J., and Kessler, R. C. (1997), "An Assessment of Weighting Methodology for the National Comorbidity Study,"*American Journal of Epidemiology,* **146,** pp. 439–449.

Little, S-J., Brook, L., Bryson, C., Jarvis, L., Park, A. and Thomson, K. (1998), *British Social Attitudes and Northern Ireland Social Attitudes, 1996 Surveys—Technical Report*, London: National Centre for Social Research.

Little, T. C. (1996), *Models for Nonresponse Adjustment in Sample Surveys,* Section 3.3, Ph.D. Thesis, Department of Statistics, University of California, Berkeley.

Littlefield, R. (1974), "Self-Disclosure among Negro, White and Mexican-American Adolescents," *Journal of Counseling Psychology,* **21,** pp. 133–136.

Locander, W. and Burton, J. P. (1976), "The Effect of Question Forms on Gatheriing Income Data by Telephone,"*Journal of Marketing Research,* **13,** pp. 189–192.

Long, J. F. (1990), "The Subnational Population Estimates Program of the U.S. Bureau of the Census: Past, Present, and Future," *Proceedings of the Government Statistics Section, American Statistical Association,* pp. 69–74.

Long, J. F. (1996), "Demographic Applications of Administrative Records," *Proceedings of the Government Statistics Section, American Statistical Association,* pp. 169–173.

Lohr, S. L. (1999), *Sampling: Design and Analysis,* Pacific Grove, CA: Brooks-Cole.

Lozar Manfreda, Katja (1999), "Participation in Web Surveys," Paper Presented at 9th International Meeting Dissertation Research in Psychometrics and Sociometrics, in Ogestgeest, The Netherlands, December 16–17, 1999.

Lozar Manfreda, K., Vehovar, V., and Batagelj, Z. (1999), "Measuring Web Site Visits," Paper Presented at the International Conference on Methodology and Statistics, Preddvor, Slovenia, September 20–22, 1999.

Luiten, A. and de Heer, W. F. (1994), "International Questionnaire and Itemlist 'Fieldwork Strategy,'" paper presented at the Fifth International Workshop on Household Survey Nonresponse, Ottawa, Canada.

Lynn, P. and Clarke, P. (2000), "Separating Refusal Bias and Non-Contact Bias: Evidence from UK National Surveys," woking paper, National Center for Social Research, UK.

Lynn, P. (1999), "Is the Impact of Respondent Incentives on Personal Interview Surveys Transmitted via the Interviewers?" unpublished manuscript, National Center for Social Research, UK.

Maas, C. F. and de Heer, W. F. (1995), "Response Developments and the Fieldwork Strategy," *Bulletin de Methodologie Sociologique,* **48,** pp. 36–51.

Mack, S., Huggins, V., Keathley, D., and Sundukchi, M. (1998), "Do Monetary Incentives Improve Response Rates in the Survey of Income and Program Participation?" *Proceedings of the Survey Methodology Section, American Statistical Association,* pp. 529–534.

Madans, J. H., Kleinman, J. C., Cox, C. S., Barbano, H. E., Feldman, J. J., Cohen, B., Finucane, F. F., and Cornoni-Huntley, J. (1986), "10 years after NHANES I: Report of Initial Followup, 1982–84," *Public Health Reports,* **101,** pp. 465–473.

Madow, W. G., Nisselson, H., and Olkin, I., (eds.), (1983), *Incomplete Data in Sample Surveys, Vol. I, Report and Case Studies,* New York: Academic Press.

Magaziner, J., Zimmerman, S. I., Gruber-Baldini, A. L., Hebel, J. R., and Fox, K. M. (1997),

"Proxy Reporting in Five Areas of Functional Status,"*American Journal of Epidemiology,* **146,** pp. 418–428.

Magaziner, J., Bassett, S., Hebel, J. R., and Gruber-Baldini, A. (1996), "Use of Proxies to Measure Health and Functional Status in Epidemiologic Studies of Community-Dwelling Women Ages 65 Years and Older,"*American Journal of Epidemiology,* **143,** pp. 283–292.

Mandell, L. (1975), "When to Weight: Determining Nonresponse Bias in Survey Data," *Public Opinion Quarterly,* **38,** pp. 247–252.

Mangione, T. W. (1995),*Mail Surveys: Improving the Quality,* Thousand Oaks, CA: Sage Publications.

Mann, L. (1980), "Cross-Cultural Studies of Small Groups," In: H. C. Triandis and R. W. Brislin (eds.), *Handbook of Cross-Cultural Psychology*, Boston: Allyn & Bacon.

Manski, C. F. and Lerman, S. (1977), "The Estimation of Choice Probabilities from Choice-Based Samples," *Econometrica,* **45,** pp. 1977–1988.

Marcus, A. C. and Telesky, C. W. (1983), *American Apartheid: Segregation and the Making of the Underclass,* Cambridge, MA: Harvard University Press.

Markus, G. B. and Converse, P. E., (1979)," A Dynamic Simultaneous Equation Model of Electoral Choice," *The American Political Science Review,* **73,** 4, pp. 1055–1070.

Marquis, K., Wetrogan, S., and Palacios, H. (1996), "Towards a U. S. Population Database from Administrative Records," *Proceedings of the Government Statistics Section, American Statistical Association,* pp. 117–122.

Martin, C. L. (1994), "The Impact of Topic Interest on Mail Survey Response Behavior," *Journal of the Market Research Society,* **36,** pp. 327–338.

Martin, E. (1986), *Report on the Development of Alternative Screening Procedures for the National Crime Survey.*Washington, D. C: Bureau of Social Science Research.

Martinez-Ebers, V. (1997), "Using Monetary Incentives with Hard-to-Reach Populations in Panel Surveys," *International Journal of Public Opinion Research,* **9,** pp. 77–86.

Mason, R., Lesser, V., and Traugott, M. W. (1996), "Weighting Converted Refusals in RDD Sample Surveys," paper presented at the annual meeting of the American Association for Public Opinion Research, Norfolk, Virginia.

Massey, D. S. and Denton, N. A. (1993), *American Apartheid: Segregation and the Making of the Underclass,* Cambridge: Harvard University Press.

Mavis, B. E. and Brocato, J. J. (1998), "Postal Surveys Versus Electronic Mail Surveys," *Evaluation and the Health Professions,* **21,** pp. 395–408.

McCarthy, P. J. (1969), "Pseudo-Replication: Half-Samples," *Review of the International Statistical Institute,* **37,** pp. 239–264.

McClendon, M. J. (1986), "Unanticipated Effects of No Opinion Filters on Attitudes and Attitude Strength," *Sociological Perspectives,* **29,** pp. 379–395.

McClendon, M. J. (1991), "Acquiescence and Recency Response-Order Effects in Interview Surveys,"*Sociological Methods and Research,* **20,** pp. 60–103.

McClendon, M. J. and Alwin, D. F. (1993), "No-Opinion Filters and Attitude Measurement Reliability,"*Sociological Methods and Research,* **21,** pp. 438–464.

McCool, S. F. (1991), "Using Probabilistic Incentives to Increase Response Rates to Mail Return Highway Intercept Diaries," *Journal of Travel Research,* **30,** pp. 17–19.

McDaniel, S. W. and Rao, C. P. (1980), "The Effect of Monetary Inducement on

Mailed Questionnaire Response Quality,"*Journal of Marketing Research,* **17,** pp. 265–268.

McLachlan, G. J. and Krishnan, T. (1997), "The EM Algorithm and Extensions," New York: Wiley.

Mehta, R. and Sivadas, E. (1995), "Comparing Response Rates and Response Content in Mail Versus Electronic Mail Surveys," *Journal of the Market Research Society,* **37,** pp. 429–439.

Meng, X-L. and Rubin, D. B. (1992), "Performing Likelihood-Ratio Tests with Multiply-Imputed Data Sets," *Biometrika,* **79,** pp. 103–111.

Meng, X-L. and Pedlow, S., (1992), "EM: A Bibliographic Review with Missing Articles," *Proceedings of the Statistical Computing Section, American Statistical Association,* 1992, pp. 24–27.

Meng, X. -L. (1994), "Multiple Imputation Inferences with Uncongenial Sources of Input," *Statistical Science,* **9,** pp. 538–573.

Meng, X-L, and Van Dyk, (1997), "The EM Algorithm, an Old Folk Song Sung to a Fast New Tune, *Journal of the Royal Statistical Society,* B, **5,** 000–000.

Meng, X. -L. (2001), "A Congenial Overview and Investigation of Imputation Iinferences under Uncongeniality," in R. Groves, R. J. A. Little, and J. Eltinge (eds.), *Survey Nonresponse.*New York: Wiley.

Merkle, D. M., Edelman, M. E., Dykeman, K., and Brogan, C. (1998), "An Experimental Study of Ways to Increase Exit Poll Response Rates and Reduce Survey Error," paper presented at the annual conference of the American Association for Public Opinion Research, St. Louis, Missouri.

Messmer, D. J. and Seymour, D. T. (1982), "The Effects of Branching on Item Nonresponse," *Public Opinion Quarterly,* **46,** pp. 270–277.

Miller, J. G., Bersoff, D. M., and Harwood, R. L. (1990), "Perceptions of Social Responsibilities in India and the United States: Moral Imperatives or Personal Decisions?" *Journal of Personality and Social Psychology,* **58,** pp. 33–47.

Miller, K. (1996), "The Influence of Different Techniques on Response Rate and Nonresponse Error in Mail Surveys," Masters thesis, Washington State University.

Miller, W. E., Kinder, D. R., and Rosenstone, S. J. (1992)," *American National Election Study, 1990: Post-Election Survey,"* Ann Arbor, MI: Inter-University Consortium for Political and Social Research.

Montaquila, J. M. and Jernigan, R. W. (1997), "Variance Estimation in the Presence of Imputed Data,"*Proceedings of the Section on Survey Research Methods, American Statistical Association,* pp. 273–278.

Moore, D. and Baxter, R. (1993), "Increasing Mail Questionnaire Completion for Business Populations: The Effects of Personalization and a Telephone Followup Procedure as Elements of the Total Design Method," *Proceedings of the International Conference on Establishment Surveys, American Statistical Association,* pp. 496–502.

Moore, J. C. and Marquis, K. H. (1989), "Using Administrative Record Data to Evaluate the Quality of Survey Estimates," Survey Methodology, **15,** pp. 129–143.

Morgan, D. L. (1998), "Practical Strategies for Combining Qualitative and Quantitative Methods: Applications to Health Research," *Qualitative Health Research,* **8,** pp. 362–376.

Morgenstein, D. and Brick, J. M. (1996), "WesVarPC: Software for Computing Variance Estimates from Complex Designs," *Proceedings of the 1996 Annual Research Conference,* Washington, D.C.: Bureau of the Census, pp. 861–866.

Morton-Williams, J. (1993), *Interviewer Approaches,* Aldershot: Dartmouth.

Much, N. C. (1991), "Determinations of Meaning: Discourse and Moral Socialization," in R. A. Shweder (ed.), *Thinking Through Cultures,* Cambridge: Harvard University Press.

Murphy, S. and Li, B. (1995), "Projected Partial Likelihood and Its Application to Longitudinal Data." *Biometrika,* **82,** pp. 399–406.

Myers, V. (1979), "Survey Methods and Socially Distant Respondents," *Social Work Research and Abstracts,* **15,** pp. 3–9.

Myrskylä, P. (1991), "Census by Questionnaire—Census by Registers and Administrative Records: The Experience of Finland," *Journal of Official Statistics,* **7,** pp. 457–474.

Nadilo, R. (1999), "Online Research: The Methodology for the Next Millennium. ARF's Online Research Day—Towards Validation," New York: Advertising Research Foundation, pp. 50–51.

Nadler, R. and Henning, J. (1998), "Web Surveys—for Knowledge, Lead Management, and Increased Traffic," Survey Tips by Perseus Development Corporation. http://perseusdevelopment. com/surveytips/thw_websurveys. html.

Nandi, P. K. (1982), "Surveying Asian Minorities in the Middle-Sized City," in W. T. Liu (ed.), *Methodological Problems in Minority Research.*Chicago: Pacific/Asian American Mental Health Research Center, pp. 81–92.

Narayan, S. and Krosnick, J. A. (1996), "Education Regulates the Magnitude of Some Response Effects in Attitude Measurement,"*Public Opinion Quarterly,* **60,** pp. 58–88.

Natarajan, R. and McCulloch, C. E. (1995), "A Note on the Existence of the Posterior Distribution for a Class of Mixed Models for Binomial Responses," *Biometrika,* **82,** pp. 639–643.

Nathan, G. (1980), "Substitution for Non-Response as a Means to Control Sample Size,"*Sankhya, The Indian Journal of Statistics, Series C,* **42,** pp. 50–55.

National Center for Health Statistics (1996), *NHANES III Reference Manuals and Reports, CD-ROM GPO 017-022-1358-4,* Washington, D.C.: United States Government Printing Office.

Nelson, T. O., Gerler, D., and Narens, L. (1984), "Accuracy of Feeling-of-Knowing Judgments for Predicting Perceptual Identification and Relearning,"*Journal of Experimental Psychology: General,* **113,** pp. 282–300.

Neugebauer, S., Perkins, R. C., and Whitford, D. C. (1996), *First Stage Evaluations of the 1995 Census Test Administrative Records Database, Technical Report DMD 1995 Census Test Results Memorandum Series No. 41,* Washington D.C.: Bureau of the Census.

Newcombe, H. B. (1988), *Handbook of Record Linkage: Methods for Health and Statistical Studies,* Oxford: Oxford University Press.

Newcombe, H. B., Kennedy, J. M., Axford, S. J., and James, A. P. (1959), "Automatic Linkage Of Vital Records," *Science,* **130,** pp. 954–959.

Newell, A. and Simon, H. A. (1972), *Human Problem Solving,* Englewood Cliffs: Prentice Hall, pp. 65–66.

Newman, S. W. (1962), "Differences Between Early and Late Respondents to a Mailed Survey,"*Journal of Advertising Research,* **2,** pp. 37–39.

Newton, R. R., Prensky, D., and Schuessler, K. (1982), "Form Effect in the Measurement of Feeling States," *Social Science Research,* **11,** pp. 301 –317.

New York Times, (1992), "The AIDS 'Plot' Against Blacks," May 12, pp. A22.

Neyman, J. (1934), "On the Two Different Aspects of the Representative Method: The Method of Stratified Sampling and the Method of Purposive Selection," *Journal of the Royal Statistical Society, Series A,* **97,** pp. 558–606.

Nichols, E. and Sedivi, B. (1998), "Economic Data Collection Via the Web: A Census Bureau Case Study," *Proceedings of the Survey Research Methods Section, American Statistical Association,* 1998.

Nichols, E., Willimack, D. K., and Sudman, S. (1999a), "Who Are the Reporters: A Study of Government Data Providers in Large Multi-Unit Companies," *Proceedings of the Survey Research Methods Section, American Statistical Association,* forthcoming.

Nichols, E., Willimack, D. K., and Sudman, S. (1999b), "Balancing Confidentiality and Burden Concerns in Censuses and Surveys of Large Businesses," paper presented to the Washington Statistical Society, U.S. Bureau of the Census, Washington, D.C., September.

Nisbett, R. E. and Wilson, T. D. (1977), "Telling More Than We Can Know: Verbal Reports on Mental Processes," *Psychological Review,* **84,** pp. 231–259.

Nixon, M. G., Kalton, G., and Brick, M. (1996), Compensating for Missing Best Values in the NIPRCS, *Proceedings of the Section on Survey Research Methods, American Statistical Association,* pp. 347–353.

Nordheim, E. V. (1984), "Inference from Nonrandomly Missing Categorical Data: An Example From a Genetic Study on Turner's Syndrome," *Journal of the American Statistical Association,* **79,** pp. 772–780.

Norman, D. A. (1973), "Memory, Knowledge, and the Answering of Questions," in R. L. Solso (ed.), *Contemporary Issues in Cognitive Psychology: the Loyola Symposium,* Washington, D.C.: Winston.

Norman, D. A. (1990), *The Design of Everyday Things,* New York: Currency Doubleday.

Nuckols, R. C. (1949), "Verbi!," *International Journal of Opinion and Attitude Research,* **3,** pp. 575–586.

Office of Management and Budget (1999), "Implementing Guidance for OMB Review of Agency Information Collection," draft, June 2, Washington, D.C.: GPO.

Office of Management and Budget (1979), "Interim Guidelines for Controlling and Reducing the Burden of Federal Reporting and Recordkeeping Requirements on the Public and for Responding to Commission of Federal Paperwork Recommendations," Washington, D.C.: GPO.

Office of Management and Budget (1976), *Federal Statistics: Coordination, Standards, Guidelines.* Washington, D.C.: GPO.

Ogbu, J. U. (1990), "Minority Status and Literacy in Comparative Perspective," *Daedalus,* **119,** pp. 141–168.

Oh, H. L. and Scheuren, F. J. (1983), "Weighting Adjustment for Unit Nonresponse," in W. G. Madow, I. Olkin, D. B. Rubin (eds.), *Incomplete Data in Sample Surveys,* Vol. 2, New York: Academic Press, pp. 143–184.

Olkin, L. and Tate, R. F. (1961), "Multivariate Correlation Models with Mixed Disorder and Continuous Variables," *Annals of Mathematical Statistics*, **32,** pp. 448–465.

O'Neil, M. J. (1979), "Estimating the Nonresponse Bias Due to Refusals in Telephonesurveys," *Public Opinion Quarterly,* **43,** pp. 219–232.

Onyshekvych, V. and McIndoe, D. (1999), "Internet Technology: Gaining Commercial Advantage," paper presented at the 1999 AAPOR Conference, St. Petersburg, Florida. http://surveys. over. net/method/Onyshkevych. ZIP.

Orr, L. L., Bloom, H. S., Bell, S. H., Doolittle, F., Lin, W., and Cave, G. (1996), *Does Training for the Disadvantaged Work?* Washington, D.C.: The Urban Institute Press.

Osmint, J. B., McMahon, P. B., and Martin, A. W. (1994), "Response in Federally Sponsored Establishment Surveys," paper presented to the American Statistical Association, Toronto, August.

Osterman, P. (1994a), "How Common is Workplace Transformation and Who Adopts It?" *Industrial and Labor Relations Review,* **47,** pp. 173–188.

Osterman, P. (1994b), "Supervision, Discretion and Work Organization," *American Economic Review,* **84,** pp. 380–384.

Pace, R. C. (1939), "Factors Influencing Questionnaire Returns from Former University Students," *Journal of Applied Psychology,* **23,** pp. 388–397.

Page, B. and Shapiro, R. (1983), "Effects of Public Opinion on Policy," *American Political Science Review,* **77,** pp. 175–190.

Paik, M. C. (1997), The Generalized Estimating Equation Approach When Data are Not Missing Completely at Random," *Journal of the American Statistical Association,* **92,** pp. 1320–1329.

Patterson, R. E., Kristal, A. R., and White E. (1996), "Do Beliefs, Knowledge and Perceived Norms About Diet and Cancer Predict Dietary Change?" *American Journal of Public Health,* **86,** pp. 1394–1400.

Paxton, M. C., Dillman, D. A. and Tarnai, J. (1995), "Improving Response to Business Mail Surveys," in B. G. Cox, D. A. Binder, B. N. Chinnappa, A. Christianson, M. J. Colledge, and P. S. Kott (eds.), *Business Survey Methods,* New York: Wiley.

Payne, S. L. (1950), "Thoughts about Meaningless Questions," *Public Opinion Quarterly,* **14,** pp. 687–696.

Pearl, D. K. and Fairley, D. (1985), "Testing for the Potential for Nonresponse Bias in Sample Surveys," *Public Opinion Quarterly,* **49,** pp. 553–560.

Petty, R. E. and Cacioppo, J. T. (1986), *Communication and Persuasion: Central and Peripheral Routes to Attitude Change,* New York: Springer Verlag.

Pfeffermann, D. (1988), The Effect of Sampling Design and Response Mechanism on Multivariate Regression-Based Prediction," *Journal of the American Statistical Association,* **83,** pp. 824–833.

Pfeffermann, D. (1993), "The Role of Sampling Weights When Modeling Survey Data," *International Statistical Review,* **61,** pp. 317–337.

Pfeffermann, D. (1996), "The Use of Sampling Weights for Survey Data Analysis," *Statistical Methods in Medical Research,* **5,** pp. 221–230.

Pfeffermann, D., Skinner, C., Goldstein, H., Holmes, D. J. and Rasbash, J. (1998), "Weighting for Unequal Selection Probabilities in Multilevel Models (with Discussion)," *Journal of the Royal Statistical Society,* **60,** pp. 23–40.

Pietsch, L. (1995), "Profiling Businesses to Define Frame Units," in B. G. Cox, D. A. Binder, B. N. Chinnappa, A. Christianson, M. J. Colledge, and P. S. Kott (eds.), *Business Survey Methods,* New York: Wiley.

Pilon, T. L. and Craig, N. C. (1988), "Disks-by-Mail: a New Survey Modality," Proceedings of the 1988 Sawtooth Software Conference on Perpetual Mapping, Conjoint Analysis and Computer Interviewing, Sun Valley: Sawtooth Software.

Pitkow, J. E. and Kehoe, C. M. (1997), Comments Submitted to the Federal Trade Commission Workshop on Electronic Privacy. http://www.cc. gatech.edu/gvu/user_surveys/papers/1997-05-ftc-privacy-supplement.pdf

Poe, G. S., Seeman, I., McLaughlin, J., Mehl, E., and Dietz, M. (1988), "Don't Know Boxes in Factual Questions in a Mail Questionnaire: Effects on Level and Quality of Response," *Public Opinion Quarterly,* **52,** pp. 212–222.

Pohl, N. E. and Albert V. B. (1978), "Reducing Item-Specific Nonresponse Bias," *Journal of Experimental Education,* **46,** pp. 57–64.

Pondman, L. M. (1998), *The Influence of the Interviewer on the Refusal Rate in Telephone Surveys,* Deventer: Print Partners Ipskamp.

Porst, R. and von Briel, C. (1995), "Waren Sie Vielleicht Bereit, Sich Gegenebenfalls Noch Einmal Befragen Zu Lassen? Oder: Gründe Für Die Teilnahme An Panelbefragungen." *ZUMA-Arbeitsbericht,* Nr. 95/04. Mannheim, Germany.

PR Newswire (1998), Recent Research Confirms Online Surveys are a Viable Means of Reaching General Population, *PR Newswire,* **1,** September 17.

Prescott-Clarke, P. and Primatesta, P. (eds) (1998b), *Health Survey for England, The Health of Young People 1995–97 Volume 2: Methodology and Documentation,* London: The Stationery Office.

Press, S. J. and Yang, C. J. (1974), "A Bayesian Approach to Second Guessing 'Undecided' Respondents," *Journal of the American Statistical Association,* **69,** pp. 58–67.

Prescott-Clarke, P. and Primatesta, P. (eds) (1998a), "Health Survey for England, 1996," *Methodology and Documentation,* 2, London: The Stationary Office.

Presser, S. (1977), *Survey Question Wording and Attitudes in the General Public,* Ph. D. dissertation, Ann Arbor: University of Michigan.

Presser, S. (1984), "The Use of Survey Data in Basic Research in the Social Sciences," in C. F. Turner and E. M. (eds.), *Surveying Subjective Phenomena,* New York: Russell Sage Foundation.

Presser, S. (1989), "Collection and Design Issues," in D. Kaspryzk, G. Duncan, G. Kalton, and M. P. Singh (eds.), *Panel Surveys,* New York: Wiley.

Price, V. (1999), "Editorial Note," *Public Opinion Quarterly,* **63,** pp. i–ii.

Purcell, N. J. and Kish, L. (1980), "Postcensal Estimates for Local Areas (or Domains)," *International Statistical Review,* **48,** pp. 3–18.

Purdon, S., Campanelli, P., and Sturgis, P, (1999), "Interviewers' Calling Strategies on Face-to-Face Interview Surveys," *Journal of Official Statistics,* **15,** 2, pp. 199–216.

Quenouille, M. (1949), "Approximation Tests of Correlation in Time Series," *Journal of the Royal Statistical Society,* pp. 18–84.

Raghunathan, T. E., and Grizzle, J. E, (1995), "A Split Questionnnaire Survey Design," *Journal of the American Statistical Association,* **90,** pp. 54–63.

Raghunathan, T. E., Lepkowski, J., Van Hoewyk, J., and Solenberger, P. (1997), "A Multivariate Technique for Imputing Missing Values Using a Sequence of Regression Models," Technical Report, Survey Methodology Program, Survey Research Center, ISR, University of Michigan.

Raimond, T. and Hensher, D. A. (1997), "A Review of Empirical Studies and Applications," in T. F. Golob, R. Kitamura and L. Long (eds.), *Panels for Transportation Planning,* Boston: Kluwer Academic Publishers, pp. 15–72.

Ramirez, C. (1996), "Respondent Selection in Mail Surveys of Establishments: Personalization and Organizational Roles," *Proceedings of the Survey Research Methods Section, American Statistical Association,* pp. 974–979.

Ramos, M., Sedivi, B. M., and Sweet, E. M. (1998), "Computerized Self-Administered Questionnaires," in M. P. Couper, R. P. Baker, J. Bethlehem, C. Z. F. Clark, J. Martin, W. L. Nicholls II, and J. M. O'Reilly (eds.), *Computer Assisted Survey Information Collection,* New York: Wiley.

Rancourt, E. and Hidiroglou, M. (1998), "The Use of Administrative Records in the Canadian Survey of Employment, Payrolls, and Hours," *Proceedings of the Survey Methods Section, Statistical Society of Canada.*

Rancourt, E. (1999), "Estimation with Nearest Neighbour Imputation at Statistics Canada," *Proceedings of the Survey Research Methods Section,* American Statistical Association, pp. 131–138.

Rancourt, E., Lee, H., and Särndal, C-E. (1994), "Bias Corrections for Survey Estimates from Data with Ratio Imputed Values for Confounded Nonresponse," *Survey Methodology*, **20,** pp.137–147.

Rancourt, E., Sarndal, C-E. and Lee, H. (1994), "Estimation of the Variance in the Presence of Nearest Neighbour Imputation," *Proceedings of the Section on Survey Research Methods,* American Statistical Association, pp. 888–893.

Rao, J. N. K. (1996), "On Variance Estimation with Imputed Survey Data," *Journal of the American Statistical Association,* **91,** pp. 499–506.

Rao, J. N. K. and Scott, A. J. (1981), "The Analysis of Categorical Data from Complex Sample Surveys: Chi-Squared Tests for Goodness of Fit and Independence in Two-Way Tables," *Journal of the American Statistical Association,* **76,** 734, pp. 221–230.

Rao, J. N. K. and Shao, J. (1992), "Jackknife Variance Estimation With Survey Data Under Hot Deck Imputation,"*Biometrika,* **79,** pp. 811–822.

Rao, J. N. K. and Shao, J. (1996), "On balanced Half-Sample Variance Estimation in Stratified Sampling, *Journal of the American Statistical Association,* **91,** pp. 343–348.

Rao, J. N. K. and Shao, J. (1999), "Modified Balanced Repeated Replication for Complex Survey Data, *Biometrika,* **86,** 2, pp. 403–415.

Rao, J. N. K. and Thomas, D. R. (1989), "Chi-Squared Tests for Contingency Tables," in C. J. Skinner, D. Holt, and T. M. F. Smith (eds.), *Analysis of Complex Survey Data,* pp. 89–114, New York: Wiley.

Rao, J. N. K. and Wu, C. F. J. (1988), "Resampling Inference with Complex Survey Data," *Journal of the American Statistical Association,* **83,** pp. 231–241.

Rao, J. N. K. and Sitter, R. R. (1995), "Variance Estimation Under Two-phase Sampling with Application to Imputation for Missing Data," *Biometrika,* **82,** pp. 453–460.

Rapoport, R. B. (1981), "The Sex Gap in Political Persuading: Where the 'Structuring Principle' Works,"*American Journal of Political Science,* **25,** pp. 32–48.

Rapoport, R. B. (1982), "Sex Differences in Attitude Expression: a Generational Explanation," *Public Opinion Quarterly,* **46,** pp. 86–96.

Rasbash, J. and Goldstein, H. (1994), "Efficient Analysis of Mixed Hierachical and Cross-Classified Random Structures Using A Multilevel Model," *Journal of Educational and Behavioral Statistics,* **19,** pp. 337–350.

Rawson, N. S. B. and D'Arcy, D. (1998), "Assessing the Validity of Diagnostic Information in Administrative Health Care Utilization Data: Experience in Saskatchewan," *Pharmacoepidemiology and Drug Safety,* **7,** pp. 389–398.

Reder, L. M. (1988), "Strategic Control of Retrieval Strategies." *The Psychology of Learning and Motivation,* **22,** pp. 227–259.

Redfern, P. (1989), "European Experience of Using Administrative Data for Censuses of Population: the Policy Issues that Must Be Addressed," *Survey Methodology,* **15,** pp. 83–99.

Redline, C. and Crowley, M. (1999), Unpublished Data, U.S. Bureau of the Census, Washington, D.C.

Reid, S. (1942), "Respondents and Nonrespondents to Mail Questionnaires," *Educational Research Bulletin,* **21,** pp. 87–96.

Reilly, C. and Gelman, A. (1999), "Post-stratification without population level information on the post-stratifying variable, with application to political polling,"

Renssen, R. H., Nieuwenbroek, N. J., and Slootbeek, G. T. (1997), "Variance Module in Bascula 3.0, Theoretical Bbackground" Research Paper 9712, Statistics Netherlands, Department for Statistical Methods, Voorburg, The Netherlands.

Rickard, W. (1999), "Pulling the Plug on Spam," *On the Internet,* May/June, 6.

Ribisl, K. M., Walton, M. A., Mowbray, C. T., Luke, D. A., Davidson, W. S., and Bootsmiller, B. J. (1996), "Minimizing Participant Attrition in Panel Studies Through the Use of Effective Retention and Tracking Strategies: Review and Recommendations," *Evaluation and Program Planning,* **19**.

RINE (1999), "Research on Internet in New Europe," http.//www.rine.org.

RIS (1996–1999), "RIS—Research on Internet in Slovenia," http://www.ris.org.

RIS (1999–2000), "Web Surveys Methodology," http://surveys. over. net/method.

Robins, J. M. (1997), "Non-Response Models for the Analysis of Non-Monotone Non-Ignorable Missing Data," *Statistics in Medicine,* **16,** pp. 21–37.

Robins, J. M. and Rotninsky, A. (1995), Semiparametric Effeciency in Multivariate Regression Models with Missing Data," *Journal of the American Statistical Association,* **90,** pp. 122–129.

Robins, J. M. and Wang, N. (2000), "Inference for Imputation Estimators,"*Biometrika,* **100,** forthcoming.

Robinson J. G., Ahmed, R., Das Gupta, and Woodrow, K.A. (1993), "Estimation of Population Coverage in the 1990 United States Census Based on Demographics Analysis," *Journal of the American Statistical Association*, **88,** pp. 1061–1071.

Rockwell, D. H. et al. (1961), "The Tuskegee Study of Untreated Syphilis: The 30th Year of Observation," *Archives of Internal Medicine*, **114,** pp. 792–798.

Roeher, G. A. (1963), "Effective Techniques in Increasing Response to Mailed Questionnaires," *Public Opinion Quarterly,* **27,** pp. 299–302.

Rogers, R. G., Carrigan, J. A., and Kovar, M. G. (1997), "Comparing Mortality Estimates Based on Different Administrative Records," *Population Research and Policy Review,* **16,** pp. 213–224.

Rogers, T. F. (1976)," Interviews by Telephone and in Person: Quality of Response and Field Performance," *Public Opinion Quarterly,* **40,** pp. 51–65.

Rommetveit, R. (1974), *On Message Structure: A Framework for the Study of Language and Communication,* New York: Wiley.

Ronitzky, A. and Robins, J. (1997), "Analysis of Semi-parametric Regression Models with Non-ignorable Nonresponse," *Statistics in Medicine,* **16,** pp. 81–102.

Ronitzky, A., Robins, J. M., and Scharfstein, D. O. (1998), "Semiparametric Regression for

Repeated Outcomes with Nonignorable Nonresponse," *Journal of the American Statistical Association,* **93,** pp. 1321–1339.

Rorschach, H. (1942), *Psychodiagnostics, a Diagnostic Test Based on Perception* (Translated by P. Lemkau and B. Kronbenerg), Bern: Huber.

Rosenbaum, P. R. and Rubin, D. B. (1983), "The Central Role of the Propensity Score in Observational Studies for Casual Effects," *Biometrica,* **70,** pp. 41–55.

Rosenbaum, P. R. and Rubin, D. B. (1984), "Reducing Bias in Observational Studies Using Subclassification on the Propensity Score," *Journal of the American Statistical Association,* **79,** pp. 516–524.

Rosenbaum, P. R. (1995), "Quantiles in Nonrandom Samples and Observational Studies," *Journal of the American Statistical Association,* **90,** pp. 1424–1431.

Rosenberg, N., Izard, C. E., and Hollander, E. P. (1955), Middle Category Response: Reliability and Relationship to Personality and Intelligence Variables,"*Educational and Psychological Measurement,* **15,** pp. 281–290.

Rowland, M. L. and Forthofer, R. N. (1993), "Investigation of Nonresponse Bias: Hispanic Health and Nutrition Examination Survey," *Vital and Health Statistics,* **2,** National Center for Health Statistics, p. 119.

Royce, D., Hardy, F., and Beelen, G. (1997), "Project to Improve Provincial Economic Statistics," Proceedings of the Statistics Canada International Symposium, Statistics Canada, Ottawa, Ontario.

Rubin, D. B. (1976), "Inference and Missing Data," *Biometrika,* **63,** pp. 581–590.

Rubin, D. B. (1977), "Formalizing Subjective Notions about the Effect of Nonrespondents in Sample Surveys," *Journal of the American Statistical Association,* **72,** pp. 538–543.

Rubin, D. B. (1978), "Multiple Imputations in Sample Surveys—A Phenomenological Bayesian Approach to Nonresponse," *Proceedings of the Survey Research Methods Section, American Statistical Association,* pp. 20–34.

Rubin, D. B. (1984), "Bayesianly Justifiable and Relevant Frequency Calculations for the Applied Statistician," *Annals of Statistics,* **12,** pp. 1151–1172.

Rubin, D. B. and Schenker, N., (1986), "Multiple Imputation for Interval Estimation from Simple Random Samples with Ignorable Nonresponse," *Journal of the American Statistical Association,* **81,** pp. 366–374.

Rubin, D. B. (1987), *Multiple Imputation for Nonresponse in Surveys,* New York: Wiley.

Rubin, D. B. (1996), "Multiple Imputation After 18+ Years," *Journal of the American Statistical Association,* **91,** pp. 473–489.

Rudolph, B. A. and Greenberg, A. G. (1994), "Surveying of Public Opinion: The Changing Shape of an Industry," NORC report to the Office of Technology Assessment, Chicago.

Sailer, P., Weber, M., and Yau, E. (1993), "How Well Can IRS Count the Population?" *Proceedings of Government Statistics Section, American Statistical Association,* pp. 138–142.

Sanchez, M. E. and Morchio, G. (1992), "Probing 'Don't Know' Answers," *Public Opinion Quarterly,* **56,** pp. 454–474.

Saris, W. E. (1998), "Ten Years of Interviewing without Interviewers: The Telepanel," in M. P. Couper, R. P. Baker, J. Bethlehem, C. Z. F. Clark, J. Martin, W. L. Nicholls II, and J. M. O'Reilly (eds.), *Computer Assisted Survey Information Collection,* New York: Wiley.

Särndal, C. E. and Swensson, B. (1987), "A General View of Estimation for Two Phases of

Selection with Applications to Two-Phase Sampling and Nonresponse," *International Statistical Review,* **55,** 279–294.

Särndal, C-E. (1990), "Methods for Estimating the Precision of Survey Estimates when Imputation has been Used," Proceedings of Statistics Canada Symposium Canada: Measurement and Improvement of Data Quality, *Statistics Canada*, pp. 337–347.

Särndal, C-E. (1992), "Methods for Estimating the Precision of Survey Estimates when Imputation has been Used," *Survey Methodology*, **18,** pp. 241–252.

Särndal, C., Swensson, B., and Wretman, J. (1992), *Model Assisted Survey Sampling,* New York: Springer-Verlag.

SAS Institute, (1992), "The Mixed Procedure in SAS/STST Software, Chages and Enhancements," Release 6. 07, Technical Report P–229, SAS Institute, Inc., Cary, North Carolina.

Schacter, D. L. (1983), "Feeling of Knowing in Episodic Memory,"*Journal of Experiment Psychology: Learning, Memory, and Cognition,* **9,** pp. 39–54.

Schaeffer, N. C. and Bradburn, N. M. (1989), "Respondent Behavior in Magnitude Estimation," *Journal of the American Statistical Association,* **84,** pp. 402–413.

Schaefer, D. R. and Dillman, D. A. (1998), "Development of a Standard E-Mail Methodology: Results of an Experiment," *Public Opinion Quarterly,* **62,** pp. 378–397. http://survey. sesrc. wsu. edu/dillman/papers/E-Mailppr. pdf.

Schaeffer, N. C. (1980), "Evaluating Race of Interviewer Effects in a National Survey," *Sociological Methods and Research,* **8,** pp. 400–419.

Schafer, J. L., Ezzati-Rice, T. M., Johnson, W., Khare, M., Little, R., and Rubin, D. (1996), "The NHANES III Multiple Imputation Project," *Proceedings of the Survey Research Methods Section, American Statistical Association,* pp. 28–37.

Schafer, J. L. (1999), "Models and Software for Multiple Imputation," Presentation at the 1999 Joint Statistical Meetings in Baltimore.

Schafer, J. L. (1997), "Analysis of Incomplete Multivariate Data," London: Chapman & Hall.

Schafer, J. L. and Schenker, N. (2000), "Inference with Imputed Conditional Means," *Journal of the American Statistical Association,* **95,** in press.

Schafer, J. L., Ezzati-Rice, T. M., Johnson, W., Khare, M., Little, R. J. A., and Rubin, D. B. (1996), "The NHANES III Multiple Iimputation Project," *Proceedings of the Survey Research Methods Section of the American Statistical Association,* pp. 28–37.

Schafer, J. L., Khare, M., and Ezzati-Rice, T. M. (1993), "Multiple Imputation of Missing Data in NHANES III, *Proceedings of the U.S. Bureau of the Census Annual Research Conference,* pp. 459–487, Bureau of the Census, Washington, D.C.

Scharfstein, D. O., Rotnitsky, A., and Robins, J. M., (1999), "Adjusting for Non-Ignorable DropOut Using SemiParametric Models, (with discussion)," *Journal of the American Statistical Association,* **94,** pp. 1096–1146.

Schechter, B., Sykes, J., and DiCarlo, J. (1997), *Creation of the January 1994 MATH SIPP Microsimulation Model and Database.*Washington, D.C.: Mathematica Policy Research, Inc.

Schlenker, B. R. (1986), "Self-Identification: Toward an Integration of the Private and Public Self," In R. Baumeister (ed.), *Public Self and Private Self,* New York: Springer Verlag.

Schenker, N. and A. H. Welsh (1988), "Asymptotic Results for Multiple Imputation,"*Annals of Statistics,* **16,** pp. 1550–1566.

Schmiedeskamp, J. W. (1962), "Reinterviews by Telephone," *Journal of Marketing,* **26,** pp. 28–34.

Schnell, R. (1997), *Nonresponse in Bevölkerungsumfragen, Ausmaß, Entwicklung und Ursachen* [In German: Nonresponse in Sample Surveys], Opladen: Leske & Budrich.

Schuman, H. and Presser, S. (1981), *Questions and Answers in Attitude Surveys: Experiments on Question Form, Wording, and Context,* New York: Academic Press.

Schwarz, N. and Clore, G. L. (1996), "Feelings and Phenomenal Experiences," in E. T. Higgins and A. Kruglanski (eds.), *Social Psychology: Handbook of Basic Principles,* New York: Guilford, pp. 433–465.

Schwarz, N., Grayson, C. E., and Knauper, B. (1998), "Formal Features of Rating Scales and the Interpretation of Question Meaning," *International Journal of Public Opinion Research,* **10,** pp. 177–183.

Schwarz, N. and Sudman, S. (1994), *Autobiographical Memeory and the Validity of Reptrospective Reports,* New York: Springer Verlag.

Schwarz, N. and Sudman, S. (1996), *Answering Questions: Methodology for Determining Cognitive and Communicative Processes in Survey Research,* San Francisco: Jossey-Bass.

Scott, C. (1961), "Research on Mail Surveys," *Journal of the Royal Statistical Society,* **124,** pp. 143–205.

Scott, A. J. and C. J. Wild (1986), Fitting Logistic Models under Case-control or Choice Based Sampling. *Journal of the Royal Statistical Society,* B, **48,** pp. 170–182.

Scott, A. J. and Holt, D. (1982), "The Effect of Two-Stage Sampling on Ordinary Least-Squares Methods," *Journal of the American Statistical Association,* **77,** pp. 848–854.

Scott, A. J. and Knott, M. (1976), "An Approximate Test for Use with AID," *Applied Statistics,* **25,** pp. 103–106.

Shah, B. V., Barnwell, B. G., and Bieler, G. S. (1996), *SUDAAN User's Manual, Version 6. 4* (2nd eds), Research Triangle Park, NC: Research Triangle Institute.

Shah, B. V., Barnwell, B. G., and Bieler, G. S. (1997), *SUDAAN User's Manual, Release 7. 5.* Research Triangle Park, NC: Research Triangle Institute.

Shao, J. and Chen, Y. (1999), "Approximate Balanced Half Samples and Related Replication Methods for Imputed Survey Data," *Sankhya,* B, Special Issue on Sample Surveys, pp. 187–201.

Shao, J. and Sitter, R. R. (1996), Bootstrap for Imputed Survey Data, *Journal of the American Statistical Association,* **91,** pp. 1278–1288.

Shao, J. and Steel, P. (1999), "Variance Estimation for Survey Data with Composite Imputation and Nonnegligible Sampling Fractions," *Journal of the American Statistical Association,* **94,** pp. 254–265.

Shao, J. and Wang, H. (1999), "Sample Correlation Coefficient Based on Survey Data under Regression Imputation," Unpublished manuscript.

Shao, J., Chen, Y., and Chen, Y. (1998), Balanced Repeated Replication for Stratified Multistage Survey Data Under Imputation," *Journal of the American Statistical Association,* **93,** pp. 819–831.

Shatos, R., Moore, D., and Dillman, D. A. (1998), "Establishment Surveys: The Effect of

Multi-Mode Sequence on Response Rate," *Proceedings of the Survey Research Methods Section, American Statistical Association,* pp. 981–987.

Sheehan, K. B. and Hoy, M. G. (1999), "Using E-mail to Survey Internet Users in the United States: Methodology and Assessment, *Journal of Computer Mediated Communication,* **4,** p. 3. http://209. 130. 1. 169/jcmc/vol4/issue3/sheehan. html.

Shettle, C. and Mooney, G. (1999), "Monetary Incentives in Government Surveys," *Journal of Official Statistics,* **15,** pp. 231–250.

Shettle, C., Guenther, P., Kaspryzk, D., and Gonzalez, M. E. (1994), "Investigating Nonresponse in Federal Surveys," *Proceedings of the Survey Research Methods Section, American Statistical Association,* Washington, D.C.

Shuttleworth, F. K. (1940), "Sampling Errors Involved in Incomplete Returns to Mail Questionnaires," *Psychological Bulletin,* **37,** pp. 437.

Siegel, P. M. (1995), "Developing Postcensal Income and Poverty Estimates for All U. S. Counties," *Proceedings of the Government Statistics Section, American Statistical Association,* Alexandria, Virginia, pp. 166–171.

Siemiatycki, J. (1979), "A Comparison of Mail, Telephone, and Home Interview Strategies for Household Health Surveys,"*American Journal of Public Health,* **69,** pp. 238–245.

Sigelman, C. K., Winer, J. L., and Schoenrock, C. J. (1982), "The Responsiveness of Mentally Retarded Persons to Questions," *Education and Training of the Mentally Retarded,* **17,** pp. 120–124.

Singer, E., Groves, R. M., and Corning, A. D. (1999a), "Differential Incentives: Beliefs About Practices, Perceptions of Equity, and Effects on Survey Participation," *Public Opinion Quarterly,* **63,** pp. 251–260.

Singer, E. and Kohnke-Aguirre, L. (1979), "Interviewer Expectation Effects: A Replication and Extension." *Public Opinion Quarterly,* **43,** pp. 245–260.

Singer, E., Van Hoewyk, J., and Maher, M. P. (1998), "Does the Payment of Incentives Create Expectation Effects?" *Public Opinion Quarterly,* **62,** pp. 152–164.

Singer, E., Van Hoewyk, J., and Maher, M. P. (2000), "Experiments with Incentives in Telephone Surveys." *Public Opinion Quarterly,* **64,** pp. 171–88.

Singer, E., Mathiowetz, N. A., and Couper, M. P. (1993), "The Impact of Privacy and Confidentiality Concerns on Survey Participation: The Case of the 1990 U. S. Census," *Public Opinion Quarterly,* **57,** pp. 465–482.

Singer, E, Van Hoewyk, J., and Gebler, N. (1999b), "The Effect of Incentives on Response Rates in Interviewer-Mediated Surveys," *Journal of Official Statistics,* **15,** pp. 217–230.

Singer, E., Frankel, M. R., and Glassman, M. B. (1983), "The Effect of Interviewers' Characteristics and Expectations on Response," *Public Opinion Quarterly,* **47,** pp. 68–83.

Sirken, M. (1975), *Evaluation and Critique of Household Sample Surveys of Substance Abuse. in Alcohol and Other Drug Use and Abuse in the State of Michigan* (Final report), prepared by the Office of Substance Abuse Services, Michigan Department of Public Health.

Sitter, R. R. (1992), "A Resampling Procedure for Complex Survey Data," *Journal of the American Statistical Association,* **87,** pp. 755–765.

Sitter, R. R. (1993), "Balanced Repeated Replications Based on Orthogonal Multi-Arrays,"*Biometrika,* **80,** 211–221.

Sitter, R. R. and Rao, J. N. K. (1997), "Imputation for Missing Values and Corresponding Variance Estimation," *Canadian Journal of Statistics,* **25,** pp. 61–73.

Skinner, C. J. and Rao, J. N. K. (1999), "Jackknife Variance Estimation for Multivariate Statistics under Hot Deck Imputation. *Journal of Statistical Planning and Inferences,* forthcoming.

Skinner, C. J. (1986), "Design Effects of Two-Stage Sampling," *Journal of the Royal Statistical Society, Series B,* **48,** pp. 89–99.

Skinner, C. J. (1989), "Introduction to Part A," in C. J. Skinner, D. Holt, and T. M. F. Smith (eds.) *Analysis of Complex Survey Data,* New York: Wiley, pp. 23–58.

Skinner, C. J. (1991), "On the Efficiency of Raking Ratio Estimation for Multiple Frame Surveys," *Journal of the American Statistical Association,* **86,** pp. 779–784.

Small, K. (1992), *Urban Transportation Economics,* Switzerland: Harwood Academic Publishers.

Smith, C. B. (1997), "Casting the Net: Surveying an Internet Population," *Journal of Computer Mediated Communication,* **3,** p. 1. http://www.ascusc.org/jcmc/vol3/issue1/smith.html.

Smith, P. B. and Bond, M. H. (1998), *Social Psychology Across Cultures,* 2nd ed., London: Prentice Hall Europe.

Smith, T. (1984), "Nonattitudes: A Review and Evaluation," in Turner, C. and Martin, E. (eds.), *Surveying Subjective Phenomena,* Vol. 2, New York: Russell Sage, pp. 215–255.

Smith, T. M. F. (1994)," Sample Surveys 1975–1990," in *Analysis of Complex Surveys,* New York: Wiley.

Smith, T. W. (1983), "The Hidden 25 Percent: An Analysis of Nonresponse on the 1980 General Social Survey," *Public Opinion Quarterly,* **47,** pp. 386–404.

Smith, T. W. (1995), "Trends in Non-Response Rates," *International Journal of Public Opinion Research,* **7,** pp. 157–171.

Smith, T. W. (1984), "Estimating Nonresponse Bias with Temporary Refusals," *Sociological Perspectives,* **27,** pp. 473–489.

Smith, T. W., (1999), "Designing Nonresponse Standards," Paper presented to the International Conference on Survey Nonresponse, Portland, Oregon, November.

Snijkers, G., Hox, J. J., and de Leeuw, E. D. (1999), "Interviewers' Tactics For Fighting Survey Nonresponse," *Journal of Official Statistics,* **15,** pp. 185–198.

SOLAS for Missing Data Analysis 1.0 (1997), Saugas, MA: Statistical Solutions.

Spaeth, J. L. and O'Rourke, D. P. (1994), "Designing and Implementing the National Organizations Study,"*American Behavioral Scientist,* **37,** pp. 872–890.

Spain, S. W. (1998), "Top–10 Web Survey Issues and How to Address Them," *Market Research Library.* http://www.researchinfo.com/library/top_10_web.shtml.

Spiegelhalter, D., Thomas, A., Best, N., and Gilks, W. (1996), *Bayesian Inference Using Gibbs Sampling Manual (Version ii).*Cambridge: MRC Biostatistics Unit, Institute of Public Health.

Stata Corporation, (2001), *Stata Reference Manual*, College Station Texas: Stata Press.

Statistics Netherlands (1998), *Integration of Household Surveys; Design, Advantages, Methods.* Netherlands Official Statistics, Vol. 13, Special Issue, Statistics Netherlands, Voorburg, The Netherlands.

Steeh, C. G. (1981), "Trends in Nonresponse Rates, 1952–1979," *Public Opinion Quarterly,* **45,** pp. 40–57.

Steeh, C., Kirgis, N., Cannon, B., and DeWitt, J. (1999), "Are They Really as Bad as They Seem? Nonresponse Rates at the End of the Twentieth Century," paper presented at the International Conference on Survey Nonresponse, Portland, Oregon, October 28–31.

Steel, D. and Boal, P. (1988), "Accessibility by Telephone in Australia: Implications for Telephone Surveys," *Journal of Official Statistics,* **4,** pp. 285–297.

Steel, P. and Fay, R. E. (1995), "Variance Estimation for Finite Populations with Imputed Data," *Proceedings of the Section on Survey Research Methods, American Statistical Association*, **94,** pp. 374-379.

Stinchcombe, A. L., Jones, C., and Sheatsley, P. (1981), "Nonresponse Bias for Attitude Questions," *Public Opinion Quarterly,* **45,** pp. 359–375.

Suchman, E. A. (1950), "The Intensity Component in Attitude and Opinion Research," in S. A. Stouffer, L. Guttman, E. A. Suchman, P. F. Lazarsfeld, S. A. Star, and J. A. Clausen (eds.), *Measurement and Prediction,* Princeton, NJ: Princeton University Press, pp. 213–276.

Suchman, E. A. and McCandless, B. (1940), "Who Answers Questionnaires," *Journal of Applied Psychology,* **24,** pp. 758–769.

Suchman, L. and Jordan, B. (1990), "Interactional Troubles in Face-toFace Survey Interviews." *Journal of the American Statistical Association,* **85,** pp. 232–241.

Sudman, E. (1966), "Probability Sampling with Quotas," *Journal of the American Statistical Association,* **61,** pp. 749–771.

Sudman, S. Bradburn, N. and Schwartz, N. (1996), *Thinking About Answers,* San Francisco: Jossey-Bass.

Sudman, S., Bradburn, N., Blair, E. and Stocking, C. (1977), "Modest Expectations: The Effects of Interviewer Prior Expectations on Responses," *Sociological Methods and Research*, **6,** pp. 171–182.

Sugiyama, M. (1992), "Responses and Nonresponses," in L. Lebart (ed), *Quality of Information in Sample Surveys,* Paris: Dunod.

Tambor, E., Chase, G., Faden, R., Geller, G., Hofman, K., and Holzman, N. (1993), "Improving Response Rates Through Incentive and Follow-Up: The Effect of a Survey of Physicians' Knowledge of Genetics,"*American Journal of Public Health,* **83,** pp. 1599–1603.

Tanner, M. A., (1996), "Tools for Statistical Inference: Methods for the Exploration of Posterior Distributions and Likelihood Functions, 3rd ed., New York: Springer-Verlag.

Tanner, M. A. and Wong, W. H., (1987), "The Calculation of Posterior Distributions by Data Augementation," *Journal of the American Statistical Association,* **82,** pp. 528–550.

Terhanian, G. and Black, G. S. (1999), Understanding the Online Population: Lessons from the Harris Poll and the Harris Poll Online, ARF's Online Research Day—Towards Validation, New York: Advertising Research Foundation, pp. 28–33.

Terrell, F. and Terrell, S. L. (1981), "An Inventory to Measure Cultural Mistrust among Blacks," *Western Journal of Black Studies,* **3,** pp. 180–185.

Thomas, S. B. and Quinn, S. C. (1991), "The Tuskegee Syphilis Study, 1932–1972: Implications for HIV Education and AIDS Risk Education Programs in the Black Community," *American Journal of Public Health*, **81,** pp. 1498–1504.

Thompson, M. E. (1997)," Theory of Sample Surveys, London: Chapman Hall.

Thomsen, I. (1973), "A Note on the Efficiency of Weighting Subclass Means to Reduce the Effects of Non-Response When Analyzing Survey Data," *Statistisk Tidskrift,* **11,** pp. 278–283.

Thomsen, I. B. and Siring, E. (1983), "On the Causes and Effects of Nonresponse: Norwegian Experiences," in W. G. Madow and I. Olkin (eds.), *Incomplete Data in Sample Surveys,* Vol. 3, New York: Academic Press.

Thurston, S. W. and Zaslavsky, A. M. (1996), "Variance Estimation in Microsimulation Models of the Food Stamp Program," *Proceedings of the Social Statistics Section, American Statistical Association,* pp. 4–9.

Titterington, D. M. and Sedransk, J. (1986), "Matching and Linear Regression Adjustment in Imputation and Observational Studies," *Sankhya, The Indian Journal of Statistics,* Series B, **48,** pp. 347–367.

Tobin, J. (1958), "Estimation of Relationships for Limited Dependent Variables," *Econometrica,* **26,** pp. 24–36.

Tollefson, M. and Fuller, W. A. (1992), "Variance Estimation for Samples with Random Imputation," *Proceedings on the Section on Survey Research Methods, American Statistical Association*, pp. 758–763.

Tomaskovic-Devey, D., Leiter, J., and Thompson, S. (1994), "Organizational Survey Nonresponse,"*Administrative Science Quarterly,* **39,** pp. 439–457.

Tomaskovic-Devey, D., Leiter, J., and Thompson, S, (1995), "Organizational Survey Nonresponse," *Administrative Science Quarterly,* **39,** pp. 439–457.

Tourangeau, R. (1992), "Attitudes as Memory Structures: Belief Sampling and Context Effects," in N. Schwartz and S. Sudman (eds.), *Context Effects in Social and Psychological Research,* New York: Springer Verlag.

Tourangeau, R. (1984), "Cognitive Science and Survey Methods," in T. Jabine et al. (eds),*Cognitive Aspects of Survey Methodology: Building a Bridge Between Disciplines,* Washington, D.C.: National Academy Press.

Tourangeau, R., and Rasinski, K. A. (1988), "Cognitive Processes Underlying Context Effects and Attitude Measurement. "*Psychological Bulletin,* **103,** pp. 299–314.

Tourangeau, R. Shapiro, G., Kearney, A., and Ernst, L. (1997), "Who Lives Here? Survey Undercoverage and Household Roster Questions," *Journal of Official Statistics,* **13,** pp. 1–18.

Tourangeau, R., Rips, L. J., and Rasinski, K. (2000), *The Psychology of Survey Response,* Cambridge, UK: Cambridge University Press.

Train, K. (1986), *Qualitative Choice Analysis: Theory, Econometrics, and an application to Automobile Demand"* Cambridge, MA: MIT Press.

Traugott, M. W. (1987), "The Importance of Persistence in Respondent Selection for Preelection Surveys," *Public Opinion Quarterly,* **51,** pp. 48–57.

Tremblay, V. (1986), "Practical Criteria for Definition of Weighting Classes," *Survey Methodology,* **12,** pp. 85–97.

Triandis, H. C. (1994), *Culture and Social Behavior,* New York: McGraw-Hill.

Triandis, H. C. (1996), "The Psychological Measurement of Cultural Syndromes," *American Psychologist,* **51,** pp. 407–415.

Triandis, H. C., Marin, G., Hui, C. H., Lisansky, J., and Ottati, V. (1984), "Role Perceptions of Hispanic Young Adults," *Journal of Personality and Social Psychology,* **47,** pp. 1363–1374.

Tuckel, P., and O'Niell, H., (1995), "A Profile of Answering Machine Owners and Screeners," *Proceedings of the Section on Survey Research Methods," American Statistical Association,* pp. 1157–1162.

Tulp, D. R. and Kusch, G. L. (1993), "Nonresponse Study: Mandatory vs. Voluntary Reporting in the PACE Survey," paper presented at the International Conference on Establishment Surveys, Buffalo, New York.

Turner, A. G. (1982), "What Subjects of Survey Research Believe About Confidentiality," in J. E. Sieber (ed.), *The Ethics of Social Research: Surveys and Experiments,* New York: Springer Verlag.

Turner, C. F., Lessler, J. T., and Gfroerer, J. (1992), *"Survey Measurement of Drug Use: Methodological Studies,"* Rockville, MD: National Institute on Drug Abuse.

Turner, C. F., Lessler, J. T., George, B. J., Hubbard, M. L., and Witt, M. B. (1992), "Effects of Mode of Administration and Wording on Data Quality," in C. F. Turner, J. T. Lessler, and J. C. Gfroerer (eds.), *Survey Measurement of Drug Use Methodological Studies,* Washington, D.C.: National Institute of Drug Abuse, U.S. Department of Health and Human Services, pp. 221–243.

Turner, C. F. and Martin, E. (1984), *Surveying Subjective Phenomena,* New York: Russell Sage Foundation.

Turner, P. A. (1993), *I Heard it Through the Grapevine,* Berkeley: University of California Press.

Tuten, T. L. (1997), "Getting A Foot in the Electronic Door: Understanding Why People Read or Delete Electronic Mail," *ZUMA—Arbeitsbericht,* **8.**

U.S. Bureau of the Census (1978), "The Current Population Survey: Design and Methodology," *Technical Report, Technical Paper 40,* U.S. Government Printing Office.

Vacca, E. A., Mulry, M., and Killion, R. A. (1996), "The 1995 Census Test: a Compilation of Results and Decision," Technical Report DMD 1995 Census Test Results Memorandum # 46, United States Department of Commerce, Bureau of the Census.

Vaillancourt, P. M. (1973), "Stability of Children's Survey Responses," *Public Opinion Quarterly,* **37,** pp. 373–387.

Van Goor, H. and Stuiver, B. (1998), "Can Weighting Compensate for Nonresponse Bias in a Dependent Variable: an Evaluation of Weighting Methods to Correct for Substantive Bias in a Mail Survey Among Dutch Municipalities," *Social Science Research,* **27,** pp. 481–499.

Van Grol, H. J. M. (1997), "Evaluating the Use Of Induction Loops For Travel Time Estimation," Presented at 8th IFAC/IFIP/IFORS Symposium on Transportation Systems, Chania, Greece, June pp. 16–18.

van Leeuwen, R. and de Leeuw, E. (1999), I Am Not Selling Anything: Experiments in Telephone Introductions, paper presented at the International Conference on Survey Nonresponse, Portland, Oregon.

Vehovar, V. (1999), "Field Substitution and Unit Non-response," *Journal of Official Statistics,* **15,** pp. 335–350.

Vehovar, V., Batagelj, Z., and Lozar, K. (1999), "Self-selected Web Surveys: Can the Weighting Solve the Problem?" paper presented at the 1999 AAPOR Conference, St. Petersburg, Florida. http://www.ris. org/si/ris98/aapor99.html.

Vehovar, V., Lozar Manfreda, K., and Batagelj., Z. (2000), "Sensitivity of E-Commerce Measurement to Survey Instrument," In Thirteenth Bled Electronic Commerce Conference, Bled, Slovenia, June 19–20, 2000, *The End of the Beginning: Proceedings,* S. Klein, B. O'Keefe, J. Gricar, and M. Podlogar (eds.), Kranj: Moderna Organizacija, pp. 528–543.

Velasco, A. (1992), "Ethnographic Evaluation of the Behavioral Causes of Undercount in the Community of Sherman Heights, San Diego, California," Ethnographic Evaluation of the 1990 Census, Report #22, prepared under Joint Statistical Agreement 89–42 with the Bureau of the Census, Washington, D.C.

Venter, P. and Prinsloo, M. (1999), "The Internet and the Changing Role of Market Research," *Proceedings of the ESOMAR Worldwide Internet Conference Net Effects,* Amsterdam: ESOMAR, pp. 215–227.

Verma, R. B. P. and Parent, P. (1985), "An Overview of The Strengths and Weaknesses of the Selected Administrative Data Files," *Survey Methodology,* **11,** pp. 171–179.

Vernon, S. W., Roberts, R. E., and Lee, E. S. (1984), "Ethnic Status and Participation in Longitudinal Health Surveys," *American Journal of Epidemiology,* **119,** pp. 99–113.

Visser, P. S., Krosnick, J. A., Marquette, J., and Curtin, M. (2000), "Improving Election Forecasting: Allocation of Undecided Respondents, Identification of Likely Voters, and Response Effects," in P. Lavarkas and M. Traugott (eds.), *Election Polls, the News Media and Democracy,* Chatham House.

Voss, D. S., Gelman, A., and King, G. (1995), "Pre-Election Survey Methodology: Details from Nine Polling Organizations, 1988 and 1992," *Public Opinion Quarterly,* **59,** pp. 98–132.

Waien, S. A. (1997), "Linking Large Administrative Databases: A Method for Conducting Emergency Medical Services Cohort Studies Using Existing Data," *Academic Emergency Medicine,* **4,** pp. 1087–1095.

Wakim, A. (1987), *Evaluation of Coverage and Response in the Manufacturers' Shipments, Inventories, and Orders Survey,* Washington, D.C.: U.S. Bureau of the Census.

Waksberg, J. (1985), "Comments: 'Some Research Issues in Random-Digit-Dialing Sampling and Estimation,'" *Proceedings of the Bureau of the Census Annual Research Conference,* Washington, D.C.: U.S. Bureau of the Census, pp. 87–92.

Wallschlaeger, C. and Busic-Snyder, C. (1992), *Basic Visual Concepts and Principles,* Dubuque, Iowa: Wm. C. Brown Publishers.

Wardman, M. (1998), The Value of Travel Time: A Review of British Evidence,"*Journal of Transportation Economics and Policy,* **32,** pp. 285–316.

Warriner, K., Goyder, J., Gjertsen, H., Hohner, P., and McSpurren, K. (1996), "Charities, No, Lotteries, No, Cash, Yes: Main Effects and Interactions in a Canadian Incentives Experiment," paper presented at the Survey Non-Response Session of the Fourth International Social Science Methodology Conference, University of Essex, Institute for the Social Sciences, Colchester, UK.

Weeks, M. F., Kulka, R. A., and Pierson, S. A., (1987), "Optimal Call Scheduling for a Telephone Survey," *Public Opinion Quarterly,* **51,** pp. 540–549.

Weinberg, E. (1971), *Community Surveys with Local Talent: A Handbook,* Chicago: National Opinion Research Center.

Weisband, S. and Kiesler, S. (1996), "Self-Disclosure on Computer Forms: Meta-Analysis and Implications," Tucson: University of Arizona. http://uainfo. arizona. edu/~weisband/chi/chi96. html.

Weisbord, R. G. (1973), "Birth Control and the Black American: A Matter of Genocide," *Demography*, **10,** pp. 571–590.

Weiss, M. S. (1977), "The Research Experience in a Chinese-American Community," *Journal of Social Issues,* **33,** pp. 120–132.

Wellman, J. D., Hawk, E. G., Roggenbuck, J. W., and Buhyoff, G. J. (1980), "Mailed Questionnaire Surveys and the Reluctant Respondent: an Empirical Examination of Differences between Early and Late Respondents," *Journal of Leisure Research,* 2nd Quarter, pp. 164–173.

Whitridge, P., Bureau, M., and Kovar, J. (1990), "Use of Mass Imputation to Estimate for Subsample Variables," *Proceedings of the Section on Business and Economic Statistics, American Statistical Association,* pp. 132–137.

Wickens, C. D. (1992), *Engineering Psychology and Human Performance Theory (2nd ed.),* New York: HarpersCollins.

Wiese, C. J. (1998), "Refusal Conversions: What Is Gained?" *National Network of State Polls Newsletter,* pp. 1–3.

Wilensky, H. L. (1964), "The Professionalization of Everyone?" *American Journal of Sociology,* **70,** pp. 137–158.

Wilkinson, G. N. and Rogers, C. E. (1973), "Symbolic Description of Factorial Models for Analysis of Variance," *Applied Statistics,* **32,** pp. 392–399.

Williams, S. R. and Folsom, R. (1977), "Bias Resulting from School Nonresponse: Methodology and Findings," prepared by the Research Triangle Institute for the National Center for Educational Statistics.

Willimack, D. K., Nichols, E., and Sudman, S. (1999a), "Understanding the Questionnaire in Business Surveys," *Proceedings of the Section on Survey Research Methods, American Statistical Association.*

Willimack, D. K., Nichols, E., Sudman, S., and Mesenbourg, T. L. (1999b), "Cognitive Research on Large Company Reporting Practices: Preliminary Findings and Implications for Data Collectors and Users," paper prepared for April meeting of the Census Advisory Committee of Professional Associations, U.S. Bureau of the Census, Washington D. C.

Willis, G., Sirken, M., and Nathan, G. (1994), "The Cognitive Aspects of Response to Sensitive Survey Questions," Cognitive Methods Staff Working Paper No. 9, Hyattsville, MD: National Center for Health Statistics.

Willke, J., Adams, C., and Girnius, Z. (1999), "Internet Testing. A Landmark Study of the Differences Between Mail Intercept and On-Line Interviewing in the United States," Proceeding of from the Worldwide Internet Seminar 1998 in Paris, France, Amsterdam: ESOMAR, pp. 145–157.

Wilmot, A. (1999), *Family Resources Survey: Annual Technical Report on the Fifth Survey Year: April 1997-March 1998,* London: Office for National Statistics.

Winglee, M., Kalton, G., Rust, K., and Kasprzyk, D. (1994), "Handling Item Nonresponse in the U. S. Component of the IEA Reading Literacy Study," in Binkley, M., Rust, K., and Winglee, M. (eds.), *Methodological Issues in Comparative Educational Studies,* NCES 94–469, U.S. Department of Education, Office of Educational Research and Improvement, National Center for Health Statistics, Washington, D.C.

Winglee, M., Ryaboy, S., and Judkins, D. (1993), "Imputation for the Income and Assets Module of the Medicare Current Beneficiary Survey," *Proceedings of the Section on Survey Research Methods of the American Statistical Association,* pp. 463–467.

Winglee, M., Kalton, G., Rust, K. and Kaspryzk, D. (in press), "Handling Item Nonresponse in the U.S. Component of the IEA Reading Literacy Study, *Journal of Educational and Behavioral Statistics.*

Winkler, W. E. (1995), "Matching and Record Linkage," in B. G. Cox, D. Binder, B. N. Chinnappa, A. Christianson, M. J. Colledge, and P. S. Kott (eds.), *Business Survey Methods,* New York: Wiley, pp. 355–384.

Wiseman, F. and McDonald, P. (1978), *The Nonresponse Problem in Consumer Telephone Surveys.*Report No. 78–116. Cambridge, MA: Marketing Science Institute.

Wiseman, F. and McDonald, P. (1980), *Towards the Development of Industry Standards of Response and Nonresponse Rates.* Report 80-101. Cambridge, MA: Marketing Science Institute.

Wiseman, F. (1983), "Editor's Preface," *Proceedings of Marketing Science Institute Workshop,* Cambridge, MA: Marketing Science Institute.

Witt, K. and Bernstein, S. (1992), "Best Practies in DBM Surveys," *Proceedings of Sawtooth Software Conference,* Evanston: Sawtooth Software.

Wolfinger, R. E. and Rosenstone, S. J. (1980), *Who Votes?* New Haven: Yale University Press.

Wolter, K. M. (1985), *Introduction to Variance Estimation.*New York: Springer-Verlag.

Woodall, G. (1998), "Market Research on the Internet." http://www.rockresearch.com/html/nmr01.htm.

Woodruff, S. I., Edwards, C. C., and Conway, T. L. (1998), "Enhancing Response Rates to a Smoking Survey for Enlisted U. S. Navy Women,"*Evaluation Review,* **22,** pp. 780–791.

Worden, G. and Hamilton, H. (1989), "The Use of Mandatory Reporting Authority to Improve the Quality of Statistics," paper prepared for April meeting of the Joint Census Advisory Committee, U.S. Bureau of the Census, Washington, D.C.

Wright, J. R., and Niemi, R. G. (1983), "Perceptions of Issue Positions,"*Political Behavior,* **5,** pp. 209–223.

Wu, C. F. J. (1991), Balanced repeated replications based on mixed orthogonal arrays, *Biometrika,* **78,** pp 181–188.

Wu, M. C. and Bailey, K. R., (1988), "Estimation and Comparison of Changes in the Presence of Informative Right Censoring: Conditional Linear Model, *Biometrics,* **45,** pp. 939–955.

Wu, M. C. and Carroll, R. J. (1988), "Estimation and Comparison of Changes in the Presence of Informative Right Censoring by Modeling the Censoring Process," *Biometrics,* **44,** pp. 175–188.

Wurdeman, K. and Pistiner, A. L. (1997), "1995 Administrative Records Evaluation-Phase II," *Technical Report DMD 1995 Census Test Results Memorandum Series # 54 (Revised),* Washington, D.C.: United States Department of Commerce, Bureau of the Census.

Wydra, D. (1999), "Online Tracking: A New Frontier. ARF's Online Research Day—Towards Validation," New York: Advertising Research Foundation, pp. 34–36.

Xie, F. and Paik, M. (1997), "Multiple Imputation and Methods for the Missing Covariates in Generalized Estimating Equations," *Biometrics,* **53,** pp. 1538–1546.

Yammarino, F. J., Skinner, S. J., and Childers, T. L. (1991), "Understanding Mail Survey Response Behavior," *Public Opinion Quarterly,* **55,** pp. 613–639.

Yansaneh, I. S., Wallace, L., and Marker, D., (1998), "Imputation Methods for Large Complex Datasets: An application to the NEHIS," *Proceedings of the Section on Survey Research Methods of the American, Statistical Association,* pp. 314–319.

Ying, Y. (1989), "Nonresponse on the Center for Epidemiological Studies-Depression Scale in Chinese Americans," *The International Journal of Social Psychiatry,* **35,** pp. 156–163.

Yu, E. (1982), "Problems in Pacific/Asian American Community Research," in W. T. Liu (ed.), *Methodological Problems in Minority Research,* Chicago: Pacific/Asian American Mental Health Research Center, pp. 93–118.

Yu, J. and Cooper, H. (1983), "A Quantitative Review of Research Design Effects on Response Rates to Questionnaires," *Journal of Marketing Research,* pp. 2036–2044.

Zabel, J. E. (1998), "An Analysis of Attrition in the Panel Study of Income Dynamics and the Survey of Income and Program Participation with an Application to a Model of Labor Market Behavior,"*The Journal of Human Resources,* **33**(2), pp. 479–506.

Zaller, J. and Feldman, S. (1992), "A Simple Theory of the Survey Response: Answering Questions versus Revealing Preferences," *American Journal of Political Science,* **36,** pp. 579–616.

Zandan, P. and Frost, L. (1989), "Customer Satisfaction Research Using Disk-By-Mail," *Preceedings of Sawtooth Software Conference,* Evanston: Sawtooth Software.

Zanutto, E. (1998), "Imputation for Unit Nonresponse: Modeling Sampled Nonresponse Follow-up, Administrative Records, and Matched Substitutes," Ph. D. thesis, Department of Statistics, Harvard University.

Zanutto, E. and Zaslavsky, A. M. (1995a), A Model for Imputing Nonsample Households with Sampled Nonresponse Follow-Up," *Proceedings of Survey Research Methods Section, American Statistical Association,* pp. 608–613.

Zanutto, E. and Zaslavsky, A. M. (1995b), "Models for Imputing Nonsample Households with Sampled Nonresponse Followup," *Proceedings of the U.S. Bureau of the Census Annual Research Conference,* **11,** pp. 673–686.

Zanutto, E. and Zaslavsky, A. M (1996), "Estimating a Population Roster from an Incomplete Census Using Mailback Questionnaires, Administrative Records, and Sampled Nonresponse Followup," *Proceedings of the U.S. Bureau of the Census Annual Research Conference,* Bureau of the Census, Washington, D.C. pp. 741–760.

Zanutto, E., and Zaslavsky, A. M. (1997), "Modeling Census Mailback Questionnaires, Administrative Records, and Sampled Nonresponse Followup, to Impute Census Nonrespondents," Proceedings of the Survey Research Methods Section, American Statistical Association, pp. 754–759.

Zaslavsky, A. M. and Wolfgang, G. S. (1993), Triple-System Modeling of Census, Post-Enumeration Survey, and Administrative-List Data," *Journal of Business and Economic Statistics,* **11,** pp. 279–288.

Zeger, S. L. and Liang, K.-Y. (1986), "Longitudinal Data Analysis for Discrete and Continuous Outcomes," *Biometrics,* **42,** pp. 121–130.

Zukerberg, A., Nichols, E., and Tedesco, H. (1999), "Designing Surveys for the Next Millennium: Internet Questionnaire Design Issues," paper presented at the 1999 AAPOR Conference, St. Petersburg, Florida. http://surveys. over. net/method/zukerberg. ZIP.

Index

WILEY SERIES IN SURVEY METHODOLOGY
Established in Part by WALTER A. SHEWHART AND SAMUEL S. WILKS

Wiley Series in Survey Methodology covers topics of current research and practical interests in survey methodology and sampling. While the emphasis is on application, theoretical discussion is encouraged when it supports a broader understanding of the subject matter.

The authors are leading academics and researchers in survey methodology and sampling. The readership includes professionals in, and students of, the fields of applied statistics, biostatistics, public policy, and government and corporate enterprises.

BIEMER, GROVES, LYBERG, MATHIOWETZ, and SUDMAN · Measurement Errors in Surveys
COCHRAN · Sampling Techniques, *Third Edition*
COUPER, BAKER, BETHLEHEM, CLARK, MARTIN, NICHOLLS, and O'REILLY (editors) · Computer Assisted Survey Information Collection
COX, BINDER, CHINNAPPA, CHRISTIANSON, COLLEDGE, and KOTT (editors) · Business Survey Methods
*DEMING · Sample Design in Business Research
DILLMAN · Mail and Telephone Surveys: The Total Design Method, *Second Edition*
DILLMAN · Mail and Internet Surveys: The Tailored Design Method
GROVES and COUPER · Nonresponse in Household Interview Surveys
GROVES · Survey Errors and Survey Costs
GROVES, DILLMAN, ELTINGE, and LITTLE · Survey Nonresponse
GROVES, BIEMER, LYBERG, MASSEY, NICHOLLS, and WAKSBERG · Telephone Survey Methodology
*HANSEN, HURWITZ, and MADOW · Sample Survey Methods and Theory, Volume I: Methods and Applications
*HANSEN, HURWITZ, and MADOW · Sample Survey Methods and Theory, Volume II: Theory
KISH · Statistical Design for Research
*KISH · Survey Sampling
KORN and GRAUBARD · Analysis of Health Surveys
LESSLER and KALSBEEK · Nonsampling Error in Surveys
LEVY and LEMESHOW · Sampling of Populations: Methods and Applications, *Third Edition*
LYBERG, BIEMER, COLLINS, de LEEUW, DIPPO, SCHWARZ, TREWIN (editors) · Survey Measurement and Process Quality
SIRKEN, HERRMANN, SCHECHTER, SCHWARZ, TANUR, and TOURANGEAU (editors) · Cognition and Survey Research
VALLIANT, DORFMAN, and ROYALL · Finite Population Sampling and Inference: A Prediction Approach

*Now available in a lower priced paperback edition in the Wiley Classics Library.